AF572370

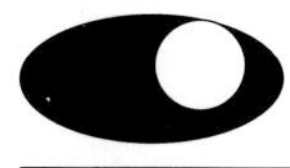

Current Topics in Membranes, Volume 39

Developmental Biology of Membrane Transport Systems

Current Topics in Membranes, Volume 39

Series Editors

Arnost Kleinzeller
Department of Physiology
University of Pennsylvania
School of Medicine
Philadelphia, Pennsylvania

Douglas M. Fambrough
Department of Biology
Johns Hopkins University
Baltimore, Maryland

Yale Series Editors

Joseph F. Hoffman and Gerhard Giebisch
Department of Cellular and Molecular Physiology
Yale University School of Medicine
New Haven, Connecticut

Murdoch Ritchie
Department of Pharmacology
Yale University School of Medicine
New Haven, Connecticut

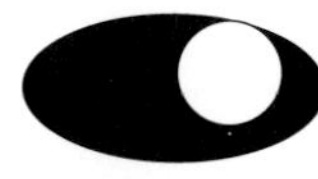

Current Topics in Membranes, Volume 39

Developmental Biology of Membrane Transport Systems

Guest Editor
Dale J. Benos
Department of Physiology & Biophysics
University of Alabama
Birmingham, Alabama

ACADEMIC PRESS, INC.
Harcourt Brace Jovanovich, Publishers
San Diego New York Boston London Sydney Tokyo Toronto

This book is printed on acid-free paper. 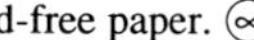

Academic Press, Inc.
San Diego, California 92101

United Kingdom Edition published by
Academic Press Limited
24–28 Oval Road, London NW1 7DX

Library of Congress Catalog Number: 70-117091

International Standard Book Number: 0-12-153339-5

PRINTED IN THE UNITED STATES OF AMERICA
91 92 93 94 9 8 7 6 5 4 3 2 1

Contents

Contributors

Numbers in parentheses indicate the pages on which the authors' contributions begin.

Kimon J. Angelides (229), Departments of Molecular Physiology and Biophysics, and Neuroscience, Baylor College of Medicine, Houston, Texas 77030

Dennis A. Ausiello (395), Renal Unit, Massachusetts General Hospital, Harvard Medical School, Boston, Massachusetts 02114

Maria Isabel Behrens (327), Centro de Estudios Cientificos de Santiago, Santiago, Casilla 16443, Chile

Dale J. Benos (121), Department of Physiology and Biophysics, University of Alabama at Birmingham, Birmingham, Alabama 35294

Paul Blount (277), Departments of Molecular Biology and Pharmacology, Washington University School of Medicine, St. Louis, Missouri 63110

M. D. Cahalan (357), Department of Physiology and Biophysics, University of California at Irvine, Irvine, California 92717

Horacio F. Cantiello (395), Renal Unit , Massachusetts General Hospital, Harvard Medical School, Boston, Massachusetts 02114

Michael J. Caplan (37), Department of Cellular and Molecular Physiology, Yale University School of Medicine, New Haven, Connecticut 06510

K. G. Chandy (357), Department of Physiology and Biophysics, University of California at Irvine, Irvine, California 92717

A. Michael Frace (3), Department of Physiology, Emory University School of Medicine, Atlanta, Georgia 30322

J. Jay Gargus (3), Department of Physiology and Section of Medical Genetics, Emory University School of Medicine, Atlanta, Georgia 30322

S. Grissmer (357), Department of Physiology and Biophysics, University of California at Irvine, Irvine, California 92717

Eun-hye Joe (229), Departments of Molecular Physiology and Biophysics, and Neuroscience, Baylor College of Medicine, Houston, Texas 77030

Hyun Dju Kim (181), Department of Pharmacology, School of Medicine, University of Missouri, Columbia, Missouri 65212

Douglas Kline (89), Department of Biological Sciences, Kent State University, Kent, Ohio 44242

Ramon Latorre (327), Centro de Estudios Cientificos de Santiago, Santiago, Chile and Departamento de Biologia Facultad de Ciencias, Universidad de Chile, Santiago, Chile

Edwin W. McCleskey (295), Department of Cell Biology and Physiology, Washington University School of Medicine, St. Louis, Missouri 63110

John Paul Merlie (277), Departments of Molecular Biology and Pharmacology, Washington University School of Medicine, St. Louis, Missouri 63110

Douglas H. Robinson (121), Department of Physiology and Biophysics, University of Alabama at Birmingham, Birmingham, Alabama 35294

Jean E. Schroeder (295), Department of Cell Biology and Physiology, Washington University School of Medicine, St. Louis, Missouri 63110

Michael W. Smith (153), AFRC Institute of Animal Physiology and Genetics Research, Babraham, Cambridge CB2 4AT, England

Foreword

This volume is one of the first to carry the new series title *Current Topics in Membranes*. The title change is a small, formal acknowledgment that "*and Transport*" has proven far too restrictive in describing what biological membranes can do. Today we can more easily appreciate the full impact of Hans Krebs' words in the Foreword to Volume 1: "membranes provide an essential framework for almost every functional activity of cells." Recent volumes and ones currently in the planning and production stages treat a very broad spectrum of biological phenomena, including membrane–cytoskeletal interactions, cell–cell recognition, biogenesis of cellular organelles, host–parasite relations, and transepithelial transport. The present volume is focused on developmental regulation of membrane functions.

When the series began in 1970, the two principal foci were membrane transport processes and membrane structure. Kinetic analyses of membrane-based processes seemed highly evolved, emboldening the original editors, Felix Bronner and Arnost Kleinzeller, to state in the Preface of the first volume: "pioneering work has made possible a rigorous description of biological transport in kinetic terms." The physicochemical properties of biological membranes and motifs of supramolecular organization that might embody these properties were not known in any detail.

The original orientation of the *Current Topics* series was rooted in the work of early pioneers of the field, including the great nineteenth century biologists, Carl Nägeli and Ernest Overton. Nägeli recognized in 1855 that the boundary layer separating plant cell protoplasm from its environment (later called the cell membrane) permitted permeation (endosmosis and exosmosis) of water but not of various solutes. Thus, membranes were characterized by what we would now denote as membrane semipermeability. In addition, he foreshadowed recognition of the membrane as a dynamic structure. The elemental role of lipids in membrane stucture was supported by the farsighted experiments and concepts of Overton (1895–1899).

By 1970, the fluid-mosaic nature of biological membranes was already gaining wide acceptance, even before its enunciation in a seminal article by Singer and Nicholson. However, there remained a complete lack of infor-

Carl Nägeli

Ernest Overton

mation about the underlying molecular details of membrane structure and functions. This was charmingly captured in Steck's vegetable patch model of the erythrocyte membrane. With proper foresight, the Preface to Volume 1 of *Current Topics in Membranes and Transport* continued, "elucidation of the underlying molecular mechanism has lagged behind [kinetic studies]. . .," and an era of scientific progress was predicted to bring forth *Current Topics* volumes on such topics as genetic determinants of membrane structure and function.

The era of marvelous revelations of molecular detail is indeed upon us. A profusion of primary structures of membrane proteins issues from molecular biological studies. Three-dimensional structures of membrane components are being determined at atomic resolution. The macromolecular machinery underlying one membrane phenomenon after another (and regulatory mechanisms thereof) is being characterized; machinery for transport processes; for secretion and endocytosis; for various mechanisms of cell locomotion; for recognition of and response to signals from other cells; for fertilization and embryogenesis; and on and on. Not only is our field transformed by application of molecular biology techniques but also by an ever broadening spectrum of conceptual and experimental approaches from various disciplines. For example, current microscopic measurements report the molecular contortions of single membrane channels. What role should *Current Topics in Membranes* play now and in the future? The primary function remains (as stated so simply in the Preface to Volume 1) "to stimulate thought and experiments." In pursuit of this goal, it is abundantly clear that we must work toward integrating the wealth of new information into cogent views of processes localized in the cell membrane, a distinct organelle of all cells. In a world rich (one might say "awash") in new information, this is an increasingly challenging task. We will continue to do our best to insure that each volume is an assemblage of key information and thoughtful perspectives, providing a timely, insightful view of each current topic in membranes.

DOUGLAS M. FAMBROUGH
ARNOST KLEINZELLER

Preface

The elucidation of the molecular nature of ion and solute transport systems is a field that has blossomed in recent years due to the availability of specific chemical and immunological probes, as well as the application of relatively straightforward molecular biological techniques. But just as important to an understanding of the transport of materials across biological membranes is defining the intrinsic and extrinsic controls signaling the appearance of specific membrane transport systems during the life cycle of a cell. This aspect of membrane biology is perhaps one of the most interesting, yet least understood. Hence, the main purpose of this volume is to summarize the current state of knowledge of the developmental aspects of membrane transport and to expose areas where critical research information is needed. Most importantly, it is the intent of this volume: first, to bring together scientists in the varied disciplines of membrane biophysics, cell and developmental biology, immunology, and molecular biology; second, to assess critically the status of the field; and third, to stimulate thought and further work in areas that have been neglected.

The book is organized into four parts: molecular biology of transport proteins and membrane protein sorting, fertilization and early embryonic development, developmental biology of ion and solute co- and counter-transport, and ion channel development. Chapters 1 (Frace and Gargus) and 2 (Caplan) [Part I] acquaint the reader with pertinent aspects of protein structure and function, provide an overview of the molecular characteristics of transport proteins, and detail membrane protein sorting mechanisms and biogenesis of membrane polarity. Part II (chapters by Kline, and Robinson and Benos) concerns the most obvious features of development: growth and differentiation of germinal cells and the nascent embryo, and the associated membrane changes coinciding with or predicting each developmental stage. Part III discusses the evolution of different co- and counter-transporters: enterocyte differentiation (Smith) and signal transduction systems influencing red blood cell maturation and volume-sensitive ion and solute transport (Kim). Part IV deals specifically with the control, regulation, and differentiation of ion channels, notably, voltage-sensitive Na^+ channels (Angelides and Joe), the nicotinic acetylcholine receptor/channel (Blount and Merlie), voltage-dependent Ca^{2+} channels

(McCleskey and Schroeder), potassium channels (chapters by Behrens and Latorre, and Cahalan, Chandy, and Grissmer), and epithelial Na^+ channels (Cantiello and Ausiello).

I would like to thank all of the authors for their outstanding and prompt contributions and Dr. Arnost Kleinzeller and Dr. Doug Fambrough for inviting me to prepare this volume, and for their advice and guidance. My sincere appreciation goes to Ms. Cathy Guy for her invaluable assistance in managing and coordinating all the manuscripts.

DALE J. BENOS

Previous Volumes in Series

Current Topics in Membranes and Transport

Volume 12 Carriers and Membrane Transport Proteins (1979)
Edited by F. Bronner and A. Kleinzeller

Volume 13 Cellular Mechanisms of Renal Tubular Ion Transport* (1980)
Edited by Emile L. Boulpaep

Volume 14 Carriers and Membrane Transport Proteins (1980)
Edited by F. Bronner and A. Kleinzeller

Volume 15 Molecular Mechanisms of Photoreceptor Transduction* (1981)
Edited by William H. Miller

Volume 16 Electrogenic Ion Pumps* (1982)
Edited by Clifford L. Slayman

Volume 17 Membrane Lipids of Prokaryotes (1982)
Edited by Shmuel Razin and Shlomo Rottem

Volume 18 Membrane Receptors (1983)
Edited by Arnost Kleinzeller and B. Richard Martin

Volume 19 Structure, Mechanism, and Function of the Na/K Pump* (1983)
Edited by Joseph F. Hoffman and Bliss Forbush III

Volume 20 Molecular Approaches to Epithelial Transport* (1984)
Edited by James B. Wade and Simon A. Lewis

Volume 21 Ion Channels: Molecular and Physiological Aspects (1984)
Edited by Wilfred D. Stein

Volume 22 The Squid Axon (1984)
Edited by Peter F. Baker

* *Part of the series from the Yale Department of Cellular and Molecular Physiology*

Volume 23 Genes and Membranes: Transport Proteins and Receptors* (1985)
Edited by Edward A. Adelberg and Carolyn W. Slayman

Volume 24 Membrane Protein Biosynthesis and Turnover (1985)
Edited by Philip A. Knauf and John S. Cook

Volume 25 Regulation of Calcium Transport Across Muscle Membranes (1985)
Edited by Adil E. Shamoo

Volume 26 Na^+-H^+ Exchange, Intracellular pH, and Cell Function* (1986)
Edited by Peter S. Aronson and Walter F. Boron

Volume 27 The Role of Membranes in Cell Growth and Differentiation (1986)
Edited by Lazaro J. Mandel and Dale J. Benos

Volume 28 Potassium Transport: Physiology and Pathophysiology* (1987)
Edited by Gerhard Giebisch

Volume 29 Membrane Structure and Function (1987)
Edited by Richard D. Klausner, Christoph Kempf and Jos van Renswoude

Volume 30 Cell Volume Control: Fundamental & Comparative Aspects in Animal Cells (1987)
Edited by R. Gilles, Arnost Kleinzeller, and L. Bolis

Volume 31 Molecular Neurobiology: Endocrine Approaches (1987)
Edited by Jerome F. Strauss III and Donald W. Pfaff

Volume 32 Membrane Fusion in Fertilization, Cellular Transport, and Viral Infection (1988)
Edited by Nejat Düzgünes and Felix Bronner

Volume 33 Molecular Biology of Ionic Channels* (1988)
Edited by William S. Agnew, Toni Claudio, and Frederick J. Sigworth

Volume 34 Cellular and Molecular Biology of Sodium Transport* (1989)
Edited by Stanley G. Schultz

Volume 35 Mechanisms of Leukocyte Activation (1990)
Edited by Sergio Grinstein and Ori D. Rotstein

Volume 36 Protein-Membrane Interactions* (1990)
Edited by Toni Claudio

Volume 37 Channels and Noise in Epithelial Tissues (1990)
Edited by Sandy I. Helman and Willy Van Driessche

Current Topics in Membranes

Volume 38 Ordering the Membrane Cytoskeleton Tri-Layer*(1991)
Edited by Mark S. Mooseker and Jon S. Morrow

PART I

Molecular Biology of Transport Proteins and Membrane Protein Sorting

CHAPTER 1

Molecular Biology of Membrane Transport Proteins

A. Michael Frace and J. Jay Gargus*
Department of Physiology and *Section of Medical Genetics, Emory University School of Medicine, Atlanta, Georgia 30322

I. INTRODUCTION

Over only the past few years there has been an explosive increase in the hard structural information available about membrane transport proteins, adding a welcome dimension to those cornerstones in cellular physiology which for so long have existed in our texts only as black boxes. These new data, coming from the application of molecular genetic techniques to the analysis of membrane proteins, make bold predictions: they suggest that all of the complexities and subtleties of an individual transporter's structure and function, as well as its fine-tuned pattern of expression during development, are determined by the nucleic acid sequence of its gene, and that the synthesis, secondary structure, membrane localization, transport selectivity, and energy requirements of the transporter, all, in turn, can be understood from the analysis of a simple ordered arrangement of four nucleotide bases in that small segment of DNA. Largely, this new information has come in a raw form, not readily yielding to methods of classical physiological data reduction. The extrapolation of physiologically relevant functions from this simple code has forced new voculabulary and concepts into mainstream physiology and has consumed an enormous amount of recent effort. The results to date, while impressive, still reveal gaping holes in our understanding of molecular information. Yet, as these types of studies promise to supply vital information about how, when, and why transport proteins work at the most fundamental of levels, it seems likely they will continue to play an expanding role in our field. For these data to be understood, not merely accepted as fact, a fair amount of background material not traditionally part of physiology must be assimilated.

The purpose of this chapter is to introduce to investigators of membrane protein function the necessary concepts and vocabulary to allow a discussion of what appear to be some fundamental structural properties these transport mechanisms share. We examine properties of protein structure that relate to a plasma membrane environment and attempt to dissect membrane protein structures to reveal how the DNA code produces them.

II. THE CODING OF MEMBRANE PROTEIN STRUCTURE

A. *Historical Perspective*

To establish some fundamentals and to become acquainted with the tools used to dissect protein structure, let us first take a backward approach and examine what we know about the higher-order structure of membrane proteins to see what it must represent in terms of DNA sequence.

Until the mid-1970s, the structure of proteins confined to a lipid environment was virtually unknown. Membrane physiologists truly worked on black boxes. The structural organization of fibrous proteins as well as some globular proteins had been examined by X-ray diffraction analysis, and exciting discoveries such as protein folding into β sheets and coiled coils of α helices had been reported. However, a critical characteristic of those proteins was that they were able to be crystallized, necessary to the application of X-ray diffraction analysis. Unfortunately, membrane proteins prove to be highly resistant to crystallization. This remains a major obstacle, but it was completely insurmountable until 1983 (Michel, 1983). Fortunately, in the interim, technical advances in the field of electron diffraction and a special, remarkably organized membrane protein allowed some progress to be made. In 1975, Henderson and Unwin (1975) applied electron diffraction technology to a naturally occurring crystal of a transmembrane protein, bacteriorhodopsin, the purple membrane protein of *Halobacterium*. At last, a rough, three-dimensional picture of a membrane protein was available. The result was consistent with a structure of seven α helices, each approximately 40 Å in length, long enough to span a biological bilayer. The next step toward deciphering membrane protein data came in 1979 when Khorana *et al.* (1979) sequenced the bacteriorhodopsin protein and demonstrated it to have seven stretches of 23 to 25 hydrophobic amino acids that had the potential to form α helices of the requisite 40-Å length. Since α helices derive stability by producing maximal intrachain H-bonding between adjacent residues, they are energetically well suited structures for spanning the lipid domain of the bilayer (which is poorly suited to such bond formation). Alternatively, β structure could only be accommodated should a large multistrand "barrel" form prior to insertion. These hydrophobic α-helical domains are the most noticeable aspect of membrane protein structure, and they are now commonly used as signatures of these proteins. Finally, in 1981, Khorana's group sequenced the cDNA and gene of bacteriorhodopsin, confirming the sequence (Dunn *et al.*, 1981). Ultimately, they performed *in vitro* mutagenesis, replacing key amino acid residues. At last a good deal was known about *a* membrane protein. From these studies on bacteriorhodopsin, and the confirmatory and extending work on the photosynthetic reaction center (where true crystallographic data first became available on membrane proteins; Deisenhofer *et al.*, 1985), a working model for membrane protein structure appeared. As recombinant DNA work accelerated, this ordered accession of protein structural data (tertiary toward primary) was turned on its head, and suddenly an imposing amount of primary sequence data became available on functional plasma membrane proteins, all of which had to be viewed through the lens of these two transporter molecules for which higher-order structure was known.

B. Eukaryotic Gene Structure

To begin an analysis of the molecular properties of transport proteins, as such data have currently become known, requires an orientation to the functional anatomy of the genes which encode these proteins. To this end, we have compiled common themes found in eukaryotic genes into a consensus cartoon of a "typical" gene. To read the cartoon we will take the perspective of the enzyme that "reads" the gene, RNA polymerase II, and discuss aspects of the DNA structure which plays a role in its translation into an mRNA chain growing in a 5′ to 3′ direction. A low-resolution cartoon reveals three basic regions to the gene (Fig. 1). At the core of the gene will be an area where the protein-coding domains are found. Here, the information content of the DNA is conferred primarily through the triplet codons used in protein synthesis. Flanking the core are regions of nucleotide sequence which function in a different manner, not as codes for protein but as structural sites which are critical in controlling the transcription and translation of the message.

Upstream, or 5′, to the coding region is an area of control elements referred to generically as the promoter. These are *cis* control elements in that they control the expression only of adjacent segments of DNA (as opposed to *trans* elements, which can exert control over genes even on different pieces of DNA, usually via the proteins they encode). The *cis* elements exert control over the transcription of the gene by serving as binding sites for specific DNA-binding proteins (*trans* regulator proteins) (Dynan and Tjian, 1985; Ptashne, 1988). Together, the *cis* elements and the *trans* regulators control the access of the RNA polymerase to the gene. The macromolecular complex of *trans* elements assembled on the *cis* elements of a given gene determines whether the nuclear transcription machinery will find this gene to be a suitable target for expression in a given cell at a given time. The genes encoding the *trans* regulators are also regulated in a very similar fashion, in essence creating a powerful hierarchy of control through the regulation of a regulator.

Several conserved functional elements can be found within the promoter. The most distant of these are usually the enhancers (Maniatis *et al.*, 1987; Dynan, 1989). These elements can be thousands of base pairs upstream from the gene, but they function independently of their exact position in that they may be found downstream from the gene, or even in the middle of it. Even the direction from which this sequence is read is not critical in that an inverted enhancer element remains effective. These aspects of enhancer function immediately suggest that enhancers work in a manner quite different from protein-encoding genetic elements. There have been many different enhancer sequences identified and they share little homology. Many function in a tissue-specific or developmental stage-specific manner, and thus seem to provide a mechanism which can activate genes in specific tissues or only at certain stages in development, providing the kind of genetic regulation clearly needed by a metazoan organism. Enhancers achieve these effects by serving as the binding sites for their

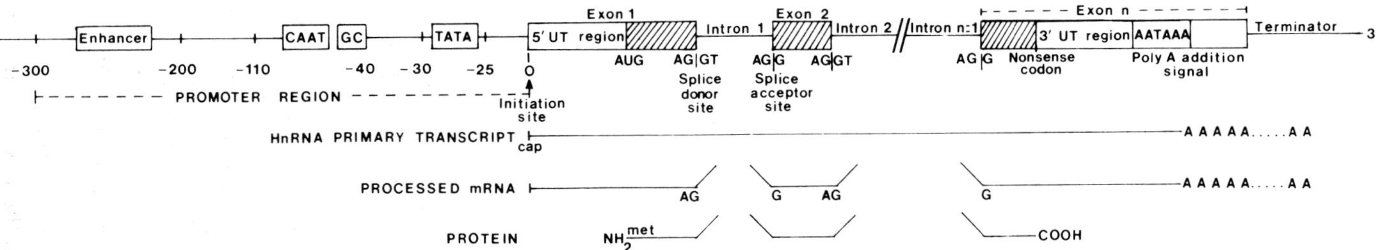

FIG. 1. Functional anatomy of a "typical" eukaryotic gene and its transcriptional products. The top line represents the crude nucleotide sequence of DNA in the region of the gene. Functional regions are labeled and discussed in the text. The second line from the top represents the crude nucleotide sequence of the primary transcript hn RNA product produced directly by RNA polymerase II. It is aligned with the gene (top line) to indicate the regions transcribed. The 5′ cap and 3′ poly(A) tail are early modifications of the transcript and are not reflected in the genome. The third line represents the crude nucleotide sequence of fully mature cytoplasmic mRNA, again aligned with the gene. Note the introns have been spliced out of the transcript and that splice donor and splice acceptor sites are now covalently ligated, converting . . . AG/GT . . . AG/G . . . sequences to a sequence of . . . AGG . . . . The 5′ cap and 3′ poly(A) tail are unaltered. The bottom line represents the crude amino acid sequence of the primary translational product as it is formed by the ribosome before any posttranslational modifications or cleavages. It too is aligned with the gene to indicate regions which are translated. Neither the 5′ nor 3′ untranslated regions encode amino acids. The protein begins with an N-terminal initiation methionine and ends with the C-terminal amino acid which is encoded by the triplet preceding the nonsense termination codon. (From Gargus, 1989, with permission.)

specific *trans* regulating DNA-binding proteins. A part of the mechanism by which these elements control gene expression is by altering the structure of the surrounding chromatin to render the DNA locally much more accessible to other protein transcription factors in the nucleoplasm. Thus, the interaction between enhancers and their *trans* factors appear critical to gene expression, as if they were tissue-specific on/off switches (Maniatis *et al.*, 1987).

The next *cis* regulatory elements the RNA polymerase will encounter are between 110 and 40 bases upstream from the start of transcription. Among others, these include the CAT box and the GC-rich elements, drawing their names from the consensus sequence of the element. Elements in this region are also binding sites for a growing family of different *trans* regulator proteins. The aggregate function of these elements seems to be to set the rate at which a gene will be transcribed once it has been activated. For instance, some developmental stage-specific genes might require very active transcription for brief periods, whereas housekeeping genes (those that operate continuously in cells) might only need a rare transcript made.

A final, highly conserved, control element in the 5′ region is encountered in a very consistent location 25 to 30 base pairs (bp) from the start of transcription. It is the TATA box, again named for its consensus nucleotide sequence. Its role seems to be to position the RNA polymerase precisely for the initiation of transcription. Synthetically repositioning this element repositions the location at which initiation occurs; elimination of the TATA box often eliminates initiation.

Many of the *trans* regulating proteins that interact at these *cis* elements have been isolated and the molecular details important in regulation are now being revealed. The overall goal of these upstream sequence elements is to allow RNA polymerase II to be attracted only to those genes destined to be expressed in a given tissue at a specific time, to initiate transcription accurately and at a rate consistent with the function of the product of the gene.

The initial DNA sequences transcribed into RNA through base-pairing still do not carry the code for protein structure. As the process of transcription begins, the base at the transcript's extreme 5′ end is enzymatically "capped" with a methyguanosine residue. This cap structure and the entire initial RNA sequence (the 5′ untranslated region) remain intact as the primary RNA transcript matures into mRNA, for these structures will play an important role as sites participating in the initiation of protein synthesis.

The 5′ untranslated region is followed by the first protein-coding region. It begins with the triplet codon AUG on the RNA strand, the methionine initiation codon. Since RNA code can be "read" in any one of its three possible triplet reading frames, and protein synthesis must begin by reading in only one specific frame, recognition of the appropriate initiation codon is critical. The first occurrence of the sequence AUG following the cap structure on the message generally sets the unique reading frame, but position alone is clearly not sufficient to define initiation. The initiation AUG is further delineated by a consensus sequence which surrounds it (Kozak, 1988). The sequence (GCC)GCCACCAUG(G) is the favored sequence, deviations from this suppressing the potency of the site for initiation. The adenine nucleotide three bases upstream from the AUG codon is the most conserved, and seemingly most important, residue of this sequence.

The AUG initiation codon may be followed by triplet code for amino acids not found in the mature protein. These amino acids are nonetheless incorporated at the amino terminus of the nascent protein. These peptide sequences may be signal sequences which play a role in localizing proteins to different membrane compartments, or they may be other parts of preproteins which are posttranslationally removed from the mature protein.

Once the RNA polymerase initiates transcription it continues to copy one strand of the DNA molecule, forming an uninterrupted transcript that grows to a great length. An average mammalian transcript is 8000 bases (8 kb) long; often they exceed 20 kb, and some seem to be hundreds of kilobases long. Obviously, this is far more sequence than is needed to encode a protein. Mature cytoplasmic mRNA carries sequences copied from the 5′ untranslated region and the first protein-encoding region of the gene. Since this part of the gene's sequence exits the nucleus, it is defined to be the first exon of the gene. The next region of the gene encountered by the polymerase, however, though transcribed in the nucleus into RNA, is not found in the mature cytoplasmic mRNA, and of course therefore does not code for protein. This is the first intron, or intervening sequence, of the gene. It can be just tens of bases long or can be tens of thousands of bases in length. Intron sequences are spliced out of the transcript and degraded in the nucleus. Most of the excess sequence found in eukaryotic transcripts consists of these introns, but their function presently remains unknown.

Following the first intron of a gene is the second exon. It, and all subsequent exons, begins with a conserved sequence element referred to as the splice acceptor site, and ends with a different element called a splice donor site. These consensus sites serve to effect an accurate splice that keeps the reading frame of the second exon exactly in frame with that of the first exon when the intron sequence is removed, through a mechanism that is described below. The alternation between exons and introns continues down the length of the gene. Most eukaryotic genes have at least one intron; some contain 50 or more. The last exon encountered by the polymerase contains a nonsense codon, one of three codons for which the cell lacks a tRNA and which serves to terminate translation of the protein. Beyond the termination codon the final exon contains the 3' untranslated region of the gene (Birnstiel *et al.,* 1985). This region carries a sequence element which on the RNA transcript will serve as a recognition element to which a new polymerase will bind. This site is called the poly(A) addition signal (AAUAA). It serves to bind poly(A) polymerase, an enzyme that cuts the transcript beyond this signal and adds a long sequence of adenosine residues, the poly(A) tail (Wickens and Stephenson, 1984). This tail seems to play a role in stabilizing the message in the cytoplasm and it is a feature that nearly all eukaryotic mRNAs have in common.

The primary RNA transcript is very short-lived and is found exclusively in the nucleus. Here, it is processed into mature mRNA by a splicing mechanism that serially removes the introns, creating perfect splices and keeping all exons in-frame by making use of the splice donor and acceptor sites (Sharp, 1987). The splicing occurs on small ribonucleoprotein complexes found in the nucleus, called snRNPs ("snurps") (Maniatis and Reed, 1987). Only after the splicing reaction has occurred can the message exit the nucleus and enter the cytoplasm.

Exons do not appear to be random divisions of protein code; rather, they seem to define functional domains of a protein, as if they were units moved about and reassociated in different combinations with one another over evolutionary time. As an example, see the genomic arrangement of the anion carrier's transmembrane domains mentioned in Section III,A. This notion perhaps explains the evolutionary significance of introns and splicing: this mechanism facilitates evolution by enhancing the shuffling of functional gene fragments. It would allow genetic rearrangements without the requirement that genes break and religate perfectly in-frame. A break could occur anywhere in the large intron targets, and the splicing mechanisms could be relied on to reestablish the reading frame (Darnell, 1978; Crick, 1979).

C. Membrane Insertion

Once the genetic code has been faithfully translated into protein, one is still left with the interesting dilemma of how to get a membrane protein into the membrane (Wickner and Lodish, 1985; Zimmermann and Meyer, 1986; Singer *et al.,* 1987; Hartmann *et al.,* 1989). Models based on proteins which are secreted across the membrane are a good starting point. These proteins make use of a leader sequence at the amino terminus that serves as a signal to direct secretion. Typically, these signal sequences consist of about 20 residues with an apolar, uncharged, central domain surrounded by polar, basic amino acids. These signal sequences are recognized by special signal recognition proteins (SRP) and docking proteins specific to the membrane target. In some instances, the protein is inserted as it is being translated, that is, cotranslationally, while still bound in its ribosomal complex. Other proteins have displayed an ability to be inserted after translation is complete, an ability termed posttranslational insertion. For secretory proteins, whose fate is simply to cross the membrane a single time, the insertion process continues until the protein is extruded or a highly hydrophilic region near the carboxyl terminus, a stop insertion sequence, is encountered. In each scenario, a leader peptidase renders insertion irreversible by cleaving the signal sequence after the membrane has been traversed.

The case of integral membrane proteins is more complex for several reasons. One is that the final membrane topology involves multiple spans, often up to a dozen. Another unsolved complexity is that quite often membrane transport proteins display no amino-terminal insertion sequence, nor is it clear whether they are inserted cotranslationally or posttranslationally. Several membrane proteins which display similar structural characteristics (no cleaved amino-terminal sequence, large extracytosolic hydrophilic domains, cytosolic amino and carboxyl termini), such as the ATPase family, band 3, the sodium channel, and the glucose transporter, have been modeled to insert in a similar fashion (Wickner and Lodish, 1985) (Fig. 2). A series of internal hydrophobic sequences is cryptically encoded in the primary amino acid sequence. SRP binds to the first hydrophobic (or signal) sequence, interacts with docking proteins, and begins insertion of the hydrophobic sequence into the endo-

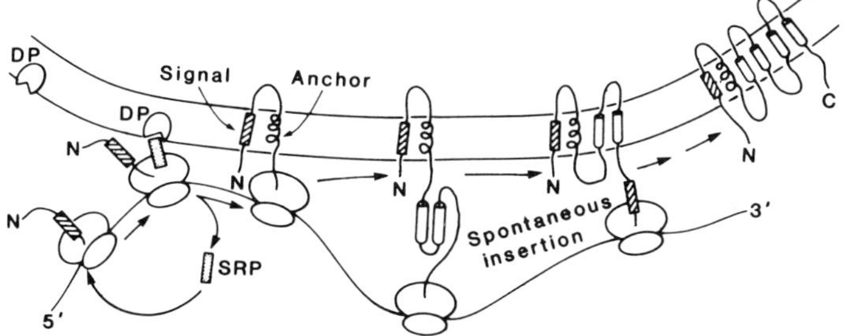

FIG. 2. Insertion of multispanning protein. See text for description. (From Wickner and Lodish, 1985, with permission.)

plasmic reticulum membrane. The orientation of this first hydrophobic span appears to be dictated by the dipole moment of its flanking sequence (Hartmann *et al.*, 1989). Loop structures or hairpins are then presumed to form and be stabilized by hydrophobic interactions in the membrane. In this manner, a series of antiparallel transmembrane helices can be formed. Further hydrophobic domains adjacent to the signal loop are thought to fold easily into the membrane, driven by favorable energetic reactions with the lipid bilayer. The process continues iteratively until a stop insertion sequence is reached. However, several variations on this general theme are necessary for the various types of transmembrane proteins, since no single model will suffice in describing a consensus insertion mechanism.

D. Prediction of Protein Structure

Once the genetic code has produced a protein's primary sequence, one is faced with a cryptic code which needs to be converted into higher-order structure. To predict secondary and tertiary structure from primary sequence information, several techniques and algorithms have been applied which take advantage of limits on structure presumably imposed by the nature of the lipid bilayer. Typical analytical methods involve the assignment of a numerical value to each residue related to each amino acid's potential contribution to particular protein substructures in lipid or aqueous environments. While the resulting data from the several algorithms are not strictly comparable, the trends produced are generally consistent. Analysis of the hydropathic nature of protein structure is a key determinant to whether fields of residues are likely to lie in an aqueous or lipid environment. The Kyte–Doolittle plot (Kyte and Doolittle, 1982) is a simple method which adopts a hydropathy scale based on various sources and assigns to each amino acid a "best guess" overall value which is likely to be more accurate for some residues than for others. The program utilizes a "moving segment" approach to average the hydropathy of a window of amino acids, usually a number sufficient to span a bilayer. Consecutive values are assigned to each window from the amino terminus to the carboxyl terminus. These values are then compared to a standard value of hydropathy averaged from many sequenced proteins. Hydropathy plots of membrane proteins exhibit well-delineated areas representing residues likely to be in or out of the bilayer. However, a caveat is that packing properties of the residues (e.g., steric interplay tendencies) are not considered. For instance leucine, valine, and isoleucine residues have substantial, but poorly understood, impact on protein folding that is not related to their hydropathy values. Such trends go unnoticed in this analysis, for the sake of simplicity.

More sophisticated analysis paradigms have been developed that begin to give weight to trends empirically observed in the body of X-ray diffraction and other structural studies of proteins. These methods offer to be more applicable in the analysis of transport proteins. Examples of more directed analysis programs are those of Engelman *et al.* (1986) and Eisenberg *et al.* (1984). Engelman *et al.*'s analysis is more directed to the study of transbilayer helices, noting whether helix residues are polar or nonpolar. They determined that when helices are composed of polar residues, a different scale of hydropathy should be applied to enhance accuracy. The analysis of Eisenberg *et al.* considers the hydrophobicity and amphiphilicity of helices to assign a value termed hydrophobic moment. This allows the tendencies of helices to be buried in the membrane or to seek the membrane surface to be considered.

Other protein conformation prediction methods consider trends in X-ray diffraction data to assign numerical values to amino acids. Chou–Fasman analysis (Chou and Fasman, 1978) assigns a hierarchical order for each amino acid based on its likelihood of being found in certain protein conformations. For instance, leucines are most frequently assigned to inner helical cores, valines are scored as strong β sheet formers, and other residues can be assigned as helix breakers or helix indifferent.

In all cases, these methods are educated guesswork, not proofs. As new trends appear in newly compiled data, the programs become more complex and, presumably, more accurate. However, at our present stage of understanding, most of these programs produce roughly compatible models in the analysis of membrane proteins and are useful, simple tools which can be used to guide experimentation.

III. MODEL STRUCTURES

The primary structures of several integral membrane transport proteins have been reported and their secondary conformations predicted. Notable trends have been established already, defining families of transport proteins, each with a characteristic fingerprint or trait. The remainder of this chapter is devoted to scanning the molecular structure of several transport proteins. Since reporting on all of the available structures is already too sizeable a task, we have chosen to pick examples of different families of transport proteins to highlight what we feel are their more interesting aspects.

A. Anion Exchanger

The band 3 anion exchanger is the most abundant protein in the erythrocyte membrane, though homologs have been found in other tissues (Wagner *et al.*, 1987; Alper *et al.*, 1989). It mediates a one-for-one exchange of chloride and bicarbonate, critical to pH buffering and systemic respiration. The organization of the murine band 3 gene and the sequence of a cDNA clone have been described (Kopito *et al.*, 1987; Kopito and Lodish, 1985). The cDNA contains a 2900 bp open reading frame which codes for 895 amino acids. Two in-frame initiation codons are found in close approximation; however, only the second one is flanked by a consensus initiation site and actually functions in this role. Based on *in situ* proteolysis data, the protein is divided structurally and functionally into two domains (Jennings, 1985). The amino-terminal half is a polar, negatively charged, cytoplasmic domain. Binding sites for hemoglobin and glycolytic enzymes are found here, localized to the first 11 residues (Walder *et al.*, 1984). Also found in the amino-terminal domain is the site for binding to ankyrin (Bennett and Stenbuck, 1980), associating band 3 with the cytoskeleton. The carboxyl-terminal half of band 3 produces anion transporter activity (Grinstein *et al.*, 1978). Twelve membrane-spanning segments, connected by short segments of polar amino acids, are modeled for this portion of the protein (Fig. 3A). Five of the membrane-spanning regions are predicted to form amphipathic helices. One of these, segment 5, contains the lysine residue(s) involved in disulfonic acid stilbene inhibitor binding.

The murine band 3 gene is a single copy gene spanning 17 kb. It is composed of 20 exons which range in size from 73 to 253 bp. The first exon and 68 bp of the second exon code for the 5′ untranslated region, while the final exon codes for only the 3′ untranslated region. All of the remaining exons code entirely for protein sequences (Kopito *et al.*, 1989). A fascinating correlation is found between the location of intron/exon splice junctions and the functional units of secondary structure predicted by the hydropathy plot (Fig. 3B). This is especially evident in the carboxyl domain, where the introns interrupt the sequence after each unit of transmembrane α helix. In this manner, it appears that studies of the genomic organization of a transporter can serve as a check on the structural model proposed by amino acid sequence analysis.

B. Glucose Transporter

In mammalian cells of nonepithelial origin, glucose transport occurs by facilitated diffusion down its concentration gradient, not coupled to me-

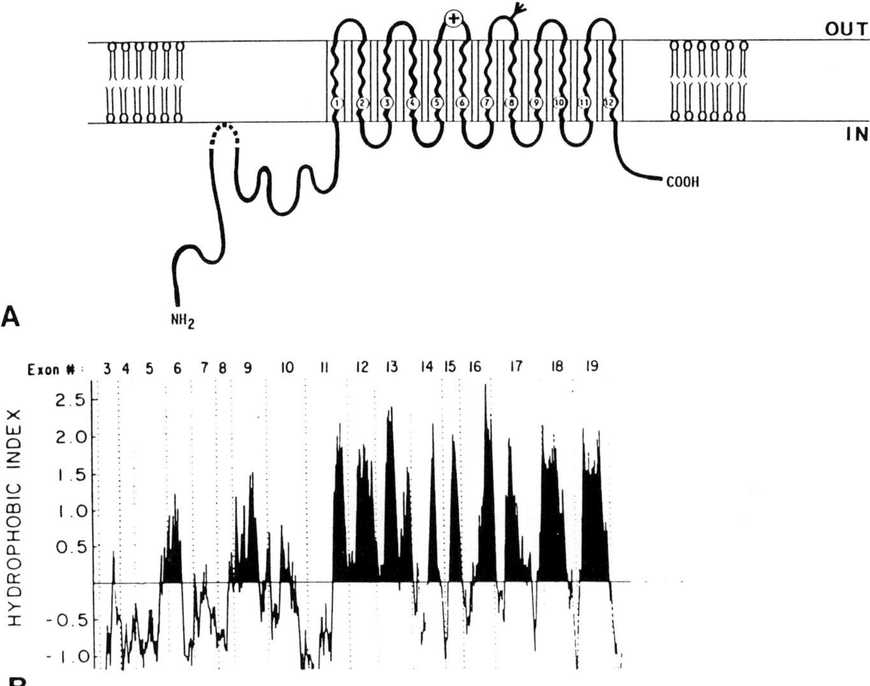

FIG. 3. (A) Proposed model of the band 3 anion exchanger oriented in the plasma membrane. The sites for isothiocyanate stilbene binding (⊕) and oligosaccharide attachment (Ψ) are indicated. Transmembrane helices 1–12 are labeled. (After Kopito and Lodish, 1985). (B) Hydropathy profile of the band 3 anion exchanger mRNA. The plot is generated using Kyte–Doolittle analysis (see text for details). The vertical dotted lines represent splice junction sites where protein coding exons are interrupted by introns. (From Kopito *et al.*, 1987, with permission.)

tabolism or to cation gradients. Specific glucose transporters from several tissues have been isolated and sequenced (Mueckler *et al.*, 1985; Fukomoto *et al.*, 1988; Thorens *et al.*, 1988). Those from erythrocytes and the HepG2 hepatoma are extremely homologous if not identical. The open reading frame of this transporter is 1,476 bp, coding for 492 amino acid residues. The 5′ and 3′ untranslated regions of the mRNA are particularly large, with the 5′ region being highly GC rich, predicting that it will form hairpin loop secondary structures of unknown function. Predictions of the polypeptide structure by Kyte–Doolittle and Eisenberg analysis propose 12 membrane-spanning segments, 5 of which are amphipathic helices. Amide and hydroxyl residues in these amphipathic helices are postulated to be hexose binding sites. Chou–Fasman analysis reveals β turns to be prominent in the hydrophilic sections that separate the helical domains. An extremely hydrophilic cytoplasmic domain is proposed for a central portion of the polypeptide, residues 207 through 272. Both the amino terminus and the carboxy terminus reside in the cytosol according to these analyses (Mueckler *et al.*, 1985), conflicting with proteolytic digestion experiments carried out on sealed and unsealed erythrocyte ghosts (Shanahan and D'Artel-Ellis, 1984), where the amino-terminal sequence could never be cleaved.

A variety of glucose transporters have been found to be distributed in a tissue-specific manner. A distinct glucose transporter has been discovered in tissues which increase glucose transport in response to insulin through the translocation of glucose transporters to the cell surface from a preformed intracellular pool (James *et al.*, 1988). The insulin-regulated glucose transporter (IRGT) is found exclusively in heart, skeletal muscle, and brown and white adipocytes (James *et al.*, 1989; Birnbaum, 1989). Slightly larger, its open reading frame is 1,680 bp, coding for 509 amino acids. The amino acid sequence is 63% identical to the HepG2 transporter and a similar secondary structure is predicted. Another glucose transporter bearing 55% amino acid homology to HepG2 is found in liver and in low abundance in kidney and small intestine. It is similar to HepG2 in size and organization but its genetic locus has been mapped to human chromosome 3, whereas the HepG2 locus is found on chromosome 1 (Fukomoto *et al.*, 1988; Shows *et al.*, 1987). Another set of hexose transporters found, surprisingly, to share high homology with HepG2 are two bacterial sugar transporters for xylose and arabinose (Maiden *et al.*, 1987). Nearly identical in size and secondary organization, these transporters are 40% identical to HepG2 yet are evolutionarily distant and functionally quite different in that they are proton-coupled transporters. All of the glucose transporters, including those distant family members of bacterial origin, are found to contain a consensus sequence of Arg-X-Gly-Arg-Arg (Arg can be replaced by Lys). This sequence is predicted to be cytosolic and to form β turns. It also resembles the peptide sequence recognized by cyclic AMP-dependent protein kinase.

C. Sodium-Coupled Carriers

1. Na^+/H^+ Antiporter

The eukaryotic Na^+/H^+ antiporter is a ubiquitous membrane transport protein that acts as a primary proton extruding system. The antiporter is driven by the Na^+ gradient produced by the Na^+/K^+-ATPase. It serves in conjunction with other pH-regulating transport proteins to control intracellular pH and to respond to periods of acid stress or mitogenic stimulation (Aronson and Boron, 1986). The eukaryotic amiloride-sensitive antiporter is encoded by a large, 5.6-kb message whose open reading frame (2,445 bp) codes for a 815-amino acid protein of molecular mass (M_r) of 99,354 (Sardet *et al.*, 1989, 1990). It has little sequence homology to any other sequenced transport protein. The 5′ untranslated region contains four potential in-frame methionine initiation codons which each could produce short peptides from the transcript. Initiation at these alternative sites is postulated to play some role in posttranscriptional control of antiporter expression. The antiporter molecule is composed of two major domains. The amino-terminal 500 amino acids make up the amphipathic membrane-spanning regions of the molecule. Depending on the hydrophobicity analysis program, 10 or 12 transmembrane α helices are found in this region, along with four large extracellular hydrophilic domains. Three potential N-glycosylation sites are found at predicted extracellular residues 75, 370, and 410. The remainder of the molecule is a large, positively charged, hydrophilic, cytoplasmic domain that contains several consensus phosphorylation sites. Activation of the antiporter by several mitogens has, in fact, been demonstrated to phosphorylate its cytoplasmic serine residues over a time course similar to the alkaline pH change induced by growth factor activation (Sardet *et al.*, 1990).

With a sequenced cDNA clone of the Na^+/H^+ antiporter in hand, the molecular studies of this transport protein promise to allow several avenues of intracellular signaling and control to be studied. With its multitude of control systems (G-proteins, kinases, growth factors, and oncogenes), the antiporter and its functional mutants are likely to be forerunners in unlocking these molecular mechanisms in a eukaryotic system.

The prokaryotic Na^+/H^+ antiporter, the product of the *ant* gene, shows little similarity with its eukaryotic counterpart in either primary sequence or cellular function (Karpel *et al.*, 1988). The prokaryotic Na^+/H^+ exchanger, isolated from *Escherichia coli,* functionally provides a sodium extrusion mechanism driven by a large H^+ gradient. The small *ant* gene ($<$ 1.58 kb) contains a 1,085 bp open reading frame which codes for a 362-amino acid protein (M_r 38,683). The antiporter is organized across the membrane in a fashion similar to the hydrophobic amino-terminal domain of its eukaryotic counterpart. Ten transmembrane helices are found in Engelman hydropathy analysis, along with two sizable extracellular domains. No cytoplasmic domain is predicted in the bacterial Na^+/H^+ antiporter, perhaps reflecting the antiporter's very different regulation in eukaryotes.

2. Sodium/Glucose Cotransporter

Sodium cotransport drives the active uptake of many organic substrates into eukaryotic cells, including sugars and amino acids. The Na^+/glucose carrier is the sole member of this class of transport proteins whose sequence and secondary structure have been examined (Hediger *et al.*, 1987). It was isolated utilizing oocyte expression to screen for an active cDNA clone. The full-length clone has an open reading frame of 2,010 bp and codes for a 662-amino acid protein of M_r 73,080. The secondary structure model (Eisenberg and Garnier analysis) includes 11 membrane-spanning sequences, 5 of which are amphipathic. Key features in the modeled structure are two large hydrophilic domains near the carboxyl terminus. One region is predicted to be extracellular and links transmembrane regions 7 and 8. The other hydrophilic domain is intracellular, and links transmembrane segments 10 and 11. These regions are mostly polar charged residues which, from previous work that implicates lysine residues in glucose binding site formation (Peerce and Wright, 1984), are hypothesized to be the extracellular and intracellular binding sites for glucose. The binding site for Na^+ is unknown; however, it is believed to be approximately 30 Å from the glucose binding site and appears to involve a tyrosine residue (Peerce and Wright, 1985, 1986). Another notable characteristic of the Na^+/glucose carrier is its lack of homology with virtually any other type of sugar transporter, either of prokaryotic or eukaryotic origin. Also, the Na^+/glucose transporter sequence does not contain the consensus Arg-X-Gly-Arg-Arg sequence conserved in all other sugar transporter sequences (Maiden *et al.*, Hediger *et al.*, 1987).

D. Ion Pumps

The maintenance of transmembrane ion gradients requires an ion translocation mechanism that is able to actively transport ions against their electrochemical gradient. Ion pumps are the primary generators of these gradients in mammalian cells since they are capable of tranducing the chemical bond energy of metabolites into the osmotic work of moving an ionic species across the plasma membrane (Skou, 1988). A large number of ion pumps responsible for the transport of a variety of ionic species have been described, most sharing several traits in common. Most utilize the hydrolysis of phosphate groups from ATP as an energy source. We will examine the sequence and secondary structure of the Na^+/K^+-ATPase as an example of this type of transport protein to be compared with other ion pumps.

The Na^+/K^+-ATPase mediates the electrogenic transport of sodium and potassium against their electrochemical gradients. The pump consists of a catalytic subunit (the α subunit) and a smaller subunit of unknown function (the β subunit). The binding site for ATP (Farley *et al.*, 1984) and the phosphorylation site (Bastide *et al.*, 1973) are located on the cytoplasmic face of the catalytic subunit. The binding of cardiac glycosides (ouabain, strophanthidin), which are potent inhibitors of the pump, occurs on the extracellular face of the α subunit. The α subunit is a 1,016 amino acid protein of M_r 110,000 (Shull *et al.*, 1985; Kawakami *et al.*, 1985). The sheep full-length cDNA clone begins with a 254-bp 5′ untranslated region that is GC-enriched. An unambiguous consensus initiation sequence begins an open reading frame of 3,084 bp. A 3′ untranslated region follows, consisting of 340 bp. Eight hydrophobic regions are modeled as transmembrane α-helical segments (H1–H8). Very little of the protein is thought to be extracellular. Both termini are modeled to be cytoplasmic and no cleaved amino-terminal insertion sequence is evident. The phosphorylated intermediate of the pump reaction is predicted to be produced by phosphorylation of an aspartyl residue at position 396. A surrounding sequence of Cys-Ser-(Asp)-Lys, which was predicted by protein radiolabeling and subsequent group-specific proteolysis (Bastide *et al.*, 1973), is found at this region of the cDNA sequence. The nucleotide binding site is modeled to lie approximately 130 amino acids away from the phosphorylation site. It is localized by tracing the site of fluorescein isothiocyanate (FITC) binding, an analog which is thought to interact with part of the nucleotide binding site. A peptide of 10 amino acids, including a reactive lysine, has been independently identified as a FITC binding peptide (Farley *et al.*, 1984; Kirley *et al.*, 1984) and a corresponding coding sequence is found in the ATPase cDNA at residues 496–506. The amino-terminal region is modeled to be the domain for selective ion translocation. This lysine-rich region is postulated to control local sodium and potassium concentrations during the E1–E2 conformational shift (Shull *et al.*, 1985). Interestingly, this amino terminus of approximately 35 amino acids is not found to have a counterpart in the Ca^{2+}-ATPase (MacLennan *et al.*, 1985, reinforcing the hypothesis that it is specific in its monovalent cation translocation function.

The β subunit of the Na^+/K^+-ATPase is a smaller protein of M_r 55,000 (Shull *et al.*, 1986b). Its 302 amino acids are modeled to form only one transmembrane segment. Contact with the α subunit is predicted to occur at the carboxyl-terminal domain of the β subunit. The β subunit has an apparent homology with a small subunit, the KdpC protein, of the homologous *E. coli* K^+-ATPase (Hesse *et al.*, 1984).

Homologies are found throughout the ATPase family. This now includes several Na^+/K^+-ATPases, Ca^{2+}-ATPases, and K^+-ATPases; and H^+-ATPases from eukaryotic as well as prokaryotic sources (Fig. 4). β subunits, however, are only found for the Na^+/K^+- and H^+/K^+-ATPases (Reuben *et al.*, 1990). All exhibit appreciable sequence homology, particularly near the phosphorylation site (Serrano *et al.*, 1986). Each is thought to form an aspartyl-phosphate intermediate conformation in the pump cycle. The H4 segment (residues 313–341 in the Na^+/K^+-ATPase) is a hydrophobic domain whose sequence is highly conserved throughout this family of ion pumps. Because of its proximity to the cytoplasmic phosphorylation site (aspartate 369), it is a likely candidate for transmitting structural conformation changes produced by phosphorylation into the intramembrane portions of the pump (i.e., it could serve as an energy

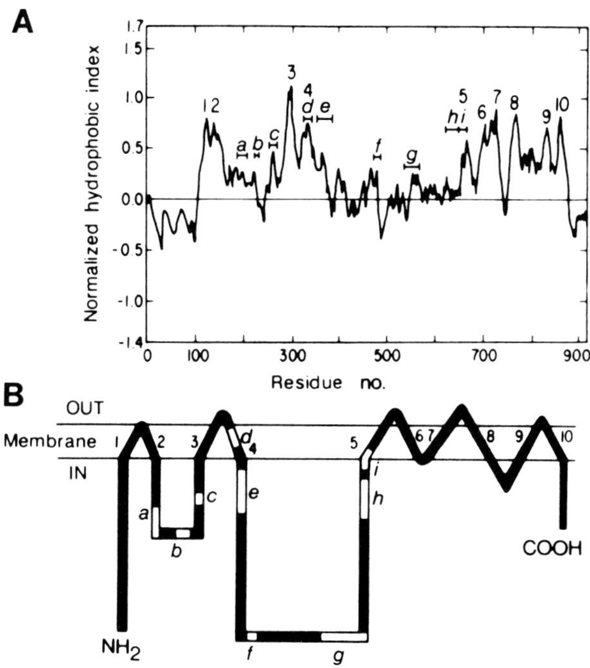

FIG. 4. (A) Hydrophobicity profile of the yeast plasma membrane ATPase. Hydrophobicity values were normalized to a mean value of 0.0 and standard deviation 1.0 and averaged over spans of 21 amino acids, as suggested by Eisenberg. Regions a–i correspond to the conserved sequences shown in C; regions 1–10 are the hydrophobic stretches that are candidates for membrane-spanning sequences. Hydrophilic regions have negative values. (B) Model of the transmembrane structure of the ATPase. The length of the bar representing the polypeptide chain is approximately proportional to the number of amino acids in every domain. The amino-terminal domain (NH_2) contains 115 amino acids and is highly hydrophilic, without any of the features of a signal peptide. Two hydrophobic stretches of ~21 amino acids each in the form of an α helix may allow the protein to cross the membrane twice, placing the amino-terminal domain and the next hydrophilic domain of ~130 amino acids on the same side of the membrane. This hydrophilic region contains the first three conserved sequences shown in C. Hydrophobic stretches 3 and 4 may represent two more membrane-

C

```
                                 a                            b
                     - *  -  **--    *  -* *  -          -* * -* **
Ca2+-ATPase      140 IKAKDIVPGDIVEIAVGDKVPADIRL 165  175 VDQSILTVES 184
Na+ +K+ ATPase   173 INAEEVVVGDLVEVKGGDRIPADLRI 198  206 VDNSSLTGES 215
H+-ATPase        191 IPANEVVPGDILQLEDGTVIPTDGRI 216  225 IDQSAITGES 234
K+-ATPase        121 VPADQLRKGDIVLVEAGDIIPCDGEV 146  153 VDESAITGES 162

                          c                        d
                         **                 -  -*  - - --
                 222 VVVATGVNTEIGK 234  304 VAAIPEGLPAVIT 316
                 250 IVVYTGDRTVMGR 262  322 VANVPEGLLATVT 334
                 259 VVTATGDNTFVGR 271  331 IIGVPVGLPAVVT 343
                 173 FASVTGGTRILSD 185  260 VCLIPTTIGGLLS 272

                                       e
                          - -      -*  -    -  *****-*
                 325 RMAKKNAIVRSLPSVETLGCTSVICSDKTGTLTTN 359
                 343 RMARKNCLVKNLEAVETLGSTSTICSDKTGTLTQN 377
                 352 YLAKKQAIVQKLSAIESLAGVEILCSDKTGTLTKN 386
                 281 RMLGANVIATSGRAVEAAGDVDVLLLDKTGTITLG 315
                                               ↑

                       f
                      **
                 513 VKGAPEGVIDRCT 525
                 500 MKGAPERILDRCS 512
                 473 VKGAPLSALKTVE 485
                 394 RKGSVDAIRRHVE 406

                                    g
                     *  -          *  *--  *-***    *
                 600 DPPRIEVASSVKLCRQAGIRVIMITGDNKGTAV 632
                 586 DPPRAAVPDAVGKCRSAGIKVIMVTGDHPITAK 618
                 534 DPPRDDTAQTVSEARHLGLRVKMLTGDAVGIAK 566
                 447 DIVKGGIKEAFAQLRKMGIKTVMITGDNRLTAA 479

                                   h
                     -    *    - * **** ** * *  *- --*
                 686 IVEFLQSFDEITAMTGDGVNDAPALKKAEIGIAM 719
                 694 IVEGCQRQGAIVAVTGDGVNDSPALKKADIGVAM 727
                 618 VVEILQNRGYLVAMTGDGVNDAPSLKKADTGIAV 651
                 500 LALIRQAEGRLVAMTGDGTNDAPALAQADVAVAM 533

                             i
                      --   - *   -
                 721 SGTAVAKTASEMVLADDNF 739
                 730 AGSDVSKQAADMILLDDNF 748
                 653 GATDAARSAADIVFLAPGL 671
                 537 SGTQAAKEAGNMVDLDSNP 555
```

spanning domains, and would place the large hydrophilic domain of ~310 amino acids on the same side of the membrane as the two hydrophilic domains that are more proximal to the amino terminus. The hydrophilic central domain contains five of the conserved sequences shown in C; as this segment also contains the phosphorylation site and the ATP binding site, it should be exposed on the cytoplasmic side of the membrane. The conserved region d (corresponding to most of hydrophobic stretch 4) is close to the phosphorylation site, and has been proposed to constitute an energy transducing pathway into the transmembrane domain that is involved in cation transport. The last six hydrophobic domains may also span the membrane. If so, the hydrophilic carboxy-terminal domain (COOH) of ~46 amino acids would be on the cytoplasmic side, leaving very little of the enzyme exposed to the external medium. (C) Conserved sequences in four different ATPases having phosphorylated intermediates. The aspartyl residue that forms such an intermediate is marked by an arrow. Only sequences that are conserved in at least three of the enzymes are shown and residues that are identical in all four enzymes are indicated by asterisks. Highly conserved replacements are indicated by a dash. Homology comparisons were made using the DIAGON program of Staden. The Na^+/K^+-ATPase sequences corresponds to the sheep kidney enzyme, the Ca^{2+}-ATPase sequence is from the cardiac muscle sarcoplasmic reticulum, K^+-ATPase refers to the product of the *KdpB* gene of *E. coli,* and the H^+-ATPase refers to the yeast plasma membrane ATPase. (From Serrano *et al.*, 1986, with permission.)

transducing mechanism). A variable characteristic of the Na^+/K^+ pumps is their sensitivity to ouabain. The isolation of cDNA sequence from ouabain-sensitive and -insensitive species and site-directed mutagenesis has thus far been unable to identify a specific primary sequence responsible for glycoside binding. Presently, incorporating several lines of evidence, the ouabain binding site is thought to be constructed from portions of the molecule at the H1–H2 extracellular junction (Shull *et al.*, 1986b; Lingrel *et al.*, 1990).

Hydropathy plots of the various ATPase sequences predict a very similar secondary structure organization, the consensus interpretation of which, unlike the anion carrier, is not borne out by the organization of exons domains. An alternative model proposing only seven transmembrane helices for the Na^+/K^+ pump, however, is quite consistent with the organization of exon domains (Ovchinnikov *et al.*, 1986, 1988; Modyanov, 1990). Additionally, secondary structure modeling for ATPases has been predicted based on high-resolution electron micrographs. Negatively stained vesicles of sarcoplasmic reticulum, enriched in Ca^{2+}-ATPase, display a visible globular formation approximately 35 Å in diameter (Greaser *et al.*, 1969). This "head" is attached to the cytoplasmic membrane face by a thin "stalk" construction. Functional correlates to these structures have been suggested (MacLennan *et al.*, 1985; Brandl *et al.*, 1986; Green *et al.*, 1988). The large cytoplasmic domains of the ATPase could comprise the globular portion of the protein containing the phosphorylation and nucleotide binding sites. A stalk region is proposed to tether the headpiece to the transmembrane helical domains. In Ca^{2+}-ATPase, this region, when modeled for helical structure, is strongly amphipathic, and is enriched with aligned glutamate residues, possibly forming a calcium binding site.

The ATPases are encoded by distinct, though ancestrally related, genes which preserve sequence homology as well as intron position (Ovchinnikov *et al.*, 1988). Several isoforms of Na^+/K^+-ATPase and the Ca^{2+}-ATPase have been isolated. The Na^+/K^+-ATPase catalytic subunit has at least three isoforms (Shull *et al.*, 1986a), which vary in size and in sensitivity to insulin and cardiac glycosides. Three β subunits have also been identified (Lingrel *et al.*, 1990). Several forms of these subunits may be found in the same tissue, albeit in variable abundance. Separate genes of 20–25 kb for each α isoform have been identified and are closely grouped on human chromosome 1 (Shull and Lingrel, 1987; Chehab *et al.*, 1987). Calcium ATPases have different isoforms as well. However, their distribution appears more dependent on tissue type, with fast twitch and slow twitch muscle exhibiting distinct isoforms. Genes for each isoform have been isolated and mapped to human chromosomes 16 and 12, respectively (MacLennan *et al.*, 1987).

E. Ion Channels

A distinct subset of membrane transport proteins are those found to produce large discrete, measurable changes in the membrane's electrical conductance. The requirement for such an observation is a transiently produced aqueous pore across the membrane bilayer that is capable of conducting ions. In some respects, ion channel transport is rather simple compared to pumps or facilitated carriers in that a simple single-state transition from closed pore to open pore is all that is required to produce a functional channel (Miller, 1989). Channel complexities arise predominantly due to temporal control of these transitions by voltage or ligand binding (gating) and also in producing ion-selective transport through an opened pore. Molecular structure/function studies of ion channels are further advanced in that they have been accompanied by advances in extremely precise functional measurements of channel activity at the single molecule level, provided by patch electrode analysis.

Several cDNAs for ion channels have been sequenced and notable trends in channel structure observed. First, channels appear to be formed by aggregation of multiple, similar (or identical) subunits. In the case of Na^+ and Ca^{2+} channels, the "subunits" are homologous repeat domains found within one large polypeptide (Noda *et al.*, 1984; Tanabe *et al.*, 1987), whereas nicotinic acetylcholine receptor channels are composed of multiple different peptide subunits (Changeux *et al.*, 1984). Hypotheses stemming from sequence and imaging studies of the acetylcholine receptor channel suggest that these subunits form transmembrane pores by aggregation. The cylindrical aggregate then supports an aqueous pore at its center (Noda *et al.*, 1982; Toyoshima and Unwin, 1988). The diameter of the various channel pores appears to be determined by the number of subunits in the aggregate, varying from about 16 Å for a hexamer to 4 Å for a tetramer (Unwin, 1986). A second trend noted in sequence structure are arrays of hydrophobic residues which appear to be arranged in helical form, producing an alignment of polar and apolar surfaces, an amphipathic helix. The key functional features these amphipathic helices confer on channels have been demonstrated in a simplified fashion by structure/function analysis of synthetic peptide ion channels. These studies serve to define the minimal requirements for channel identity (Lear *et al.*, 1988). Extremely simple peptides were designed with several features in mind: (1) the peptides were of minimal size to form an α helix that would span a lipid bilayer; (2) the helix makeup was amphipilic to provide aggregation of polar faces and retain a hydrophobic shell to be stabilized by the lipid bilayer; (3) the helix was composed of only two amino acid residues for simplicity, leucine to produce hydrophobic and helical interactions, and serine for its polar side chain, which, in proper orientation, can donate and accept hydrogen bonds with water, thereby creating a hydrated environment. When *three* repeats of seven residues, NH_2-$(LSSLLSL)_3$-$CONH_2$, were inserted into lipid bilayers, spontaneous channel events that were very similar to those of the acetylcholine receptor were found. Replacing one serine with leucine [$(LSLLLSL)_3$] was found to alter the channel's selectivity from one preferring small cations to one selective exclusively for protons. These results, as well as computer modeling of energetically likely conformations that these helical segments could produce, test several predictions of channel structure. Based on this work, it seems likely that polar faces of α helices from neighboring subunits can aggregate to form ion-conducting pores. It also demonstrates that very minor changes in amino acid sequence can produce dramatic functional changes. It is possible that changing amino acids to vary their contribution to helical energy minima (leucines and prolines, for example) may produce dramatic changes in interhelical packing and thereby result in corresponding changes in pore size and selectivity.

The following sections highlight molecular aspects of several ion channels. An in-depth analysis is not attempted; however, highlights exhibited by specific channels or channel types are presented. Further descriptions are available for some channels in other chapters of this volume.

1. Sodium and Calcium Channels

A key property of some ion channels of excitable membranes is a dependence on changes in membrane potential to vary their conformation and conductive states. Sequences and structures for several Na^+ channels (Noda *et al.*, 1986a) and for a dihydropyridine receptor/Ca^{2+} channel (Tanabe *et al.*, 1987) are presently available, along with two types of voltage-dependent K^+ channels: a well-studied A-type channel from *Drosophila* (Papazian *et al.*, 1987; Kamb *et al.*, 1987), and delayed rectifier channels from mammalian brain (Baumann *et al.*, 1988; Frech *et al.*, 1989).

Sodium channels from eel electroplaque (Noda *et al.*, 1982), rat and rabbit muscle (Barchi, 1983; Barchi *et al.*, 1984), and three channel cDNAs from rat brain (Noda *et al.*, 1986a) have been isolated and sequenced. In each case, the channel is a product of a separate gene, but they all exhibit identical biophysical properties. In all cases, the primary polypeptide produced is large, approximately 1800–2000 amino acids. The sequence contains four homologous internal repeats (I–IV), with each of these possessing up to six possible transmembrane hydrophobic segments (Fig. 5). These six repeats were designated S1–S6 by Noda *et al.* (1984) and they have come to act as signposts for channels of this type. S1 and S3 are composed primarily of nonpolar residues and they exhibit a net negative

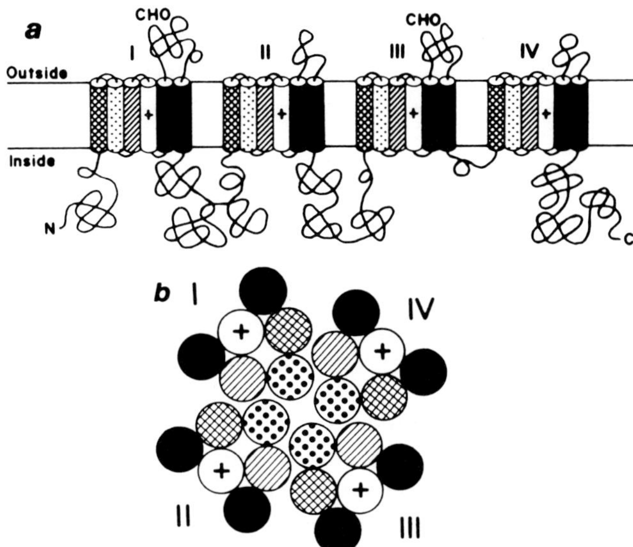

FIG. 5. (a) Proposed transmembrane topology of the sodium channels; (b) proposed arrangement of the transmembrane segments viewed in the direction perpendicular to the membrane. In a, the four units of homology spanning the membrane are displayed linearly. Segments S1–S6 in each repeat (I–IV) are indicated by cylinders as follows: S1, cross-hatched; S2, stippled; S3, hatched; S4, indicated by a plus sign; S5 and S6, solid. Putative sites of N-glycosylation (CHO) are indicated. In b, the ionic channel is represented as a central pore surrounded by the four units of homology. Segments S1–S6 in each repeat (I–IV) are represented by circles indicated as in a. (From Noda *et al.,* 1986a, with permission.)

charge. S2 is also predominantly nonpolar, but with a neutral charge. If a helical structure is assumed for each of these segments, the nonpolar residues are aligned on one side of the helix, with the opposite side containing the polar charged residues, thus forming amphipathic helices. S5 and S6 are strongly hydrophobic and uncharged and presumed to be helical structures. A signature for voltage-dependent channels is apparent in segment S4. Here, residues are largely nonpolar, but they are interrupted at every third residue by a postively charged lysine or arginine residue. Site-directed mutagenesis has verified the S4 segment to be the voltage sensor for channel activation (Stuhmer *et al.,* 1989). In repeat I, replacement of only one positive residue with an uncharged glutamine residue in segment S4 was found to affect greatly the activation kinetics (Hodgkin–Huxley *m* parameter) without affecting inactivation kinetics. Mutations have been directed toward corresponding residues of the other internal repeats; however, these mutants fail to express functional channels, so the role of these sites remains unclear. In a similar vein, mutations directed toward intracellular residues between repeats III and IV have implicated this region as critical in channel inactivation (Stuhmer *et al.*, 1989). This finding seems confirmed by studies with an antibody raised against the peptide sequence between repeat III and IV. The antibody has been found to alter channel inactivation kinetics (Vassilev *et al.*, 1988). Whether the specific residues changed in mutation studies are singularly responsible for the altered feature of channel function or if the altered function results from some more global structural effect remains the hardest question to resolve, and is largely unknown.

The DHP receptor/Ca^{2+} channel is found to be organized very similarly to the Na^+ channel (Tanabe *et al.*, 1987). It is of similar size (1873 residues) and also contains four homologous internal repeats, each with six transmembrane segments (Fig. 6). Each of these is highly homologous to those of the Na^+ channel, including the voltage-sensitive S4 segment. The homology in sequence and organization of the structural elements suggests that the Na^+ and Ca^{2+} channel genes evolved from a common ancestral channel gene, a gene which itself underwent tandem duplications of the basic S1–S6 unit to produce a pseudomultimeric structure.

2. Potassium Channels

Two types of voltage-dependent K^+ channels have been isolated and sequenced. The most thoroughly studied is the A-type channel, or fast-inactivating K^+ channel, from the Shaker locus of *Drosophila*. Two delayed rectifier K^+ channels have also been cloned and sequenced.

The Shaker gene was identified by cloning its chromosomal locus, taking advantage of its linkage to known genetic markers. Genomic sequence subsequently allowed cDNA clones to be isolated and studied (Kamb *et al.*, 1987; Papazian *et al.*, 1987). An immediately noticeable difference between the Shaker clones and other channel clones is the small size of Shaker cDNAs, being only about one-quarter to one-third the size of those of other voltage-dependent ion channels. It is therefore hypothesized that the Shaker K^+ channel is a multimer made up of several individual subunits. Thus, instead of multiple repeats of subunits within the same polypeptide, its subunits are individual peptides, potentially each encoded by different mRNAs. In common with other voltage-gated channels, each subunit has six putative transmembrane segments, including the voltage sensor S4 region (Fig. 6). Several different cDNA clones from the Shaker locus have been isolated and sequenced and each has its own characteristic current profile when expressed alone in oocytes (Timpe *et al.*, 1988a,b). When traced back to their genomic origins, it is found that all the different cDNAs of the various clones come from a single gene that possesses 12

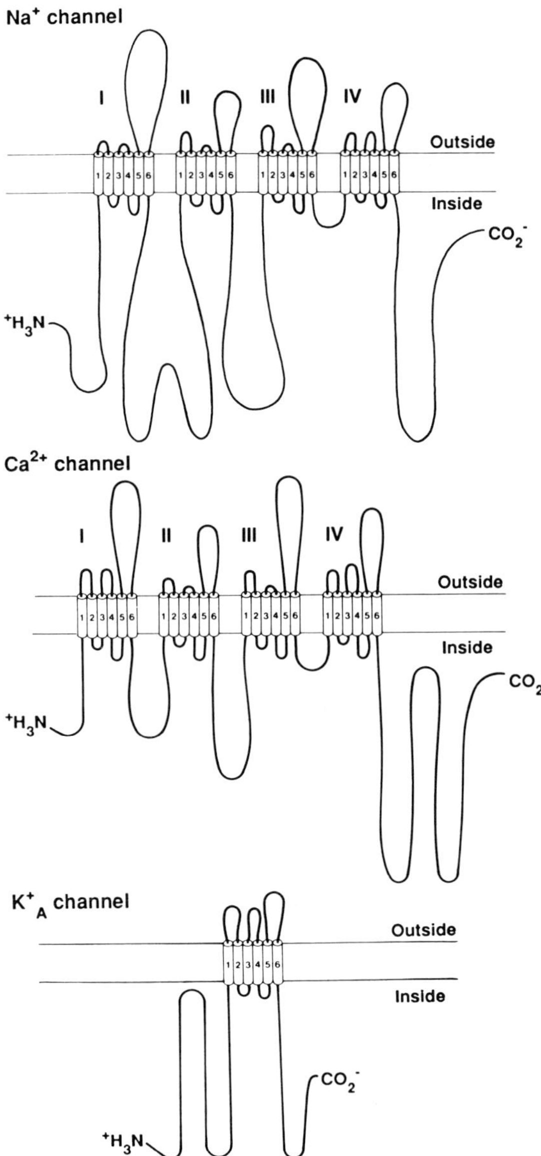

FIG. 6. Proposed transmembrane arrangements of the principal subunits of Na^+, Ca^{2+}, and A-current K^+ channels (K^+_A). The protein folding models for rat brain Na^+ channels, rabbit skeletal muscle Ca^{2+} channels, and *Drosophila* A-current K^+ channels are presented to illustrate overall sequence similarities. Segment H5, which is proposed to be transmembrane, is illustrated as part of an extracellular segment between S5 and S6. (From Catterall, 1988, with permission.)

different exons ranging over 65 kb of DNA—the Shaker locus (Schwarz *et al.*, 1988). Each cDNA is found to share a central region of some 1163 nucleotides. This central region is encoded by eight exons (about 10 kb of genomic DNA) that form a basic channel core. Each clone, however, possesses distinct 5′ and 3′ exons that come from various parts of the Shaker locus. These additional coding regions modify the "prototype" channel function. The suggestion is that the multiple cDNA clones are the result of alternative splicing of the large primary transcript (Schwarz *et al.*, 1988). Therefore, the Shaker locus can be viewed as a large transcription unit capable of coding for several different peptides. These peptides can be mixed and matched to produce distinct types of A-type K^+ channels, perhaps in a tissue-specific manner. It seems that the evolutionary strategy of the K^+ channel progenitor was to increase diversity, taking advantage of the combinatorials available through alternative splicing and independent subunits in a multimer. This strategy strongly contrasts with that taken by the Na^+ channel. Here, evolutionary pressure seemed to call for a fixed and ordered combination of subunits to produce a constant finely tuned stereotypical channel event. Random combination of subunits could not be chanced, so instead chromosomal duplication of the locus produced one specific ordered array of repeats I through IV.

Other loci (Shab, Shaw, Shal) coding for K^+ channels homologous to Shaker clones have been isolated in *Drosophila* strains which lack a functional Shaker locus (Butler *et al.*, 1989). The diversity of K^+ channels thus is not defined only by alternative splicing at a single locus but is additionally complex in that other loci coding for similar channels may be present in the same system.

Two delayed rectifier channel clones, rck1 and drk1, have been isolated from rat brain (Baumann *et al.*, 1988; Frech *et al.*, 1989). The drk1 clone is a 3.4-kb cDNA producing a K^+ channel of 853 amino acids. Analysis of the peptide sequence shows that the amino-terminal half of the protein resembles other voltage-gated channels in that it contains six putative transmembrane segments, including a typical S4 region. The carboxyl-terminal half of the protein is a large cytoplasmic domain of unrelated sequence, though it does contain two sites for potential cyclic AMP-dependent phosphorylation.

The rck1 clone is a delayed rectifier clone notable because it was isolated by hybridization with a Shaker cDNA probe, indicating substantial homology with the A-type channel. The peptide sequence of the channel shows 50% homology to the amino acid sequence of Shaker-type channels. Interestingly, the rck1 channel retains a similar conductance, selectivity, and voltage dependence to the A-type currents, and varies only in its inactivation profile. Analysis of the sequence data shows the cDNA to be highly homologous to Shaker in the core region but to vary in a carboxyl-terminal extension sequence. It is therefore likely that the Shaker locus, as well as the Shab and Shaw loci, are capable of coding for different types of K^+ channels, not simply A-type channels. Notably, a 7-amino acid sequence occurring at the beginning of the "core" sequence of Shaker has been found in every K^+ channel clone isolated thus far (Frech *et al.*, 1989).

3. Acetylcholine Receptor Channel

The nicotinic acetylcholine receptor (AChR) channel is composed of four types of homologous subunites: α, β, γ, and δ. The α subunit has been demonstrated to carry the ACh binding site and bungarotoxin binding site. The stoichiometry and functional organization of the channel are presented by Blount and Merlie (Chapter 8, this volume) and are not repeated here. However, site-directed mutagenesis experiments have been performed on this channel system that reveal properties of the channel pore lining and indicate residues which determine its conductive properties. As with the previously discussed channels, the AChR channel is composed of multisubunits, each with obvious hydrophobic, transmembrane segments, designated M1–M4 (Noda *et al.*, 1982) and MA (Finer-Moore and Stroud, 1984). The residues in areas M1–M4 are predominantly uncharged, nonpolar residues; however, polar residues are found in each segment. If the MA segment is modeled as an α helix, the polar residues align, forming a classical amphipathic helix. For this reason, each subunit was initially hypothesized to contribute its MA region to the pore lining. However, deletion and site-directed mutagenesis experiments have clearly demonstrated an alternative arrangement. By deleting the MA region (Mishina *et al.*, 1985) or by mutating negatively charged and glutamine residues exclusively in regions closely flanking the M2 segment (Imoto *et al.*, 1988), it has been demonstrated that the M2 segment, and not the amphipathic MA helix, dominates the permeation properties of the pore. The conductance properties of the pore were shown by Imoto *et al.* (1988) to be determined by a negatively charged ring structure composed of glutamine residues which flank the M2 region. The failure of similar mutations in other transmembrane regions (M1, M3, M4) to produce variations in conductance properties suggests that they do not contribute to the pore lining. More critical to the modeling of channel structure, these works point out that the role long presumed for amphipathic helices, that of forming channel pore linings, is not universal and may not be correct.

IV. CONCLUSIONS AND PERSPECTIVES

In assessing the structures already determined for membrane transport proteins, it is apparent that Nature has chosen common themes for these molecules. At the most basic level is the structure contributed by the hydrophobic α helix. This hydrogen-bonded structure appears ideally suited to spanning the lipid bilayer: all of the H bonds occur intramolecularly and a hydrophobic face is left to interact with the lipid. For this reason, α helices are primarily utilized as the membrane-spanning domains of the transporters. Several helices appear to be required to form an aggregate which can provide a transmembrane substrate pathway sequestered within a hydrophobic shell. To some extent, the amphipathic nature of certain helices may facilitate their aggregation, since hydrophilic faces must come to be sequestered. The transmembrane pathway appears to be remarkably similar for all the very different physiological transport mechanisms, from carriers to pumps to channels. It is composed of side chains from key amino acids found in the helices. The physiologically distinct mechanisms seem to arise from apparently subtle differences in the kinetics of conformational changes within these protein pore structures.

As the studies of Lear *et al.* (1988) demonstrate, extremely simple peptides *can* suffice to translocate ions across a membrane, much simpler than those found in naturally occurring transport proteins. Surprisingly minor alterations can radically change the nature of the transported species. The variations now found to complicate what must have been a simple primordial theme likewise appear to define the core evolutionary families to be the transport *mechanisms*. Nature appears to have discovered a variation useful, creating a distinct physiological function we now recognize in the manifestation of pumps, channels, or carriers. For instance, it is clear that all channels are much more similar than are all Na^+ transporting proteins (Na^+ channel, Na^+/H^+ antiporter, Na^+/glucose cotransporter, Na^+/K^+ pump). Apparently, once the primordial mechanisms arose, it was relatively easy to diversify that prototype to allow it to serve other substrate species. It is not obvious that this needed to be the case. If it were very hard to devise a mechanism which could transport Na^+, it might well have been that the different mechanisms of Na^+ transport reflected minor variations in the primordial Na^+ pathway. Certainly, nucleotide binding sites, regulatory phosphorylation sites, and "energizing" phosphorylation sites have proved to be just that type of conserved genetic subroutine more easily disseminated (as an exon unit?) to a variety of different proteins than reinvented *de novo* in each. It is also clear that the diversification of these mechanisms has not been limited to these changes occurring within key residues of the transmembrane helices, but that entire domains could be found appended to this cardinal structure, such as the cytoskeleton binding domain of band 3, the regulatory domain of the eukaryotic Na^+/H^+ antiporter, or the headpiece of the pumps.

Molecular biological techniques have invigorated studies elucidating the structures and functions of membrane transport proteins. A beginning of a molecular basis for physiological function is being defined. Equally important to the structural descriptions we have discussed is that molecular techniques are also beginning to describe the structures and sites responsible for the regulation of transcription and expression of the transporters, opening the vista to understanding a new layer of physiological control.

References

Alper, S. L., Natale, J., Gluck, S., Lodish, H. F., and Brown D. (1989). Subtypes of intercalated cells in rat kidney collecting duct defined by antibodies against erythroid band 3 and renal vacuolar H^+-ATPase. *Proc. Natl. Acad. Sci. U.S.A.* **86,** 5429–5433.

Aronson, P., and Boron, W. F., eds. (1986). "Na-H Exchange, Intracellular pH and Cell Function," Current Topics in Membranes and Transport, Vol. 26. Academic Press, New York.

Barchi, R. (1983). Protein components of the purified sodium channel from rat skeletal muscle. *J. Neurochem.* **40,** 1377–1385.

Barchi, R., Tanaka, J. C., and Furman, R. (1984). Molecular characteristics and functional reconstitution of muscle voltage-sensitive sodium channels. *J. Cell. Biochem.* **26,** 135–146.

Bastide, F., Meissner, G., Fleischer, S., and Post, R. L. (1973). Similarity of the active site of phosphorylation of adenosine triphosphate for transport of sodium and potassium ions in kidney to that of transport of calcium ions in the sarcoplasmic reticulum of muscle. *J. Biol. Chem.* **248,** 8385–8391.

Baumann, A., Grupe, A., Ackerman, A., and Pongs, O. (1988). Structure of the voltage-dependent potassium channel is highly conserved from Drosophila to vertebrate central nervous systems. *Embo J.* **7,** 2457–2463.

Bennett, V., and Stenbuck, P. J. (1980). Association between ankyrin and the cytoplasmic domain of band 3 isolated from the human erythrocyte membrane. *J. Biol. Chem.* **255,** 6424–6432.

Birnbaum, M. (1989). Identification of a novel gene encoding an insulin-responsive glucose transporter protein. *Cell* **57,** 305–315.

Birnstiel, M., Busslinger, M., and Strub, K. (1985). Transcription termination and 3′ processing. The end is in site. *Cell* **41,** 349–359.

Brandl, C. J., Green, N. M., Korczak, B., and MacLennan, D. (1986). Two Ca ATPase genes: Homologies and mechanistic implications of deduced amino acid sequences. *Cell* **44,** 597–607.

Butler, A., Wei, A., Baker, K., and Salkoff, L. (1989). A family of putative potassium channel genes in Drosophila. *Science* **243,** 943–947.

Catterall, W. A. (1988). Structure and function of voltage-sensitive ion channels. *Science* **242,** 50–61.

Changeux, J. P., Devillers-Thiery, A., and Chermouille, P. (1984). Acetylcholine receptor: An allosteric protein. *Science* **225,** 1335–1345.

Chehab, F., Kan, Y., Lau, M., Hartz, J., Kao, F., and Blotstein, R. (1987). Human placental Na,K-ATPase α subunit: cDNA cloning, tissue expression, DNA polymorphism, and chromosomal localization. *Proc. Natl. Acad. Sci. U.S.A.* **84,** 7901–7905.

Chou, P. Y., and Fasman, G. O. (1978). Empirical predictions of protein conformation. *Annu. Rev. Biochem.* **47,** 251–276.

Crick, F. (1979). Split genes and RNA splicing. *Science* **204,** 264–271.

Darnell, J., Lodish, H., and Baltimore, D. (1986). "Molecular Cell Biology." Freeman, San Francisco, California.

Darnell, J. R. (1978). Implications of RNA-RNA splicing in evolution of eukaryotic cells. *Science* **202,** 1257–1260.

Deisenhofer, J., Epp, O., Miki, K., Huber, R., and Michel, H. (1985). Structures of the protein subunits in the photosynthetic reaction center of Rhodopseudomonas viridis at 3 Å resolution. *Nature (London)* **318,** 618–624.

Dunn, R., McCoy, J., Simsek, M., Majumdar, A., Chang, S., Rajbandary, U., and Khorana, H. G. (1981). The bacteriorhodopsin gene. *Proc. Natl. Acad. Sci. U.S.A.* **78,** 6744–6748.

Dynan, W. S. (1989). Modularity in promoters and enhancers. *Cell* **518,** 1–4.

Dynan, W., and Tjian, R. (1985). Control of eukaryotic messenger RNA synthesis by sequence-specific DNA-binding proteins. *Nature (London)* **316,** 774–778.

Eisenberg, D., Schwarze, E., Komaromy, M., and Wall, R. (1984). Analysis of membrane and surface protein sequences with the hydrophobic moment plot. *J. Mol. Biol.* **179,** 125–142.

Engelman, D. M., Steitz, T., and Goldman, A. (1986). Identifying nonpolar transbilayer helices in amino acid sequences of membrane proteins. *Annu. Rev. Biophys. Biophys. Chem.* **5,** 321–353.

Farley, R., Tran, C., Carilli, C., Hawke, D., and Shively, J. (1984). The amino acid sequence of a fluorescein-labelled peptide from the active site of (Na,K)-ATPase. *J. Biol. Chem.* **259,** 9532–9535.

Finer-Moore, J., and Stroud, R. M. (1984). Amphipathic analysis and possible formation of the ion channel in an acetylcholine receptor. *Proc. Natl. Acad. Sci. U.S.A.* **81,** 155–159.

Frech, G., VanDongen, A. M., Schuster, G., Brown, A., and Joho, R. (1989). A novel potassium channel with delayed rectifier properties isolated from rat brain by expression cloning. *Nature (London)* **340,** 642–645.

Fukumoto, H., Sieno, S., Imura, H., Seino, Y., Eddy, R., Fukushima, Y., Byers, M., Shows, T., and Bell, G. (1988). Sequence, tissue distribution, and chromosomal localization of mRNA encoding a human glucose transporter-like protein. *Proc. Natl. Acad. Sci. U.S.A.* **85,** 5434–5438.

Gargus, J. J. (1989). Tools for the molecular analysis of gene structure and function. *In* "Molecular Biology in Physiology" (S. Chien, ed.), pp. 19–34. Raven, New York.

Greaser, M. L., Cassens, R., Hoekstra, W., and Briskey, E. (1969). Purification and ultrastructure properties of the calcium accumulating membranes in isolated sarcoplasmic reticulum preparations from skeletal muscle. *J. Cell. Physiol.* **74,** 37–50.

Green, N. M., Taylor, M., and MacLennan, D. (1988). A consensus structure for cation pumps. *In* "The Ion Pumps: Structure, Function, and Regulation," (W. Stein, ed.), pp. 15–24. Alan R. Liss, New York.

Grinstein, S., Ship, S., and Rothstein, A. (1978). Anion transport in relation to proteolytic dissection of band 3 protein. *Biochim. Biophys. Acta* **507,** 294–304.

Hartmann, E., Rapoport, T. A., and Lodish, H. F. (1989). Predicting the orientation of eukaryotic membrane-spanning proteins. *Proc. Natl. Acad. Sci. U.S.A.* **86,** 5786–5790.

Hediger, M. A., Coady, M. J., Ikeda, T. S., and Wright, E. M. (1987). Expression cloning and cDNA sequencing of the Na/glucose co-transporter. *Nature (London)* **330,** 379–381.

Henderson, R., and Unwin, P. N. T. (1975). Three-dimensional model of purple membrane obtained by electron microscopy. *Nature (London)* **257,** 28–32.

Hesse, J., Wieczorek, L., Altendorf, K., Reicin, A. S., Dorus, E., and Epstein, W. (1984). Sequence homology between two membrane transport ATPases, the Kdp-ATPase of Escherichia coli and the Ca-ATPase of sarcoplasmic reticulum. *Proc. Natl. Acad. Sci. U.S.A.* **81,** 4746–4750.

Imoto, K., Busch, C., Sakmann, B., Mishina, M., Konno, T., Nakai, J., Bujo, H., Mori, Y., Fukuda, K., and Numa, S. (1988). Rings of negatively charged amino acids determine the acetylcholine receptor channel conductance. *Nature (London)* **335,** 645–648.

James, D., Brown, R., Navarro, J., and Pilch, P. (1988). Insulin-regulatable tissues express a unique insulin-sensitive glucose transport protein. *Nature (London)* **333,** 183–185.

James, D., Strube, M., and Mueckler, M. (1989). Molecular cloning and characterization of an insulin-regulatable glucose transporter. *Nature (London)* **338,** 83–87.

Jennings, M. (1985). Kinetics and mechanism of anion transport in red blood cells. *Annu. Rev. Physiol.* **47,** 519–533.

Kamb, A., Iverson, L. E., and Tanouye, M. A. (1987). Molecular characterization of *Shaker,* a *Drosophila* gene that encodes a potassium channel. *Cell* **50,** 405–413.

Karpel, R., Olami, Y., Taglicht, D., Schuldiner, S., and Padan, E. (1988). Sequencing of the gene *ant* which affects the Na/H antiporter activity in *Escherichia coli. J. Biol. Chem.* **263,** 10408–10414.

Kawakami, K., Noguchi, S., Noda, M., Takahashi, H., Ohta, T., Kawamura, M., Nojima, H., Nagano, K., Hirose, T., Inayama, S., Hayashida, H., Miyata, T., and Numa, S. (1985). Primary structure of the α-subunit of *Torpedo californica* (Na+K)ATPase deduced from cDNA sequence. *Nature (London)* **316,** 733–736.

Khorana, H. G., Gerber, G. E., Herlihy, W., Gray, C., Anderegg, R., Nihei, K., and Biemann, K. (1979). Amino acid sequence of bacteriorhodopsin. *Proc. Natl. Acad. Sci. U.S.A.* **78,** 2225–2229.

Kirley, T., Wallick, E., and Lane, L. (1984). The amino acid sequence of the fluorescein isothiocyanate reactive site of lamb and rat kidney Na and K dependent ATPase. *Biochem. Biophys. Res. Commun.* **125,** 767–773.

Kopito, R. R., and Lodish, H. F. (1985). Primary structure and transmembrane orientation of the murine anion exchange protein. *Nature (London)* **316,** 234–238.

Kopito, R. R., Anderson, M. and Lodish, H. (1987). Structure and organization of the murine band 3 gene. *J. Biol. Chem.* **262,** 8035–8040.

Kopito, R. R., Anderson, M., and Lodish, H. (1989). Molecular genetics of the mouse anion exchanger. *In* "Molecular Biology in Physiology" (S. Chien, ed.), pp. 35–46. Raven, New York.

Kozak, M. (1988). An analysis of 5′-noncoding sequences from 699 vertebrate messenger RNAs. *Nucleic Acids Res.* **15,** 8125–8148.

Kyte, J., and Doolittle, R. F. (1982). A simple method for displaying the hydropathic character of a protein. *J. Mol. Biol.* **157,** 105–132.

Lear, J., Wasserman, Z., and DeGrado, W. F. (1988). Synthetic amphiphilic peptide models for protein ion channels. *Science* **240,** 1177–1181.

Lingrel, J. B., Orlowski, J., Price, E., Jewell, E., Szamraj, O., Pathak, B., and Tyson, P. (1990). Structure function studies of the Na,K-ATPase and regulation of the α-subunit genes. *J. Gen. Physiol.* **96,** 1a.

MacLennan, D., Brandl, C., Korczak, B., and Green, N. M. (1985). Amino-acid sequence of a Ca^{++}/Mg^{++}-dependent ATPase from rabbit muscle sarcoplasmic reticulum, deduced from its complementary DNA sequence. *Nature (London)* **316,** 696–700.

MacLennan, D., Brandl, C., Champaneria, S., Holland, P., Powers, V., and Willard, H.

(1987). Fast-twitch and slow-twitch/cardiac Ca ATPase genes map to human chromosomes 16 and 12. *Somatic Cell Mol. Genet.* **13,** 341–346.

Maiden, M., Davis, E., Baldwin, S., Moore, D., and Henderson, P. (1987). Mammalian and bacterial sugar transport proteins are homologous. *Nature (London)* **325,** 641–643.

Maniatis, T., and Reed, R. (1987). The role of small nuclear ribonucleoprotein particles in pre-mRNA splicing. *Nature (London)* **325,** 673–678.

Maniatis, T., Goodbourn, S., and Fischer, J. A. (1987). Regulation of inducible and tissue-specific gene expression. *Science* **236,** 1237–1245.

Michel, H. (1983). Crystallization of membrane proteins. *TIBS* **8,** 56–59.

Miller, C. (1989). Genetic manipulation of ion channels: A new approach to structure and mechanism. *Neuron* **2,** 1195–1205.

Mishina, M., Tobimatsu, T., Imoto, K., Tanaka, K., Fujita, Y., Fukuda, K., Kurasaki, M., Takahashi, H., Morimoto, Y., Hirose, T., Inayama, S., Takahashi, T., Kuno, M., and Numa, S. (1985). Location of functional regions of acetylcholine receptor α-subunit by site-directed mutagenesis. *Nature (London)* **313,** 364–369.

Modyanov, N. N. (1990). Probing the folding of the sodium pump subunits. *J. Gen. Physiol.* **96,** 2a.

Mueckler, M., and Lodish, H. (1986). The human glucose transporter can insert posttranslationally into microsomes. *Cell* **44,** 629–637.

Mueckler, M., Caruso, C., Baldwin, S., Panico, M., Blench, I., Morris, H., Allard, W. J., Lunhard, G., and Lodish, H. (1985). Sequence and structure of a human glucose transporter. *Science* **229,** 941–945.

Noda, M., Takahashi, H., Tanabe, T., Toyosato, M., Furutani, Y., Hirose, T., Asai, M., Inayama, S., Miyata, T., and Numa, S. (1982). Primary structure of α-subunit precursor of *Torpedo californica* acetylcholine receptor deduced from cDNA sequence. *Nature (London)* **299,** 793–797.

Noda, M., Shimizu, S., Tanabe, T., Takai, T., Kayano, T., Ikeda, T., Takahashi, H., Nakayama, H., Kanaoka, Y., Minamino, N., Kangawa, K., Matsuo, H., Raftery, M. A., Hirose, T., Inayama, S., Hayashida, H., Miyata, T., and Numa, S. (1984). Primary structure of *Electrophorus electricus* sodium channel deduced from cDNA sequence. *Nature (London)* **312,** 121–127.

Noda, M., Takayuki, I., Kayano, T., Suzuki, H., Takeshima, H., Kurasaki, M., Takahashi, H., and Numa, S. (1986a). Existence of distinct sodium channel messenger RNAs in rat brain. *Nature (London)* **320,** 188–192.

Noda, M., Ikeda, T., Suzuki, H., Takeshima, H., Takahashi, T., Kuno, M., and Numa, S. (1986b). Expression of functional sodium channels from cloned cDNA. *Nature (London)* **322,** 826–828.

Ovchinnikov, Y. A., Modyanov, N. N., Broude, N. E., Petrukhin, K. E., Grishin, A. V., Arzamazova, N. M., Aldanova, N. A., Monastyrskaya, G. S., and Sverdlov, E. D. (1986). Pig kidney Na^+,K^+-ATPase: Primary structure and spatial organization. *FEBS Lett.* **201,** 237–245.

Ovchinnikov, Y. A., Monastryskaya, G. S., Broude, N. E., Ushkaryov, Y. A., Melkov, A. M., Smirnov, Y. V., Malyshev, I. V., Allikmets, R. L., Kostina, M. B., Dulubova, I. E., Kiyatkin, N. I., Grishin, A. V., Modyanov, N. N., and Sverdlov, E. D. (1988). Family of human Na^+,K^+-ATPase genes: Structure of the gene for the catalytic subunit (αIII-form) and its relationship with structural features of the protein. *FEBS Lett.* **233,** 87–94.

Papazian, D. M., Schwarz, T. L., Tempel, B. L., and Jan, Y. N. (1987). Cloning of genomic and complementary DNA from *Shaker*, a putative potassium channel gene from *Drosophila*. *Science* **237,** 749–753.

Peerce, B., and Wright, E. M. (1984). Sodium-induced conformational changes in the glucose transporter of intestinal brush borders. *J. Biol. Chem.* **259,** 14105–14112.

Peerce, B., and Wright, E. M. (1985). Evidence for tyrosyl residues at the Na site on the intestinal Na/glucose cotransporter. *J. Biol. Chem.* **260,** 6026–6031.

Peerce, B., and Wright, E. M. (1986). Distance between substrate sites on the Na-glucose cotransporter by fluorescence energy transfer. *Proc. Natl. Acad. Sci. U.S.A.* **83,** 8092–8096.

Ptashne, M. (1988). How eukaryotic transcriptional activators work. *Nature* (*London*) **335,** 683–689.

Reuben, M. A., Lasater, L. S., and Sachs, G. (1990). Characterization of a β subunit of the gastric H/K-transporting ATPase. *Proc. Natl. Acad. Sci. U.S.A.* **87,** 6767–6771.

Sardet, C., Franchi, A., and Pouysségur, J. (1989). Molecular cloning, primary structure and expression of the human growth factor-activatable Na/H antiporter. *Cell* **56,** 271–280.

Sardet, C., Councillon, L., Franchi, A., and Pouysségur, J. (1990). Growth factors induce phosphorylation of the Na/H antiporter, a glycoprotein of 110 kD. *Science* **247,** 723–726.

Schwarz, T., Tempel, B., Papazian, D., Jan, Y. N., and Jan. L. Y. (1988). Multiple potassium channel components are produced by alternative splicing at the *Shaker* locus in Drosophila. *Nature* (*London*) **331,** 137–142.

Serrano, R., Kielland-Brandt, M., and Fink, G. (1986). Yeast plasma membrane ATPase is essential for growth and has homology with (Na+K), K- and Ca-ATPases. *Nature* (*London*) **319,** 689–693.

Shanahan, M., and D'Artel-Ellis, J. (1984). Orientation of the glucose transporter in the human erythrocyte membrane: Investigation by *in situ* proteolytic dissection. *J. Biol. Chem.* **259,** 13878–13884.

Sharp, P. (1987). Splicing of messenger RNA precursors. *Science* **235,** 766–771.

Shows, T., Eddy, R. L., Byers, M. G., Fukushima, Y., Dehaven, C., Murray, J. C., and Bell, G. (1987). Polymorphic human glucose transporter gene (GLUT) is on chromosome 1p31.3-p35. *Diabetes* **36,** 546–549.

Shull, G., Schwartz, A., and Lingrel, J. (1985). Amino-acid sequence of the catalytic subunit of the (Na+K)ATPase deduced from a complementary DNA. *Nature* (*London*) **316,** 691–695.

Shull, G., Greeb, J., and Lingrel, J. (1986a). Molecular cloning of three distinct forms of the Na,K-ATPase α-subunit from rat brain. *Biochemistry* **25,** 8125–8132.

Shull, G., Lane, L., and Lingrel, J. (1986b). Amino-acid sequence of the β-subunit of the (Na+K)ATPase deduced from a cDNA. *Nature* (*London*) **321,** 429–431.

Shull, M., and Lingrel, J. (1987). Multiple genes encode the human Na,K-ATPase catalytic subunit. *Proc. Natl. Acad. Sci. U.S.A.* **84,** 4039–4043.

Singer, S. J., Maher, P. A., and Yaffe, M. P. (1987). On the transfer of integral proteins into membranes. *Proc. Natl. Acad. Sci. U.S.A.* **84,** 1960–1964.

Skou, J. C. (1988). Overview: The Na/K pump. *Method Enzymol.* **156,** 1–28.

Stuhmer, W., Conti, F., Suzuki, H., Wang, X., Noda, M., Yahagi, N., Kubo, H., and Numa, S. (1989). Structural parts involved in activation and inactivation of the sodium channel. *Nature* (*London*) **339,** 597–603.

Tanabe, T., Takeshima, H., Mikami, A., Flockerzi, V., Takahashi, H., Kangawa, K., Kojima, M., Matsuo, H., Hirose, T., and Numa, S., (1987). Primary structure of the receptor for calcium channel blockers from skeletal muscle. *Nature* (*London*) **328,** 313–318.

Thorens, B., Sarkar, H., Kaback, H. R., and Lodish, H. (1988). Cloning and functional expresion in bacteria of a novel glucose transporter present in liver, intestine, kidney, and β-pancreatic islet cells. *Cell* **55,** 281–290.

Timpe, L. C., Schwarz, T. L., Tempel, B. L., Papazian, D. M., Jan, Y. N., and Jan, L. Y. (1988a). Expression of functional potassium channels from *Shaker* cDNA in *Xenopus* oocytes. *Nature* (*London*) **331,** 143–145.

Timpe, L. C., Jan, Y. N., and Jan, L. Y. (1988b). Four cDNA clones from the *Shaker* locus of Drosophila induce kinetically distinct A-type potassium currents in Xenopus oocytes. *Neuron* **1,** 659–667.

Toyoshima, C., and Unwin, N. (1988). Ion channel of acetylcholine receptor reconstructed from images of postsynaptic membranes. *Nature* (*London*) **336,** 247–250.

Unwin, N. (1986). Is there a common design for cell membrane channels? *Nature* (*London*) **323,** 12–13.

Vassilev, P., Scheuer, T., and Catterall, W. A. (1988). Identification of an intracellular peptide segment involved in sodium channel inactivation. *Science* **241,** 1658–1661.

Wagner, S., Vogel T., Lietzke, R., Koob, R., and Drenkhahan, D. (1987). Immunochemical localization of a band 3-like anion exchanger in collecting duct of human kidney. *Am. J. Physiol.* **253,** F213–F221.

Walder, J., Chatterjee, R., Steck, T. L., Low, P. S., Musso, G. F., Kaiser, E. T., Rogers, P. H., and Arnone, A. (1984). The interaction of hemoglobin with the cytoplasmic domain of band 3 of the human erythrocyte membrane. *J. Biol. Chem.* **259,** 10238–10246.

Wickens, M., and Stephenson, P. (1984). Role of the conserved AAUAAA sequence: Four AAUAAA point mutations prevent messenger RNA 3′ end formation. *Science* **226,** 1045–1057.

Wickner, W. T., and Lodish, H. F. (1985). Multiple mechanisms of protein insertion into and across membranes. *Science* **230,** 400–408.

Zimmermann, R., and Meyer, D. I. (1986). A year of new insights into how proteins cross membranes. *TIBS* **11,** 512–515.

CHAPTER 2

Biogenesis and Sorting of Plasma Membrane Proteins

Michael J. Caplan
Department of Cellular and Molecular Physiology, Yale University School of Medicine, New Haven, Connecticut 06510

I. INTRODUCTION

The plasma membrane constitutes an animal cell's link with and protection from the outside world. The cell surface membrane serves as the selective permeability barrier which supports the nonequilibrium distribution of solutes necessary for the maintenance of homeostasis. The plasmalemma also possesses the elements which endow a cell with the capacity to converse with its environment. It is now widely accepted that the membrane's protein components deserve most of the credit for these manifold capabilities. As is discussed throughout this volume, it is channel, pump, and cotransport proteins embedded in the cell surface lipid bilayer which control the composition of the intracellular milieu. Plasmalemmal receptor and transducer proteins allow the cell to recognize and respond to various external influences. Membrane-associated proteins anchor cells to their substrata and mediate their integration into tissues. These examples suggest that many of the properties of a given cell type may be attributed to the protein composition of its plasma membrane (Guidotti, 1986). It is not surprising, therefore, that most cells go to great lengths to control the nature and distribution of polypeptides which populate their plasmalemmas.

There are a number of ways in which the cell can exert this control. As is true for almost any polypeptide, cells carefully regulate the expression of genes encoding plasma membrane proteins (Darnell, 1982). Transcriptional regulation probably accounts for most of the individuality associated with the cell surface membrane. The extent to which the plasmalemmal constituents of two cells differ from one another probably reflects the idiosyncracies of their respective genetic programs more than any other single influence. Posttranscriptional control of synthesis may also play a role in defining the composition of the plasma membrane. Recent evidence from a number of systems suggests that the cell may not diligently apply itself to translating each and every mRNA as soon as it is transcribed. Mechanisms exist through which messenger RNAs (especially those specifying secretory and membrane proteins) can be warehoused in the cytosol until the proteins which they encode are called for (Walter and Blobel, 1981b; Wolin and Walter, 1988, 1989).

Although transcriptional and translational controls are extremely important, the cell's ability to oversee the assembly of its plasmalemma does not end with protein synthesis. Proteins destined for insertion into the plasma membrane pass through a complex system of processing organelles prior to arriving at their site of ultimate functional residence. Each of these organelles makes a unique contribution to the maturation of these proteins as they transit through them. Each of these organelles also serves as a potential decision point, at which the cell can make choices about a protein's delivery, distribution, and life span.

This chapter focuses on the postsynthetic steps involved in the biogenesis of plasma membrane proteins. It begins with a discussion of some of the events common to all plasmalemmal polypeptides, with special emphasis on those which contribute directly to the character of the cell surface. The second half of the chapter is devoted to the specializations associated with cell types possessing differentiated cell surface subdomains. Epithelial cells (among many other cell types) are characterized by plasma membranes which can be morphologically and functionally divided into two or more distinct compartments (Simons and Fuller, 1985; Matlin, 1986; Caplan and Matlin, 1989; Rodriguez-Boulan and Nelson, 1989). The pathways through which membrane proteins are sorted to and retained in these differentiated regions of the cell surface are just beginning to be understood.

The mechanisms through which the plasma membrane is generated and maintained have bearing on all of the processes discussed in this volume. Our current understanding of these mechanisms is, for the most part, rudimentary and rapidly evolving. The goal of the synopsis presented here, therefore, is to not to provide an exhaustive review. Rather, its aim is to acquaint its readers with some of the important and fascinating questions confronting investigators interested in the cell biology of the plasma membrane.

II. BIOGENESIS OF MEMBRANE PROTEINS

A. *The Processing Pathway*

In the late 1960s, Jamieson and Palade demonstrated that newly synthesized secretory proteins follow a rigorously prescribed route through the cell (Jamieson and Palade, 1967a,b, 1968a,b; Palade, 1975). Autoradiographic experiments at the electron microscopic level revealed that, on completion of their synthesis, the zymogens of pancreatic exocrine cells are first associated with the rough endoplasmic reticulum (RER). Several minutes later, the newly synthesized polypeptides are transported, via vesicular carriers, to the Golgi complex. The newly synthesized proteins are delivered to the cis-most (i.e., ER-facing) stack of the Golgi and subsequently progress through each stack to the terminal or trans-most cisterna. Condensation of the secretory proteins begins in the trans Golgi stack and continues as the zymogens pass through condensing vacuoles to secretory granules, where they remain until they are discharged from the cell in response to a secretagogue. Although aspects of this pathway reflect specializations unique to endocrine and exocrine secretory cells, in general outline it has proved applicable to virtually any type of animal cell. Furthermore (and most importantly for this discussion), the pathway delineated in these studies is also pursued by newly synthesized membrane proteins (Bergmann and Singer, 1983).

Over the ensuing years it has been shown that each step along the route which a membrane protein follows to the cell surface, i.e., ER, cis Golgi, trans Golgi, is associated with specific processing reactions. Thus, the postsynthetic maturation of secretory and membrane proteins is temporally and spatially compartmentalized. The reactions which occur in each organelle build on those which were carried out at the previous station. Furthermore, the processing steps associated with a given locus are frequently absolute prerequisites for the further processing events which will occur in subsequent locations. With regard to these features, the posttranslational course pursued by membrane and secretory proteins is analogous to an assembly line. Like an assembly line, the process is carried out vectorially and in strict series. There is no evidence for skipping or repeating steps under normal circumstances. The flux of proteins from the ER to the cell surface is, for the most part, unidirectional (Palade, 1975; Rothman, 1987). Each step of the process is also endowed with some capacity for quality control. The cell can monitor the degree to which processing steps have been successfully completed and, if it is not satisfied, it can activate mechanisms to detain or destroy those proteins which it judges to be imperfect (Lippincott-Schwartz, 1988).

Successful passage through this endomembraneous network is required of every protein which will ultimately reside in the plasmalemma. Our understanding of the mechanism and purpose underlying each of the modifications associated with stations along this route is far from complete. It is clear, however, that the biosynthetic pathway contributes a number of possible points at which control of the composition of cell surface membrane can be exerted. The sections which follow describe the steps which constitute this pathway in greater detail.

B. The Role of the Endoplasmic Reticulum: Membrane Insertion

The machinery responsible for protein synthesis resides in the cytoplasm (Palade, 1975). Ribosomes, tRNAs, mRNAs, elongation factors, etc. are all soluble components of the cytoplasmic space. In synthesizing a membrane or secretory protein, therefore, the cell is faced with the problem of producing a polypeptide which must ultimately be separated by a lipid bilayer from the molecules required for its production. The pathways which cells have evolved to overcome this difficulty have been explored during the course of over two decades of elegant research.

It is now generally accepted that the vast majority of animal cell membrane and secretory proteins are inserted into or across the membrane of the RER cotranslationally (Walter and Lingappa, 1986). Ribosomes translating secretory or membrane proteins become bound to the cytoplasmic surface of the ER. The polypeptide emerging from the bound ribosome enters and traverses the ER membrane as it is elongated (Blobel and Dobberstein, 1975). When translation is complete, the protein is released from the ribosome and the ribosome is released from the RER membrane. In the case of secretory proteins, release from the ribosome results in the discharge of the polypeptide across the membrane and into the lumen of the RER. Transmembrane proteins are released in what will presumably be their final topology, with endodomains facing the cytosol, ectodomains exposed at the lumenal face of the RER, and membrane-spanning domains embedded in the lipid bilayer (Blobel, 1980).

Proteins destined for cotranslational insertion into or across the RER generally carry within their primary structure a contiguous stretch of ~ 15 amino acids which is required to initiate the membrane interaction. These "signal sequences" tend to be fairly hydrophobic and are most commonly located at a protein's amino terminus (von Heijne, 1985). Upon emerging from the ribosome, the signal sequence interacts specifically with a cytosolic complex of proteins and RNA called signal recognition particle (SRP) (Walter and Blobel, 1981a,b; Walter *et al.,* 1981).

The interaction with SRP serves two purposes. Evidence from *in vitro* studies indicates that association with SRP often causes a temporary translation arrest (Walter and Blobel, 1981b; Wolin and Walter, 1989). The nascent polypeptide remains bound to the polysome, but elongation ceases. The bound SRP also acts as a guide which directs a polysome to the membrane of the RER. Targeting of the SRP–polysome complex to the RER requires the intercession of the SRP receptor or docking protein, a heterodimeric transmembrane component of the RER which acts as a receptor for SRT-bearing polysomes (Meyer *et al.,* 1982; Gilmore *et al.,* 1982a,b). Interestingly, one of the polypeptides of the SRP receptor shares a region of homology with one of the protein components of SRP itself, prompting the speculation that this domain corresponds to a signal sequence binding site (Bernstein *et al.,* 1989; Romisch *et al.,* 1989).

Once bound to docking protein, the SRP–polysome complex dissociates. Another transmembrane component of the RER appears to bind the signal sequence after its release from SRP (Weidmann *et al.,* 1987). Discharge of SRP from the ribosome requires GTP, which appears to interact with the docking protein (Connolly and Gilmore, 1989; Hoffman and Gilmore, 1988) and perhaps with the 54-kDa subunit of SRP itself (Bernstein *et al.,* 1989; Romisch *et al.,* 1989). On release of SRP, the translation block is relieved and synthesis of the nascent polypeptide continues in association with the (as yet poorly understood) machinery responsible for actually translocating elongating polypeptides across the ER membrane (Krieg *et al.,* 1989). This translocation is not driven by translation and seems to require energy in the form of hydrolyzable nucleotides (Perara *et al.*, 1986; Chen and Tai, 1987; Hoffman and Gilmore, 1988). By halting translation until a free docking protein site on the RER is encountered, SRP may prevent secretory and membrane proteins from being completely synthesized and discharged into the cytosol. If no translocation sites on the RER are available, SRP may serve to route the translation complex into what might be characterized as a holding pattern. Resumption of active translation is predicated upon the translocation machinery signaling its readiness to process one of those waiting in line.

Once it has succeeded in initiating translocation, the signal sequence is of no further use. Signal peptidase, an enzyme resident in the membrane of the RER, cleaves most amino-terminal signal sequences from their parent proteins, leaving the mature polypeptide ~ 15 amino acids shorter than the protein encoded by the mRNA (Evans *et al.*, 1986). Interestingly, it is precisely this cotranslational removal of the signal sequence which greatly facilitated the discovery and characterization of the signal sequence-mediated membrane insertion pathway (Milstein *et al.*, 1972).

It must be pointed out that the pathway described above, although elegant, is not all encompassing. A large number of deviations from this model have been documented. Most dramatic of these, perhaps, is the situation which prevails for proteins inserted into the membranes of mitochondria and chloroplasts. It is now clear that these polypeptides are synthesized in their entirety in the cytosol (Attardi and Schatz, 1988; Schmidt and Mishkind, 1986). Following completion of their translation, they undergo signal sequence-mediated import into or across the relevant membrane. Posttranslational membrane insertion has also been demonstrated in yeast and bacteria (Hansen *et al.*, 1986; Waters and Blobel, 1986; Rothblatt and Meyer, 1986; Wickner, 1988). The situation in yeast is especially interesting, since these organisms possess an endomembranous system which is extremely similar to that found in higher eukaryotes. Yeast secretory and membrane proteins can be synthesized on free, cytoplasmic ribosomes and discharged into the cytoplasm as soluble proteins. In order to retain the ability to be inserted into or across the membrane of the ER, however, these newly synthesized polypeptides must associate with members of a class of cytosolic proteins which serve to prevent them from folding into their final conformations. In bacteria the GroEL family of proteins serves this purpose (Hemingson *et al.*, 1988), whereas in yeast the hsc70 complex mediates this function (Deshaies *et al.*, 1988; Chirico *et al.*, 1988). These "chaperonins," which are related to the heat-shock family, serve a function somewhat analogous to that of SRP. Their associations with newly synthesized proteins maintain the competence of these polypeptides for membrane insertion until this operation can be accomplished. Dissociation of the unfolded protein from the chaperonin is energy dependent (Rothman, 1989).

Signal sequences need not always be amino terminal. It is becoming clear that stretches of hydrophobic amino acids can serve as signal sequences when placed at a number of different positions in a protein's linear sequence. Although this phenomenon has been demonstrated for secretory proteins (Lingappa *et al.*, 1979; Perara and Lingappa, 1985), it appears to be far more common among membrane proteins. It is now apparent that the hydrophobic membrane-spanning sequences which anchor proteins in the bilayer can also serve as their "start" or "stop" transfer sequences (Yost *et al.*, 1983; Rothman *et al.*, 1988). A large number (perhaps even the majority) of membrane proteins have no amino-terminal signal peptide. This is especially true of polytopic proteins, that is, those which span the bilayer several times. The transmembrane domains of these multiloop proteins can apparently serve to start or stop translocation, thus mediating the SRP-dependent and cotranslational weaving of the nascent polypeptide into the membrane fabric of the RER (Anderson *et al.*, 1983). The extent to which the interaction of transmembrane domains with SRP can cause translation arrest remains to be established.

The paradigm presented here for the insertion of membrane proteins into the RER is very much an evolving one. The component parts of the translocation machinery are just beginning to be identified (Krieg *et al.*, 1989) and the biophysical parameters governing the insertion of a protein into or through a lipid bilayer remain mysterious (Engelman and Steitz, 1981). Furthermore, exceptions to this general rule continue to surface. Of special relevance to the topics discussed in this volume is the case of the erythrocyte glucose transporter. Recent evidence from *in vitro* translation experiments indicates that this protein, which spans the bilayer as many as 12 times, can be synthesized in its entirety on free cytosolic ribosomes and inserted into microsomes derived from the RER posttranslationally (Mueckler and Lodish, 1986). This behavior, although apparently commonplace in yeast and bacteria, is unprecedented among the plasmalemmal proteins of higher eukaryotes. It is not yet clear whether the glucose transporter is actually intercalated posttranslationally *in vivo*. The degree to which a chaperonin (related to the bacterial GroEL or yeast hsc70 class of proteins) might participate in this posttranslational insertion has also not been investigated. This is at least a possibility, since members of the chaperonin family have been shown to be present in the cytoplasm of mammalian cells. For instance, hsc70 is present in the cytoplasm of animal cells, where it is known to mediate the uncoating of clathrin-coated vesicles (Chappell *et al.*, 1986). The prevalence of the glucose transporter's renegade conduct—among other transporters or other membrane proteins in general—needs to be further investigated. Independent of its applicability, however, the example of the glucose transporter echoes a theme which has recurred several times during the preceding discussion: that the initial step in the biosynthesis of a plasma membrane protein is complex and offers the cell several options for variation and regulation.

Incorporation into the membrane of the RER constitutes the first in a series of sorting steps which will be encountered by a newly synthesized plasma membrane protein. This first step separates all membrane proteins from the rest of the polypeptides being synthesized coincidently. In a sense, the RER is a staging area, in which membrane proteins destined for a number of different subcellular locales are gathered in order to begin a course of common processing operations. Subsequent sorting events will separate these membrane proteins into subclasses destined for different destinations or modifications.

C. The Role of the Endoplasmic Reticulum: Processing and Transport to the Golgi Complex

1. Cotranslational Processing

The processing steps which contribute to the maturation of a plasmalemmal protein begin as soon as its membrane insertion is initiated. As portions of the protein are woven into and through the membrane, they undergo covalent modifications and start to assume a tertiary structure. In addition to cleavage of the signal sequence (Evans *et al.*, 1986), alterations generally associated with a membrane protein's residence in the ER include folding (Hurtley and Helenius, 1989), disulfide bond formation (Freedman, 1989), N-linked glycosylation (Hirschberg and Snider, 1987), and oligomerization (Hurtley and Helenius, 1989). Departure from the RER frequently requires that these operations be successfully completed (Rose and Doms, 1988).

A great deal has been learned about the biochemical processes involved in N-linked glycosylation. Although the purposes served by the addition of sugar groups to proteins remain rather mysterious, the mechanics of their addition are well worked out. The sugar structure characteristic of N-linked glycosylation is a branched tree containing (among other sugar types) nine mannose residues and an *N*-acetylglucosamine at its stem (Lennarz, 1987). While it is being preassembled in the ER, the sugar structure is attached to the membrane through a linkage between its initial *N*-acetylglucosamine residue and a molecule of dolichol phosphate lipid. The completed sugar tree is transfered en bloc from the dolichol lipid to lumenally disposed asparagine acceptor sites on nascent polypeptides.

Addition of the sugar occurs while the proteins are still in the process of being cotranslationally inserted into and across the bilayer. Sugars are added only to those asparagine residues which appear in the appropriate context, that is, as part of the sequence Asn-X-Ser or Asn-X-Thr. In many proteins, only a subset of those asparagines appearing in this configuration actually get glycosylated. The basis for this selectivity is not fully understood.

Disulfide bond formation is the responsibility of the lumenally disposed enzyme protein disulfide isomerase (PDI) (Freedman, 1989). As with glycosylation, disulfide bond formation proceeds cotranslationally. The PDI complex may actually be involved in several functions, including aspects of *N*-linked glycosylation.

On emerging into the lumen of the RER, the nascent polypeptide begins immediately to fold and acquire tertiary structure. This appears to be a complicated and multifactorial process, involving both the biophysical properties of the peptide chain itself and the intervention of protein complexes which catalyze folding. It has been elegantly demonstrated that, in mitochondria, for example, the folding of proteins delivered to the matrix space is controlled through an interaction with hsc60 (Ostermann *et al.*, 1989). This protein complex, composed of 14 identical subunits, is very similar in structure and function to the bacterial GroEl family (Hemingson *et al.*, 1988). These "foldases" appear to interact with nascent polypeptides as they traverse membranes and, through controlled binding and unbinding, play a role in orchestrating their folding (Rothman, 1989). It has been suggested that immunoglobin binding protein (see below) might serve a similar function for proteins inserted into the RER.

Finally, it must be noted that the addition of sugars and the formation of disulfide bonds will affect the balance of biophysical forces and thus contribute to the realization of the protein's ultimate conformation. Similarly, the acquisition of structure driven by the chemical nature of the primary sequence will influence the placement of sugars and disulfide bonds by making certain asparagine and cysteine residues more or less available to their respective modifying enzymes. Thus, a membrane protein's final tertiary conformation is formed as the result of a dynamic interplay among the proteins which facilitate folding, and the enzymatic reactions and biophysical forces which are brought to bear during cotranslational membrane insertion.

2. Oligomeric Assembly

Many plasmamembrane proteins are multimeric. This is certainly true of ion transporters, as exemplified by band III (a homodimer) (Jay and Cantley, 1986), the Na^+/K^+-ATPase (a heterodimer) (Jorgensen, 1982), and the nicotinic acetylcholine receptor (a heteropentamer) (Popot and Changeux, 1984). Recent evidence indicates that assembly of newly synthesized monomers into the appropriate higher order structures occurs during their period of residence in the RER.

A number of studies illustrating this point have been performed on the trimeric spike glycoproteins of enveloped viruses. Studies combining pulse-chase labeling with density gradient centrifugation reveal that assembly occurs very early after the completion of protein synthesis (Copeland *et al.*, 1986, 1988; Gething *et al.*, 1986). These experiments also indicate that those monomers which do not get incorporated into trimers never pass from the ER to the Golgi. Temperature-sensitive mutants of spike glycoproteins which are barred from trimerizing at elevated temperatures can only depart the ER at a lower, permissive temperature (Doms *et al.*, 1988).

Retention in the ER of incompletely folded or assembled polypeptides may perhaps be mediated by a protein called immunoglobin binding protein (BiP) (Bole *et al.*, 1986; Hurtley *et al.*, 1989). This 72-kDa polypeptide is a soluble component of the ER lumenal space. It is a member of the family of glucose-regulated proteins, which are a subset of polypeptides encoded by the heat-shock response genes (Munro and Pelham, 1986). BiP binds to newly synthesized proteins whose folding or oligomerization is incomplete or has not been completed correctly (Hurtley and Helenius, 1989; Kassenbrock *et al.*, 1988). As was suggested above, an interaction with BiP may facilitate the folding of ER inserted proteins. BiP binding is apparently reversible and, at least *in vitro*, can be terminated through an interaction with ATP. The BiP polypeptide is itself retained in the ER by virtue of its four carboxy-terminal amino acids, whose sequence is KDEL (Munro and Pelham, 1987). It has been shown that proteins bearing this sequence at their carboxy termini are retained in the ER through an interaction with a recently identified lumenally disposed receptor (Vaux *et al.*, 1990). It has been suggested that this retention is active, that is, that KDEL-bearing proteins are capable of escaping to the Golgi or a pre-Golgi compartment, from which they are recaptured via a KDEL-mediated mechanism (Pelham, 1988).

Many misfolded or unassembled proteins retained in the ER are ultimately degraded in a compartment distinct both from the ER and the lysosome. This degradation has been best studied for the case of the T cell antigen receptor, which is composed of seven subunits and which apparently assembles inefficiently under normal circumstances (Clevers *et al.*, 1988). Only about 5% of this protein complex passes from the ER to the Golgi, while the remaining 95% is rapidly turned over by some process which cannot be interrupted by inhibitors of lysosomal enzymes (Klausner *et al.,* 1990). The nature and subcellular location of this degredative activity have yet to be established.

Similar observations on oligomeric assembly have been gathered for endogenous cellular proteins involved in membrane transport. Studies on the Na^+/K^+-ATPase indicate that the assembly of the α- and β-subunits to form the heterodimer occurs almost as soon as the proteins are released from the ribosomes (Tamkun and Fambrough, 1986). Studies in which the subunit polypeptides are expressed individually in fibroblasts (Takeyasu *et al.,* 1987, 1988) or in *Xenopus* oocytes (Geering *et al.,* 1989; Caplan, 1990) suggest that the α-subunit cannot leave the ER unless it is appropriately complexed with a β-subunit. These observations are especially interesting in light of the potential mechanism they offer the cell for regulating cell surface Na/K^+-ATPase levels. It has recently been shown, for example, that in chick myocytes the α-subunit is produced in excess of the β polypeptide (Taormino and Fambrough, 1990). Those α-subunits unable to assemble remain in the ER and are probably degraded through mechanisms similar to those discussed above. Conditions which induce the cell to increase its complement of cell surface sodium pump result in an increase in the transcription of only the mRNA encoding the β-subunit. The resultant increase in the quantity of newly synthesized β-subunit polypeptide in the membrane of the RER allows more heterodimer to assemble and more Na^+/K^+-ATPase to be delivered to the cell surface.

In the case of the Na^+/K^+-ATPase, time spent in the ER may effect more than folding and oligomerization. Recent evidence suggests that at least some aspects of the sodium pump's catalytic activity may be initiated during this early postsynthetic period (Caplan *et al.,* 1990; Geering *et al.,* 1987). Ouabain is a specific inhibitor of the Na^+/K^+-ATPase, which achieves its inhibition by binding to the sodium pump's α-subunit. Ouabain binds with greatest affinity to only one of the conformations through which the Na^+/K^+-ATPase passes during the course of its enzymatic cycle (Forbush, 1983). *In vitro,* the sodium pump can be driven into this conformation in two ways, one of which requires enzymatic hydrolysis of ATP and one of which does not. Experiments employing the photoactivatible NAB derivative of ouabain and anti-ouabain antibodies were performed on cultured kidney cells which had been briefly pulse labeled with {^{35}S]methionine (Caplan *et al.,* 1990). It was found that, immediately upon completion of the pulse period, newly synthesized radiolabeled Na^+/K^+-ATPase was capable of binding ouabain under the ATP-independent conditions. Ouabain binding requiring ATP hydrolysis, however, was not observed until the newly synthesized sodium pumps were at least 10 min old. Results from other studies suggest that the newly synthesized sodium pump still resides in the ER at this early time point (Tamkun and Fambrough, 1986). These observations suggest that the sodium pump may undergo some form of activation or conformational maturation during its passage through the ER. Similar conclusions have been generated in studies using controlled proteolysis to examine the conformational repertoire of the newly synthesized sodium pump (Geering *et al.*, 1987). The nature of this putative activation step remains to be elucidated. It is also not clear whether other ion pumps or, more generally, other plasma membrane enzymes undergo some manner of activation during the course of their postsynthetic processing in the ER. It is interesting to speculate, however, that intracellular activation might play a role in regulating the cell's plasmalemmal levels of functional Na^+/K^+-ATPase.

3. Transport to the Golgi

Newly synthesized proteins depart from the RER in vesicular carriers which bud from the ER membrane. Morphological evidence suggests that budding occurs from the transitional elements, specialized smooth subdomains of the RER which are closely apposed to the cis-most cisterna of a Golgi complex (Jamieson and Palade, 1967a). The vesicles appear to be coated with some matrix which is morphologically and immunologically distinct from clathrin (Orci *et al.*, 1986; Pfeffer and Rothman, 1987). Delivery of proteins to the Golgi can be reversibly blocked by incubating cells at 16°C (Saraste *et al.*, 1986). The formation of the vesicles and their subsequent targeting to the Golgi appears to be a complex, multistage process which has recently become susceptible to *in vitro* analysis. By examining the processing of newly synthesized membrane proteins in permeabilized cells, investigators have been able to define some of the soluble, cytosolic components which are necessary for the maintenance of ER to Golgi vesicular traffic. Once the vesicles have budded, GTP is required to activate them for the fusion step with the Golgi acceptor (Beckers and Balch, 1989; Beckers *et al.*, 1989). A cytosolic protein called NSF (*N*-ethylmaleimide sensitive factor) is also required at this stage. A cytosolic calcium concentration in the range of 0.1 μM appears to be critically required for fusion of the vesicular carrier with the cis Golgi stack.

The precise roles of these soluble effectors have yet to be defined. It is interesting to note, however, that GTP and NSF may be common elements in all of the subsequent vesicular transport events involved in carrying proteins to the plasmalemma. NSF has been isolated and found to be identical with an NEM-sensitive protein necessary for the movement of vesicles between succesive Golgi stacks (Beckers *et al.*, 1989; Malhorta *et al.*, 1988). GTP and small, ras-like GTP-binding proteins appear to be involved in each stage of the secretory pathway (Bourne, 1988), as is discussed below. It is rather aesthetically pleasing to consider that the cell might exploit similar mechanisms to accomplish the transport of vesicular carriers between the numerous organelles in the secretory pathway. The mechanical aspects of budding and fusing vesicles are probably quite similar whether these operations are occurring in the ER or in the Golgi. It would not be surprising, therefore, if a considerable amount of the cellular machinery which has evolved to solve these problems in one setting were to be brought to bear in another. The extent to which this generalized machinery exists remains to established, as do the identities of components which could endow specificity to each pathway served by this putative general scheme.

Finally, it must be pointed out that the existence of vesicular flux from the ER to the Golgi suggests the existence of an equal flux of membrane from the Golgi to the ER. The idea that membrane delivered from the ER to Golgi must be retrieved arises from steady-state considerations and from the observation that the biochemical compositions of the ER and Golgi are discrete (Palade, 1975; Pfeffer and Rothman, 1987). Although it has long been postulated, evidence for this return traffic has only recently been gathered in experiments employing brefeldin A. This drug results in a dissolution of the Golgi complex and the redistribution of its membrane protein markers to the ER (Lippincott-Schwartz *et al.*, 1990; Doms *et al.*, 1989). In order to understand these observations, it has been suggested that a flux of vesicles normally carries membrane from the Golgi to the ER and that this flux is able to exclude, via some sorting process, polypeptides native to the Golgi cisternae. It is thought that brefeldin A may disturb this putative sorting process, thus allowing proteins normally associated exclusively with the Golgi membranes to enter the back-flux and thus become associated with the ER. According to this interpretation, the Golgi proteins which arrive in the ER are simply allowed to enter a pathway which normally exists in the cell and from which they are normally excluded. The validity of this assertion remains to be established. Some manifestation of the sorting mechanism which it postulates must exist, however, since newly synthesized membrane proteins pass from ER to Golgi to the cell surface, while the resident proteins of these compartments are not swept along in this flow.

D. The Role of the Golgi Complex: Covalent Modifications and Stabilization of Conformation

1. Sugar Modifications

The best studied functions associated with the membranous stacks of the Golgi complex relate to the processing of *N*-linked oligosaccharides. The sugar structure transferred from dolichol phosphate to the asparagine residues of nascent polypeptides contains nine mannose and three glucose residues (Hirschberg and Snider, 1987). The glucose residues and one mannose residue are removed very shortly after the sugar chain is transferred from its dolichol intermediate to the acceptor polypeptide. This initial step in the pathway of sugar processing reactions occurs while the newly synthesized protein is still resident in the ER. All subsequent modifications are carried out by enzymes associated with elements of the Golgi complex.

The early, high-mannose sugar structures are susceptible to the activity of endoglycosidase H, an enzyme which specifically cleaves these oligosaccharides from their associated polypeptides (Kobata, 1979). Arrival of polypeptides at the cis cisterna of the Golgi complex is generally heralded by the action of two α-mannosidase activities specifically localized to this compartment which function in a pair of tandem reactions to trim five of the mannose residues from the sugar structures of most newly synthesized proteins. The loss of these mannose residues is accompanied by a loss of sensitivity to endoglycosidase H (Kornfeld and Kornfeld, 1985). Consequently, endoglycosidase H has proved extremely useful in studies on the biosynthesis of membrane proteins by serving as a biochemical marker for the progression of newly synthesized polypeptides from the ER through the early Golgi.

Once they have been divested of six mannose residues, the newly shorn oligosaccharides are remodeled by enzymes associated with the medial and trans elements of the Golgi stack. Transport of proteins among Golgi stacks is accomplished by vesicles bearing a non-clathrin coat, which is presumably identical to that associated with the vesicles which mediate ER to Golgi traffic (Orci *et al.*, 1986). Sugar transferases add *N*-acetylglucosamine, galactose, and sialic acid residues in strict series, with the action of each transferase being rigorously dependent on the biosynthetic history of the oligosaccharide structure up to that point (Kornfeld and Kornfeld, 1985). While all N-linked sugars begin with the same dolichol intermediate, the final complex oligosaccharide produced by the actions of the Golgi enzymes varies from protein to protein. Apparently, different proteins have different affinities for or susceptibilities to the individual subsets of lumenally disposed sugar transferases which they encounter as they transit the Golgi complex. Since each stage of the modification is in part determined by its predecessor, small variations in the degree of processing can produce large compositional and structural alterations in the final constellation of sugars sculpted by the Golgi processing enzymes. Once again, the function of these sugars and the role that this compositional variation plays in dictating the various properties of membrane proteins remain largely unknown.

2. Other Covalent Modifications

The Golgi complex contributes a number of other posttranslational modifications to newly synthesized membrane proteins. The addition and maturation of O-linked sugars (which are attached to the hydroxyl groups of serine and threonine residues) are thought to occur entirely within the confines of the Golgi stacks (Hanover *et al.*, 1982; Cummings *et al.*, 1983). Similarly, phosphorylation and sulfation of sugars, proteins, and proteoglycans have been conclusively associated with the Golgi apparatus (Kornfeld and Kornfeld, 1985; Huttner, 1988). While it is beyond the scope of this review to discuss the mechanisms of each of these covalent modifications in detail, it is important to point out that, as with the processing of N-linked sugars, these reactions are thought to occur along a temporal and spatial assembly line. The enzymes responsible for the individual steps in the modification are compartmentalized, that is, restricted in their distribution to only a subset of the Golgi cisternae. Thus, the action of each enzyme occurs in a spatial and temporal sequence. At each station in its saltatory progress through the Golgi stacks the newly synthesized protein encounters one component of the processing machinery at a time, in isolation from the other enzymes of this system.

3. Stabilization of Conformation

Recent evidence from a number of laboratories suggests that the modifications proteins experience while transiting the Golgi complex are not limited to the covalent variety discussed above. Studies on the trimeric hemagglutinin (HA) protein of the influenza virus indicate that it undergoes a "stabilization" of its tertiary structure during or soon after its residence in the Golgi. Copeland *et al.* (1986) found that influenza HA protein forms its characteristic homotrimers while still in the membranes of the RER. Biochemical analysis, however, reveals that these early trimers are not bound together as tightly as those which populate the cell surface membrane. Solubilization of the newly synthesized trimers under fairly gentle detergent conditions can lead to the dissolution of the complex, whereas the mature plasmalemmal trimers are resistant to this treatment. These investigators found that the transition from the unstable to the stable conformer occurs while the protein is passing through or out of the Golgi complex. The nature of the modification (if any) which brings about this stabilization remains unknown. It is interesting to note, however, that Skibbens *et al.* (1989) have found that an intramolecular disulfide bond in the HA protein is lost at roughly the same point in this protein's posttranslational processing that stabilization is achieved. It seems likely, therefore, that during its time in the Golgi (or not long thereafter) the protein modifies its folding pattern in a manner which affects both its own tertiary structure and its affinity for oligomerization.

E. The Role of the Golgi Complex: The Trans Golgi Network and Delivery to the Cell Surface

1. The Functions of the Trans Golgi Network

For most plasma membrane proteins, the staging area for delivery to the cell surface is the final cisterna of the Golgi apparatus, which has come to be referred to as the trans Golgi network, or TGN. The TGN is morphologically discernable from the other Golgi stacks and appears to be endowed with a number of properties not shared by other Golgi elements (Griffiths and Simons, 1986). For example, addition of sialic acid appears to occur entirely within the confines of the TGN (Fuller *et al.*, 1985). Furthermore, newly synthesized proteins accumulate within the TGN when their progress to the cell surface is impeded by lowering the temperature to 20°C (Matlin and Simons, 1983; Griffiths *et al.*, 1985; Saraste and Kuismanen, 1984). Most importantly for the purposes of this review, the TGN appears to be the site at which proteins bound for different subcellular compartments become segregated from one another. Biochemical and immunocytochemical experiments performed on a number of systems have revealed that polypeptides destined for different subcellular locales remain together as far as the TGN and become separated from one another within its confines (Fuller *et al.*, 1985; Rindler *et al.*, 1984; Orci *et al.*, 1987; Tooze *et al.*, 1987). This sorting function of the TGN is discussed at greater length in the next section of this review.

Proteins depart the TGN in clathrin-coated vesicles, which tend to bud from the lateral boundaries of this structure's membranous stack (Orci *et al.*, 1984). Vesicles emanating from the TGN carry newly synthesized polypeptides to a large number of cellular destinations, including prelysosomal endosomes (Griffiths *et al.*, 1988), secretory granules, and the cell surface. As has been discussed above, the trafficking of these vesicles to their appropriate subcellular targets is likely to be dependent on cytosolic factors such as NSF and SNAPs [soluble NSF attachment proteins, which appear to link NSF to the membranes of vesicular carriers (Clary *et al.*, 1990)] as well as GTP and small ras-like GTP-binding proteins. It has been shown in yeast, for example, that fusion of secretory granules with the cell surface is prevented by mutations in a ras-like GTP-binding protein encoded by the *Sec 4* gene (Salminen and Novick, 1987).

2. Cell Surface Delivery

After departing the TGN, the secreted products of cells are frequently packaged into secretory granules which are stored in the cytoplasm until some physiological stimulus initiates their fusion with the plasmalemma and the discharge of their content (Burgess and Kelly, 1987). In contrast, no such regulated delivery pathway exists for most plasmalemmal polypeptides. The vast majority of plasma membrane proteins are delivered directly and constitutively from the TGN to the cell surface. They are not accumulated in any storage compartment within the cell and their arrival at the cell surface is not triggered by a physiological stimulus.

There are, however, at least four exceptions to this general rule. The gastric H^+/K^+-ATPase (Urushidani and Forte, 1987), the H^+-ATPase of renal intercalated cells (Schwartz and Al-Awqati, 1986), the insulin-sensitive glucose transporter of adipocytes (Blok *et al.*, 1988), and the water channels of the renal collecting duct (Handler, 1988) are all resident in cytosolic vesicles whose fusion with the cell surface membrane is under rigid physiological control. Thus, stimulation of gastric acid secretion by histamine is associated with the insertion of an enormous intracellular pool of H^+/K^+-ATPase into the gastric parietal cell's apical plasma membrane (Urushidani and Forte, 1987). Similarly, ADH induces the fusion of water channel-containing vesicles with the apical membrane of collecting duct epithelial cells (Handler, 1988), and insulin initiates the delivery of intracellular stores of glucose channels into the adipocyte plasma membrane (Blok *et al.*, 1988). In this manner, the cells are able to carefully and rapidly influence the activity of membrane transport processes without directly modulating the catalytic activities of the transporters themselves. This adaptation may have evolved to control the function of transport systems whose structural characteristics render them, for some reason, poor candidates for regulation by covalent or allosteric modification.

Removal of the stimulus for membrane insertion results in the retrieval of these transport proteins from the cell surface via an endocytic process. This endocytosis does not, however, appear to involve traditional acidic endosomes and lysosomes (Lencer *et al.*, 1990). Instead, the recycled proteins appear to be returned to their storage vesicles or, alternatively, those storage vesicles are regenerated through the endocytic process. The mechanisms through which these "stored" membrane proteins are targeted from the TGN (and subsequently the cell surface) to their intracellular pools remain to be established. Furthermore, the generality of this pathway—its applicability to other transport proteins or to membrane proteins in general in other cell types—is not at all clear. This specialization is, however, an excellent example of the cellular capacity discussed in the Introduction, that is, the ability to control temporally and spatially the composition of the plasma membrane.

III. SORTING AND EPITHELIAL POLARITY

In Section II of this review, a general outline of the postsynthetic pathway pursued by newly synthesized plasma membrane proteins was presented. The generic pathway described above is more or less applicable to any plasmalemmal polypeptide in any animal cell type. It can be fairly well encompassed by the observation that a newly synthesized membrane protein's progress to the cell surface is marked by its saltatory passage through distinct subcellular processing stations (see Fig. 1). While essentially correct, however, this summation does not appropriately emphasize a critically important theme which is common to each step in this pathway. During each phase in a membrane protein's processing, it must be examined by some mechanism capable of determining whether the organelle it currently occupies is in fact its site of ultimate functional residence. Those proteins which are identified by this mechanism as belonging to the organelle in which they reside must be retained, while those deemed inappropriate must be allowed to travel to the next processing locus. In other words, the RER must be able to hold on to newly synthesized components of the RER membrane while allowing proteins destined for the Golgi and the cell surface to proceed. Similarly, each subcompartment of the Golgi must be able to recognize and retain its constituent proteins from among the traffic of newly synthesized membrane proteins which passes through them.

The nature of this mechanism is beginning to be elucidated. Sequences have been identified on the carboxy termini of resident ER proteins which mediate their retention in the ER (Nilsson *et al.*, 1989). A sequence in the transmembrane domain of the Golgi protein has been shown to be important in its Golgi localization (Machamer and Rose, 1987). Clearly, some primary, secondary, or tertiary structural determinants of membrane proteins must contain the information that is used by the cell to target them to the appropriate destination. Furthermore, the cell must possess some machinery which is capable of interpreting and acting on these signals. Every newly synthesized membrane protein which enters the cell's endomembranous processing network must be sorted, actively or passively, and targeted to the correct subcellular destination.

The sorting problem is compounded in polarized epithelial cells. Epithelial cells line body cavities, essentially forming the barrier between the inside and the outside of an organism. It is the selective permeability

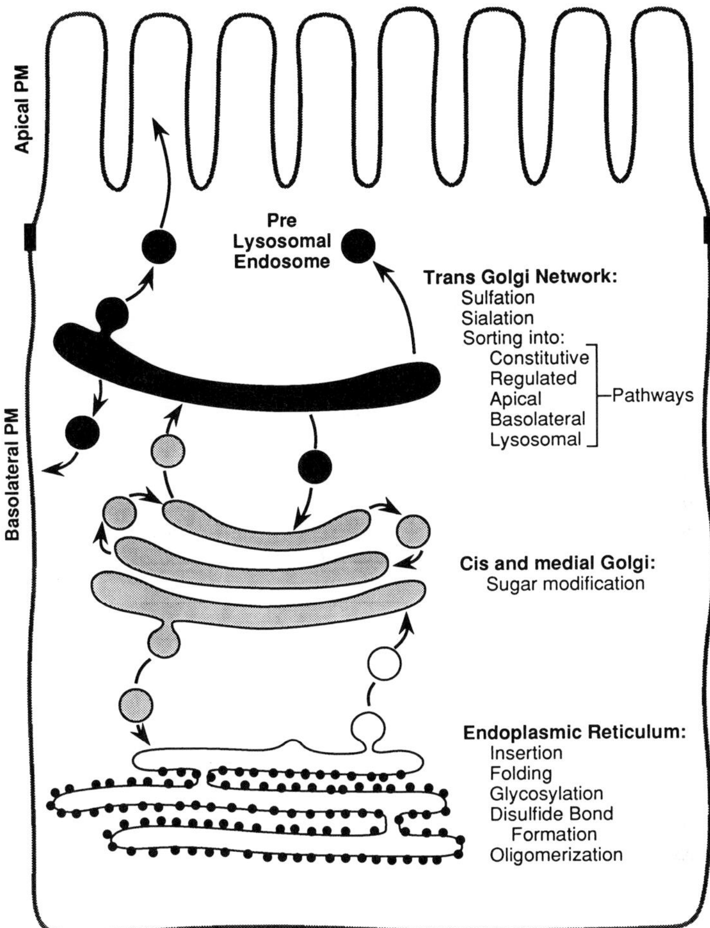

FIG. 1. Newly synthesized plasma membrane (PM) proteins undergo a wide variety of modifications and sorting operations en route to the cell surface. As described in the text, each stage in the postsynthetic maturation of a plasmalemmal protein is associated with unique processing functions. Each stage is also endowed with the capacity to both determine a protein's subsequent destination and mediate its appropriate targeting. In this sketch, drawn to represent a typical epithelial cell, notable events in a plasma membrane protein's postsynthetic course are indicated next to the organelles in which they occur. See text for details.

barriers provided by epithelia which control the exchange of solutes and fluid between an organism and its environment. The plasma membranes of polarized epithelial cells are divided into two morphologically and biochemically distinct domains (Simons and Fuller, 1985; Matlin, 1986; Caplan and Matlin, 1989; Rodriguez-Boulan and Nelson, 1989). The apical membrane, which is frequently endowed with microvilli, generally faces the lumen of a body compartment which is topologically continuous with the extracorporeal space. The basolateral plasmalemma rests on the epithelium's basement membrane and is in contact with the extracellular fluid space. Both the phospholipid and polypeptide compositions of the two domains are markedly different. The boundary between the apical and basolateral cell surfaces is delineated by tight junctions, which impede the passage of molecules between the two extracellular fluid compartments (see Fig. 2).

Perhaps the most important physiological feature of epithelial cells is their capacity to carry out vectorial transport, that is, the net movement of substances from one extracellular fluid compartment to the other, frequently in the face of steep unfavorable concentration gradients. The ability of transporting epithelia to mediate vectorial transport is explained by the asymmetric distribution of transport proteins among their two plasmalemmal surfaces (Schultz, 1986). Movement of a substance from one compartment to the other requires the participation of both apical and basolateral transporters working in series. The differential placement of pumps, cotransporters, and channels in the apical and basolateral plasma membranes bestows on epithelia their ability to carry out unidirectional and uphill transport. The value of this asymmetry is nicely illustrated by the principal cell of the renal collecting duct (O'Neil, 1987; Koeppen and Giebisch, 1985). The basolateral plasma membrane of the principal cell is extremely rich in Na^+/K^+-ATPase. This complement of sodium pump functions to keep the cytosolic concentration of sodium low. A sodium channel in the apical membrane allows sodium to flow down its concentration gradient from the tubule lumen into the cytoplasm. Apically entering sodium is expelled from the cytoplasm via the basolateral Na^+/K^+-ATPase and is thus transported from the lumenal to the extracellular fluid compartment. The energy of the sodium gradient generated by the Na^+/K^+-ATPase is exploited to drive the energetically unfavorable resorption of sodium from the renal tubule. This scheme is critically dependent on its geometry. It only works if the sodium pump and the sodium channel are present in different membrane surfaces which are separated by relatively sodium-impermeable tight junctions. Clearly, the cell must possess some mechanisms which are capable of organizing and maintaining this geometry.

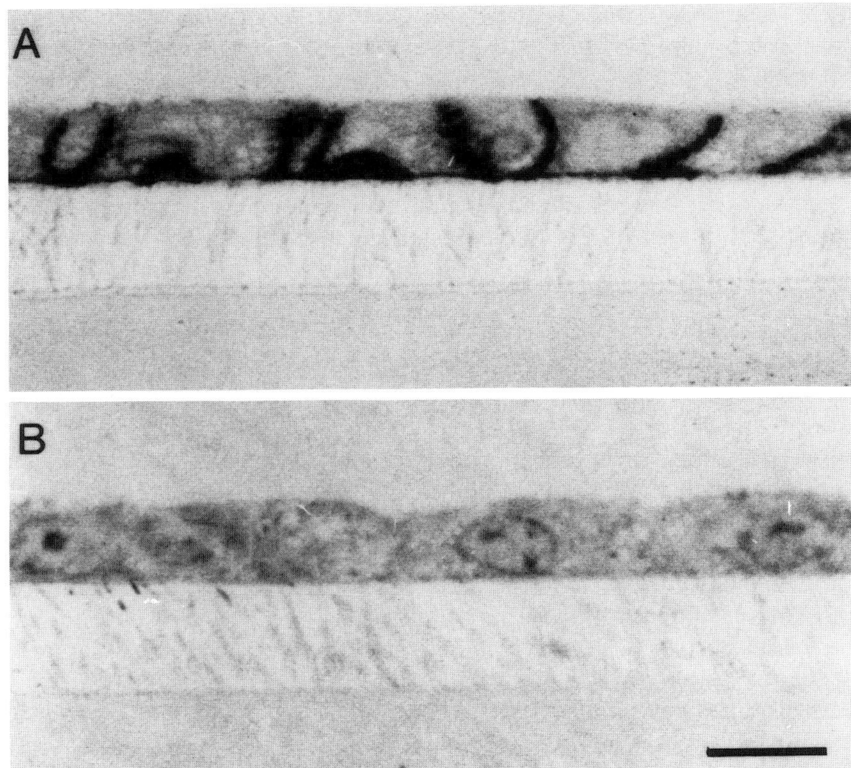

FIG. 2. The plasma membranes of polarized epithelial cells are divided into two biochemically distinct domains. The protein compositions of the apical and basolateral membranes of an epithelial cell are essentially distinct. In A, cultured canine kidney epithelial cells (MDCK) grown on permeable filter supports were labeled for immunoelectron microscopy with an antibody directed against the Na^+/K^+-ATPase α-subunit followed by a secondary reagent coupled to horseradish peroxidase. Dense reaction product is associated with only the lateral and basal membranes. No sodium pump can be detected on the apical surface. No staining is detected when a non-immune IgG is employed (B). Bar, 10 μm. (From Caplan *et al.*, 1986.)

Asymmetric distributions are not limited to the epithelial polypeptides which participate in ion transport. The vast majority of plasmalemmal proteins in polarized epithelial cells are restricted in their distributions to one or the other cell surface domain. For the purposes of this discussion, therefore, it is useful to think of an epithelial cell's plasma membrane as two distinct organelles which, although physically contiguous, are equipped for specific and unique functions.

The existence of this plasmalemmal compositional anisotrophy implies that the cell possesses mechanisms to both generate and maintain it. The cell must be able to recognize newly synthesized membrane proteins bound for one or the other cell surface domain, target them to their appropriate destination, and retain them there following their arrival. As was discussed in the context of the resident proteins of the ER and the Golgi, this postulated cellular sorting capability must, in some form or another, be based on the concepts of sorting signals and sorting machinery. We define a sorting signal as that information encoded in some aspect of a protein's structure which the cell uses to determine the protein's site of ultimate functional residence. Sorting machinery, in this context, refers to all of the cellular components involved in recognizing, interpreting, and acting on the information contained in sorting signals (Caplan and Matlin, 1989).

Research into the nature of epithelial sorting signals and the cellular sorting machinery is now over a decade old. A tremendous amount of work in this field was inspired by a seminal observation reported in 1978 by Rodriguez-Boulan and Sabatini, which made it clear that this problem could be made accessible to the techniques of cell and molecular biology. These investigators studied the budding of enveloped viruses from polarized epithelial cells in culture. They took advantage of the fact that a line of polarized epithelial cells derived from the dog renal tubule (Madin–Darby canine kidney, or MDCK; Madin and Darby, 1975) is susceptible to infection with both the influenza virus and the vesicular stomatitis virus (VSV). Through an electon microscopic analysis, they noticed that the influenza virus buds predominantly from the apical surface of infected cells, while VSV buds predominantly from the basolateral plasmalemma.

The encapsulating lipid bilayer membranes of enveloped viruses are notable for an extremely high density of transmembrane "spike" glycoproteins. In immunoelectron microscopic studies, Rodriguez-Boulan and Sabatini found that, prior to viral budding, the viral spike glycoproteins synthesized by infected cells accumulate in the plasma membrane domain from which budding will occur. Thus, influenza HA protein is targeted to the apical plasma membrane, where it awaits incorporation into virions. Similarly, the VSV G protein behaves as a basolateral plasma membrane protein until it becomes associated with the departing viral envelope. This observation was immensely important in that it provided investigators in the field of epithelial polarity with the first model system for membrane protein sorting whose components could be readily manipulated to suit an experimental design.

Within a few years after these initial observations, the genes encoding the influenza HA and VSV G proteins were cloned and expressed by transfection in MDCK cells. In a number of studies it was found that the spike glycoproteins synthesized by transfected cells still accumulate on their characteristic membrane domains, even in the absence of all of the other viral components that are normally synthesized during the course of infection (Stephens *et al.*, 1986; Gottlieb *et al.*, 1986b; Roth *et al.*, 1983; Jones *et al.*, 1985). For all intents and purposes, it could be said that the influenza HA protein behaves like a bona fide apical protein and the VSV G protein like a bona fide component of the basolateral plasmalemma. From these experiments it was clear that all of the information necessary to target these polypeptides to their respective cell surface domains resides within the proteins themselves. Their sorting is not dependent on any contribution from the viral genome. These studies comprised the first direct demonstrations that epithelial sorting signals exist and are wholly determined by the characteristics of the sorted proteins themselves.

Over the ensuing years, a tremendous amount of effort has been devoted to uncovering the nature of these signals and of the sorting machinery which interprets them. At this point in time, it is safe to say that neither of these elemental components of the epithelial sorting system is well understood. Identifying putative sorting signals and components of the cellular sorting machinery has proved extremely difficult for both technical and theoretical reasons. While the mechanisms underlying sorting have yet to be elucidated, however, a tremendous amount has been learned which offers clues into the nature of the sorting process. A synopsis of this literature is presented below, divided somewhat arbitrarily into discussions of sorting pathways, sorting signals, sorting machinery, and the biogenesis and maintenance of the polarized state.

A. Sorting Pathways

The first, and perhaps most accessible, question explored in the field of epithelial polarity relates to where, within the cell, sorting occurs. Three experimentally distinguishable alternatives were proposed to describe the route taken by newly synthesized plasmalemmal polypeptides on their way to the cell surface (Evans, 1980). These models can be identified as vectorial sorting, random sorting, and obligate missorting. The vectorial sorting paradigm predicts that all of the operations required to target a protein to the appropriate cell surface domain occur prior to that protein's arrival at the cell surface. According to this scheme, sorting is an intracellular process and transport of newly synthesized plasmalemmal proteins to the cell surface is vectorial in the sense that a polypeptide's first appearance at the cell surface is coincident with its arrival at the membrane domain in which it rightly belongs.

According to the random sorting model, no sorting occurs prior to cell surface delivery. Apical and basolateral proteins depart the TGN together and are delivered without preference to both cell surface domains. In this formulation, sorting is the product of selective endocytosis, which removes misplaced proteins from the cell surface and reroutes them to the appropriate domain. Obligate missorting is something of a compromise between the mutally exclusive vectorial and random proposals. This model predicts that apically and basolaterally directed proteins depart the TGN together, perhaps in the same vesicular carrier, and are delivered together to one of the two plasmalemmal surfaces. The subset of proteins for which this delivery process is correct, i.e., those which arrive directly at their proper destination, are allowed to remain in place. The polypeptides which find themselves in the incorrect membrane domain as a result of the initial delivery step are retrieved by endocytosis and shuttled transcytotically to the opposite surface. In theory, the obligate misdelivery model is equally compatible with the initial membrane insertion step occurring at either the apical or the basolateral pole of the cell. In practice, evidence has been gathered supporting only the latter possibility.

The first studies of sorting pathways were performed on MDCK cells infected with either the influenza virus or VSV. Cells were metabolically labeled with [^{35}S]methionine and exposed at their apical or basolateral surfaces to either trypsin or to antibodies directed against the ectodomains of the viral spike glycoproteins (Misek *et al.,* 1984; Matlin and Simons, 1984; Pfeiffer *et al.,* 1985). The susceptibility of these proteins to proteolysis or to antibody binding was assessed by immunoprecipitation followed by gel electrophoresis and fluorography. It was found that the VSV G protein was never even transiently available for interaction with apically added antibody or vulnerable to the action of apically added trypsin (Pfeiffer *et al.,* 1985). Similarly, the influenza HA protein was only affected by these treatments when they were applied from the apical side (Misek *et al.,* 1984; Matlin and Simons, 1984). These findings were most consistent with the vectorial sorting model, which would predict that the newly synthesized spike glycoproteins would appear first and only at their appropriate membrane surfaces and would thus never be available to reagents added at the opposite side.

Subsequent studies have since verified that, in MDCK cells, vectorial sorting applies to endogenous proteins as well. The pathway followed by

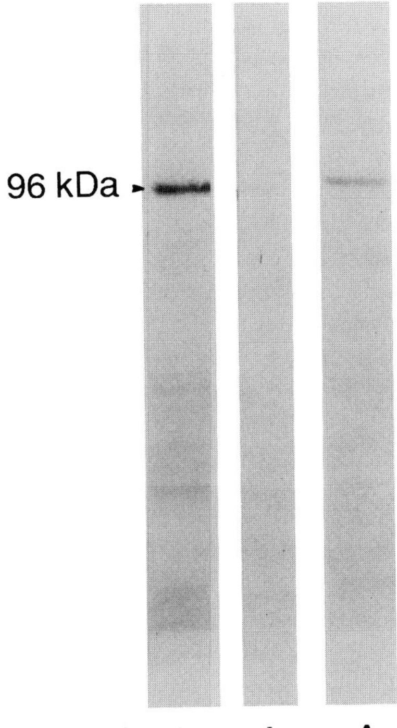

FIG. 3. The newly synthesized sodium pump is sorted vectorially to the basolateral plasma membrane in MDCK cells. MDCK cells were grown on a filter, as depicted in Fig. 2. Following pulse labeling with [^{35}S]methionine, they were exposed to NAB-ouabain at their apical or basolateral surfaces, photolyzed, solubilized, and subjected to immunoprecipitation with anti-ouabain antibodies as described in the text. As can be seen in the figure, NAB-ouabain had access to the newly synthesized Na^+/K^+-ATPase when it was added to the basolateral (B) but not the apical (A) medium compartment. Rupturing intercellular tight junctions through calcium chelation allowed apically added NAB-ouabain to interact with the newly synthesized sodium pump (EDTA +). These experiments indicate that the sodium pump does not appear, even briefly, on the apical surface prior to its arrival at the basolateral plasmalemma. (From Caplan *et al.*, 1986.)

the newly synthesized Na^+/K^+-ATPase was assessed using a protocol similar to that described above in the experiments relating to intracellular activation of the sodium pump (Caplan *et al.*, 1986, 1990) (see Fig. 3). MDCK cells were grown on permeable filter supports in order to allow for simultaneous and independent access to both membrane surface domains. Following a brief pulse labeling with [^{35}S]methionine, the photoactivatable NAB derivative of ouabain was added to the apical or basolateral medium compartment. After a 90-min chase incubation in the presence of this photoaffinity reagent, bound NAB-ouabain was covalently incorporated into the Na^+/K^+-ATPase α-subunit through exposure to UV light. Membranes prepared from these cells were subjected to immunoprecipitation with an anti-ouabain antibody and immunoprecipitates were analyzed by gel electrophoresis followed by fluorography. If newly synthesized sodium pump appeared, even briefly, on the apical surface, then radiolabeled α-subunit would have been present in immunoprecipitates from cells exposed to apical NAB-ouabain during the chase. Analysis of the fluorographs, however, revealed that newly synthesized Na^+/K^+-ATPase was only accessible to NAB-ouabain when this compound was added to the basolateral medium. It was possible to conclude from these experiments that less than 5% of the newly synthesized sodium pump appeared, stably or transiently, in the apical plasma membrane.

Similar experiments examining the sorting of endogenous MDCK proteins have taken advantage of biotin coupled to the amine-reactive NHS group to label selectively proteins expressed at the apical and basolateral surfaces of filter grown cells (Lisanti *et al.*, 1989b; Le Bivic *et al.*, 1990). Metabolically labeled proteins accessible to the biotin probe were isolated on avidin-conjugated Sephrarose beads and analyzed by gel electrophoresis followed by fluorography. Proteins which are normally resident in the basolateral plasma membrane were never accessible to apically added NHS-biotin throughout the course of their posttranslational processing. Apical proteins were similarly invulnerable to conjugation with basolaterally applied reagent. Taken together, all of these results provide a strong demonstration that sorting in MDCK cells occurs intracellularly and is complete prior to the arrival of newly synthesized membrane proteins at the cell surface.

The observation that MDCK cell sorting is vectorial invites curiosity about the intracellular site at which sorting occurs. Rindler *et al.*, examined this question by performing immunoelectron microscopy on MDCK cells which had been doubly infected with both the influenza virus and VSV (Rindler *et al.*, 1984). These investigators found that the influenza HA and VSV G proteins could be colocalized in all of the intracellular membranous structures involved in membrane protein processing up through the TGN. Segregation of the apical from the basolateral proteins seemed to happen at or beyond this point.

Similar observations were made in a biochemical investigation performed by Fuller *et al.* (1985). Their experiment took advantage of the fact that, in addition to the HA protein, the influenza viral genome encodes a transmembrane neuraminidase which is incorporated into the apical plasma membrane and subsequently into viral envelopes. As was mentioned in the previous section, sialic acid is added in the TGN (Fuller *et al.*, 1985; Griffiths and Simons, 1986). Furthermore, it must be noted that incubation of cells at 20°C allows newly synthesized proteins to progress only as far as the TGN (Matlin and Simons, 1983; Saraste and Kuismanen, 1984; Griffiths *et al.*, 1985). Warming the cells to 37°C relieves the temperature block and allows the accumulated polypeptides to exit the TGN and proceed to the cell surface. Fuller *et al.*, found that when VSV-infected cells were incubated at 20°C, the G protein, which is normally multiply sialated, became hypersialated. This is not surprising, since the 20°C block traps the G protein in the compartment containing the sialyl transferase. Interestingly, however, when MDCK cells doubly infected with VSV and influenza virus were incubated at 20°C, the G protein was found to be markedly hyposialated. This observation was taken as an indication that the VSV G protein and the influenza neuraminidase had physical access to one another during their temperature-induced confinement in the TGN. This conclusion fit well with Rindler *et al.*'s (1984) morphological assessment that apically and basolaterally directed proteins were still intermingled at the TGN stage of their processing.

Immunoelectron microscopic studies have also been performed on nonpolarized endocrine cells which manifest regulated and constitutive secretory pathways. It was noted that, in the TGN, proteins destined for packaging into regulated secretory granules were physically segregated from membrane and secretory proteins as well as from intraluminal viral particles, all of which depart the Golgi for the cell surface via the constitutive route (Orci *et al.*, 1987; Tooze *et al.*, 1987). Taken together with the results discussed above, these observations suggest that sorting begins in the TGN and is completed by the time (or certainly not long after) newly synthesized polypeptides depart this final recognizable locus of the intracellular processing pathway.

While the vectorial model almost certainly applies to MDCK cells, it must be noted that other epithelial cell types appear to target membrane proteins to their cell surfaces via a different sorting pathway. In cell fractionation studies, Bartles *et al.*, (1987) followed the postsynthetic route pursued by two apical proteins in hepatocytes. They concluded that both proteins appeared in a fraction derived principally from the basolateral plasma membrane prior to being detectable in membranes cosedimenting with the apical plasmalemma. This observation suggested that these polypeptides might be targeted from the Golgi first to the basolateral surface and subsequently transcytosed to the apical side. This mode of cell surface delivery would conform to the obligate missorting model presented above. Interestingly, this route also mirrors the course followed by the polymeric immunoglobulin receptor of hepatocytes (Geuze *et al.*, 1984; Hoppe *et al.*, 1985). This transmembrane protein is delivered from the Golgi to the basolateral cell surface, where it binds IgA circulating in the blood. The IgA–receptor complex is endocytosed and transported to the apical membrane, at which point the receptor's ectodomain is cleaved from its transmembrane anchor. The ectodomain fragment, referred to as secretory component, is released into the bile in association with the bound IgA.

Cell fractionation-based and cell surface labeling studies of the sorting pathways in intestinal cells have also provided evidence for some manner of obligate missorting pathway. Metabolic pulse labeling and cell fractionation were used to follow the brush border enzymes aminopeptidase N (Massey *et al.*, 1987) and sucrase-isomaltase (Hauri *et al.*, 1979) in small intestinal enterocytes. Analysis of immunoprecipitates from the various membrane fractions suggested that this protein complex was associated with the basolateral plasma membrane prior to arriving at the apical surface. While all of these results are extremely interesting, cell fractionation studies are always open to the criticism that difficult-to-control-for and difficult-to-measure cross-contamination may artifactually skew the results. Similar experiments in other laboratories suggest that microvillar proteins are vectorially sorted in intestinal cells (Danielsen and Cowell, 1985).

More recent experiments have taken advantage of the biotin labeling protocol outlined above to observe the behavior of plasma membrane proteins synthesized by a polarized human intestinal adenocarcinoma cell line SK-CO-15 (Le Bivic *et al.*, 1989) and by the intestinal carcinoma cell line Caco-2 (Matter *et al.*, 1990). These experiments revealed that, while basolateral proteins were targeted vectorially in both cell systems, the behavior of apical proteins was somewhat more complicated. Apical proteins were vectorially targeted in the SK-CO-15 cells, whereas in the Caco-2 cells only a portion of the newly synthesized apical proteins were vectorially sorted and made their initial plasmalemmal appearance at the apical cell surface. The remainder were initially accessible to basolaterally added NHS-biotin and were subsequently transcytosed to the apical plasma membrane. Apparently, these cells handle apical plasmalemmal proteins via some combination of the vectorial and obligate missorting pathways.

It is perhaps somewhat surprising that the sorting pathway, which one might be tempted to regard as a fundamental property of all epithelial cells, appears to vary from one cell type to another. A possibile explanation for this puzzling variability, however, might be found in the tissue-specific secretory behavior manifest by epithelial cells. It has been demonstrated that MDCK cells release secretory proteins into both the apical and basolateral media compartments. (Kondor-Koch *et al.*, 1985; Gottlieb *et al.*, 1986a; Caplan *et al.*, 1987). Furthermore, it has been shown that the default pathway for secretory proteins, that is, the route pursued by secretory proteins incapable of interacting with the cellular sorting machinery, is apical and basolateral (Kondor-Koch *et al.*, 1985; Gottlieb *et al.*, 1986a; Caplan *et al.*, 1987). The default pathway is thought to reflect to some extent the relative volume, or capacity, of apically and basolaterally directed membrane carriers which are available to bulk flow cargo.

In contrast to the example of MDCK cells, hepatocytes seem to have no direct secretory pathway to the apical surface. While a huge volume of secretory proteins is released at the basolateral membrane into the hepatic sinusoids, no proteins have been identified which are secreted directly at the apical surface into the bile canaliculi. Furthermore, studies on the default pathway associated with Caco-2 cells indicated that the bulk of unsorted polypeptides are released basolaterally by this cell line (Hughson *et al.*, 1989; Rindler and Traber, 1988). Taken together, these observations suggest that the membrane protein sorting model associated with a given cell type may reflect the relative activities of its apical and basolateral secretory pathways. In hepatocytes, for example, the absence of an apical secretory pathway might reflect, or be responsible for, an absence of membranous traffic from the Golgi to the apical cell surface. Thus, apically bound proteins in hepatocytes may have no choice but to depart the TGN in basolaterally bound vesicles. The same could be said for Caco-2 cells, in which the basolaterally directed default pathway may reflect a paucity of apically directed vesicular carriers.

This model is currently favored, since it seems to correlate two separate cellular behaviors. It also raises a number of interesting and difficult to answer questions. For example, is the existence of an apically directed secretory cargo required in order to maintain vesicular traffic from the TGN to the apical surface? Conversely, is there no direct apical secretory cargo in certain epithelial cell types precisely because there is no direct route to the apical surface? In those cells which employ obligate misdelivery of apical proteins to the basolateral surface, do the apical and basolateral proteins occupy the same vesicular carriers, or are they sorted from one another in the TGN and carried to the basolateral surface in separate, segregated shuttles? What is the endocytic sorting mechanism that allows these cells to recognize apical proteins in the basolateral surface and remove them for transort to their appropriate destination? These questions, although somewhat arcane, are of much more than academic interest. Their solutions, if obtainable, will tell us a tremendous amount about the control of intracellular membrane traffic and the mechanisms of membrane protein targeting.

B. Sorting Signals

As was mentioned above, the demonstration that the VSV G and influenza HA proteins, expressed by transfection in MDCK cells, were sorted with high fidelity to the appropriate cell surface domains provided the first solid evidence that sorting signals are wholly contained within some aspect of the structure of the sorted molecule itself. Experiments by a number of groups had also shown that blocking the addition of N-linked sugars to the viral spike glycoproteins [either through the action of the glycosylatin inhibitor tunicamycin (Green *et al.*, 1981; Roth *et al.*, 1979) or by infecting strains of MDCK cells defective in glycosylation (Green *et al.*, 1981)] had no effect on the sorting of these polypeptides. In light of these observations, it seems certain that the information required to specify sorting is encoded in the amino acid sequence of a polypeptide, either directly or by virtue of the tertiary structure this sequence effects. A large body of literature has now developed out of the attempts to identify, or at least narrow, the search for these sorting signals. As of this writing, it is safe to say that nothing even remotely approximating an answer is yet available, although a few general rules may be emerging. The most popular approaches in the hunt for sorting signals have involved the construction of chimeric or truncated versions of the viral spike glycoproteins, whose sorting could be analyzed in transfected cells. These experiments have produced very complex and frequently contradictory results. Since this field is rather tangled and has been reviewed fairly recently (Caplan and Matlin, 1989), it is simply summarized here.

Truncated forms of the influenza HA protein which lack cytosolic and transmembrane domains are secreted from transfected epithelial cells. Analysis of this secretion reveals that it is predominantly apical (Roth *et al.*, 1987). It should be noted, however, that another group has found that a similar construct may be released into both the apical and basolateral compartments (Gonzalez *et al.*, 1987). Similar anchor-minus constructs of the VSV G protein (and of the basolaterally directed spike glycoprotein of the Friend mink leukemia virus) are released from MDCK cells into both the apical and basolateral media (Gonzalez *et al.*, 1987; Stephens and Compans, 1986). Chimeras composed of the influenza HA ectodomain and the VSV G transmembrane and endodomains are sorted apically (Roth *et al.*, 1987; McQueen *et al.*, 1986). The complementary constructs, bearing the VSV G ectodomain wedded to the influenza HA transmembrane and endodomains, are probably basolaterally targeted (McQueen *et al.*, 1987; Puddington, *et al.*, 1987; Compton *et al.*, 1989). These results seem to suggest that the ectodomains are important for the sorting of both apical and basolateral proteins, although the random secretion of the anchor-minus VSV G truncation is not entirely consistent with this formulation.

Recent studies have demonstrated that proteins anchored to the membrane via glycophosopholipids are expressed by epithelial cells essentially exclusively in the apical plasmalemma (Lisanti *et al.*, 1988, 1990). The members of this fascinating class of membrane proteins, which includes within its roster alkaline phosphatase, 5′-nucleotidase, and trypanosomal surface antigens, are intially synthesized on bound polysomes as transmembrane proteins (Cross, 1987). While still in the membrane of the RER they are cleaved from their transmembrane portion and transferred covalently to a lumenally facing glycosyl-phosphatidylinositol molecule (GPI). Two groups have tested the possibility that attachment to a GPI anchor is in itself a sufficient signal to ensure apical targeting. Brown *et al.* (1989) engineered a construct in which the VSV G ectodomain was joined with the lipid acceptor site of Thy-1, a GPI-coupled lymphocyte antigen. When expressed in MDCK cells, the resultant lipid-linked G protein was sorted to the apical plasmalemma. When placental alkaline phosphatase (PLAP) was expressed in MDCK cells, this GPI-linked protein was delivered to the apical membrane. An anchor-minus form of PLAP, which lacks the lipid association, was secreted both apically and basolaterally in a roughly 2 : 1 ratio. Attachment of PLAP to the VSV G transmembrane domain and cytosolic tail resulted in a chimeric protein that was targeted basolaterally. Similar results were gathered by Lisanti *et al.* (1989a). These investigators coupled the GPI-linked tail of decay-accelerating factor (DAF) to the ectodomain of herpes simplex glycoprotein D (a basolateral protein) and to human growth hormone (normally targeted for regulated secretion). Both constructs were delivered to the apical surface of transfected MDCK cells. These observations seem to support the notion that lipid attachment might serve to convey an apical sorting signal; however, it is difficult to synthesize all of these experiments into a single coherent picture.

A cDNA encoding the polymeric immunoglobulin receptor has been transfected into MDCK cells. Remarkably, this protein retraces in these cultured kidney cells the complicted course it follows in hepatocytes (Mostov and Deitcher, 1986). The receptor first appears at the basolateral surface, from which it is endocytosed and transported to the apical membrane. Immediately before or after its arrival at the apical surface the receptor undergoes a proteolytic cleavage which releases the secretory component into the apical media. The fact that this protein follows its rather indirect course in a cell type which normally manifests vectorial sorting has prompted the suggestion that the polymeric immunoglobulin receptor might possess hierarchical sorting signals, that is, two sorting signals whose expression is temporally or spatially determined. According to this line of thought, a basolateral sorting signal predominates during the protein's initial voyage from the TGN to the basolateral cell surface. After basolateral delivery, however, an apical signal somehow gains sway and the protein is transcytosed. Interestingly, the receptor undergoes phosphorylation on a serine residue at approximately the time that it is delivered to the basolateral surface (Larkin *et al.*, 1986).

An anchor-minus construct of the polymeric immunoglobulin receptor was secreted from transfected MDCK cells into the apical medium (Mostov *et al.*, 1987). Another receptor construct, which lacks only the cytoplasmic tail, was also delivered directly to the apical membrane (Mostov *et al.*, 1986). Site-directed mutagenesis experiments which converted the phosphorylated serine residue into an alanine resulted in a protein which remained at the basolateral surface and was transcytosed extremely slowly (Casanova *et al.*, 1990). Converting the serine to an aspartic acid residue, however, resulted in a protein which was subject to rapid transcytosis.

Anyone encountering this hodgepodge of chimeras and truncations for the first time can be forgiven for feeling a pronounced urge to skip this section of the review. It is certainly not immediately obvious that any single thread might unify this rather disparate collection of experimental results. There are, however, theories which can explain many, if not all, of these observations and which are, therefore, worth considering. Perhaps the most aesthetically pleasing of these invokes the concept of a default pathway for membrane proteins. As was discussed above, the default, or unsorted, pathway need not necessarily lead to both cell surfaces. One can imagine, for example, a scenario in which only incorporation into an apically directed vesicle required a special sorting interaction for membrane proteins. According to this model, any protein not specifically pulled out of the stream of newly synthesized membrane proteins flowing through the TGN would be carried by bulk flow to the basolateral surface. Were this the case, the absence of a sorting signal would in itself serve as a sorting signal, directing basolateral localization. It should be noted that a cell which manifest a basolateral default pathway for membrane proteins could still possess an apical and basolateral default pathway for secreted proteins, since soluble polypeptides might be able to gain access to the lumenal space of apically directed carriers without need of a special signal.

If we go forward with the idea that the basolateral route corresponds to the default pathway (as has recently been proposed as well by Simons and

Wandinger-Ness, 1990), we can generate a scheme which is at least sufficiently self-consistent to produce testable hypotheses. The postulates of this proposal are as follows: (1) the VSV G protein contains no sorting signal and is carried to the basolateral surface via the default pathway; (2) the influenza HA protein contains an apical sorting signal in its ectodomain; and (3) a glycolipid anchor is sufficient to mediate apical targeting and, therefore, glycolipid-anchored proteins carry no other sorting signals (as has been suggested by Brown *et al.*, 1989). Were all of these axioms to be true, one would expect that anchor-minus HA (which contains the ectodomain's apical sorting signal) would be released apically, while anchor-minus VSV G protein (which lacks a sorting signal) would be released via the secretory default pathway to both surfaces. Furthermore, an HA ectodomain–VSV G tail chimera would carry only an apical signal and be delivered to the apical surface, whereas a G ectodomain–HA tail chimera would bear no sorting signal and would follow the membrane protein default pathway to the basolateral plasmalemma. Lipid-linked VSV G ectodomain would travel to the apical surface by virtue of the signal embedded in the lipid linkage. Anchor-minus PLAP, however, which carries no signal, would be a candidate for the apically and basolaterally oriented secretory default release. Similarly, VSV G tail–PLAP ectodomain hybrids would be signal-less and thus destined for basolateral insertion.

Finally, this model can be expanded to integrate the polymeric immunoglobulin receptor observations if it is assumed that the ectodmain of this protein possesses an apical sorting signal which is unrecognizable so long as the cytoplasmic tail exerts some inhibitory influence. Phosphorylation of a serine residue on the tail would remove that inhibition and permit expression of the apical targeting information. According to this formulation, an intact and unphosphorylated receptor would display no recognizable signal and would thus be shuttled by default to the basolateral surface. Removal of the inhibitory tail, in either a tail-minus or anchor-minus truncation, would result in the apical targeting of the remainder of the polypeptide. Altering the phosphorylation site would prevent the tail's inhibitory influence from being reversed and would condemn the mutant protein to permanent residence in the basolateral plasmalemma. It should be noted, however, that the results of experiments (described above) in which the phosphorylated serine residue is converted to an aspartate (which may preserve the negative charge) are difficult to reconcile with this model. It may be that the signals and mechanisms involved in biosynthetic sorting differ from those associated with postendocytic sorting. Were this the case, the complicated situation of the polymeric immunoglobulin receptor could be explained more readily.

Evidence for this latter possibility comes from experiments in which the B1 and B2 isoforms of the macrophage Fc receptor were expressed in MDCK cells (Hunziker and Mellman, 1989). These two proteins are identical except for a 47-amino acid in-frame insertion in the cytoplasmic tail of the B1 polypeptide, which appears to diminish its capacity for endocytosis. The B1 form was found mostly apically, whereas B2 was predominantly detected in the basolateral domain. Both receptors could mediate IgG transcytosis in only the apical to basolateral direction. These results suggest that the steady-state distributions of these transcytotic receptors may reflect not only their handling by the biosynthetic sorting machinery but also their affinity for and interactions with elements of the endocytic apparatus.

The preceding scheme, although certainly prolix, is at least more or less consistent with many of the facts. Furthermore, the supposition of a basolateral default pathway is not without some theoretical justification. Most of the proteins commonly found in epithelial basolateral membranes are also present in the plasma membranes of nonpolarized cells. In contrast, apical membranes are generally endowed with "epithelia-specific" proteins. It could be argued that economy-minded Nature would design a system that would require sorting signals for only those proteins whose expression is limited to cells with more than one plasmalemmal destination. Confirmation or rejection of the model presented above will require elucidation of the membrane protein default pathway. Further clarification will require the identification and characterization of several membrane protein sorting signals as well as the identification of at least some of the cellular components which recognize and interpret these signals. Until these advances are achieved, models such as the one presented above, although unsatisfying, will have to suffice.

C. Sorting Mechanisms

If little is known of sorting signals, less is known about the mechanisms of sorting. None of the cellular components which participate in the targeting of newly synthesized epithelial membrane proteins have been identified. Very little has been learned or can be inferred about the routines which these putative components employ in carrying out their functions. Those who labor in the sorting field are inclined to think that this paucity of information derives from the inherent difficulty of the problem rather than from the inherent abilities of its investigators. In defense of this position, it should be noted that sorting has, until very recently, only been accessible to study in intact cells. *In vitro* (or semi-*in vitro* systems) capable of carrying out bona fide sorting are only just being developed (Tooze and Huttner, 1990). It is not surprising, therefore, that the biochemical correlates of the sorting process have proved difficult to dissect.

The best understood sorting mechanism is the one which functions to target newly synthesized lysosomal enzymes from the Golgi complex to a prelysosomal compartment. A number of very elegant experiments have demonstrated that a family of receptors exists which recognize a mannose 6-phosphate residue added to the sugar structure of newly synthesized hydrolases during their passage through the Golgi (for review see Kornfeld, 1987). On binding these enzymes, the receptors transport them from the Golgi to a prelysosomal compartment whose acidic pH induces the dissolution of the receptor–hydrolase complex. Receptors unburdened of their ligand are free to return to the Golgi and to participate in another round of sorting. Weak bases such as NH_4Cl elevate the pH of the prelysosomal compartment and thus prevent release of the delivered enzyme. Since the complexed receptors cannot participate in further sorting, the cell's complement of available receptors is quickly exhausted and the newly synthesized hydrolases pass through the Golgi without being diverted to the lysosomal pathway. Thus, in the presence of this drug, newly synthesized lysosomal enzymes are secreted from the cell.

There is some reason to believe that an analogous model may be applicable to at least some aspects of other sorting phenomena. The requirement for sorting signals would certainly suggest the involvement of receptors which, by analogy with the lysosome system, might be expected to divert proteins into the appropriate pathways. Furthermore, the sorting of secretory proteins between the regulated and constitutive pathways in endocrine cells has been shown to require the participation of intracellular low-pH compartments (Kelly, 1985). Elevation of the pH of such compartments through the addition of weak bases such as NH_4Cl prevents sorting to the regulated pathway and results in the constitutive release of proteins which are normally stored intracellularly. It is tempting to think that low-pH compartments might subserve the same receptor regenerating function for regulated pathway sorting that they perform for the lysosmal enzymes.

Similar evidence has been gathered for proteins secreted constitutively from the basolateral surface of MDCK cells (Caplan *et al.*, 1987). The basement membrane components laminin and heparan sulfate proteoglycan (HSPG) are normally released into the basolateral medium. In the presence of NH_4Cl, both of these proteins are released almost equally into both media compartments. Interestingly, the sorting of an apical secretory protein is unaffected by NH_4Cl. The basolateral targeting of the sodium pump is also not perturbed by this drug (Caplan *et al.*, 1986) (see Fig. 4).

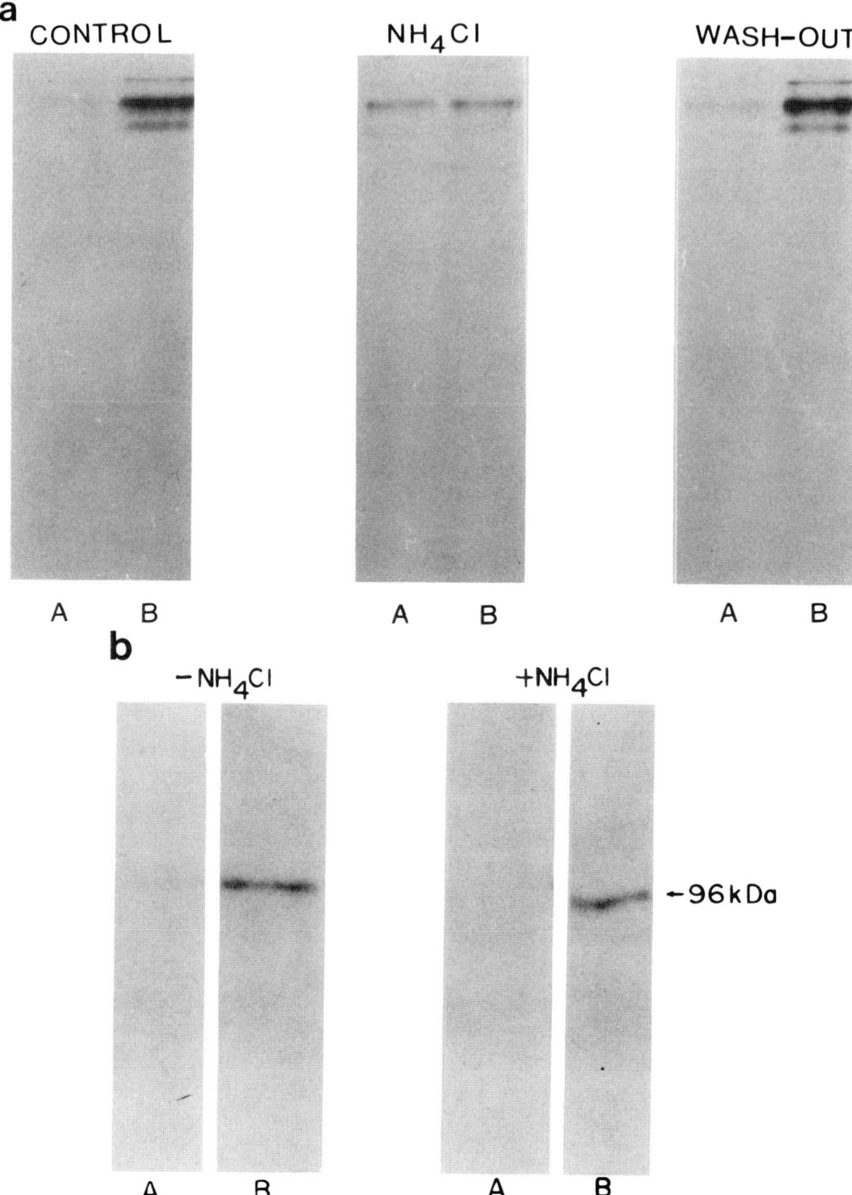

FIG. 4. A low-pH compartment is required for basolateral secretory, but not membrane, protein sorting in MDCK cells. MDCK cells grown on filters, as depicted in Fig. 2, secrete the basement membrane protein laminin into the basolateral medium (a, Control). In the presence of weak bases which elevate the pH of intracellular acidic compartments, laminin is released into both the apical and basolateral media (a, NH_4Cl). Removal of the weak base restores normal secretion (a, Wash-out). In contrast, the targeting of the newly synthesized Na^+/K^+-ATPase (assessed according to the protocol outlined in the text and Fig. 3) is unaffected by the presence of weak bases (b). Thus, distinct and pharmacologically separable mechanisms must operate in the sorting of these two proteins to the basolateral surface of MDCK cells. See text for details. (From Caplan *et al.*, 1986, 1987.)

If a lysosomal sorting-like model actually applies, it should be possible to identify the pH-sensitive receptor. A candidate for this function has been isolated from endocrine and exocrine cells (Chung *et al.,* 1989). This 25-kDa polypeptide appears to interact in a pH-sensitive fashion with only those proteins destined for packaging into regulated secretory granules. The degree to which this protein actually participates in sorting, however, remains to be established. Much less progress has been made in attempts to identify putative epithelial sorting receptors.

Finally, it should be noted that microtubules have been implicated as cellular components important in the epithelial sorting process. MDCK cells treated with microtubule-dissolving drugs deliver apical membrane and secretory proteins to both cell surface domains (Parczyk *et al.,* 1989; Rindler *et al.,* 1987). Interestingly, the basolateral delivery of membrane proteins is unaffected by the disruption of the microtubular network. Similar results have been gathered with intestinal epithelial cells (Achler *et al.,* 1989). It remains to be learned whether microtubules are functioning as tracks along which vesicles powered by microtubule motors are directed to their appropriate destinations or if they are subserving some other, as yet undefined function.

D. Generation and Maintenance of Epithelial Polarity

The discussion up to this point has focused on the sorting of membrane proteins in an already established epithelium. A closely related and extremely interesting problem is associated with the initial formation of a polarized epithelium *de novo* from nonpolarized cells. This process occurs during embryogenesis, where the nonepithelial early embryo gives rise to the highly polarized blastodermal epithelium. It also happens during the routine passage of polarized tissue culture cells, which involves the disruption of an established epithelial monolayer by trypsinization followed by the reorganization of a polarized cell layer subsequent to replating. A thorough discussion of the fascinating literature relating to these processes is beyond the scope of this review. Furthermore, it has been extremely well reviewed elsewhere (Rodriguez-Boulan and Nelson, 1989; Nelson, 1989). It is useful, however, to take note of some of the general themes which seem to be emerging from research in this field.

The generation of epithelial polarity requires cell–cell contact. MDCK cells plated on a substrate but prevented from forming cell–cell contacts are unable to generate fully polarized plasmalemmal domains (Gonzalez-Mariscal *et al.,* 1985; Nelson and Veshnock, 1987a; Vega-Salas *et al.,* 1987a). Conversely, MDCK cells grown in suspension but allowed to form cell–cell contacts are able to organize at least partially polarized plasmalemmal domains (Rodriguez-Boulan *et al.*, 1983; Wang *et al.*, 1990). In MDCK cells, the initiation of cell–cell contact has been shown to induce at least two effects. The first effect is the insertion into the plasma membrane of an intracellular structure called VAC, or vacuolar apical compartment. Vega-Salas *et al.* (1987b) have found that in MDCK cells denied cell–cell contact, apical membrane proteins accumulate in an intracellular compartment which is morphologically distinct from the organelles of the processing pathway. These fairly large vacuolar structures are frequently endowed with microvilli, which protrude into their lumena. The formation of cell–cell contacts induces the exocytosis of these structures, resulting in the insertion of pre–formed apical membrane at the cell surface. These observations suggest that, prior to the formation of an epithelium, the uniquely epithelial apical membrane cannot be expressed at the cell surface. Membrane traffic within the individual cells of a developing epithelium, therefore, appears to be directly responsive to influences from neighboring cells.

The second effect which has been recognized is the assembly of the cytoskeleton. The cytosolic surface of the basolateral plasma membrane of many epithelial cells (including MDCK) is covered by a cytoskeletal meshwork whose composition is remarkably similar to that of the erythrocyte cytoskeleton (Nelson, 1989). This web of ankyrin and fodrin (non-erythroid spectrin) appears to be bound to the membrane through interactions with trans-membrane proteins. The Na^+/K^+-ATPase has been shown to be one of the membrane anchors for this complex (Nelson and Veshnock, 1987b; Nelson and Hammerton, 1989; Morrow *et al.*, 1989), as has the cell adhesion molecule uvomorulin (Nelson *et al.*, 1990). In isolated MDCK cells denied cell–cell contact, the protein components of the cytoskeleton are disorganized and unassembled (Nelson and Veshnock, 1987a). Very quickly after the initiation of cell–cell contact, however, the basolateral cytoskeleton is formed. Nelson *et al.* (1990) have suggested that uvomorulin, which is involved in establishing cell–cell contact, communicates this event to the cytoplasm through its interaction with the cytoskeleton. According to this model, uvomorulin aggregates at sites of cell–cell contact, which in turn leads to the assembly of the cytoskeleton in these regions. The forming cytoskeleton essentially traps those proteins, such as the sodium pump, which are capable of interacting with it, leading to the generation of a biochemically differentiated membrane domain (McNeill *et al.*, 1990). Evidence from detergent extraction experiments (Salas *et al.*, 1988) suggests that interactions with an assembled, insoluble cytoskeleton may exist for both apical and basolateral proteins. While as yet unproven, this scheme offers an interesting and potentially testable explanation for the epithelial cell's capacity to generate or regenerate a polarized state on interaction with its neighbors.

Once a polarized state is formed, the epithelial cell must be able to maintain it. Basolateral proteins must be prevented from diffusing into the apical domain and vice versa. The most important mechanism for preventing this intermixing is almost certainly the tight junction. Tight junctions have been shown to serve as barriers to the two-dimensional diffusion of both lipids and proteins (Cereijido *et al.*, 1989; van Meer, 1989; Dragsten *et al.*, 1981). Furthermore, disruption of tight junctions has been shown to allow randomization of epithelial plasmalemmal domains. The cytoskeleton may also play a role in preserving biochemical polarity. The involvement of the sodium pump with the cytoskeleton, for example, almost certainly curtails its mobility in the plane of the membrane (Jesaitis and Yguerabide, 1986).

Finally, endocytosis and postendocytic sorting must contribute to the stability of the polarized state. Fuller and Simons (1986) found that, following endocytosis, the basolateral transferrin receptor of MDCK cells was recycled to the basolateral membrane with greater than 99% fidelity. Matlin *et al.* (1983) found that basolateral proteins inserted into the apical surface (by fusing the VSV into the apical membrane) were efficiently removed and transcytosed to the basolateral plasmalemma. It seems clear, therefore, that the cell possesses mechanisms for continuously monitoring the compositions of its cell surface domains and for undertaking corrections should they be required.

IV. CONCLUSION

The cell surface membrane is the boundary between a cell and its environment. In the case of polarized epithelial cells, the apical plasma membrane is frequently the boundary between an organism and its environment. Consequently, the proteins that populate the plasmalemma play a tremendous role in determining the properties of individual cells and of whole tissues. Cells go to great trouble to regulate the compositions of their plasma membranes and to organize those membranes into subdomains capable of specialized functions. The processes through which this control and organization are created are the subjects of extremely active investigation. As promised in the introduction, this review has been long on description and speculation and short on the presentation of definitive mechanisms. I hope, however, that I have conveyed some of the excitement and complexity that attend this field and that I have succeeded in demonstrating its relevance to the membrane transport phenomena which are the subject of the remainder of this volume.

Acknowledgments

The author is supported by NIH GM42136 and a fellowship from the David and Lucille Packard Foundation.

References

Achler, C., Filmer, D., Merte, C., and Drenckhahn, D. (1989). Role of microtubules in polarized delivery of apical membrane proteins to the brush border of the intestinal epithelium. *J. Cell Biol.* **109,** 179–189.

Anderson, D. J., Mostov, K. E., and Blobel, G. (1983). Mechanisms of integration of *de novo*-synthesized polypeptides into membranes: Signal recognition particle is required for integration into microsomal membranes of calcium ATPase and of lens MP26 but not of cytochrome b_5. *Proc. Natl. Acad. Sci.* (*U.S.A.*) **80,** 7249–7253.

Attardi, G., and Schatz, G. (1988). Biogenesis of mitochondria. *Annu. Rev. Cell Biol.* **4,** 289–333.

Bartles, J. R., Ferracci, H. M., Steiger, G., and Hubbard, A. L. (1987). Biogenesis of the rat hepatocyte plasma membrane *in vivo:* Comparison of the pathways taken by apical and basolateral proteins using subcellular fractionation. *J. Cell Biol.* **105,** 1241–1251.

Beckers, C. J., and Balch, W. E. (1989). Calcium and GTP: Essential components in vesicular trafficking between the endoplasmic reticulum and Golgi apparatus. *J. Cell Biol.* **108,** 1245–1256.

Beckers, C. J., Block, M. R., Glick, B. S., Rothman, J. E., and Balch, W. E. (1989). Vesicular transport between the endoplasmic reticulum and the golgi stack requires the NEM-sensitive fusion protein. *Nature* (*London*) **339,** 397–398.

Bergmann, J. E., and Singer, S. J. (1983). Immunoelectron microscopic studies of the intracellular transport of the membrane glycoprotein (G) of vesicular stomatitis virus in infected Chinese hamster ovary cells. *J. Cell Biol.* **97,** 1777–1787.

Bernstein, H. D., Poritz, M. A., Strub, K., Hoben, P. J., Brenner, S., and Walter, P. (1989). Model for signal sequence recognition from amino-acid sequence of 54 K subunit of signal recognition particle. *Nature* (*London*) **340,** 482–486.

Blobel, G. (1980). Intracellular protein topogenesis. *Proc. Natl. Acad. Sci.* (*U.S.A.*) **77,** 1496–1500.

Blobel, G., and Dobberstein, B. (1975). Transfer of proteins across membranes. *J. Cell Biol.* **67,** 852–862.

Blok, J., Gibbs, E. M., Lienhard, G. E., Slot, J. W., and Geuze, H. J. (1988). Insulin-induced translocation of glucose transporter from post-Golgi compartments to the plasma membrane of 3T3-L1 adipocytes. *J. Cell Biol.* **106,** 69–76.

Bole, D. G., Hendershot, L. M., and Kearney, J. F. (1986). Posttranslational association of immunoglobulin heavy chain binding protein with nascent heavy chains in non-secreting and secreting hybridomas. *J. Cell Biol.* **102,** 1558–1566.

Bourne, H. R. (1988). Do GTPases direct membrane traffic in secretion? *Cell* **53,** 669–671.

Brown, D. A., Crise, B., and Rose, J. K. (1989). Mechanism of membrane anchoring affects polarized expression of two proteins in MDCK cells. *Science* **245,** 1499–1501.

Burgess, T. L., and Kelly, R. B. (1987). Constitutive and regulated secretion of proteins. *Annu. Rev. Cell Biol.* **3,** 243–293.

Caplan, M. J. (1990). Biosynthesis and sorting of the sodium, potassium-ATPase. *In* "Regulation of Potassium Transport Across Biological Membranes" (L. Reuss, J. M. Russell, and G. Szabo, eds.), pp. 77–101. Univ. of Texas Press, Austin.

Caplan, M. J., and Matlin, K. S. (1989). Sorting of membrane and secretory proteins in polarized epithelial cells. *In* "Functional Epithelial Cells in Culture" (K. S. Matlin and J. D. Valentich, eds.), pp. 71–127. Alan R. Liss, New York.

Caplan, M. J., Anderson, H. C., Palade, G. E., and Jamieson, J. D. (1986). Intracellular sorting and polarized cell surface delivery of NA,K-ATPase, an endogenous component of MDCK cell basolateral plasma membranes. *Cell* **46,** 623–631.

Caplan, M. J., Stow, J. L., Newman, A. P., Madre, J., Anderson, H. C., Farquhar, M. G., Palade, G. E., and Jamieson, J. D. (1987). Dependence on pH of polarized sorting of secreted proteins. *Nature (London)* **329,** 632–635.

Caplan, M. J., Forbush, B., III, Palade, G. E., and Jamieson, J. D. (1990). Biosynthesis of the Na,K-ATPase in Madin–Darby canine kidney cells: Activation and cell surface delivery. *J. Biol. Chem.* **265,** 3528–3534.

Casanova, J. E., Breitfeld, P. P., Ross, S. A., and Mostov, K. E. (1990). Phosphorylation of the polymeric immunoglobulin receptor required for its efficient transcytosis. *Science* **248,** 742–745.

Cereijido, M., Ponce, A., and Gonzalez-Mariscal, L. (1989). Tight junctions and apical/basolateral polarity. *J. Membr. Biol.* **110,** 1–9.

Chappell, T. G., Welch, W. J., Schlossman, D. M., Palter, K. B., Schlessinger, M. J., and Rothman, J. E. (1986). Uncoating ATPase is a member of the 70 kilodalton family of stress proteins. *Cell* **45,** 3–13.

Chen, L., and Tai, P. C. (1987). Evidence for the involvement of ATP in co-translational translocation. *Nature (London)* **328,** 164–168.

Chirico, W. J., Waters, M. G., and Blobel, G. (1988). 70K heat shock related proteins stimulate protein translocation into microsomes. *Nature (London)* **332,** 805–810.

Chung, K.-N., Walter, P., Aponte, G., and Moore, H.-P. H. (1989). Molecular sorting in the secretory pathway. *Science* **243,** 192–197.

Clary, D. O., Griff, I. C., and Rothman, J. E. (1990). SNAPs, a family of NSF attachment proteins involved in intracellular membrane fusion in animals and yeast. *Cell* **61,** 709–721.

Clevers, H., Alarcon, B., Wileman, T., and Terhorst, C. (1988). The T cell receptor/CD3 complex: A dynamic protein ensemble. *Annu. Rev. Immunol.* **6,** 629–662.

Compton, T., Ivanov, I. E., Gottlieb, T., Rindler, M. J., Adesnik, M., and Sabatini, D. D. (1989). A sorting signal for the basolateral delivery of the vesicular stomatitis virus (VSV) G protein lies in its lumenal domain: Analysis of the targeting of VSV G-influenza Hemagglutinin chimeras. *Proc. Natl. Acad. Sci. U.S.A.* **86,** 4112–4116.

Connolly, T.,and Gilmore, R. (1989). The signal recognition particle receptor mediates the GTP-dependent displacement of SRP from the signal sequence of the nascent polypeptide. *Cell* **57,** 599–610.

Copeland, C. S., Doms, R. W., Bolzau, E. M., Webster, R. G., and Helenius, A. (1986). Assembly of influenza hemagglutinin trimers and its role in intracellular transport. *J. Cell Biol.* **103,** 1179–1191.

Copeland, C. S., Zimmer, K.-P., Wagner, K. R., Healey, G. A., Mellman, I., and Helenius, A. (1988). Folding, trimerization and transport are sequential events in the biogenesis of influenza virus hemagglutinin. *Cell* **53,** 197–209.

Cross, G. A. M. (1987). Eukaryotic protein modification and membrane attachment via phosphatidylinositol. *Cell* **48,** 179–181.

Cummings, R. D., Kornfeld, S., Schneider, W. J., Hobgood, K. K., Tolleshaug, H., Brown, M. S., and Goldstein, J. L. (1983). Biosynthesis of *N*- and *O*-linked oligosaccharides of the low density lipoprotein receptor. *J. Biol. Chem.* **258,** 15261–15273.

Danielsen, E. M., and Cowell, G. M. (1985). Biosynthesis of intestinal microvillar proteins: Evidence for an intracellular sorting taking place in, or shortly after, exit from the Golgi complex. *Eur. J. Biochem.* **152,** 493–499.

Darnell, J. E., Jr. (1982). Variey in the level of gene control in eukaryotic cells. *Nature (London)* **297,** 365–371.

Deshaies, R. J., Koch, B. D., Werner-Washburne, M., Craig, E. A., and Shekman, R. (1988). A sub-family of stress proteins facilitates translocation of secretory and mitochondrial precursor polypeptides. *Nature (London)* **332,** 800–805.

Doms, R. W., Ruusala, A., Machamer, C., Helenius, J., Helenius, A., and Rose, J. K. (1988). Differential effects of mutations in three domains on folding, quaternary structure, and intracellular transport of VSV G protein *J. Cell Biol.* **107,** 89–99.

Doms, R. W., Russ, G., and Yewdell, J. W. (1989). Brefeldin A redistributes resident and itinerant Golgi proteins to the endoplasmic reticulum. *J. Cell Biol.* **109,** 61–72.

Dragsten, P. R., Blumenthal, R., and Handler, J. S. (1981). Membrane asymmetry in epithelia: Is the tight junction a barrier to diffusion in the plasma membrane? *Nature (London)* **294,** 718–722.

Engleman, D. M., and Steitz, T. A. (1981). The spontaneous insertion of proteins into and across membranes: The helical hairpin hypothesis. *Cell* **23,** 411–422.

Evans, E., Gilmore, R., and Blobel, G. (1986). Purification of microsomal signal peptidase as a complex. *Proc. Natl. Acad. Sci. U.S.A.* **83,** 581–585.

Evans, W. H. (1980). A biochemical dissection of the functional polarity of the plasma membrane of the hepatocyte. *Biochim. Biophys. Acta* **604,** 27–64.

Forbush, B., III (1983). Cardiotonic steroid binding to Na,K-ATPase. *Curr. Top. Membr. Transp.* **19,** 167–201.

Freedman, R. B. (1989). Protein disulfide isomerase: Multiple roles in the modification of nascent secretory proteins. *Cell* **57,** 1069–1072.

Fuller, S. D., and Simons, K. (1986). Transferrin receptor polarity and recycling accuracy in "tight" and "leaky" strains of Madin–Darby canine kidney cells. *J. Cell Biol.* **103,** 1767–1779.

Fuller, S. D., Bravo, R., and Simons, K. (1985). An enzymatic assay reveals that proteins destined for the apical or basolateral domains of an epithelial cell line share the same late Golgi compartments. *EMBO J.* **4,** 297–307.

Geering, K., Kraehenbuhl, J.-P., and Rossier, B. C. (1987). Maturation of the catalytic α-subunit of the Na,K-ATPase during intracellular transport. *J. Cell Biol.* **105,** 2613–2619.

Geering, K., Theulaz, I., Verray, F., Hauptle, M. T., and Rossier, B. C. (1989). A role for the beta-subunit in the expression of functional Na,K-ATPase in *Xenopus* oocytes. *Am. J. Physiol.* **257,** C851–C858

Gething, M. J., McCammon, K., and Sambrook, J. (1986). Expression of wild-type and mutant forms of influenza hemagglutinin: The role of folding in intracellular transport. *Cell* **46,** 939–950.

Geuze, H. J., Slot, J. W., Strous, G. J. A. M., Peppard, J., von Figura, K., Hasilik, A., and Schwartz, A. L. (1984). Intracellular receptor sorting during endocytosis: Comparative immunoelectron microscopy of multiple receptors in rat liver. *Cell* **37,** 195–204.

Gilmore, R., Blobel, G., and Walter, P. (1982a). Protein translocation across the endoplasmic reticulum I. Detection in the microsomal membrane of a receptor for the signal recognition particle. *J. Cell Biol.* **95,** 463–469.

Gilmore, R., Walter, P.,and Blobel, G. (1982b). Protein translocation across the endoplasmic reticulum II. Isolation and characterization of the signal recognition particle receptor. *J. Cell Biol.* **95,** 470–477.

Gonzalez, A., Rizzolo, L., Rindler, M., Adesnik, M., Sabatini, D. D., and Gottlieb, T. (1987) Nonpolarized secretion of truncated forms of the influenza hemagglutinin and the vesicular stomatitis virus G protein from MDCK cells. *Proc. Natl. Acad. Sci. U.S.A.* **84,** 3738–3742.

Gonzalez-Mariscal, L., Chavez de Ramirez, B.,and Cereijido, M. (1985). Tight junction formation in cultured epithelial cells (MDCK). *J. Membr. Biol.* **86,** 113–125.

Gottlieb, T. A., Beaudry, G., Rizzolo, L., Colman, A., Rindler, M., Adesnik, M., and Sabatini, D. D. (1986a). Secretion of endogenous and exogenous proteins from polarized MDCK cell monolayers. *Proc. Natl. Acad. Sci. U.S.A.* **83,** 2100–2104.

Gottlieb, T. A., Gonzalez, A., Rizzolo, L., Rindler, M. J., Adesnik, M., and Sabatini, D. D. (1986b). Sorting and endocytosis of viral glycoproteins in transfected polarized epithelial cells. *J. Cell Biol.* **102,** 1242–1255.

Green, R. F., Meiss, H. K., and Rodriguez-Boulan, E. J. (1981). Glycosylation does not determine segregation of viral envelope proteins in the plasma membrane of epithelial cells. *J. Cell Biol.* **89,** 230–239.

Griffiths, G., and Simons, K. (1986). The tans Golgi network: Sorting at the exit site of the Golgi complex. *Science* **234,** 438–443.

Griffiths, G., Pfeiffer, S., Simons, K., and Matlin, K. S. (1985). Exit of newly synthesized membrane proteins from the trans cisterna of the Golgi complex to the plasma membrane. *J. Cell Biol.* **101,** 949–964.

Griffiths, G., Hoflack, B., Simons, K., Mellman, I., and Kornfeld, S. (1988). The mannose-6-phosphate receptor and the biogenesis of lysosomes. *Cell* **52,** 329–341.

Guidotti, G. (1986). Membrane proteins: Structure, arrangement and disposition in the membrane. *In* "Physiology of Membrane Disorders" (T. E. Andreoli, J. F. Hoffman, D. D. Fanestil, and S. G. Shultz, eds.), pp. 45–55. Plenum, New York.

Handler, J. S. (1988). Antidiuretic hormone moves membranes. *Am. J. Physiol.* **255,** F375–F382.

Hanover, J. A., Elting, J., Mintz, G. R., and Lennarz, W. J. (1982). Temporal aspects of the N- and O-glycosylation of human chorionic gonadotrophin. *J. Biol. Chem.* **257,** 10172–10177.

Hansen, W., Garcia, P., and Walter, P. (1986). *In vitro* protein translocation across the yeast endoplasmic reticulum: ATP-dependent post-translational translocation of preproalpha factor. Cell **45,** 397–406.

Hauri, H. P., Quaroni, A., and Isselbacher, J. (1979). Biosynthesis of intestinal plasma membrane: Post translational route and cleavage of sucrase-isomaltase. *Proc. Natl. Acad. Sci. U.S.A.* **76,** 5183–5186.

Hemingson, S. M., Woolford, C., van der Vies, S. M., Tilly, K., Dennis, D. T., Georgopoulos, C. P., Hendrix, R. W., and Ellis, R. J. (1988). Homologous plant and bacterial proteins chaperone oligomeric protein assembly. *Nature (London)* **333,** 330–334.

Hirschberg, C. B., and Snider, M. D. (1987). Topography of glycosylation in the rough endoplasmic reticulum and Golgi apparatus. *Annu. Rev. Biochem.* **56,** 63–87.

Hoffman, K. E., and Gilmore, R. (1988). Guanosine triphosphate promotes the post-translational integration of opsin into the endoplasmic reticulum membrane. *J. Biol. Chem.* **263,** 4381–4385.

Hoppe, C. A., Connolly, T. P., and Hubbard, A. L. (1985). Transcellular transport of polymeric IgA in the rat helpatocyte: Biochemical and morphological characterization of the transport pathway. *J. Cell Biol.* **101,** 2113–2123.

Hughson, E. J., Cutler, D. F., and Hopkins, C. R. (1989). Basolateral secretion of kappa light chain in the polarised epithelial cell line Caco-2. *J. Cell Sci.* **94,** 327–332.

Hunziker, W., and Mellman, I. (1989). Expression of macrophage-lymphocyte Fc receptors in Madin–Darby canine kidney cells: Polarity and transcytosis differ for isoforms with or without coated pit localization domains. *J. Cell Biol.* **109,** 3291–3302.

Hurtley, S. M., and Helenius, A. (1989). Protein oligomerization in the endoplasmic reticulum. *Annu. Rev. Cell Biol.* **5,** 277–307.

Hurtley, S. M., Bole, D. G., Hoover-Litty, H., Helenius, A., and Copeland, C. S. (1989). Interactions of misfolded influenza virus hemagllutinin with binding protein (BiP). (1989). *J. Cell Biol.* **108,** 2117–2126.

Huttner, W. B. (1988). Tyrosine sulfation and the secretory pathway. *Annu. Rev. Physiol.* **50,** 363–376.

Jamieson, J. D., and Palade, G. E. (1967a). Intracellular transport of secretory proteins in the pancreatic exocrine cell I. Role of the peripheral elements of the Golgi complex. *J. Cell Biol.* **34,** 577–596.

Jamieson, J. D., and Palade, G. E. (1967b). Intracellular transport of secretory proteins in the pancreatic exocrine cell II. Transport to condensing vacuoles and zymogen granules. *J. Cell Biol.* **34,** 597–615.

Jamieson, J. D., and Palade, G. E. (1968a). Intracellular transport of secretory proteins in the pancreatic exocrine cell III. Dissociation of intracellular transport from protein synthesis. *J. Cell. Biol.* **39,** 580–588.

Jamieson, J. D., and Palade, G. E.)1968b). Intracellular transport of secretory proteins in the pancreatic exocrine cell IV. Metabolic requirements. *J. Cell Biol.* **39,** 589–603.

Jay, D., and Cantley, L. (1986). Structural aspects of the red cell anion exchange protein. *Annu. Rev. Biochem.* **55,** 511–538.

Jesaitis, A. J., and Yguerabide, J. (1986). The lateral mobility of the Na,K-dependent ATPase in Madin–Darby canine kidney cells. *J. Cell Biol.* **102,** 1256–1263.

Jones, L. V., Compans, R. W., Davis, A. R., Bos, T. J., and Nayak, D. P. (1985). Surface expression of influenza virus neuraminidase, an amino-terminally anchored viral membrane glycoprotein, in polarized epithelial cells. *Mol. Cell. Biol.* **5,** 2181–2189.

Jorgensen, P. L. (1982). Mechanism of the Na,K pump: Protein structure and conformations of the pure Na,K-ATPase. *Biochim. Biophys. Acta* **694,** 27–68.

Kassenbrock, C. K., Garcia, P. D., Walter, P., and Kelly, R. B. (1988). Heavy-chain binding protein recognizes aberrant polypeptides translocated *in vitro*. *Nature* (*London*) **333,** 90–93.

Kelly, R. B. (1985). Pathways of protein secretion in eukaryotes. *Science* **230,** 25–32.

Klausner, R. D., Lippincott-Schwartz, J., and Bonifacino, J. S. (1990). Architectural editing: Regulating the surface expression of the T-cell antigen receptor. *Curr. Top. Membr. Transp.* **36,** 31–51.

Kobata, A. (1979). Use of endo- and exoglycosidases for structural studies of glycoconjugates. *Anal. Biochem.* **100,** 1–14.

Koeppen, B. M., and Giebisch, G. H. (1985). Mineralocorticoid regulation of sodium and potassium transport by the cortical collecting duct. *In* "Regulation and Development of Membrane Transport Processes" (J. S. Graves, ed.), pp. 89–104. Wiley, New York.

Kondor-Koch, C., Bravo, R., Fuller, S., Cutler, D., and Garoff, H. (1985). Exocytic pathways exist to both the apical and the basolateral cell surface of the polarized epithelial cell MDCK. *Cell* **43,** 297–306.

Kornfeld, R., and Kornfeld, S. (1985). Assembly of asparagine-linked oligosaccharides. *Annu. Rev. Biochem.* **54,** 631–664.

Kornfeld, S. (1987). Trafficking of lysosomal enzymes. *FASEB J.* **1,** 462–468.

Krieg, U. C., Johnson, A. E., and Walter, P. (1989). Protein translocation across the endoplasmic reticulum membrane: Identification by photocross-linking of a 39 kD integral membrane glycoprotein as part of a putative translocation tunnel. *J. Cell Biol.* **109,** 2033–2043.

Larkin, J. M., Sztul, E. S., and Palade, G. E. (1986). Phosphorylation of the rat hepatic polymeric IgA receptor. *Proc. Natl. Acad. Sci. U.S.A.* **83,** 4759–4763.

Le Bivic, A., Real, F. X., and Rodriguez-Boulan, E. J. (1989). Vectorial targeting of apical and basolateral plasma membrane proteins in a human adenocarcinoma epithelial cell line. *Proc. Natl. Acad. Sci. U.S.A.* **86,** 9313–9317.

Le Bivic, A., Sambuy, Y., Mostov, K., and Rodriguez-Boulan, E. J. (1990). Vectorial targeting of an endogenous apical sialoglycoprotein and uvomorulin in MDCK cells. *J. Cell Biol.* **110,** 1533–1539.

Lencer, W. I., Verkman, A. S., Arnaout, M. A., Ausiello, D. A., and Brown, D. (1990). Endocytic vesicles from renal papilla which retrieve the vasopressin-sensitive water channel do not contain a functional H^+-ATPase. *J. Cell Biol.* **111,** 379–389.

Lennarz, W. J. (1987). Protein glycosylation in the endoplasmic reticulum: Current topological issues. *Biochemistry* **26,** 7205–7210.

Lingappa, V. R., Lingappa, J. R., and Blobel, G. (1979). Chicken ovalbumin contains an internal signal sequence. *Nature (London)* **281,** 117–121.

Lippincott-Schwartz, J., Bonifacino, J. S., Yuan, L. C., and Klausner, R. D. (1988). Degradation from the endoplasmic reticulum: Disposing of newly synthesized proteins. *Cell* **54,** 209–220.

Lippincott-Schwartz, J., Donaldson, J. G., Schweizer, A., Berger, E. G., Hauri, H.-P., Yuan, L. C., and Klausner, R. D. (1990). Microtubule-dependent retrograde transport of proteins into the ER in the presence of brefeldin A suggests an ER recycling pathway. *Cell* **60,** 821–836.

Lisanti, M. P., Sargiacomo, M., Graeve, L., Saltiel, A. R., and Rodriguez-Boulan, E. J. (1988). Polarized apical distribution of glycosyl-phosphatidylinositol-anchored proteins in a renal epithelial cell line. *Proc. Natl. Acad. Sci. U.S.A.* **85,** 9557–9561.

Lisanti, M. P., Caras, I. W., Davitz, M. A., and Rodriguez-Boulan, E. J. (1989a). A glycophospholipid membrane anchor acts as an apical targeting signal in polarized epithelial cells. *J. Cell Biol.* **109,** 2145–2156.

Lisanti, M. P., Le Bivic, A., Sargiacomo, M., and Rodriguez-Boulan, E. J. (1989b). Steady-state distribution and biogenesis of endogenous MDCK glycorproteins: Evidence for intracellular sorting and polarized cell surface delivery. *J. Cell Biol.* **109,** 2117–2127.

Lisanti, M. P., Le Bivic, A., Saltiel, A. R., and Rodriguez-Boulan, E. J. (1990). Preferred apical distribution of glycosyl-phosphatidylinositol (GPI) anchored proteins: A highly conserved feature of the polarized epithelial cell phenotype. *J. Membr. Biol.* **113,** 155–167.

Machamer, C. E., and Rose, J. K. (1987). A specific transmembrane domain of a coronavirus E1 glycoprotein is required for its retention in the Golgi region. *J. Cell Biol.* **105,** 1205–1214.

Madin, S. H., and Darby, N. B. (1975). *In* "American Type Culture Collection Catalogue of Strains II," 1st Ed., p. 47. ATCC, Rockville, Maryland.

Malhorta, V., Orci, L., Glick, B. S., Block, M. R., and Rothman, J. E. (1988). Role of an *N*-ethylmaleimide-sensitive transport component in promoting fusion of transport vesicles with cisternae of the Golgi stack. *Cell* **54,** 221–227.

Massey, D., Ferracci, H. M., Gorvel, J. P., Rigal, A., Soulie, J. M., and Maroux, S. (1987). Evidence for the transit of aminopeptidase *N* through the basolateral membrane before it reaches the brush border of enterocytes. *J. Membr. Biol.* **96,** 19–25.

Matlin, K. S. (1986). The sorting of proteins to the plasma membrane in epithelial cells. *J. Cell Biol.* **103,** 2565–2568.

Matlin, K. S, and Simons, K. (1983). Reduced temperature prevents transfer of a membrane glycoprotein to the cell surface but does not prevent terminal glycosylation. *Cell* **34,** 233–243.

Matlin, K. S., and Simons, K. (1984). Sorting of an apical plasma membrane glycoprotein occurs before it reaches the cell surface in cultured epithelial cells. *J. Cell Biol.* **99,** 2131–2139.

Matlin, K. S., Bainton, D. F., Pesonen, M., Louvard, D., Genty, N., and Simons, K. (1983). Transepithelial transport of a viral membrane glycoprotein implanted into the apical plasma membrane of MDCK cells. I. Morphological evidence. *J. Cell Biol.* **97,** 627–637.

Matter, K., Brauchbar, M., Bucher, K., and Hauri, H. P. (1990). Sorting of endogenous plasma membrane proteins occurs from two sites in cultured human intestinal epithelial cells (Caco-2). *Cell* **60,** 429–437.

McNeill, H., Ozawa, M., Kemler, R., and Nelson, W. J. (1990). Novel function of the cell adhesion molecule uvomorulin as an inducer of cell surface polarity. *Cell* **62,** 309–316.

McQueen, N., Nayak, D. P., Stephens, E. B., and Compans, R. W. (1986). Polarized expression of a chimeric protein in which the transmembrane and cytoplasmic domains of influenza virus hemagglutinin have been replaced by those of the vesicular stomatitis G protein. *Proc. Natl. Acad. Sci. U.S.A.* **83,** 9318–9322.

Meyer, D. I., Krause, E., and Dobberstein, B. (1982). Secretory protein translocation across membranes—the role of the "docking protein". *Nature (London)* **297,** 647–650.

Milstein, C., Brownlee, G., Harrison, T., and Mathews, M. B. (1972). A possible precursor of immunoglobin light chains. *Nature (London), New Biol.* **239,** 117–120.

Misek, D. E., Bard, E., and Rodriguez-Boulan, E. J. (1984). Biogenesis of epithelial cell polarity: Intracellular sorting and vectorial exocytosis of an apical plasma membrane glycoprotein. *Cell* **39,** 537–546.

Morrow, J. S., Cianci, C. D., Ardito, T., Mann, A. S., and Kashgarian, M. (1989). Ankyrin links fodrin to the alpha subunit of Na,K-ATPase in Madin–Darby canine kidney cells and in intact renal tubule cells. *J. Cell Biol.* **108,** 455–465.

Mostov, K. E., and Deitcher, D. L. (1986). Polymeric immunoglobulin receptor expressed in MDCK cells transcytoses IgA. *Cell* **46,** 613–621.

Mostov, K. E., de Bruyn Kops, A., and Deitcher, D. L., (1986). Deletion of the cytoplasmic domain of the polymeric immunoglobulin receptor prevents basolateral localization and endocytosis. *Cell* **47,** 359–364.

Mostov, K. E., Breitfeld, P. P., and Harris, J. M. (1987). An anchor-minus form of the polymeric immunoglobulin receptor is secreted predominantly apically in Madin–Darby canine kidney cells. *J. Cell Biol.* **105,** 2031–2036.

Mueckler, M., and Lodish, H. F. (1986). Post translational insertion of a fragment of the glucose transporter requires phosphoanhydride bond cleavage. *Nature (London)* **322,** 459–462.

Munro, S., and Pelham, H. R. B. (1986). An HSP 70-like protein in the ER: Identity with the 78 kD glucose regulated protein and immunoglobulin heavy chain binding protein. *Cell* **46,** 291–300.

Munro, S., and Pelham, H. R. B. (1987). A C-terminal signal prevents secretion of luminal ER proteins. *Cell* **48,** 899–907.

Nelson, W. J. (1989). Development and maintenance of epithelial polarity: A role for the submembranous cytoskeleton. *In* "Functional Epithelial Cells in Culture" (K. S. Matlin and J. D. Valentich, eds.), pp. 3–42. Alan R. Liss, New York.

Nelson, W. J., and Hammerton, R. W. (1989). A membrane-cytoskeletal complex containing Na,K-ATPase, ankyrin and fodrin in Madin–Darby canine kidney cells: Implications for the biogenesis of epithelial cell polarity. *J. Cell Biol.* **108,** 893–902.

Nelson, W. J., and Veshnock, P. J. (1987a). Modulation of fodrin (membrane skeleton) stability by cell–cell contact in Madin–Darby canine kidney epithelial cells. *J. Cell Biol.* **104,** 1527–1537.

Nelson, W. J., and Veshnock, P. J. (1987b). Ankyrin binding to Na,K-ATPase and implications for the organization of membrane domains in polarized cells. *Nature (London)* **328,** 533–536.

Nelson, W. J., Shore, E. M., Wang, A. Z., and Hammerton, R. W. (1990). Identification of a membrane-cytoskeletal complex containing the cell adhesion molecule uvomorulin (E-cadherin), ankyrin and fodrin in Madin–Darby canine kidney epithelial cells. *J. Cell Biol.* **110,** 349–357.

Nilsson, T., Jackson, M., and Peterson, P. A. (1989). Short cytoplasmic sequences serve as retention signals for transmembrane proteins in the endoplasmic reticulum. *Cell* **58,** 707–718.

O'Neil, R. G. (1987). Adrenal steroid regulation of potassium transport. *Curr. Top. Membr. Transp.* **28,** 185–206.

Orci, L., Halban, P., Amherdt, M., Ravazzola, M., Vassalli, J.-D., and Perrelet, A. (1984). A clathrin-coated, Golgi-related compartment of the insulin secreting cell accumulates proinsulin in the presence of monensin. *Cell* **39,** 39–47.

Orci, L., Glick, B. S., and Rothman, J. E. (1986). A new type of coated vesicular carrier that appears not to contain clathrin: Its possible role in protein transport within the Golgi stack. *Cell* **46,** 171–184.

Orci, L., Ravazzola, M., Amherdt, M., Perrelet, A., Powell, S. K., Quinn, D. L., and Moore, H.-P. H. (1987). The trans-most cisternae of the Golgi complex: A compartment for sorting of secretory and plasma membrane proteins. *Cell* **51,** 1039–1051.

Ostermann, J., Horwich, A. L., Neupert, W., and Hartl, F.-U. (1989). Protein folding in mitochondria requires complex formation with HSP60 and ATP hydrolysis. *Nature (London)* **341,** 125–130.

Palade, G. E. (1975). Intracellular aspects of the process of protein synthesis. *Science* **189,** 347–358.

Parczyk, K., Haase, W., and Kondor-Koch, C. (1989). Microtubules are involved in the secretion of proteins at the apical cell surface of the polarized epithelial cell, Madin–Darby canine kidney. *J. Biol. Chem.* **264,** 16837–16846.

Pelham, H. R. B. (1988). Evidence that lumenal ER proteins are sorted from secreted proteins in a post ER compartment. *EMBO J.* **7,** 913–918.

Perara, E., and Lingappa, V. R. (1985). A former amino terminal signal sequence engineered to an internal location directs translocation of both flanking protein domains. *J. Cell Biol.* **101,** 2292–2301.

Perara, E., Rothman, R. E., and Lingappa, V. R. (1986). Uncoupling translation from translocation: Implications for transport of proteins across membranes. *Science* **232,** 348–352.

Pfeffer, S. R., and Rothman, J. E. (1987). Biosynthetic protein transport and sorting by the endoplasmic reticulum and Golgi. *Annu. Rev. Biochem.* **56,** 829–852.

Pfeiffer, S., Fuller, S. D., and Simons, K. (1985). Intracellular sorting and basolateral appearance of the G protein of vesicular stomatitis virus in MDCK cells. *J. Cell Biol.* **101,** 470–476.

Popot, J.-L., and Changeux, P.-J. (1984). Nicotinic receptor of acetylcholine: Structure of an oligomeric integral membrane protein. *Physiol. Rev.* **64,** 1162–1239.

Puddington, L., Woodgett, C., and Rose, J. K. (1987). Placement of the cytoplasmic domain alters sorting of a viral glycoprotein in polarized cells. *Proc. Natl. Acad. Sci. U.S.A.* **84,** 2756–2760.

Rindler, M. J., and Traber, M. G. (1988). A specific sorting signal is not required for the polarized secretion of newly synthesized proteins from cultured intestinal epithelial cells. *J. Cell Biol.* **107,** 471–479.

Rindler, M. J., Ivanov, I. E., Plesken, H., Rodriguez-Boulan, E. J., and Sabatini, D. D. (1984). Viral glycoproteins destined for the apical or basolateral plasma membrane domains traverse the same Golgi apparatus during their intracellular transport in double infected Madin–Darby canine kidney cells. *J. Cell Biol.* **98,** 1304–1319.

Rindler, M. J., Ivanov, I. E., and Sabatini, D. D. (1987). Microtubule-acting drugs lead to the nonpolarized delivery of the influenza hemagglutinin to the cell surface of the polarized Madin–Darby canine kidney cells. *J. Cell Biol.* **104,** 231–241.

Rodriguez-Boulan, E., and Nelson, W. J. (1989). Morphogenesis of the polarized epithelial cell phenotype. *Science* **245,** 718–725.

Rodriguez-Boulan, E. J., and Sabatini, D. D. (1978). Asymmetric budding of viruses in epithelial monolayers: A model system for study of epithelial polarity. *Proc. Natl. Acad. Sci. U.S.A.* **75,** 5071–5075.

Rodriguez-Boulan, E. J., Paskiet, K. T., and Sabatini, D. D. (1983). Assembly of enveloped viruses in MDCK cells: Polarized budding from single attached cells and from clusters of cells in suspension. *J. Cell Biol.* **96,** 866–874.

Romisch, K., Webb, J., Herz, J., Prehn, S., Frank, R., Vingron, M., and Dobberstein, B. (1989). Homology of 54 K protein of signal recognition particle, docking protein and two *E. coli* proteins with putative GTP binding domains. *Nature* (*London*) **340,** 478–482.

Rose, J. K., and Doms, R. W. (1988). Regulation of protein export from the endoplasmic reticulum. *Annu. Rev. Cell Biol.* **4,** 257–288.

Roth, M. G., Fitzpatrick, J. P., and Compans, R. W. (1979). Polarity of influenza and vesicular stomatitis virus maturation in MDCK cells: Lack of requirement for glycosylation of viral glycoproteins. *Proc. Natl. Acad. Sci. U.S.A.* **76,** 6430–6434.

Roth, M. G., Compans, R. W., Giusti, L., Davis, A. R., Nayak, D. P., Gething, M. J., and Sambrook, J. S. (1983). Influenza virus hemagglutinin expression is polarized in cells infected with recombinant SV40 viruses carrying cloned hemagglutinin DNA. *Cell* **33,** 435–443.

Roth, M. G., Gunderson, D., Patil, N., and Rodriguez-Boulan, E. J. (1987). The large external domain is sufficient for the correct sorting of secreted or chimeric influenza virus hemagglutinins in polarized monkey kidney cells. *J. Cell Biol.* **104,** 769–782.

Rothblatt, J. A., and Meyer, D. I. (1986). Secretion in yeast: Translocation and glycosylation of preproalpha factor *in vitro* can occur via an ATP-dependent post-translational mechanism. *EMBO J.* **5,** 1031–1036.

Rothman, J. E. (1987). Protein sorting by selective retention in the endoplasmic reticulum and Golgi stack. *Cell* **50,** 521–522.

Rothman, J. E. (1989). Polypeptide chain binding proteins: Catalysts of protein folding and related processes in cells. *Cell* **59,** 591–601.

Rothman, R. E., Andrews, D. W., Calayag, C. M., and Lingappa, V. R. (1988). Construction of defined polytopic integral transmembrane proteins: The role of signal and stop transfer sequence permutations. *J. Biol. Chem.* **263,** 10470–10480.

Salas, P. J., Vega-Salas, D. E., Hochman, J., Rodriguez-Boulan, E. J., and Edidin, M. (1988). Selective anchoring in the specific plasma membrane domain: A role in epithelial cell polarity. *J. Cell Biol.* **107,** 2363–2376.

Salminen, A., and Novick, P. J. (1987). A *ras*-like protein is required for a post-Golgi event in yeast secretion. *Cell* **49,** 527–538.

Saraste, J., and Kuismanen, E. (1984). Pre- and post-Golgi vacuoles operate in the transport of Semliki Forest virus membrane glycoproteins to the cell surface. *Cell* **38,** 535–549.

Saraste, J., Palade, G. E., and Farquhar, M. G. (1986). Temperature-sensitive steps in the transport of secretory proteins through the Golgi complex in exocrine pancreatic cells. *Proc. Natl. Acad. Sci. U.S.A.* **83,** 6425–6429.

Schlutz, S. G. (1986). Cellular models of epithelial ion transport. *In* "Physiologic of Membrane Disorders" (T. E. Andreoli, J. F. Hoffman, D. D. Fanestil, and S. G. Schultz, eds.), pp. 519–534. Plenum, New York.

Schmidt, G. W., and Mishkind, M. L. (1986). The transport of proteins into chloroplasts. *Annu. Rev. Biochem.* **55,** 879–912.

Schwartz, G. J., and Al-Awqati, Q. (1986). Regulation of transepithelial H^+ transport by exocytosis and endocytosis. *Annu. Rev. Physiol.* **48,** 153–161.

Simons, K., and Fuller, S. D. (1985). Cell surface polarity in epithelia. *Annu. Rev. Cell Biol.* **1,** 295–340.

Simons, K., and Wandinger-Ness, A. (1990). Polarized sorting in epithelia. *Cell* **62,** 207–210.

Skibbens, J. E., Roth, M. G., and Matlin, K. S. (1989). Differential extractibility of influenza virus hemagglutinin during intracellular transport in polarized epithelial cells and nonpolar fibroblasts. *J. Cell Biol.* **108,** 821–832.

Stephens, E. B., and Compans, R. W. (1986). Nonpolarized expression of a secreted murine leukemia virus glycoprotein in polarized epithelial cells. *Cell* **47,** 1053–1059.

Stephens, E. B., Compans, R. W., Earl, P., and Moss, B. (1986). Surface expression of viral glycoproteins is polarized in epithelial cells infected with recombinant vaccinia viral vectors. *EMBO J.* **5,** 237–245.

Takeyasu, K., Tamkun, M., Siegel, N., and Fambrough, D. M. (1987). Expression of hybrid Na,K-ATPase molecules after transfection of mouse Ltk^- cells with DNA Encoding the β-subunit of an avian brain sodium pump. *J. Biol. Chem.* **262,** 10733–10740.

Takeyasu, K., Tamkun, M., Renaud, K. J., and Fambrough, D. M. (1988). Ouabain-sensitive Na,K-ATPase activity expressed in mouse L cells by transfection with DNA encoding the α-subunit of an avian sodium pump. *J. Biol. Chem.* **263,** 4347–4354.

Tamkun, M., and Fambrough, D. M. (1986). The Na,K-ATPase of chick sensory neurons: Studies on biosynthesis and intracellular transport. *J. Biol. Chem.* **261,** 1009–1019.

Taormino, J. P., and Fambrough, D. M. (1990). Pre-translational regulation of the Na,K-ATPase in response to demand for ion transport in cultured chicken skeletal muscle. *J. Biol. Chem.* **265,** 4116–4123.

Tooze, J., Tooze, S. A., and Fuller, S. D. (1987). Sorting of progeny coronavirus from condensed secretory proteins at the exit from the *trans* Golgi network of AtT 20 cells. *J. Cell Biol.* **105,** 1215–1226.

Tooze, S. A., and Huttner, W. B. (1990). Cell-free protein sorting to the regulated and constitutive secretory pathways. *Cell* **60,** 837–847.

Urushidani, T., and Forte, J. G. (1987). Stimulation-associated redistribution of H,K-ATPase activity in isolated gastric glands. *Am. J. Physiol.* **252,** G458–G465.

van Meer, G. (1989). Polarity and polarized transport of membrane lipids in a cultured epithelium. *In* "Functional Epithelial Cells In Culture" (K. S. Matlin and J. D. Valentich, eds.), pp. 43–69. Alan R. Liss, New York.

Vaux, D., Tooze, J., and Fuller, S. (1990). Identification by anti-idiotype antibodies of an intracellular membrane protein that recognizes a mammalian endoplasmic reticulum retention signal. *Nature (London)* **345,** 495–502.

Vega-Salas, D. E., Salas, P. J. I., Gunderson, D., and Rodriguez-Boulan, E. J. (1987a). Formation of the apical pole of epithelial (Madin–Darby canine kidney) cells: polarity of an apical protein is independent of tight junctions while segregation of a basolateral marker requires cell–cell interactions. *J. Cell Biol.* **104,** 905–916.

Vega-Salas, D. E., Salas, P. J. I., and Rodriguez-Boulan, E. J. (1987b). Modulation of the expression of an apical plasma membrane protein of Madin–Darby canine kidney cells: Cell–Cell interactions control the appearance of a novel intracellular storage compartment. *J. Cell Biol.* **104,** 1249–1259.

von Heijne, G. (1985). Signal sequences: The limits of variation. *J. Mol. Biol.* **184,** 99–105.

Walter, P., and Blobel, G. (1981a). Translocation of proteins across the endoplasmic reticulum II. Signal recognition protein (SRP) mediates the selective binding to microsomal membranes of *in vitro* assembled polysomes synthesizing secretory protein. *J. Cell Biol.* **91,** 551–556.

Walter, P., and Blobel, G. (1981b). Translocation of proteins across the endoplasmic reticulum III. Signal recognition protein (SRP) causes signal sequence-dependent and site-specific arrest of chain elongation that is released by microsomal membranes. *J. Cell Biol.* **91,** 557–561.

Walter, P., and Lingappa, V. R. (1986). Mechanism of protein translocation across the endoplasmic reticulum membrane. *Annu. Rev. Cell Biol.* **2,** 499–516.

Walter, P., Ibrahimi, I., and Blobel, G. (1981). Translocation of proteins across the endoplasmic reticulum I. Signal recognition protein (SRP) binds to *in vitro* assembled polysomes synthesizing secretory protein. *J. Cell Biol.* **91,** 545–550.

Wang, A. Z., Ojakian, G. K., and Nelson, W. J. (1990). Steps in the morphogenesis of a polarized epithelium: Uncoupling the roles of cell–cell and cell–substratum contact in establishing plasma membrane polarity in multicellular epithelial (MDCK) cysts. *J. Cell Sci.* **95,** 137–151.

Waters, M. G., and Blobel, G. (1986). Secretory protein translocation in a yeast cell-free system can occur post-translationally and requires ATP hydrolysis. *J. Cell Biol.* **102,** 1543–1550.

Wichner, W. (1988). Mechanisms of membrane assembly: General lessons from the study of M13 coat protein and *Escherichia coli* leader peptidase. *Biochemistry* **27,** 1081–1086.

Wiedmann, M., Kurzchalia, T. V., Hartmann, E., and Rapoport, T. A. (1987). A signal sequence receptor in the endoplasmic reticulum membrane. *Nature (London)* **328,** 830–833.

Wolin, S. L., and Walter, P. (1988). Ribosome pausing and stacking during translation of a eukaryotic mRNA. *EMBO J.* **7,** 3559–3569.

Wolin, S. L., and Walter, P. (1989). Signal recognition particle mediates a transient elongation arrest of preprolactin in reticulocyte lysate. *J. Cell Biol.* **109,** 2617–2622.

Yost, C. S., Hedgpeth, J., and Lingappa, V. R. (1983). A stop transfer sequence confers predictable transmembrane orientation to a previously secreted protein in cell-free systems. *Cell* **34,** 759–766.

PART II

Fertilization and Early Embryonic Development

CHAPTER 3

Electrical Characteristics of Oocytes and Eggs

Douglas Kline
Department of Biological Sciences, Kent State University, Kent, Ohio 44242

I. INTRODUCTION

Study of the electrical properties of oocytes and eggs has interested embryologists and electrophysiologists for some time. Accurate recordings of the membrane potential of unfertilized eggs were first made by Tyler and co-workers (1956), who examined starfish eggs. Some time later, Miyazaki *et al.* (1972) reported that the egg of an ascidian was electrically excitable and produced an action potential when a small depolarizing current was injected. Subsequently, studies of voltage-sensitive channels in oocytes and eggs in a number of species were made to determine how the channels in these cells might be related to the ion channels found at later stages of development (Hagiwara and Jaffe, 1979; Hagiwara, 1983). In this chapter, I discuss some of the more recent findings and highlight some of the relationships between the electrical characteristics of animal oocytes and eggs and the cellular events of oocyte growth, maturation, and fertilization.

Differentiation and growth of the female gamete occur before the completion of meiosis (Fig. 1). After yolk deposition and cell enlargement, development of the oocyte is usually arrested in prophase of the first meiotic division and the oocyte contains an enlarged nucleus or germinal vesicle. In some species, release from meiotic arrest at the germinal vesi-

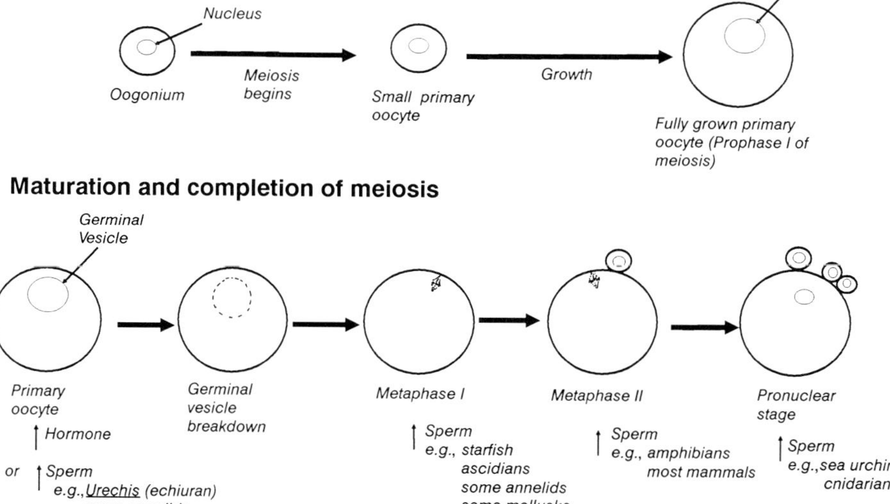

FIG. 1. The timing of oocyte growth and sperm entry with respect to the meiotic divisions. Oocyte growth occurs during an extended phase of meiotic prophase and the fully grown primary oocyte is arrested at this stage. Resumption of meiosis in the primary oocyte may be triggered by sperm in some species or by a maturation-inducing hormone in other species. Oocyte maturation, induced by hormonal stimulation, results in a mature egg that may be arrested, depending on the species, at metaphase of the first or second meiotic division, and resumption of meiosis at these stages is triggered by sperm during fertilization. In some species, meiosis is completed and sperm enter the egg at the pronuclear stage. A few of the animal groups are indicated to illustrate the different stages at which sperm entry normally occurs.

cle stage is triggered by sperm and meiosis is completed after sperm entry. In many species, resumption of meiosis is triggered by a hormonal stimulus. The resumption of meiosis, referred to as maturation, is accompanied by morphological and physiological changes, including changes in electrical properties. Maturation and ovulation furnish a mature, fertilizable egg which, depending on the species, may be arrested at metaphase of the first or second meiotic division or have completed meiosis and be arrested at the pronuclear stage (haploid interphase nucleus). Fertilization of the mature egg results in the completion of meiosis (if not already complete), resumption of the cell cycle, and the start of embryogenesis. In many cases, the response of the egg to sperm at fertilization includes dramatic changes in the electrical characteristics of the egg.

II. ION CHANNELS IN OOCYTES AND MATURE EGGS

A. Action Potentials in Invertebrates

The oocytes and eggs of a number of invertebrate species produce action potentials (reviewed in Hagiwara and Jaffe, 1979). These are generally longer in duration than action potentials produced in nerve and muscle cells. For example, the action potential in the unfertilized egg of the sea urchin lasts about 9 sec (Chambers and de Armendi, 1979), and the action potential in the unfertilized egg of the nemertean worm *Cerebratulus* lasts about 9 min (Jaffe *et al.*, 1986).

The early component of the action potential in these invertebrate eggs is due to a voltage-activated Ca^{2+} conductance and the second phase, or plateau phase, is primarily sodium dependent. The ionic conductances during the action potential of the mature egg of *Cerebratulus* serve to illustrate some of the features of the action potential in a marine egg. The initial peak of the action potential reaches + 50 mV and is calcium dependent. A Na^+-dependent plateau phase at a potential of about + 25 mV follows, and it is maintained for several minutes (Fig. 2A).

Voltage–clamp studies of the *Cerebratulus* egg revealed that two separate inward currents account for the action potential. The fast inward current, responsible for the initial peak of the action potential, is a rapidly inactivating Ca^{2+} current (the current indicated by the numeral 1 in Fig. 2B). A second inward current, which is carried by Na^+ ions, starts a few seconds after the Ca^{2+} current; it develops slowly and lasts many minutes (the current indicated by the numeral 2 in Fig. 2B). In Ca^{2+}-free sea water, neither the fast nor slow inward current can be activated. If the egg is injected with the calcium chelator ethylene glycol bis(β-aminoethyl ether)-

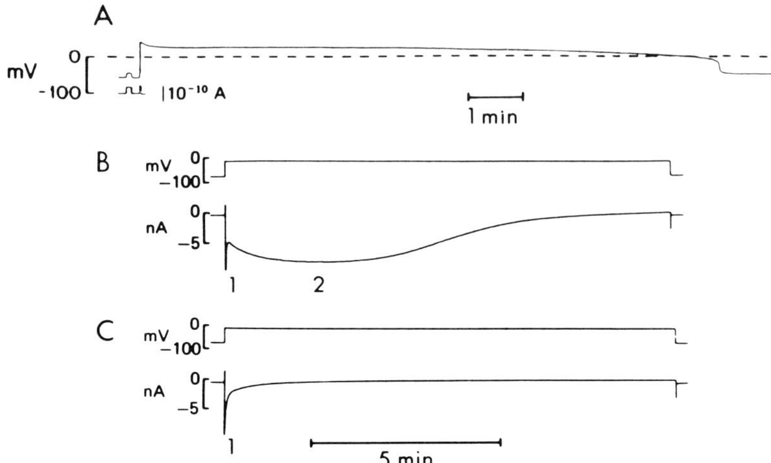

FIG. 2. Action potential and voltage-clamp currents in the egg of the nemertean *Cerebratulus lacteus*. (A) Action potential recorded in artificial sea water. The top trace is the voltage record and the dotted line indicates 0 mV. The lower trace indicates the magnitude and duration of the injected current which initiated the action potential. (B) Inward currents recorded during voltage clamp. The upper trace indicates the clamped membrane potential and the lower trace is the current recorded after a voltage-clamp step from −75 to −10 mV. The numeral 1 indicates the inward Ca^{2+} current and the numeral 2 indicates the inward Na^+ current (see text). (C) Inward current recorded during voltage clamp in an egg injected with EGTA. (From Jaffe *et al.*, 1986.)

N,N, N′, N′-tetraacetic acid (EGTA), the slow inward current is abolished while the fast inward current remains (Fig. 2C). Therefore, the action potential in the *Cerebratulus* egg is produced by activation of a voltage-sensitive Ca^{2+} conductance followed by a Ca^{2+}-activated Na^+ conductance (Jaffe *et al.*, 1986).

There is evidence to suggest that a Ca^{2+}-activated Na^+ conductance also contributes to the action potential in oocytes and eggs of some other species (for references, see Jaffe *et al.*, 1986). In some starfish oocytes, the early part of the action potential is, at least in part, calcium dependent, while the plateau phase is sodium dependent. Voltage-clamp studies of these starfish oocytes suggest that the Na^+ conductance is activated by Ca^{2+}. The unfertilized sea urchin egg and ascidian egg also produce action potentials having a Ca^{2+}-dependent initial peak and a Na^+-dependent plateau. Raising intracellular Ca^{2+} increases a Na^+-dependent inward current in the ascidian egg (Takahashi and Yoshii, 1978). It would be informative to inject sea urchin eggs and ascidian eggs with a calcium chelator to test the hypothesis that the action potential in these eggs depends, in part, on a Ca^{2+}-activated Na^+ conductance. The mechanism by which Ca^{2+} causes these channels to open is not known. Channel opening may be due to a direct interaction of Ca^{2+} with the channel or some other Ca^{2+}-dependent step may be involved.

B. Calcium-Activated Channels in Amphibians

Amphibian oocytes and mature eggs do not produce Ca^{2+}- and Na^+-dependent action potentials like those found in invertebrate species; however, frog oocytes and eggs contain a number of interesting ion channels that are sensitive to voltage or Ca^{2+}. Miledi (1982) and Barish (1983) reported a Ca^{2+}-dependent Cl^- current in the immature *Xenopus* oocyte that was activated by membrane depolarization. They proposed that the current they observed was the result of a small amount of Ca^{2+} influx through a voltage-sensitive Ca^{2+} channel followed by current through a Ca^{2+}-dependent Cl^- channel. This Cl^- current is inhibited by intracellular injection of the calcium chelator EGTA, and a similar current can be activated by injection of Ca^{2+}, injection of the Ca^{2+}-mobilizing agent inositol 1,4,5-trisphosphate ($InsP_3$), application of a Ca^{2+} ionophore (A23187), and certain external divalent cations (Miledi and Parker, 1984; Dascal *et al.*, 1985; Gillo *et al.*, 1987; Berridge, 1988; Boton *et al.*, 1989; Miledi *et al.*, 1989). These observations indicate that the *Xenopus* oocyte contains a Ca^{2+}-activated Cl^- channel.

In a recent study, Boton *et al.* (1989) found evidence for two distinct Cl^- currents, each having different sensitivities to Ca^{2+} and to a Cl^- channel blocker. It is not yet known if these currents are due to two separate ion channels or to different activation mechanisms for the same channel. Single channel recordings made from *Xenopus* oocytes with patch electrodes demonstrate the presence of low-conductance (3–4 pS) Cl^- channels (Takahashi *et al.*, 1987; Oosawa and Yamagishi, 1989). The fertilization potential in the *Xenopus* egg (Section IV,B) is produced by a Ca^{2+}-dependent Cl^- conductance. The relationship between the chloride channel(s) found in the immature oocyte and the Cl^- conductance found later in the mature egg is not known.

Kusano *et al.* (1977) reported that the membrane of the *Xenopus* oocyte depolarized in response to the application of acetylcholine. This cholinergic muscarinic response is due to the activation of a Cl^- current that is now thought to be activated by an internal signaling system involving hydrolysis of inositol phospholipids and mobilization of intracellular Ca^{2+} (Dascal, 1987). The acetylcholine-induced current is blocked by intracellular injection of EGTA (Dascal *et al.*, 1985) and it is similar to the chloride conductance induced by injection of $InsP_3$ or Ca^{2+}. The physiological function of the acetylcholine response in the oocyte is not known. Acetylcholine does not induce maturation but it does shorten the maturation time (Dascal *et al.*, 1984). Oocytes do not respond to acetylcholine after maturation (Kusano *et al.*, 1982), and fertilization is not affected by acetylcholine or an acetylcholine antagonist (Robbins and Molenaar, 1981).

The same Cl^- current can be activated by a variety of agonists following expression of exogenous receptors that are known to stimulate phospholipase C (phosphatidylinositol 4,5-bisphosphate phosphodiesterase) and cause Ca^{2+} release (Dascal, 1987; McIntosh and Catt, 1987; Nomura *et al.*, 1987). There is also evidence that exogenous receptors interact with a guanine nucleotide-binding protein (G-protein) present in the oocyte membrane which in turn stimulates phospholipase C and subsequent release of intracellular Ca^{2+} (Moriarty *et al.*, 1988, 1989).

A separate cAMP-dependent K^+ current is recorded from follicle-enclosed *Xenopus* oocytes after application of catecholamines, adenosine, gonadotropins, and several other substances (Dascal, 1987; Woodward and Miledi, 1987). There is little response to these agents after defolliculation, suggesting that the K^+ currents recorded in the oocyte are generated in the follicle cells which are coupled by gap junctions to the oocyte (Miledi and Woodward, 1989). The physiological function of this K^+ conductance is not completely known.

C. Voltage-Sensitive Sodium Channels in Amphibians

The oocyte of the frog *Xenopus laevis* contains interesting voltage-sensitive Na+ channels that are not normally detected. It is only after a prolonged depolarization to a potential more positive than about +30 mV that these channels appear (Baud *et al.*, 1982; Baud, 1983; Baud and Kado, 1984). This process has been described as the "induction" of Na^+ channels. Once induced, these channels can be opened by a smaller depolarization from the resting potential (to about − 20 mV). This Na^+ channel resembles Na^+ channels of nerve and muscle cells, but the oocyte channel shows little or no inactivation and is relatively insensitive to tetrodotoxin.

The mechanism by which these Na^+ channels are induced and their function remain unknown. Induction probably does not depend on new protein synthesis (Baud and Kado, 1984). This suggests that channels are present in the membrane of the oocyte before induction. Under physiological conditions, the oocyte membrane would not experience depolarizations strong enough to induce the channels. Baud (1983) has suggested that the channels might normally be opened in some other way, and therefore be a mechanism for the influx of Na^+ that occurs during oogenesis; however, there is no evidence for this in the *Xenopus* oocyte. There is some evidence to suggest that intracellular second messengers are not involved in the induction of the Na^+ channel (Baud and Kado, 1984), but it would be informative to examine further the possible role of intracellular messengers, including Ca^{2+}, in induction of this channel.

Spontaneous action potentials occur during maturation of the oocyte of the frog *Rana pipiens* (Schlichter, 1983a); this is the only known example of spontaneous generation of action potentials in an oocyte or egg. The action potential is due to the opening of Na^+ channels with properties similar to the Na^+ channels found in *Xenopus* oocytes; however, the Na^+ channels in *Rana* do not need to be induced by a depolarizing current. A voltage-sensitive Cl^- channel opens during the action potential and the outward Cl^- current causes the falling phase of the action potential (Schlichter, 1989).

D. Voltage-Sensitive Channels in Mammals

A voltage-activated Ca^{2+} channel is present in growing oocytes, eggs, and early cleavage-stage embryos of the mouse, but there is no evidence to suggest that the mouse egg has functional Na^+ channels (Eusebi *et al.*, 1983; Yoshida, 1985; Peres, 1987), nor does there seem to be an "inducible" Na^+ channel in the mouse oocyte or egg (Yoshida, 1985). In Ca^{2+}-free medium, monovalent cations (Na^+ or Li^+) will pass through the Ca^{2+} channel (Yoshida, 1985). Action potentials as such are not usually recorded from mouse oocytes or eggs, since these calcium channels are inactivated at the normal recording potential of −40 mV, which is probably close to the true resting potential (Peres, 1986). The developmental function of these Ca^{2+} channels is not known.

III. ELECTRICAL CHARACTERISTICS OF OOCYTES

A. The Growing Oocyte

As part of a study of fertilization, Holland and Gould-Somero (1981) examined immature, coelomic oocytes of the echiuran worm *Urechis caupo*. All oocytes, from the smallest (20 μm diameter) to the largest (120 μm diameter), had comparable resting potentials and a two-component action potential similar to the action potential in the egg.

Therefore, little modification in the electrical properties could be detected during growth of the oocytes. However, these growing oocytes had lower membrane resistance and more negative resting potentials than full-grown eggs due, at least in part, to a greater K^+-selective conductance. The reason for the decrease in K^+ conductance at the end of oogenesis in *Urechis* is not known; there may be a decrease in channel density, channel permeability, or both.

The possible changes in ion channel density and permeability during oogenesis have been explored in more detail in oocytes of the starfish *Leptasterias*. Voltage-clamp studies (Moody and Lansman, 1983) showed that the full-grown starfish oocyte displays a fast transient K^+ current (A-current) and an inwardly rectifying K^+ current. A third current is responsible for the generation of action potentials in these oocytes and represents a Ca^{2+} current through a voltage-dependent Ca^{2+} channel followed by a Ca^{2+}-activated current carried mostly by Na^+ (see Section II,A).

Moody (1985) examined these currents during oogenesis. During a rapid growth period in the second year of oogenesis, when the oocyte diameter increases fourfold, the magnitudes of the A-current and inwardly rectifying K^+ currents increase due to the addition of K^+ channels along with new membrane. However, the kinetic properties and current densities of both K^+ currents remain unchanged. The inward current carried by Ca^{2+} and Na^+ also increases in proportion to the increase in surface area during the growth period, but then abruptly increases much more after the oocyte reaches full size. The increase in amplitude of the inward current is associated with a more prolonged current (slower inactivation). The cause of this increase is not known, but, since the later part of the current is probably a Ca^{2+}-activated Na^+ current, changes in the current might be due to changes in the Ca^{2+} sensitivity of the Na^+ channel or a change in the magnitude of the intracellular rise in Ca^{2+} during the action potential. Thus, as in the oocytes of *Urechis,* the electrical properties of the starfish oocyte change little during the growth period, but do change significantly at the end of the growth period.

A similar period of abrupt change in electrical properties occurs in the oocyte of the frog *Xenopus laevis* near the end of the growth phase. The voltage-sensitive Na^+ channels that are induced by membrane depolarization appear to be present in the membrane of small, growing oocytes; however, during a short growth interval just before the oocyte reaches full size (Stage V), the Na^+ current increases dramatically. This increase in current is not due just to the addition of new membrane; the current is about five times greater while surface area increases only by about 25% (Baud, 1983).

From these examples, it appears that the addition of new membrane during oocyte growth does not usually result in extensive modification of the electrical properties of the oocyte. As new membrane is added to the growing oocyte, ion channels are also added, but the resting potential and other electrical properties remain relatively unmodified. Instead, changes occur at or near the end of oocyte growth. It is likely that these changes are associated in some way with preparation of the egg for maturation or fertilization, so it will be important to determine how these changes are regulated.

The electrical properties of full-grown, immature oocytes of a number of species have been examined (Hagiwara and Jaffe, 1979; Moreau *et al.*, 1985). Most oocytes of marine species have a resting potential around -70 mV since the membrane is predominantly permeable to K^+. Immature oocytes have a much lower membrane resistance than mature eggs. The resting potential of immature frog oocytes is between -40 and -60 mV, and the membrane resistance of frog oocytes is also less than mature eggs (Dascal, 1987).

B. Transcellular Ion Currents

In immature oocytes of the frog (Robinson, 1979) and fish (Nuccitelli, 1988), a steady transcellular current enters the animal hemisphere and leaves the vegetal hemisphere. In the full-grown oocyte of the frog *Xenopus laevis,* the inward current of about 1 $\mu A/cm^2$ in the animal hemisphere is carried mainly by Cl^- efflux, and is reduced by application of agents that block Ca^{2+} influx, suggesting that the chloride current is activated by Ca^{2+}. The localized inward current suggests a larger density of Cl^- channels in the animal hemisphere. The relationship between the Cl^- channel that carries the steady inward current and the Cl^- conductance stimulated by membrane depolarization, Ca^{2+} injection, $InsP_3$, or acetylcholine (Section II,B) is not known. Chloride currents induced by Ca^{2+} injection are also larger in the animal hemisphere (Miledi and Parker, 1984). It has been reported that acetylcholine produces larger current when applied to the animal hemisphere (Kusano *et al.*, 1982); however, Oron *et al.* (1988) measured larger currents when acetylcholine was applied to the vegetal hemisphere. The activation current at fertilization is certainly larger in the animal hemisphere (see Section IV,C).

Although this transcellular current was described over 10 years ago, its role in oogenesis is still not known. The current gradually ceases when maturation is induced by progesterone. Moreover, maturation is initiated by Ca^{2+} channel blockers that halt the transcellular current, inferring that the current might be involved in preservation of the immature state (see Section III,C); however, there is no direct evidence for this hypothesis.

Steady transcellular currents may be an indicator of animal–vegetal polarity or they might have a role in the establishment or maintenance of the animal–vegetal polarity in the oocyte. Preliminary observations of Nuccitelli (1988) indicate that the animal–vegetal current can be detected in smaller oocytes, even before they show morphological asymmetries. A steady transcellular current, acting over a long time period of oogenesis, would produce an extracellular voltage gradient. Such a voltage gradient might act to segregate charged, mobile proteins in the plain of membrane by lateral electrophoresis (Jaffe, 1977), contributing to the formation of the animal–vegetal polarity in the *Xenopus* oocyte; this idea has not yet been tested.

Steady transcellular currents do seem to be involved in establishing a voltage gradient within the cytoplasm between cells in insect ovarian complexes. Such voltage gradients appear to result in the *inter*cellular electrophoresis of charged molecules between nurse cells and oocytes. For example, movement of fluorescently labeled proteins or other fluorescent molecules across the cytoplasmic bridge between nurse cells and the oocyte in the silk moth *Hyalophora* and in *Drosophila* is clearly charge dependent (Woodruff and Telfer, 1980; Woodruff, 1989). This process would appear to have a role in the growth and development of the oocyte. Thus far, there is no evidence of *intra*cellular electrophoresis of molecules in an oocyte or egg.

C. Maturation

Maturation (the resumption of meiosis) is a complex process involving cytoplasmic maturation-promoting factor (MPF) (see reviews in Masui and Clarke, 1979; Kanatani, 1985; Masui and Shibuya, 1987; Eckberg, 1988; Smith, 1989). MPF causes germinal vesicle breakdown and chromosome condensation in the oocyte and, furthermore, is responsible for the regulation of the cell cycle in somatic cells (reviewed in Cross *et al.*, 1989).

The membrane electrical properties of oocytes of several species have been examined during maturation with the goal of learning how changes in membrane potential, ion conductance, and intracellular ion concentrations (particularly Ca^{2+}) might initiate or regulate maturation events (reviewed in Moreau *et al.*, 1985; Cork *et al.*, 1987; Cicirelli and Smith, 1987). It appears that alterations in the electrical properties or changes in intracellular ion concentrations per se may not be involved in the natural progesterone-stimulated transmembrane signaling pathway to activate MPF (Smith, 1989). However, changes in the electrical properties of the oocyte during maturation are of interest because they determine the electrical characteristics of the mature egg. The oocytes of starfish and amphibian are particularly amenable to study because maturation can be induced *in vitro* by application of 1-methyladenine (starfish) and progesterone (amphibian).

The oocyte of the starfish undergoes a change in membrane properties during maturation. As described in Section III,A, the *Leptasterias* oocyte displays three voltage-sensitive currents: two K^+ currents and an inward current carried by Ca^{2+} and Na^+ which is responsible for the action potential. During maturation, both K^+ currents decrease, while the inward Ca^{2+}-Na^+ current increases slightly (Moody and Lansman, 1983). The consequence of these changes is to increase the egg's membrane resistance and make the egg more electrically excitable. The decrease in outward K^+ current relative to the inward Ca^{2+}-Na^+ current results in an increase in amplitude and rate of rise of the action potential. By simultaneously examining ionic currents and membrane capacitance, Moody and Bosma (1985) determined that the decrease in K^+ currents occurs at the same time as membrane surface area decreases during maturation as microvilli are reabsorbed. This suggests that K^+ channels are removed as membrane is reabsorbed. Despite the loss of membrane and K^+ channels, the Ca^{2+}-Na^+ current responsible for the action potential is only slightly changed, suggesting that the Ca^{2+} and Na^+ channels might not be susceptible to loss. Moody and Bosma suggest that they might be localized to a region of the membrane not subject to endocytosis.

Similar electrophysiological changes are correlated with a removal of membrane in another starfish (*Henricia*), but there is no correlation with membrane loss in *Asterina* (Simoncini and Moody, 1990). In the smaller *Asterina* oocyte there is no decrease in surface area during maturation, although there is a decrease in both K^+ currents. This observation necessitates formulation and testing of alternative hypotheses for regulation of ionic currents during oocyte maturation. For example, in some species, loss of K^+ currents may involve a recycling of membrane to remove channels without a net loss of membrane, or existing channels in the membrane might some how be permanently inactivated or blocked.

In the amphibian oocyte, little change in membrane potential or resistance occurs immediately after application of progesterone. However, the membrane depolarizes and the resistance increases beginning at a time about halfway between application of progesterone and germinal vesicle breakdown (Wallace and Steinhardt, 1977; Kado *et al.*, 1981). Depolarization probably results from a decrease in K^+ permeability or an increase in

Na^+ permeability or both (Dascal, 1987). These changes in membrane potential and resistance are coincident with a decrease in membrane surface area due to the reabsorption of microvilli (Kado *et al.*, 1981). A systematic examination of changes in specific ion currents with respect to the membrane removal in the *Xenopus* ooctye has not been made. A change in a specific ion channel activity or density is indicated by the observation that the inducible Na^+ channel in the *Xenopus* oocyte cannot be induced after maturation (Baud and Kado, 1984).

Schlichter (1983a,b,1989) has shown that currents through voltage-sensitive Na^+, K^+, and Cl^- channels in *Rana pipiens* change during maturation. Na^+-dependent action potentials can be elicited in oocytes at metaphase of the first meiotic division. In oocytes at this stage, the action potential lasts several seconds and repolarization is produced by currents through the voltage-sensitive K^+ and Cl^- channels. Spontaneous action potentials begin at about the time of the first polar body formation. During maturation, the K^+ currents and then the Cl^- currents cease and, because the Na^+ channels do not inactivate, the action potential lasts many minutes. During this period of maturation, the Na^+ current in the *Rana* oocyte dominates and the recorded oocyte potential is always positive. In the mature egg, action potentials do not occur spontaneously, but they can be elicited by current injection, indicating that the Na^+ channels remain functional.

Preventing the spontaneous action potentials in the *Rana* oocyte delayed maturation, and Schlichter (1983b) proposed that the Na^+-dependent action potential functions to increase the concentration of intracellular Na^+, but it is not known how this might influence maturation. Schlichter (1989) has suggested a number of possible processes that could be influenced by an increase in the intracellular Na^+ concentration. For example, the increase in Na^+ might activate the Na^+/K^+-ATPase and thus alter the intracellular K^+ concentration, or an increase in intracellular Na^+ might affect Na^+-dependent transporters such as a Na^+/H^+ or Na^+/Ca^{2+} exchanger, thereby altering H^+ or Ca^{2+} concentrations. These ion changes or the change in membrane potential itself could regulate some aspect of maturation.

The capacity of the amphibian egg to produce a long-lasting fertilization potential (see Section IV,B) develops during maturation. In maturing oocytes of *Rana* and *Bufo* (between metaphase I and metaphase II), sperm induce low-amplitude, chloride-dependent depolarizations lasting less than 1 min (Schlichter and Elinson, 1981; Iwao, 1987). In contrast, the chloride-dependent fertilization potential of a mature egg reaches a more positive potential and lasts 10–20 min. During maturation, the potential changes induced by sperm gradually take on the form of the fertilization potential characteristic of the mature egg. This transformation is probably due, in part, to the decrease in K^+ conductance associated with maturation. The larger K^+ conductance in the oocyte would tend to reduce the amplitude and duration of a depolarizing current. The amplitude and duration of the fertilization potential in the mature egg might also reflect an enhancement in the ability of the mature egg to release intracellular Ca^{2+} (Charbonneau and Grey, 1984). Similar transient depolarizations are observed when immature (germinal vesicle stage) sea urchin oocytes are fertilized (McCulloh *et al.*, 1987).

There are several reports on the electrical properties of mammalian oocytes during maturation. Maturation of mammalian eggs occurs spontaneously when oocytes are removed from the follicle. No continuous electrical recordings have been made in the mouse oocyte during maturation; however, Peres (1986) found immature oocytes have a somewhat more negative resting potential than mature, ovulated eggs. Radioisotope flux measurements suggest that this decrease in the resting potential is due to a decrease in K^+ permeability (Powers, 1982). McCulloh and Levitan (1987) reported that the membrane potential of the rabbit oocyte changes little during maturation but the membrane resistance increases.

In general, as a consequence of maturation, the electrical properties of the oocyte change; K^+ conductance decreases and the resistance of the membrane increases. As a result, the mature marine egg is more excitable and small depolarizations, such as those initiated during the initial interaction of sperm and egg, produce a rapidly rising, large-amplitude action potential which is an important part of the fertilization potential (see Section IV,B). A similar increase in membrane resistance following maturation of the frog oocyte also results in a faster change in membrane potential produced by sperm-induced opening of ion channels.

IV. FERTILIZATION

A. *The Fertilization Potential and Polyspermy Prevention*

Fertilization of eggs of a number of species results in a transient change in membrane potential referred to as the fertilization potential (Fig. 3). Tyler *et al.* (1956) first reported that a change in membrane potential accompanied fertilization of the starfish egg, confirming earlier indications that the electrical properties of the egg changed at fertilization (reviewed in Hagiwara and Jaffe, 1979). Subsequent studies revealed that a change in membrane potential occurs in many species, and that positive-going fertilization potentials function in the prevention of polyspermy (Jaffe, 1976;

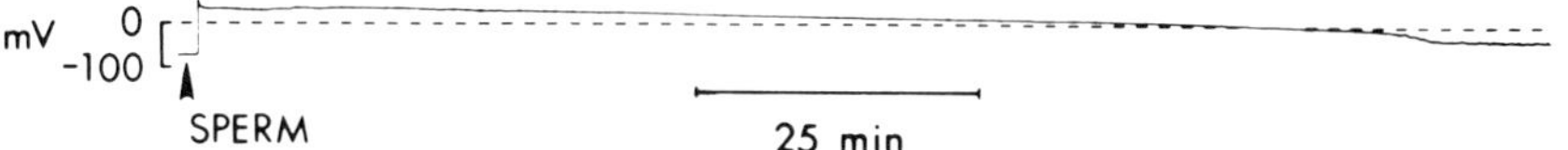

FIG. 3. Fertilization potential in the egg of *Cerebratulus*. The 0 mV potential is indicated by the dashed line. (From Kline *et al.*, 1986.)

reviewed in Jaffe and Cross, 1986). Cross-species fertilization experiments have revealed that the voltage-dependent component contributing to the electrically mediated barrier to sperm–egg fusion is contained in the sperm membrane rather than the egg membrane (see Iwao and Jaffe, 1989, and references therein).

An electrically mediated block to polyspermy occurs at fertilization in eggs of sea urchin, starfish, an echiuran worm (*Urechis caupo*), a nemertean worm (*Cerebratulus lacteus*), anuran amphibians, and possibly several other species (see Fig. 4). The electrical block to polyspermy is a transient process that usually lasts until a permanent block is established by exocytosis of cortical granules and subsequent modification of the egg's extracellular coats, or by an unknown modification of the egg's plasma membrane (reviewed in Jaffe and Gould, 1985). The fertilization potential in the sea urchin egg lasts about 1 min, during which a permanent polyspermy block is established (Jaffe, 1976). In the nemertean egg, a long-lasting fertilization potential is required since it takes about 1 hr to establish a permanent block fully (Kline *et al.*, 1985).

Goudeau and Goudeau (1989) have proposed that a sustained (5 hr or more) hyperpolarization is responsible for an electrically mediated polyspermy block in the crab egg. Thus far, this is the only example in which polyspermy may be prevented by a negative-going potential. In some species, such as the teleost fish *Oryzias* (Nuccitelli, 1980), the cnidarian *Hydractinia echinata* (Berg *et al.*, 1986), and probably most mammals (Miyazaki and Igusa, 1982; Jaffe *et al.*, 1983; McCulloh *et al.*, 1983), there may be some change in membrane potential but polyspermy prevention is not electrically mediated and these species rely on other mechanisms to prevent polyspermy. These species depend on a nonelectrical change in the egg plasma membrane and/or modification of extracellular coats to prevent sperm entry. Physiological polyspermy occurs in a number of species, including salamanders, elasmobranch fish, reptiles, birds, and some invertebrates (Jaffe and Gould, 1985). In these cases, more than one sperm enters the egg but only one male pronucleus fuses with the female pronucleus. It has been demonstrated that an electrical block to sperm–egg fusion does not occur in the salamander (Charbonneau *et al.*, 1983).

Two distinct strategies for polyspermy prevention are utilized in am-

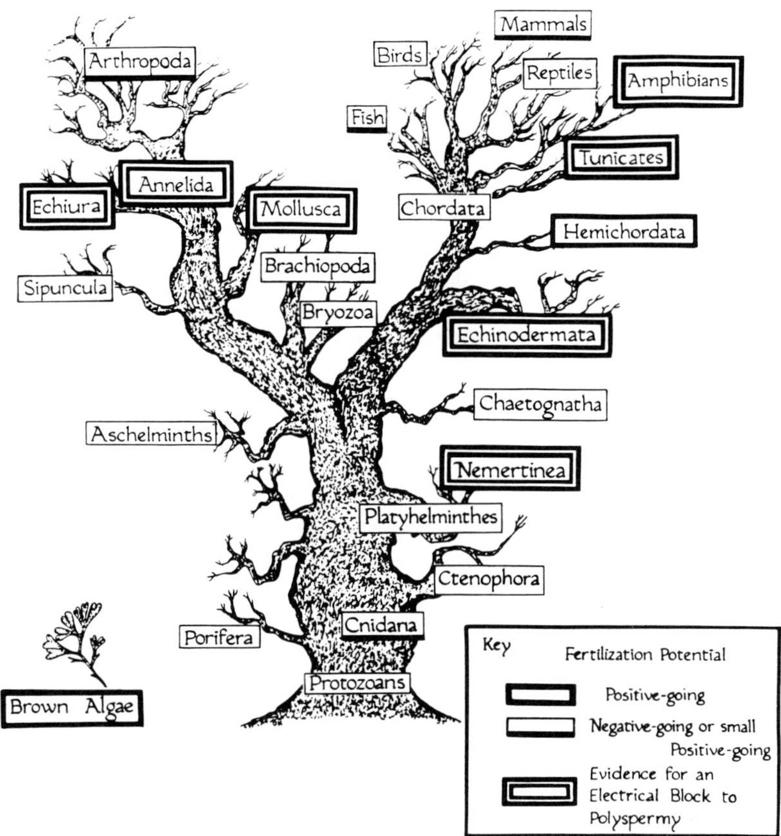

FIG. 4. A phylogenetic tree of the animal kingdom (and the brown algae) showing groups in which the electrical characteristics of eggs during fertilization have been examined. Some of the groups not yet studied are indicated by a light box. A heavy box indicates groups in which at least some species in the group have positive-going potentials. A double box indicates groups having species with positive-going potentials and for which there is evidence of an electrical block to polyspermy. The fertilization potential in the four underlined groups consists of a negative-going or a very small positive-going potential change. These notations do not imply that all species in the group share these traits. (For references see Kline *et al.*, 1985, and references indicated in the text of this chapter.) More recent evidence suggests that an electrical block to polyspermy may not occur in at least one species of mollusk even though the egg has a positive-going fertilization potential (Moreau *et al.*, 1989). There is now some evidence to suggest that an electrical mechanism may prevent polyspermy in two species of brown algae (Brawley, 1987). (From Kline *et al.*, 1985.)

phibians: in general, species of the order Anura (frogs and toads) rely on an electrical block as well as a permanent block produced by cortical granule exocytosis, while species of the order Urodela (salamanders) are physiologically polyspermic and do not rely on an electrical block (Elinson, 1986). However, recent examination of some primitive anurans and urodeles has revealed some exceptions. Talevi (1989) has examined fertilization in the egg of the primitive anuran *Discoglossus pictus*. His evidence suggests that this egg may be polyspermic (and develop normally), yet it has a positive-going fertilization potential. Moreover, depolarizing the egg by current injection during fertilization does not appear to inhibit sperm entry. Iwao (1989) examined fertilization of the egg of the primitive urodele *Hynobius nebulosis*. This egg produces an anuran-type fertilization potential, fertilization is monospermic, and sperm–egg fusion is voltage dependent.

In addition to the role of the positive-going fertilization potential in polyspermy prevention (prevention of sperm–egg fusion), an additional function for the change in membrane potential has been discovered in voltage-clamp studies of fertilization in sea urchin eggs. If the membrane potential of the egg is maintained by voltage clamp at a potential more negative that −20 mV, sperm entry is inhibited. Therefore, the membrane depolarization occurring at fertilization is necessary for sperm penetration into the egg following fusion of sperm and egg membranes (Lynn and Chambers, 1984; McCulloh *et al.*, 1987; Lynn *et al.*, 1988). It is not apparent how membrane potential regulates sperm entry.

B. Sperm-Induced Ion Conductances

1. Invertebrates

The fertilization potential of many marine species resembles the action potentials that occur in oocytes and eggs of the same species; however, the fertilization potential is usually much longer in duration, lasting for about 1 min or in some cases much longer. The calcium-dependent action potential contributes to the initial part of the fertilization potential and this is followed by a sperm-induced Na^+ or cation conductance which produces a long, positive plateau potential.

In the sea urchin egg, the sperm-induced Na^+ conductance appears to be the consequence of a rise in intracellular calcium that occurs at fertilization. In the absence of sperm, application of a calcium ionophore or the injection of $InsP_3$ to release Ca^{2+} from intracellular stores causes a change in membrane potential and inward Na^+ current similar to the responses initiated by sperm (Chambers *et al.*, 1974; Steinhardt and Epel, 1974; Slack *et al.*, 1986; Crossley *et al.*, 1988; David *et al.*, 1988). The time course of the conductance change initiated by sperm is similar to the time course of the sperm-induced rise in intracellular Ca^{2+} (Eisen *et al.*, 1984). If the fertilization potential is calcium sensitive, it should be blocked by injection of a calcium chelator prior to insemination. Swann *et al.* (1987) reported that injection of the calcium chelator EGTA inhibited most of the fertilization potential in the sea urchin egg. In addition to the increase in Na^+ conductance at fertilization, a K^+ conductance develops. This K^+ conductance may be activated by the increase in cytoplasmic pH that follows fertilization (Shen and Steinhardt, 1980).

The sperm-induced conductance change in starfish is similar to that of the sea urchin. A Ca^{2+}-activated Na^+ or cation conductance is present in the unfertilized egg of the starfish (Lansman, 1983a; Moody, 1985) and it appears to contribute to the fertilization potential (Lansman, 1983a,b). A Ca^{2+} rise at fertilization also occurs in the starfish egg (Eisen and Reynolds, 1984).

Direct evidence for a Ca^{2+}-activated Na^+ conductance contributing to the fertilization potential in the egg of *Cerebratulus* has been obtained by injecting a calcium chelator into the egg prior to fertilization (Kline *et al.*, 1986). Injection of EGTA or BAPTA/CaBAPTA buffers having a free Ca^{2+} concentration of less than 0.1 μM reduced the amplitude of the Na^+-dependent plateau potential of the *Cerebratulus* egg. Injection of a BAPTA buffer having a free Ca^{2+} concentration of about 1 μM caused a prolonged potential change very similar to that initiated by sperm. The plateau potential depends little on the concentration of external calcium, suggesting that internal Ca^{2+} stores might be involved in producing the fertilization potential. Measurements of intracellular calcium concentrations have not been made in this egg; it would be very interesting to know if Ca^{2+} rises and remains high for the duration of the fertilization potential (more than 1 hr) or declines. It is now known if a high Ca^{2+} concentration is required only to open the channels or to keep the channels open as well.

The conclusion drawn from these studies is that the prolonged positive plateau phase of the fertilization potential in eggs of sea urchin, starfish, nemertean, and possibly other marine species is probably due to the opening of a Ca^{2+}-activated cation conductance, which in some cases may show a relatively greater selectivity for Na^+ over K^+ and other cations (Lansman, 1983a,b; Kline *et al.*, 1986).

The fertilization potential of the egg of the echiuran worm *Urechis caupo* resembles the fertilization potential in eggs of other marine species in that there are both Ca^{2+} and Na^+ components; the initial rapid depolarization is due to the activation of a voltage-dependent Ca^{2+} current and the positive plateau potential is due primarily to Na^+ influx (Jaffe *et al.*, 1979).

The hypothesis that the Na^+ channels in this egg are activated by Ca^{2+} has not been tested.

Less is known about the electrical changes that occur at fertilization in marine annelid and mollusk eggs. Positive-going fertilization potentials occur in eggs of the polychaete *Chaetopterus* (Jaffe, 1983), the mollusks *Spisula* and *Dentalium* (Finkel and Wolf, 1980; Moreau *et al.*, 1989), and in oyster and abalone (Gould and Stephano, 1989). In a few marine species (crab and lobster), the fertilization potential is negative-going and due primarily to the opening of K^+ channels in the egg membrane (Goudeau and Goudeau, 1985, 1986). The role of Ca^{2+}-activated conductances in producing the fertilization potential in annelids, mollusks, and crustaceans has not been examined.

Among the lower chordates, the hemichordate *Saccoglossus* also has a positive-going potential change (Jaffe, 1983), but no further information about the fertilization potential in this species has been obtained. The fertilization potential in eggs of several ascidian species is positive-going (Dale *et al.*, 1983) and resembles the fertilization potential of the sea urchin egg. An activation potential consisting mainly of a Ca^{2+} and a Na^+ current, which resembles the fertilization potential, can be induced by application of the Ca^{2+} ionophore A23187. There is also a large increase in intracellular free Ca^{2+} at fertilization in at least two ascidian species (Speksnijder *et al.*, 1989). This suggests that a Ca^{2+}-activated Na^+ or cation conductance might be involved in generation of the fertilization potential. Dale (1987) has reported that the sperm-induced conductance at fertilization in the egg of the ascidian *Ciona* is not activated by Ca^{2+}. It would be useful to examine this again using BAPTA/CaBAPTA buffers made to buffer Ca^{2+} at different levels in the egg to determine if there are any Ca^{2+}-activated conductances in this species.

2. Vertebrates

In anuran amphibians, the fertilization potential is positive-going and lasts 10–20 min. The amphibian egg is fertilized in fresh water and the ionic conductances needed to reach a positive potential are different from marine species; for these freshwater species, the fertilization potential is primarily due to the opening of Cl^- channels in the membrane, resulting in an efflux of Cl^-. The fertilization potential has been described for various species of *Rana* (Schlichter and Elinson, 1981; Jaffe and Schlichter, 1985) and *Bufo* (Iwao *et al.*, 1981), and in *Xenopus laevis* (Webb and Nuccitelli, 1985) and *Discoglossus pictus* (Talevi *et al.*, 1985). The channels responsible for the fertilization potential are permeable to halides other than Cl^- (Webb and Nuccitelli, 1985), but under physiological conditions, Cl^- is the principal anion and the anion conductance activated at fertilization is referred to as a Cl^- conductance.

The amplitude of the fertilization potential in the frog egg depends on the concentration of extracellular Cl^-; however, the equilibrium potential for Cl^- is more positive than the fertilization potential, indicating that other conductances are activated at fertilization. A voltage-activated Na^+ conductance is activated during depolarization of the egg at fertilization (Jaffe and Schlichter, 1985; Peres and Mancinelli (1985) and a large K^+ conductance is activated at fertilization (Webb, 1984; Jaffe and Schlichter, 1985). Under the ionic conditions in which the egg is fertilized, an increase in K^+ and Na^+ conductances would tend to limit the amplitude of the fertilization potential and contribute to the repolarization of the egg.

The fertilization potential in the frog egg is Ca^{2+} dependent. The potential change is initiated by artificially increasing intracellular Ca^{2+} (Cross, 1981; Schlichter and Elinson, 1981) and by injection of $InsP_3$, which releases Ca^{2+} from cortical or subcortical stores (Busa *et al.*, 1985). A propagated wave of increased intracellular Ca^{2+} occurs in the *Xenopus* egg at fertilization (Busa and Nuccitelli, 1985; Kubota *et al.*, 1987). The sperm-induced conductance change is prevented by injection of a Ca^{2+} chelator to prevent the rise in intracellular Ca^{2+} (Kline, 1988).

In those species of mammals (hamster, mouse, and rabbit) so far studied, the potential of the egg during fertilization changes little or consists of negative-going hyperpolarizing responses (Miyazaki and Igusa, 1981; Igusa *et al.*, 1983; Jaffe *et al.*, 1983; McCulloh *et al.*, 1983). The largest responses are seen in hamster eggs, in which transient, large hyperpolarizations occur at regular intervals of 40–120 sec. The transients are superimposed on a gradual negative-going shift in potential. Each transient is due to the opening of Ca^{2+}-activated K^+ channels in the egg membrane; the responses are inhibited by EGTA injection before insemination, are initiated by Ca^{2+} injection, and coincide with transient increases in intracellular Ca^{2+} (Miyazaki and Igusa, 1982; Igusa and Miyazaki, 1983, 1986).

To review, the fertilization potential in a variety of species appears to result from the opening of Ca^{2+}-activated ion channels. The specific ion channel that is activated by calcium depends on the species; for many marine species, it is probably a Na^+ channel, for frogs and toads it is a Cl^- channel, and for hamsters it is a K^+ channel. The importance of Ca^{2+} in producing the fertilization potential is indicated by three observations. First, in many species, an increase in the concentration of intracellular free Ca^{2+} occur at fertilization. Second, artificial elevation of Ca^{2+} causes a potential change that mimics the fertilization potential. Third, in a few species, it has been demonstrated that injection of Ca^{2+} chelators prevents the potential change induced by sperm.

C. Activation Current Waves

The chloride channels responsible for the fertilization potential in the amphibian egg do not open uniformly over the egg, but rather, a ring-shaped wave of inward current (the activation current, a Cl^- ion efflux) spreads over the egg from the site of sperm entry or site of artificial activation to the opposite side of the egg. The inward current is preceded and followed by an outward current (K^+ efflux), except at the initial site of activation, where the current is first inward. Such a wave of channel opening has been described for eggs of *Xenopus* (Kline and Nuccitelli, 1985) and *Rana* (Jaffe *et al.*, 1985). The activation current in the egg of the medaka fish *Oryzias latipes* also spreads across the egg in a ring-shaped wave (Nuccitelli, 1987), and there is some evidence to suggest that an activation current wave occurs in the sea urchin egg (reviewed in Chambers, 1989).

The activation current probably reflects the wave of intracellular calcium release that occurs in the eggs of these species, and the opening of Ca^{2+}-dependent channels. Although simultaneous measurements of the calcium rise and inward current have not been made, the activation current wave in the medaka egg and *Xenopus* egg follows about the same time course as the wave of increased free Ca^{2+} detected with aequorin (Gilkey *et al.*, 1978; Kubota *et al.*, 1987).

The magnitude of the inward current is 6–7 times greater in the animal hemisphere of the *Xenopus* or *Rana* egg (Jaffe *et al.*, 1985; Kline and Nuccitelli, 1985); this suggests a localization of Cl^- channels in the animal hemisphere of the egg. A similar localization of Cl^- channels has been noted in immature *Xenopus* oocytes, and localization of Cl^- channels in the animal hemisphere is also indicated by the steady inward current entering the animal hemisphere in the immature *Xenopus* oocyte (see Section III,B). Fertilization channels appear to be even more localized in the egg of the frog *Discoglossus pictus* (Nuccitelli *et al.*, 1988). In this species, sperm entry occurs only at the animal pole in a specialized region referred to as the animal dimple. The activation current enters only the dimple region; there is no wave of ion channel opening despite the fact that a wave of increased free Ca^{2+} appears to pass through the egg. The Cl^- channels in this egg appear to be the most localized in any vertebrate egg.

The activation current wave precedes the wave of cortical granule exocytosis in the *Xenopus* and medaka egg by 10–15 sec (Kline and Nuccitelli, 1985; Nuccitelli, 1987). This indicates that the ion channels responsible for the fertilization potential are present in the membrane of the unfertilized egg and are not inserted with new membrane from the exocytosis of cortical granules. This is in agreement with the observations that the maximal opening of ion channels induced by sperm precedes the increase in membrane capacitance in sea urchin and frog (Jaffe *et al.*, 1978; Jaffe and Schlichter, 1985). In addition, in *Xenopus*, a membrane depolarization and activation current can be elicited under certain conditions that inhibit cortical granule exocytosis (Peres and Bernardini, 1985; Charbonneau and Webb, 1986; Charbonneau *et al.*, 1986).

Ion channel opening during fertilization of the *Urechis* egg is localized in a patch near the site of sperm entry (Gould-Somero, 1981). This might reflect the fact that an activating Ca^{2+} wave may not occur in eggs of protostomes (Jaffe, 1985). It would be informative to examine whether fertilization channels in other protostome species are also opened locally.

D. The Pathway to Electrical Changes at Fertilization

Calcium is an important signal, not only for ion channel opening, but for other events of egg activation (reviewed in Jaffe, 1985; Whitaker and Steinhardt, 1985). There is evidence to indicate a role for the phosphoinositide messenger system in the initiation of intracellular calcium release at fertilization. For example, in the sea urchin egg, $InsP_3$ and diacylglycerol (DAG) concentrations rise within 10 sec after fertilization (Ciapa and Whitaker, 1986). Injection of $InsP_3$ into the egg causes Ca^{2+} release and subsequent ion channel activation in a number of species (see previous discussion; reviewed in Whitaker, 1989). Guanine nucleotide-binding proteins (G-proteins) are present in sea urchin eggs (Oinuma *et al.*, 1986; Turner *et al.*, 1987) and frog eggs (Kline *et al.*, 1991). The pathway leading to Ca^{2+} release in the egg may involve a G-protein in the egg membrane (Turner and Jaffe, 1989). In other cells, G-proteins serve to couple receptor and effector molecules such as phospholipase C, which produces $InsP_3$ and DAG from the membrane lipid phosphatidylinositol 4,5-bisphosphate (Cockcroft and Gomperts, 1985; Harden, 1989).

A role for G-proteins in activating the phosphoinositide messenger system and producing electrical changes in eggs is indicated by experiments using the hydrolysis-resistant analog of GTP, GTP-γ-S, which activates G-proteins. In the sea urchin, activation of the egg's G-protein by GTP-γ-S causes an increase in intracellular Ca^{2+} and cortical vesicle exocytosis (Turner *et al.*, 1986). Additional evidence for the role of G-proteins in fertilization comes from experiments using cholera toxin, which activates certain G-proteins; injection of cholera toxin into sea urchin eggs causes cortical granule exocytosis (Turner *et al.*, 1987). The electrical responses to GTP-γ-S or cholera toxin injection in the sea urchin egg have not been examined.

In the frog, injection of GTP-γ-S causes a change in membrane potential that mimics the sperm-induced change in potential (Kline *et al.*, 1991). Similarly, in the hamster egg, GTP-γ-S causes a rise in Ca^{2+} and a change in membrane potential like that initiated by sperm (Miyazaki, 1988). The sea urchin and frog responses to GTP-γ-S can be blocked by injection of EGTA or BAPTA prior to injection of GTP-γ-S. Therefore, cortical granule exocytosis and ion channel opening are not regulated directly by a G-protein (or $InsP_3$), but are initiated by the subsequent rise in Ca^{2+}.

Recent experiments demonstrate that G-proteins in eggs can be activated by neurotransmitters following injection of mRNA for neurotransmitter receptors known to function through G-proteins. Exogenous neurotransmitter receptors, known to act by way of a G-protein to stimulate phospholipase C, were introduced into the *Xenopus* egg (Fig. 5). Application of acetylcholine to mature frog eggs, after expression of acetylcholine M1 receptors, caused a change in membrane potential very similar to that initiated by sperm as well as causing other activation responses (cortical granule exocytosis, endocytosis, and cortical pigment contraction) (Kline *et al.*, 1988). Eggs not injected with mRNA did not respond to the neurotransmitters. Starfish oocytes express neurotransmitter receptors after injection of mRNA for the serotonin 1c receptor. Following maturation, application of serotonin to such eggs causes cortical vesicle exocytosis (Shilling *et al.*, 1990). These results suggest that, as in the frog, receptor activation of a G-protein in the egg membrane will produce responses like those occurring at fertilization.

The results of these experiments with sea urchin, starfish, frog, and hamster eggs support the hypothesis that egg activation and subsequent ion channel opening occur as a result of activation of a G-protein in the egg membrane. Activation of the G-protein might occur through interaction of sperm with a receptor molecule in the egg membrane or the sperm might activate the egg's G-protein directly. There are alternative hypotheses for the mechanism of egg activation which would not involve activation of a G-protein in the egg membrane (Dale *et al.*, 1985; Gould and Stephano, 1989; Whitaker *et al.*, 1989). For example, the sperm might insert an activating molecule into the egg membrane or into the egg cytoplasm during fusion of egg and sperm. The process of egg activation needs to be examined in order to understand how sperm initiate the first electrical events in the egg.

The sperm-induced electrical changes that accompany fertilization in the sea urchin egg have been examined extensively by Chambers and associates. In particular, they have recorded currents in eggs which are voltage-clamped to prevent currents associated with the action potential. The currents induced by sperm consist of an inward current of abrupt

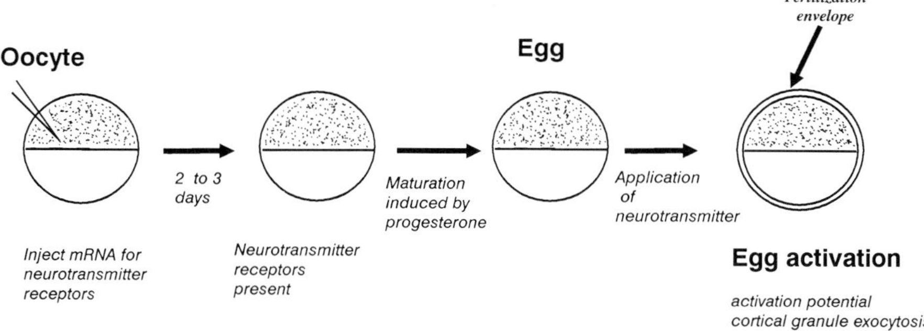

FIG. 5. Experimental design and results of an experiment to examine egg activation in the frog egg by introduction of exogenous neurotransmitter receptors.

onset which increases slowly, and which is followed by the major phase of the inward Na^+ current, coinciding with the large increase in intracellular free calcium (Lynn *et al.*, 1988). Sperm–egg fusion, as indicated by electron microscopy (Longo *et al.*, 1986) or by transfer of a DNA dye (Hinkley *et al.*, 1986), is not seen until 5–10 sec after the beginning of the first sperm-induced current. These experiments suggest that the sperm-induced ion conductances are activated just before fusion, perhaps by a receptor-mediated mechanism. However, because of the inherent limitations of these detection techniques, examination of the relationship between the electrical events and sperm–egg fusion by another method would be useful.

The initial fertilization current in the sea urchin egg appears to be localized near the site of sperm–egg interaction and might be due to the local activation of ion channels in the egg membrane or to the introduction of sperm channels following fusion of egg and sperm (Lynn *et al.*, 1988). Preliminary experiments (McCulloh and Chambers, 1986; reviewed in Chambers, 1989) suggest that fusion of egg and sperm might be coincident with the initial sperm-induced electrical responses. An increase in egg membrane capacitance, in a patch of membrane under a patch electrode containing sperm, is coincident with the first sperm-induced currents. McCulloh and Chambers propose that the initial conductance change might be due to the introduction of sperm ion channels into the egg membrane during fusion of sperm and egg membranes. This model would also be consistent with the idea that the sperm might introduce an activating substance after fusion, but it does rule out the possibility of receptor-induced activation being involved in generation of the fertilization potential. Further experimentation in sea urchin and other eggs will be necessary to determine the relationships between the early electrical events and the mechanism of egg activation.

V. CONCLUSIONS

The evidence presented here points to an important role for Ca^{2+}-activated ion conductances in the action potential of marine occytes and eggs. Moreover, it appears that Ca^{2+}-activated ion conductances may be responsible for the fertilization potential in many species. In some species, the fertilization potential has an important role in polyspermy prevention, but in other species, including mammals, the functions of the potential changes during fertilization are not known. The changes in electrical properties of oocytes during growth and maturation appear to prepare the egg for fertilization. Many interesting aspects of oogenesis, maturation, and fertilization deserve further study, a few of which include the mechanism by which Ca^{2+} regulates ion channel opening, the process by which ion conductances change during oogenesis and maturation, the function of the Ca^{2+}-activated Cl^- conductance in the frog oocyte and its relationship to the Cl^- current at fertilization, the mechanism by which ion channels in the frog oocyte and egg are localized in the animal hemisphere, the mode of ion channel activation and production of the fertilization potential in eggs of protostomes, the function of the fertilization potential in regulating sperm entry, and the molecular mechanism by which the egg is activated.

References

Barish, M. E. (1983). A transient calcium-dependent chloride current in the immature *Xenopus* oocyte. *J. Physiol.* (*London*) **342,** 309–325.

Baud, C. (1983). Developmental change of a depolarization-induced sodium permeability in the oocyte of *Xenopus laevis*. *Dev. Biol.* **99,** 524–528.

Baud, C., and Kado, R. T. (1984). Induction and disappearance of excitability in the oocyte of *Xenopus laevis:* A voltage-clamp study. *J. Physiol.* (*London*) **356,** 275–289.

Baud, C., Kado, R. T., and Marcher, K. (1982). Sodium channels induced by depolarization of the *Xenopus laevis* oocyte. *Proc. Natl. Acad. Sci. U.S.A.* **79,** 3188–3192.

Berg, C., Kirby, C., Kline, D., and Jaffe, L. A. (1986). Fertilization potential and polyspermy prevention in the egg of the hydrozoan *Hydractinia echinata*. *Biol. Bull.* (*Woods Hole, Mass.*) **171,** 485.

Berridge, M. J. (1988). Inositol trisphosphate-induced membrane potential oscillations in *Xenopus* oocytes. *J. Physiol.* (*London*) **403,** 589–599.

Boton, R., Dascal, N., Gillo, B., and Lass, Y. (1989). Two calcium-activated chloride conductances in *Xenopus laevis* oocytes permeabilized with the ionophore A23187. *J. Physiol.* (*London*) **408,** 511–534.

Brawley, S. H. (1987). A sodium-dependent, fast block to polyspermy occurs in eggs of fucoid algae. *Dev. Biol.* **124,** 390–397.

Busa, W. B., and Nuccitelli, R. (1985). An elevated free cytosolic Ca^{2+} wave follows fertilization in eggs of the frog, *Xenopus laevis*. *J. Cell Biol.* **100,** 1325–1329.

Busa, W. B., Ferguson, J. E., Joseph, S. K., Williamson, J. R., and Nuccitelli, R. (1985). Activation of frog (*Xenopus laevis*) eggs by inositol trisphosphate. I. Characterization of Ca^{2+} release from intracellular stores. *J. Cell Biol.* **101,** 677–682.

Chambers, E. L. (1989). Fertilization in voltage-clamped sea urchin eggs. *In* "Mechanisms of Egg Activation" (R. Nuccitelli, G. N. Cherr, and W. H. Clark, eds.), pp. 1–18. Plenum, New York.

Chambers, E. L., and de Armendi, J. (1979). Membrane potential, action potential and activation potential of eggs of the sea urchin, *Lytechinus variegatus*. *Exp. Cell Res.* **122,** 203–218.

Chambers, E. L., Pressman, B. C., and Rose, B. (1974). The activation of sea urchin eggs by the divalent ionophores A23187 and X-537A. *Biochem. Biophys. Res. Commun.* **60,** 126–132.

Charbonneau, M., and Grey, R. D. (1984). The onset of activation responsiveness during maturation coincides with the formation of the cortical endoplasmic reticulum in oocytes of *Xenopus laevis*. *Dev. Biol.* **102,** 90–97.

Charbonneau, M., and Webb, D. J. (1986). Multiple activation currents can be evoked in *Xenopus laevis* eggs when cortical granule exocytosis is inhibited by weak bases. *Pfleugers Arch.* **407,** 370–376.

Charbonneau, M., Moreau, M., Picheral, B., Vilain, J. P., and Guerrier, P. (1983). Fertilization of amphibian eggs: A comparison of electrical responses between anurans and urodeles. *Dev. Biol.* **98,** 304–318.

Charbonneau, M., Dufresne-Dube, L., and Guerrier, P. (1986). Inhibition of the activation reaction of *Xenopus laevis* eggs by the lectins WGA and SBA. *Dev. Biol.* **114,** 347–360.

Ciapa, B., and Whitaker, M. (1986). Two phases of inositol polyphosphate and diacylglycerol production at fertilisation. *FEBS Lett.* **195,** 347–351.

Cicirelli, M. F., and Smith, L. D. (1987). Do calcium and calmodulin trigger maturation in amphibian oocytes? *Dev. Biol.* **121,** 48–57.

Cockcroft, S., and Gomperts, B. D. (1985). Role of guanine nucleotide binding protein in the activation of polyphosphoinositide phosphodiesterase. *Nature (London)* **314,** 534–536.

Cork, R. J., Cicirelli, M. F., and Robinson, K. R. (1987). A rise in cytosolic calcium is not necessary for maturation of *Xenopus laevis* oocytes. *Dev. Biol.* **121,** 41–47.

Cross, F., Roberts, J., and Weintraub, H. (1989). Simple and complex cell cycles. *Annu. Rev. Cell Biol.* **5,** 341–395.

Cross, N. L. (1981). Initiation of the activation potential by an increase in intracellular calcium in eggs of the frog, *Rana pipiens*. *Dev. Biol.* **85,** 380–384.

Crossley, I., Swann, K., Chambers, E., and Whitaker, M. (1988). Activation of sea urchin eggs by inositol phosphates is independent of external calcium. *Biochem. J.* **252,** 257–262.

Dale, B. (1987). Fertilization channels in ascidian eggs are not activated by Ca. *Exp. Cell Res.* **172,** 474–480.

Dale, B., De Santis, A., and Ortolani, G. (1983). Electrical response to fertilization in ascidian oocytes. *Dev. Biol.* **99,** 188–193.

Dale, B., DeFelice, L. J., and Ehrenstein, G. (1985). Injection of a soluble sperm fraction into sea-urchin eggs triggers the cortical reaction. *Experientia* **41,** 1068–1070.

Dascal, N. (1987). The use of *Xenopus* oocytes for the study of ion channels. *CRC Crit. Rev. Biochem.* **22,** 317–387.

Dascal, N., Yekuel, R., and Oron, Y. (1984). Acetylcholine promotes progesterone-induced maturation of *Xenopus* oocytes. *J. Exp. Zool.* **230,** 131–135.

Dascal, N., Gillo, B., and Lass, Y. (1985). Role of calcium mobilization in mediation of acetylcholine-evoked chloride currents in *Xenopus laevis* oocytes. *J. Physiol.* (*London*) **366,** 299–313.

David, C., Halliwell, J., and Whitaker, M. (1988). Some properties of the membrane currents underlying the fertilization potential in sea urchin eggs. *J. Physiol.* (*London*) **402,** 139–154.

Eckberg, W. R. (1988). Intracellular signal transduction and amplification mechanisms in the regulation of oocyte maturation. *Biol. Bull.* (*Woods Hole, Mass.*) **174,** 95–108.

Eisen, A., and Reynolds, G. T. (1984). Calcium transients during early development in single starfish (*Asterias forbesi*) oocytes. *J. Cell Biol.* **99,** 1878–1882.

Eisen, A., Kiehart, D. P., Weiland, S. J., and Reynolds, G. T. (1984). Temporal sequence and spatial distribution of early events of fertilization in single sea urchin eggs. *J. Cell Biol.* **99,** 1647–1654.

Elinson, R. P. (1986). Fertilization in amphibians: The ancestry of the block to polyspermy. *Int. Rev. Cytol.* **101,** 59–100.

Eusebi, F., Colonna, R., and Mangia, F. (1983). Development of membrane excitability in mammalian oocytes and early embryos. *Gamete Res.* **7,** 39–47.

Finkel, T., and Wolf, D. P. (1980). Membrane potential, pH and the activation of surf clam oocytes. *Gamete Res.* **3,** 299–304.

Gilkey, J. C., Jaffe, L. F., Ridgway, E. B., and Reynolds, G. T. (1978). A free calcium wave traverses the activating egg of the medaka, *Oryzias latipes*. *J. Cell Biol.* **76,** 448–466.

Gillo, B., Lass, Y., Nadler, E., and Oron, Y. (1987). The involvement of inositol 1,4,5-trisphosphate and calcium in the two-component response to acetylcholine in *Xenopus* oocytes. *J. Physiol.* (*London*) **392,** 349–361.

Goudeau, H., and Goudeau, M. (1985). Fertilization in crabs: IV. The fertilization potential consists of a sustained egg membrane hyperpolarization. *Gamete Res.* **11,** 1–17.

Goudeau, H., and Goudeau, M. (1986). Electrical and morphological responses of the lobster egg to fertilization. *Dev. Biol.* **114,** 325–335.

Goudeau, H., and Goudeau, M. (1989). A long-lasting electrically mediated block, due to the egg membrane hyperpolarization at fertilization, ensures physiological monospermy in eggs of the crab *Maia squinado*. *Dev. Biol.* **133,** 348–360.

Gould, M., and Stephano, J. L. (1989). How do sperm activate eggs in *Urechis* (as well as in polychaetes and molluscs)? *In* "Mechanisms of Egg Activation" (R. Nuccitelli, G. N. Cherr, and W. H. Clark, eds.), pp. 201–214. Plenum, New York.

Gould-Somero, M. (1981). Localized gating of egg Na^+ channels by sperm. *Nature* (*London*) **291,** 254–256.

Hagiwara, S. (1983). "Membrane Potential-Dependent Ion Channels in Cell Membrane. Phylogenetic and Developmental Approaches," Distinguished Lecture Series of the Society of General Physiologists, Vol. 3. Raven, New York.

Hagiwara, S., and Jaffe, L. A. (1979). Electrical properties of egg cell membranes. *Annu. Rev. Biophys. Bioeng.* **8,** 385–416.

Harden, T. K. (1989). The role of guanine nucleotide regulatory proteins in receptor-selective direction of inositol lipid signalling. *In* "Inositol Lipids in Cell Signalling" (R. H. Michell, A. H. Drummond, and C. P. Downes, eds.), pp. 113–133. Academic Press, London.

Hinkley, R. E., Wright, B. D., and Lynn, J. W. (1986). Rapid visual detection of sperm-egg fusion using the DNA-specific fluorochrome Hoechst 33342. *Dev. Biol.* **118,** 148–154.

Holland, L., and Gould-Somero, M. (1981). Electrophysiological response to insemination in oocytes of *Urechis caupo*. *Dev. Biol.* **83,** 90–100.

Igusa, Y., and Miyazaki, S. (1983). Effects of altered extracellular and intracellular calcium concentration on hyperpolarizing responses of the hamster egg. *J. Physiol.* (*London*) **340,** 611–632.

Igusa, Y., and Miyazaki, S. (1986). Periodic increase of cytoplasmic free calcium in fertilized hamster eggs measured with calcium-sensitive electrodes. *J. Physiol.* (*London*) **377,** 193–205.

Igusa, Y., Miyazaki, S., and Yamashita, N. (1983). Periodic hyperpolarizing responses in hamster and mouse eggs fertilized with mouse sperm. *J. Physiol.* (*London*) **340,** 633–647.

Iwao, Y. (1987). The spike component of the fertilization potential in the toad, *Bufo japonicus:* Changes during meiotic maturation and absence during cross-fertilization. *Dev. Biol.* **123,** 559–565.

Iwao, Y. (1989). An electrically mediated block to polyspermy in the primitive urodele *Hynobius nebulosus* and phylogenetic comparison with other amphibians. *Dev. Biol.* **134,** 438–445.

Iwao, Y., and Jaffe, L. A. (1989). Evidence that the voltage-dependent component in the fertilization process is contributed by the sperm. *Dev. Biol.* **134,** 446–451.

Iwao, Y., Ito, S., and Katagiri, C. (1981). Electrical properties of toad oocytes during maturation and activation. *Dev. Growth Differ.* **23,** 89–100.

Jaffe, L. A. (1976). Fast block to polyspermy in sea urchin eggs is electrically mediated. *Nature* (*London*) **261,** 68–71.

Jaffe, L. A. (1983). Fertilization potentials from eggs of the marine worms *Chaetopterus* and *Saccoglossus*. *In* "The Physiology of Excitable Cells" (A. D. Grinnell and W. J. Moody, eds.), pp. 211–218. Alan R. Liss, New York.

Jaffe, L. A., and Cross, N. L. (1986). Electrical regulation of sperm–egg fusion. *Annu. Rev. Physiol.* **48,** 191–200.

Jaffe, L. A., and Gould, M. (1985). Polyspermy-preventing mechanisms. *In* "Biology of Fertilization" (C. B. Metz and A. Monroy, eds.), Vol. 3, pp. 223–250. Academic Press, Orlando, Florida.

Jaffe, L. A., and Schlichter, L. C. (1985). Fertilization-induced ionic conductances in eggs of the frog, *Rana pipiens*. *J. Physiol.* (*London*) **358,** 299–319.

Jaffe, L. A., Hagiwara, S., and Kado, R. T. (1978). The time course of cortical vesicle fusion in sea urchin eggs observed as membrane capacitance changes. *Dev. Biol.* **67,** 243–248.

Jaffe, L. A., Gould-Somero, M., and Holland, L. (1979). Ionic mechanism of the fertilization potential of the marine worm, *Urechis caupo* (Echiura). *J. Gen. Physiol.* **73,** 469–492.

Jaffe, L. A., Sharp, A. P., and Wolf, D. P. (1983). Absence of an electrical polyspermy block in the mouse. *Dev. Biol.* **96,** 317–323.

Jaffe, L. A., Kado, R. T., and Muncy, L. (1985). Propagating potassium and chloride conductances during activation and fertilization of the egg of the frog, *Rana pipiens*. *J. Physiol.* (*London*) **368,** 227–242.

Jaffe, L. A., Kado, R. T., and Kline, D. (1986). A calcium-activated sodium conductance produces a long-duration action potential in the egg of a nemertean worm. *J. Physiol.* (*London*) **381,** 263–278.

Jaffe, L. F. (1977). Electrophoresis along cell membranes. *Nature* (*London*) **265,** 600–602.

Jaffe, L. F. (1985). The role of calcium explosions, waves, and pulses in activating eggs. *In* "Biology of Fertilization" (C. B. Metz and A. Monroy, eds.), Vol. 3, pp. 127–165. Academic Press, Orlando, Florida.

Kado, R. T., Marcher, K., and Ozon, R. (1981). Electrical membrane properties of the *Xenopus laevis* oocyte during progesterone-induced meiotic maturation. *Dev. Biol.* **84,** 471–476.

Kanatani, H. (1985). Oocyte growth and maturation in starfish. *In* "Biology of Fertilization" (C. B. Metz and A. Monroy, eds.), Vol. 1, pp. 119–140. Academic Press, Orlando, Florida.

Kline, D. (1988). Calcium-dependent events at fertilization of the frog egg: Injection of a calcium buffer blocks ion channel opening, exocytosis, and formation of pronuclei. *Dev. Biol.* **126,** 346–361.

Kline, D., and Nuccitelli, R. (1985). The wave of activation current in the *Xenopus* egg. *Dev. Biol.* **111,** 471–487.

Kline, D., Jaffe, L. A., and Tucker, R. P. (1985). Fertilization potential and polyspermy prevention in the egg of the nemertean, *Cerebratulus lacteus*. *J. Exp. Zool.* **236,** 45–52.

Kline, D., Jaffe, L. A., and Kado, R. T. (1986). A calcium-activated sodium conductance contributes to the fertilization potential in the egg of the nemertean worm *Cerebratulus lacteus*. *Dev. Biol.* **117,** 184–193.

Kline, D., Simoncini, L., Mandel, G., Maue, R. A., Kado, R. T., and Jaffe, L. A. (1988). Fertilization events induced by neurotransmitters after injection of mRNA in *Xenopus* eggs. *Science* **241,** 464–467.

Kline, D., Kopf, G. S., Muncy, L. F., and Jaffe, L. A. (1991). Evidence for the involvement of a pertussis toxin-insensitive G-protein in egg activation of the frog, *xenopus laevis*. *Dev. Biol.* **143,** 218–229.

Kubota, H. Y., Yoshimoto, Y., Yoneda, M., and Hiramoto, Y. (1987). Free calcium wave upon activation in *Xenopus* eggs. *Dev. Biol.* **119,** 129–136.

Kusano, K., Miledi, R., and Stinnakre, J. (1977). Acetylcholine receptors in the oocyte membrane. *Nature* (*London*) **270,** 739–741.

Kusano, K., Miledi, R., and Stinnakre, J. (1982). Cholinergic and catecholaminergic receptors in the *Xenopus* oocyte membrane. *J. Physiol.* (*London*) **328,** 143–170.

Lansman, J. B. (1983a). Components of the starfish fertilization potential: Role of calcium and calcium-dependent inward current. *In* "The Physiology of Excitable Cells" (A. D. Grinnell and W. J. Moody, eds.), pp. 233–246. Alan R. Liss, New York.

Lansman, J. B. (1983b). Voltage-clamp study of the conductance activated at fertilization in the starfish egg. *J. Physiol.* (*London*) **345,** 353–372.

Longo, F. J., Lynn, J. W., McCulloh, D. H., and Chambers, E. L. (1986). Correlative ultrastructural and electrophysiological studies of sperm–egg interactions of the sea urchin, *Lytechinus variegatus*. *Dev. Biol.* **118,** 155–166.

Lynn, J. W., and Chambers, E. L. (1984). Voltage clamp studies of fertilization in sea urchin eggs. I. Effect of clamped membrane potential on sperm entry, activation, and development. *Dev. Biol.* **102,** 98–109.

Lynn, J. W., McCulloh, D. H., and Chambers, E. L. (1988). Voltage clamp studies of fertilization in sea urchin eggs. II. Current patterns in relation to sperm entry, nonentry, and activation. *Dev. Biol.* **128,** 305–323.

Masui, Y., and Clarke, H. J. (1979). Oocyte maturation. *Int. Rev. Cytol.* **57,** 185–282.

Masui, Y., and Shibuya, E. K. (1987). Development of cytoplasmic activities that control chromosome cycles during maturation of amphibian oocytes. *In* "Molecular Regulation of Nuclear Events in Mitosis and Meiosis" (R. A. Schregel, M. S. Halleck, and P. N. Rao, eds.), pp. 1–42. Academic Press, New York.

McCulloh, D. H., and Chambers, E. L. (1986). Fusion and "unfusion" of sperm and egg are voltage dependent in the sea urchin *Lytechinus variegatus*. *J. Cell Biol.* **103,** 236a.

McCulloh, D. H., and Levitan, H. (1987). Rabbit oocyte maturation: Changes of membrane resistance, capacitance, and the frequency of spontaneous transient depolarizations. *Dev. Biol.* **120,** 162–169.

McCulloh, D. H., Rexroad, C. E., and Levitan, H. (1983). Insemination of rabbit eggs is associated with slow depolarization and repetitive diphasic membrane potentials. *Dev. Biol.* **95,** 372–377.

McCulloh, D. H., Lynn, J. W., and Chambers, E. L. (1987). Membrane depolarization facilitates sperm entry, large fertilization cone formation, and prolonged current responses in sea urchin oocytes. *Dev. Biol.* **124,** 177–190.

McIntosh, R. P., and Catt, K. J. (1987). Coupling of inositol phospholipid hydrolysis to peptide hormone receptors expressed from adrenal and pituitary mRNA in *Xenopus laevis* oocytes. *Proc. Natl. Acad. Sci. U.S.A.* **84,** 9045–9048.

Miledi, R. (1982). A calcium-dependent transient outward current in *Xenopus laevis* oocytes. *Proc. R. Soc. London, Ser. B* **215,** 491–497.

Miledi, R., and Parker, I. (1984). Chloride current induced by injection of calcium into *Xenopus* oocytes. *J. Physiol. (London)* **357,** 173–183.

Miledi, R., and Woodward, R. M. (1989). Effects of defolliculation on membrane current responses of *Xenopus* oocytes. *J. Physiol. (London)* **416,** 601–621.

Miledi, R., Parker, I., and Woodward, R. M. (1989). Membrane currents elicited by divalent cations in *Xenopus* oocytes. *J. Physiol. (London)* **417,** 173–195.

Miyazaki, S. (1988). Inositol 1,4,5-trisphosphate-induced calcium release and guanine nucleotide-binding protein-mediated periodic calcium rises in golden hamster eggs. *J. Cell Biol.* **106,** 345–353.

Miyazaki, S., and Igusa, Y. (1981). Fertilization potential in golden hamster eggs consists of recurring hyperpolarizations. *Nature (London)* **290,** 702–704.

Miyazaki, S., and Igusa, Y. (1982). Ca-mediated activation of a K current at fertilization of golden hamster eggs. *Proc. Natl. Acad. Sci. U.S.A.* **79,** 931–935.

Miyazaki, S., Takahashi, K., and Tsuda, K. (1972). Calcium and sodium contributions to regenerative responses in the embryonic excitable cell membrane. *Science* **176,** 1441–1443.

Moody, W. J. (1985). The development of calcium and potassium currents during oogenesis in the starfish, *Leptasterias hexactis*. *Dev. Biol.* **112,** 405–413.

Moody, W. J., and Bosma, M. M. (1985). Hormone-induced loss of surface membrane during maturation of starfish oocytes: Differential effects on potassium and calcium channels. *Dev. Biol.* **112,** 396–404.

Moody, W. J., and Lansman, J. B. (1983). Developmental regulation of Ca^{2+} and K^{+} currents during hormone-induced maturation of starfish oocytes. *Proc. Natl. Acad. Sci. U.S.A.* **80,** 3096–3100.

Moreau, M., Guerrier, P., and Vilain, J. P. (1985). Ionic regulation of oocyte maturation. *In* "Biology of Fertilization" (C. B. Metz and A. Monroy, eds.), Vol. 1, pp. 299–345. Academic Press, Orlando, Florida.

Moreau, M., Guerrier, P., and Dufresne, L. (1989). Absence of an electrical block to polyspermy in the scaphopod mollusk *Dentalium vulgare*. *J. Exp. Zool.* **249,** 76–84.

Moriarty, T. M., Gillo, B., Carty, D. J., Premont, R. T., Landau, E. M., and Iyengar, R. (1988). $\beta\gamma$ Subunits of GTP-binding proteins inhibit muscarinic receptor stimulation of phospholipase C. *Proc. Natl. Acad. Sci. U.S.A.* **85,** 8865–8869.

Moriarty, T. M., Sealfon, S. C., Carty, D. J., Roberts, J. L., Iyengar, R., and Landau, E. M. (1989). Coupling of exogenous receptors to phospholipase C in *Xenopus* oocytes through pertussis toxin-sensitive and -insensitive pathways. Cross-talk through heterotrimeric G-proteins. *J. Biol. Chem.* **264,** 13524–13530.

Nomura, Y., Kaneko, S., Kato, K., Yamagishi, S., and Sugiyama, H. (1987). Inositol phosphate formation and chloride current responses induced by acetylcholine and serotonin through GTP-binding proteins in *Xenopus* oocyte after injection of rat brain messenger RNA. *Mol. Brain Res.* **2,** 113–123.

Nuccitelli, R. (1980). The fertilization potential is not necessary for the block to polyspermy or the activation of development in the medaka egg. *Dev. Biol.* **76,** 499–504.

Nuccitelli, R. (1987). The wave of activation current in the egg of the medaka fish. *Dev. Biol.* **122,** 522–534.

Nuccitelli, R. (1988). Ionic currents in morphogenesis. *Experientia* **44,** 657–666.

Nuccitelli, R., Kline, D., Busa, W. B., Talevi, R., and Campanella, C. (1988). A highly localized activation current yet widespread intracellular calcium increase in the egg of the frog, *Discoglossus pictus*. *Dev. Biol.* **130,** 120–132.

Oinuma, M., Katada, T., Yokosawa, H., and Ui, M. (1986). Guanine nucleotide-binding protein in sea urchin eggs serving as the specific substrate of islet-activating protein, pertussis toxin. *FEBS Lett.* **207,** 28–34.

Oosawa, Y., and Yamagishi, S. (1989). Rat brain glutamate receptors activate chloride channels in *Xenopus* oocytes coupled by inositol trisphosphate and Ca^{2+}. *J. Physiol. (London)* **408,** 223–232.

Oron, Y., Gillo, B., and Gershengorn, M. C. (1988). Differences in receptor-evoked membrane electrical responses in native and mRNA-injected *Xenopus* oocytes. *Proc. Natl. Acad. Sci. U.S.A.* **85,** 3820–3824.

Peres, A. (1986). Resting membrane potential and inward current properties of mouse ovarian oocytes and eggs. *Pfluegers Arch.* **407,** 534–540.

Peres, A. (1987). The calcium current of mouse egg measured in physiological calcium and temperature conditions. *J. Physiol. (London)* **391,** 573–588.

Peres, A., and Bernardini, G. (1985). The effective membrane capacity of *Xenopus* eggs: its relations with membrane conductance and cortical granule exocytosis. *Pfluegers Arch.* **404,** 266–272.

Peres, A., and Mancinelli, E. (1985). Sodium conductance and the activation potential in *Xenopus laevis* eggs. *Pfleugers Arch.* **405,** 29–36.

Powers, R. D. (1982). Changes in mouse oocyte membrane potential and permeability during meiotic maturation. *J. Exp. Zool.* **221,** 365–371.

Robbins, N., and Molenaar, P. C. (1981). Investigation of possible cholinergic mechanisms in fertilization of *Xenopus* eggs. *Proc. R. Soc. London, Ser. B* **213,** 59–72.

Robinson, K. R. (1979). Electrical currents through full-grown and maturing *Xenopus* oocytes. *Proc. Natl. Acad. Sci. U.S.A.* **76,** 837–841.

Schlichter, L. C. (1983a). Spontaneous action potentials produced by Na and Cl channels in maturing *Rana pipiens* oocytes. *Dev. Biol.* **98,** 47–59.

Schlichter, L. C. (1983b). A role for action potentials in maturing *Rana pipiens* oocytes. *Dev. Biol.* **98,** 60–69.

Schlichter, L. C. (1989). Ionic currents underlying the action potential of *Rana pipiens* oocytes. *Dev. Biol.* **134,** 59–71.

Schlichter, L. C., and Elinson, R. P. (1981). Electrical responses of immature and mature *Rana pipiens* oocytes to sperm and other activating stimuli. *Dev. Biol.* **83,** 33–41.

Shen, S. S., and Steinhardt, R. A. (1980). Intracellular pH controls the development of new potassium conductance after fertilization of the sea urchin egg. *Exp. Cell Res.* **125,** 56–61.

Shilling, F., Mandel, G., and Jaffe, L. A. (1990). Activation by serotonin of starfish eggs expressing the rat serotonin 1c receptor. *Cell Regul.* **1,** 465–469.

Simoncini, L., and Moody, W. J. (1990). Changes in voltage-dependent currents and membrane area during maturation of starfish oocytes: Species differences and similarities. *Dev. Biol.* **138,** 194–201.

Slack, B. E., Bell, J. E., and Benos, D. J. (1986). Inositol-1,4,5-trisphosphate injection mimics fertilization potentials in sea urchin eggs. *Am. J. Physiol.* **250,** C340–C344.

Smith, L. D. (1989). The induction of oocyte maturation: transmembrane signaling events and regulation of the cell cycle. *Development* **107,** 685–699.

Speksnijder, J. E., Corson, D. W., Sardet, C., and Jaffe, L. F. (1989). Free calcium pulses following fertilization in the ascidian egg. *Dev. Biol.* **135,** 182–190.

Steinhardt, R. A., and Epel, D. (1974). Activation of sea-urchin eggs by a calcium ionophore. *Proc. Natl. Acad. Sci. U.S.A.* **71,** 1915–1919.

Swann, K., Ciapa, B., and Whitaker, M. (1987). Cellular messengers and sea urchin egg activation. *In* "Molecular Biology of Invertebrate Development" (D. O'Connor, ed.), pp. 45–69. Alan R. Liss, New York.

Takahashi, K., and Yoshii, M. (1978). Effects of internal free calcium upon the sodium and calcium channels in the tunicate egg analysed by the internal perfusion technique. *J. Physiol.* (*London*) **279,** 519–549.

Takahashi, T., Neher, E., and Sakmann, B. (1987). Rat brain serotonin receptors in *Xenopus* oocytes are coupled by intracellular calcium to endogenous channels. *Proc. Natl. Acad. Sci. U.S.A.* **84,** 5063–5067.

Talevi, R. (1989). Polyspermic eggs in the anuran *Discoglossus pictus* develop normally. *Development* **105,** 343–349.

Talevi, R., Dale, B., and Campanella, C. (1985). Fertilization and activation potentials in *Discoglossus pictus* (Anura) eggs: A delayed response to activation by pricking. *Dev. Biol.* **111,** 316–323.

Turner, P. R., and Jaffe, L. A. (1989). G-proteins and the regulation of oocyte maturation and fertilization. *In* "The Cell Biology of Fertilization" (H. Schatten and G. Schatten, eds.), pp. 297–318. Academic Press, San Diego, California.

Turner, P. R., Jaffe, L. A., and Fein, A. (1986). Regulation of cortical vesicle exocytosis in sea urchin eggs by inositol 1,4,5-trisphosphate and GTP-binding protein. *J. Cell Biol.* **102,** 70–76.

Turner, P. R., Jaffe, L. A., and Primakoff, P. (1987). A cholera toxin-sensitive G-protein stimulates exocytosis in sea urchin eggs. *Dev. Biol.* **120,** 577–583.

Tyler, A., Monroy, A., Kao, C. Y., and Grundfest, H. (1956). Membrane potential and resistance of the starfish egg before and after fertilization. *Biol. Bull.* (*Woods Hole, Mass.*) **111,** 153–177.

Wallace, R. A., and Steinhardt, R. A. (1977). Maturation of Xenopus oocytes II. Observations on membrane potential. *Dev. Biol.* **57,** 305–316.

Webb, D. J. (1984). Ion conductance changes during amphibian early development. *Bioelectrochem. Bioenerg.* **13,** 429–438.

Webb, D. J., and Nuccitelli, R. (1985). Fertilization potential and electrical properties of the *Xenopus laevis* egg. *Dev. Biol.* **107,** 395–406.

Whitaker, M. (1989). Phosphoinositide second messengers in eggs and oocytes. *In* "Inositol Lipids in Cell Signalling" (R. H. Michell, A. H. Drummond, and C. P. Downes, eds.), pp. 459–483. Academic Press, London.

Whitaker, M. J., and Steinhardt, R. A. (1985). Ionic signaling in the sea urchin egg at fertilization. *In* "Biology of Fertilization" (C. B. Metz and A. Monroy, eds.), Vol. 3, pp. 167–221. Academic Press, Orlando, Florida.

Whitaker, M., Swann, K., and Crossley, I. (1989). What happens during the latent period at fertilization. *In* "Mechanisms of Egg Activation" (R. Nuccitelli, G. N. Cherr, and W. H. Clark, eds.), pp. 157–171. Plenum, New York.

Woodruff, R. I. (1989). Charge-dependent molecular movement through intercellular bridges in *Drosophila* follicles. *Biol. Bull.* (*Woods Hole, Mass.*) **176(S),** 71–78.

Woodruff, R. I., and Telfer, W. H. (1980). Electrophoresis of proteins in intercellular bridges. *Nature* (*London*) **286,** 84–86.

Woodward, R. M., and Miledi, R. (1987). Hormonal activation of ionic currents in follicle-enclosed *Xenopus* oocytes. *Proc. Natl. Acad. Sci. U.S.A.* **84,** 4135–4139.

Yoshida, S. (1985). Action potentials dependent on monovalent cations in developing mouse embryos. *Dev. Biol.* **110,** 200–206.

CHAPTER 4

Ion and Solute Transport in Preimplantation Mammalian Embryos

Douglas H. Robinson* and Dale J. Benos
Department of Physiology and Biophysics, University of Alabama at Birmingham, Birmingham, Alabama 35294

I. INTRODUCTION

The developing preimplantation mammalian embryo undergoes rapid transformation from a single cell to a system with multiple cell types during the variable (4 to about 30 days in the mouse and horse, respectively) preimplantation period. After fertilization, the embryo experiences a phase of cell division (3 days in the rabbit) with little or no increase in size to form the morula stage embryo. The outer cells of the morula then form tight junctions and, following this compaction event, blastulation occurs (Fig. 1). The mammalian blastocyst consists of fundamentally two differ-

*Present Address: Molecular Medicine, Beth Israel Hospital, Boston, Massachusetts 02215

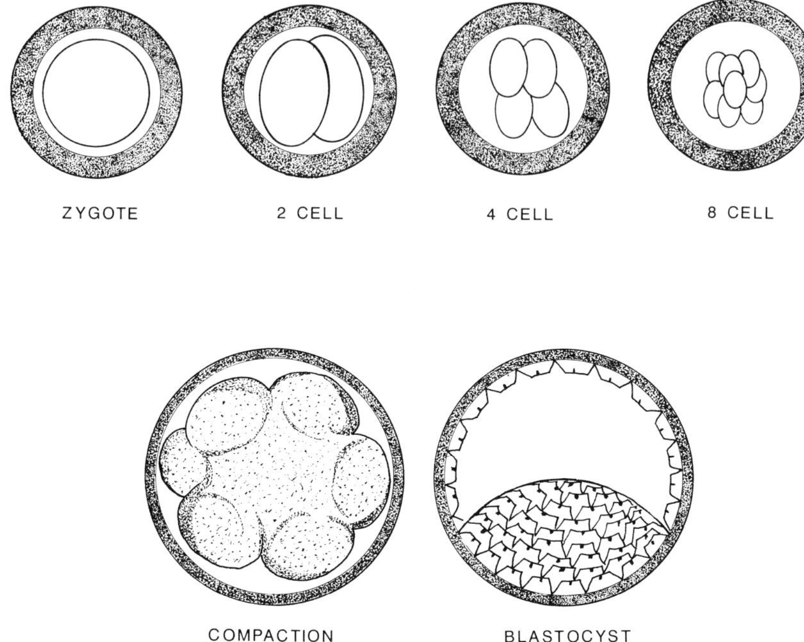

FIG. 1. General developmental sequence of the embryo during the preimplantation period. The cells of the embryo form tight junctions at the compaction event. After this time, vectorial ion transport begins, followed by expansion of the blastocoel.

ent cell types: those comprising the trophectoderm and those forming the inner cell mass (ICM). The trophectoderm, which forms the bulk of the preimplantation blastocyst, consists of cells that are squamate in appearance and, as will be seen, form a polarized epithelium. Following implantation, these cells differentiate to form the nonmaternal components of the placenta and other extrafetal tissues. The ICM is a small disk of nonpolar cells at one pole of the blastocyst which eventually forms the fetus. In the mouse and human, the ICM subtends about 20–50% of the trophectoderm, depending on gestational age, prior to implantation. The ICM in the rabbit, on the other hand, occupies only about 5% of the total trophectodermal surface at days 6 and 7 postcoitus (p.c.).

The degree of blastocyst expansion is clearly species dependent (Table I). In rodents and anthropods the embryo is minimally expanding; that is, the volume of the blastocoel (see Fig. 1) is small when compared with the total embryo size. For example, a fully expanded mouse embryo contains about 400 pl of fluid in the blastocoel (Dickson, 1966). In contrast, lago-

TABLE I

Mammals with Maximally or Minimally Expanding Embryos

Maximally expanding	Minimally expanding
Rabbit	Mouse
Pig	Rat
Cow	Hamster
Horse	Human
	Primate

morphs and ungulates have embryos that are maximally expanding. The rabbit blastocoel contains approximately 70 μl of fluid at day 7 (Biggers *et al.*, 1988), while the volume of the pig blastocoel may be in excess of 10 ml at day 10 p.c. (Overstrom, 1987). Thus, the blastocoel fluid in maximally expanding embryos constitutes a major fraction of the blastocyst's total volume.

The size differences of the commonly studied rabbit and mouse embryos promote their usefulness in experiments using different methodologies. The mouse follows a very regular developmental pattern during the early cleavage divisions (2-cell, 4-cell, etc.) and so transport-associated events at a synchronous stage can be observed. Additionally, when superovulated, a single mouse can produce 20–40 viable embryos. The mouse becomes disadvantageous in transport studies at the blastocyst stage because it is difficult to determine unequivocally the developmental age of the blastocysts and, because of the small size, the blastocoel is not easily accessible.

The rabbit has at least two advantages with respect to transport studies when compared to other species. First, the rabbit is a reflex ovulator, that is, the doe will ovulate between 10 and 11 hr after copulation, thus allowing for experiments on blastocysts of a similar developmental age. Second, the large size of the blastocyst allows easy access to the blastocoel, and large amounts of fluid can be collected from this extracellular compartment. Influx measurements can be easily made and injections of tracer materials or inhibitors into the blastocoel can be done to measure either efflux rates or characteristics of trophectodermal membrane transport.

The development of trophectodermal polarity during embryogenesis is associated with the initiation of vectorial transport of various ions and metabolic substrates. Transport of substrates such as amino acids and sugars has been studied in different embryo types, and occurs by both ion-dependent and independent mechanisms. Metabolic substrates that accumulate within embryonic cells are presumably utilized by the blastocyst to fulfill its energy requirements. Ion transport may not only be associated with regulation of cell division (O'Brien and Prettyman, 1987), but may also be involved with formation of blastocoel fluid and the subsequent increase in blastocoelic volume (Borland *et al.*, 1976; Biggers *et al.*, 1978).

In addition to substrate and ion transport, it is increasingly evident that the preimplantation embryo acquires mechanisms that are involved in the transport of large molecular substances (for example, proteins) from the uterine lumen into the blastocoel. The purpose of these mechanisms is largely unknown. Possible functions of this system may include providing the ICM with growth-regulatory hormones, and/or providing colloids necessary to generate an osmotic pressure within the blastocoel.

This review examines the transport mechanisms that are known to exist in developing mammalian embryos and the potential roles they may play in embryogenesis.

II. ION TRANSPORT

A. *Epithelial Differentiation*

The trophectoderm of the mammalian embryo functions as an epithelium and, as such, shares a number of characteristics common to other epithelia (e.g., toad bladder, nephron segments). However, during the preblastulation stages, the embryonic cells are not in an epithelial arrangement and must first polarize into an epithelial-like structure.

The first evidence of polarization is the compaction event. In the mouse embryo, compaction occurs at the 8-cell stage. During this process, numerous cell stuctures, such as microvilli and actin filaments, become localized to distinct regions of the cell (Ducibella and Anderson, 1975; Lehtonen and Bradley, 1980). Ziomek and Johnson (1980), examining concanavalin A (Con A) binding, sought to determine whether polarization was independent of cell-to-cell contact. They found that individual blastomeres from the 4-cell or 8-cell stage did not polarize *in vitro,* whereas when these cells were incubated in culture as doublets, polarization occurred. This requirement for cell-to-cell contact suggests that polarization events in the preimplantation embryo share some common characteristics with the polarization events examined in cell culture systems (reviewed in Rodriguez-Boulan and Nelson, 1989).

Early studies by Cross (1973) noted that the trophectoderm of the day-6 p.c. rabbit embryo had a resistance of about 2 k-Ω cm^2, and thus could be considered to be a tight epithelium. [Tight epithelia are classified as those having a transepithelial resistance greater than 300 Ω cm^2 (Macknight *et al.*, 1980).] Benos (1981a) extended these physiological studies by examining the lanthanum permeability in day-4 p.c. through day-6 p.c. rabbit embryos and correlated this permeability with the transtrophectodermal resistance. He found that lanthanum influx was high on day 4 p.c. but not at day 5 p.c., which implies that functional tight junctions are forming at this time. The transepithelial electrical resistance was low at day 4 (8.4 Ω cm^2), but by day 6 p.c. the resistance rose to 1.8 k-Ω cm^2.

Processes of epithelial differentiation have also been examined using microscopic techniques. Ducibella *et al.* (1975) found the early mouse blastocyst had lanthanum-impeding tight junctions. They also observed gap junctions between trophectodermal cells. The presence of gap junctions suggests that trophectodermal cells are electrically coupled. Similar findings were also found in the rabbit embryo in the same study. Magnuson *et al.* (1977) found evidence for macula occludens and gap junction formation between the 8- and 16-cell stage of mouse embryogenesis. Those embryos at the 16-cell stage, in general, had a greater level of complexity of cell junctions. As embryogenesis proceeded to the morula stage, zona occludens were found between presumptive trophectoderm cells with between 2 and 8 ridges per groove.[1] At the blastocyst stage, macula adherens were noted below the zona occludens. In addition, gap junctions were still found between trophectodermal cells. These data suggest that formation of trophectoderm junctional complexes is stage specific and occurs concomitantly with the electrical differentiation of the epithelium.

Formation of rabbit trophectodermal junctional complexes was studied by Hastings and Enders (1974b). The day-5 blastocyst had 2 to 3 ridges per zona occludens lattice and the day-6 p.c. embryo had 5 to 6 ridges per lattice. Evidently, 2 to 3 ridges are sufficient to produce characteristics of an electrically tight epithelium because the rabbit embryo expresses permeability properties of a tight epithelium at this time. In addition, Hastings and Enders (1974b) also observed that the day-5 and day-6 p.c. embryos had gap junction-like intramembranous particles, implying that the trophectoderm functions electrically as a syncytial unit.

[1] It is generally accepted that the number of strands at the tight junction is indicative of the electrical resistance of the epithelium. For example, epithelia with electrically leaky tight junctions, such as the proximal tubule, usually have only a single continuous strand in the area of the zona occludens. The toad urinary bladder, which is electrically very tight, can have up to eight strands at the tight junction (reviewed in Gumbiner, 1987).

B. Na^+/K^+-ATPase

Na^+,K^+-ATPase functions as a sodium and potassium exchanger which, under normal cellular conditions, translocates 3 Na^+ from the interior of the cell to the exterior, while 2 K^+ are moved into the cell from the external environment. As implied by the name, this protein hydrolyzes one ATP molecule per cycle and is thus thought to be the ultimate event in most energy-dependent transport processes. In epithelial tissues engaged in vectorial transport of ions and water, this enzyme is preferentially localized in high concentration within the basolateral membrane of transporting cells. In the rabbit blastocyst, tritiated ouabain binding and autoradiographic studies have demonstrated that this enzyme is localized to the basolateral (or juxtacoelic) border of the trophectoderm cell (Benos *et al.*, 1985; Robinson *et al.*, submitted). Immunofluorescence studies in the mouse embryo using an antibody directed against Na^+/K^+-ATPase subunits indicated that the enzyme was located in the basolateral membrane of the trophectodermal cell at the time of cavitation (Watson and Kidder, 1988). Interestingly, the enzyme was found within the cytoplasm of the late mouse morula (Watson and Kidder, 1988) and was localized to the cell membrane of the late morula in the rabbit (Robinson *et al.*, submitted).

Efforts to quantify Na^+ pump number have been performed utilizing different methodologies. Benos (1981b), in experiments measuring tritiated ouabain binding to rabbit embryos, found that there was a large increase in the number of ouabain binding sites between days 4 and 5 p.c., followed by a further increase between days 6 and 7 p.c. The author speculated, and later confirmed (Benos *et al.*, 1985), that this increase in binding between days 6 and 7 p.c. might be a result of ouabain binding to enzyme sites located on cells which formed the primitive endoderm and thus, were not involved in ion transport processes directly related to blastocyst expansion.

A recent study examined the rate of Na^+/K^+-ATPase synthesis by quantitating [^{35}S]methionine incorporation into the enzyme during an *in vitro* incubation period. A 90-fold increase in enzyme synthesis occurred between days 4 and 5 p.c., while there was little change between days 5 and 7 p.c. (Overstrom *et al.*, 1989). This increase in enzyme sythesis is likely a direct result of an increase in messenger number because Gardiner *et al.* (1990) found a large increase in the levels of mRNA coding for the α-subunit of Na^+/K^+-ATPase between days 4 and 5 p.c. This finding correlated with recent work that measured ouabain binding on a per-cell basis using tritiated ouabain autoradiography (Robinson *et al.*, submitted). Autoradiographic techniques have the advantage that quantitation is performed only on those cells of interest (i.e., trophectodermal cells). The autoradiographic data also indicated that a large increase in pump number occurred between days 4 and 5 p.c., with no significant change during the remainder of the preimplantation period. Interestingly, at days 6 and 7 p.c., the heaviest grain density was found in the intercellular space between the trophectodermal cells (Robinson *et al.*, submitted). Thus, the pattern of enzyme synthesis correlated well with the expression of the protein on the basolateral membrane.

If the Na^+ pump is involved in the processes of blastocyst formation and expansion, it follows that ouabain should inhibit expansion of the embryo. In fact, ouabain (10^{-3} *M*) inhibited reexpansion of cytochalasin B collapsed mouse blastocysts (DiZio and Tasca, 1977), and 10^{-4} *M* ouabain inhibited cavitation in the mouse embryo (Wiley, 1984). Curiously, in the same study, a 10-fold lower concentration of ouabain accelerated cavitation. An adequate explanation for this phenomenon is not apparent. An active Na^+ pump also seems to be required for the rabbit embryo to accumulate fluid within the blastocoel. Biggers *et al.* (1978) found that when ouabain was injected into the blastocoel of the day-6 p.c. rabbit embryo, expansion of the embryo was inhibited. External exposure to ouabain did not inhibit expansion (Biggers *et al.*, 1978), nor did ouabain bind to external sites on the day-6 p.c. rabbit embryo (Benos, 1981b).

C. Ion Channels

Our understanding of transport mechanisms in the developing embryo comes primarily from studies done in the rabbit because, as stated earlier, the embryo's large size facilitates transport measurements. The earliest electrophysiological studies were done by Cross and co-workers (Cross and Brinster, 1969, 1970; Cross, 1974), in which they examined the voltage across the trophectoderm by inserting glass microelectrodes into the blastocoel. Notably, the potential of the blastocoel was negative with respect to the outer solution at days 5 and 6 p.c. (Cross and Brinster, 1969). At day 7 p.c., the magnitude of the transtrophectodermal potential difference decreased. Powers *et al.* (1977) confirmed this finding regarding the day-6 embryo and, in addition, found that the potential difference became positive by 21 mV at day 7 p.c. These authors reported that this positive potential could be reversed to −3 mV by application of 10^{-5} *M* amiloride, a compound which is an inhibitor of apical Na^+ channels at this concentration (Benos, 1982). Amiloride had no effect on the potential difference at day 6 p.c. This finding suggested that an amiloride-sensitive Na^+ channel appeared in the apical membrane of the rabbit embryo between days 6 and 7 p.c. Amiloride does, however, have other effects, one of which is to act as a weak base (Benos *et al.*, 1983; Dubinsky and Frizzell, 1983). Thus, this change in potential may be a result of changing the cellular pH, and not a direct effect of amiloride on a channel.

Efforts to confirm this effect of amiloride through $^{22}Na^+$ influx studies have yielded conflicting results. Benos *et al.* (1985) found that there was a large increase in Na^+ influx between days 6 and 7 p.c. and that this influx at day 7 p.c. became amiloride inhibitable. Benos (1981b) also observed that 10^{-4} *m* amiloride inhibited the rate of ouabain binding in day-7 but not day-6 p.c. embryos. This finding implied that the rate of Na^+ entry decreased because the Na^+ pump can only bind ouabain in intact cells when it is actively transporting Na^+ and K^+ (Skou, 1975). It should be noted, however, that amiloride at this high concentration also inhibits both the Na^+ pump and the Na^+/H^+ exchanger (Benos, 1982; Soltoff and Mandel, 1983). These effects could also decrease, either directly or indirectly via changes in Na^+ entry or intracellular pH, the rate of ouabain binding to the Na^+ pump.

Nielsen *et al.* (1987) found that Na^+ influx changed little between days 6 and 7 p.c., and amiloride (10^{-4} *M*) had no effect on the influx. Despite the fact that the rabbit is a reflex ovulator and will ovulate within 10.5 hr of copulation, there may be a considerable variation in the functional expression of the Na^+ channel at day 7 p.c., thereby obscuring development of amiloride sensitivity. On the other hand, there may be other ion-transporting systems that obscure amiloride inhibition of Na^+ influx, or the embryo may have to be in a time-dependent or hormonally stimulated state in order to express Na^+ channel activity.

The pig blastocyst also possesses a ouabain inhibitable $^{22}Na^+$ influx during its period of expansion from days 7 to 10 p.c. (Overstrom *et al.*, 1984; Overstrom, 1987). On a per cm basis, the influx of Na^+ in the pig blastocyst from days 7 to 9 p.c. is comparable to that of the rabbit blastocyst. In day-10 p.c. blastocysts, however, the Na^+ influx increases nearly 5-fold. Interestingly, amiloride at 10^{-4} *M* inhibits the Na^+ influx only at day 9 p.c. The exact characteristics of this Na^+ entry pathway remain to be elucidated.

We (Robinson *et al.*, 1991) have recently examined the question of Na^+ channel development in the day-6 and day-7 p.c. rabbit embryo using whole-cell patch-clamp techniques and immunogold labeling of ultrathin sections with an antibody raised against epithelial apical sodium channels (Sorscher *et al.*, 1988). The immunogold studies localized the channel in day-7 p.c. embryos but not day-6 p.c. embryos to the apical membrane, and no gold particles were observed at the basolateral aspect of the cell (Fig. 2). Sodium currents were observed in both the day-6 and day-7 p.c. trophectoderm cell. Evidently, the current was carried through a similar

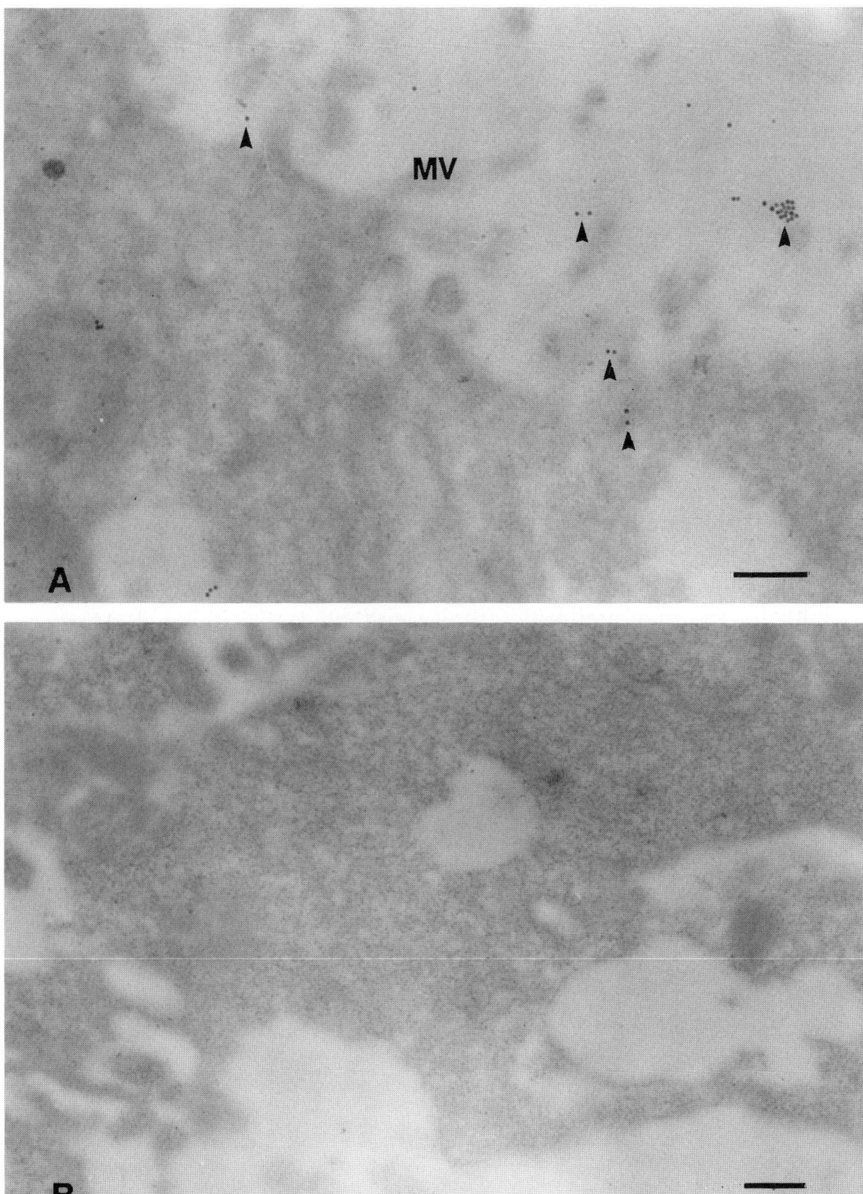

FIG. 2. Labeling of a day-7 p.c. rabbit trophectodermal cell with a polyclonal antibody raised against the apical epithelial sodium channel. Binding of the antibody is localized to only the apical membrane of the trophectodermal cell (A), with no binding seen along the basolateral membrane (B). At day 6 p.c., there is no observable binding. MV, Microvilli. Bar, 0.25 μm. (Micrograph courtesy of Dr. Peter Smith.)

pathway at these stages because the current–voltage relationships at days 6 and 7 p.c. were superimposable. However, an amiloride inhibition was observed only in the day-7 p.c. embryo (Fig. 3). The K_i for this inhibition was 10 μM, which indicates that this channel is similar to a low-amiloride-affinity channel that has recently been described in a number of tissues (Moran *et al.*, 1988; Bridges *et al.*, 1989). Thus, it appears that the majority of the Na^+ influx is carried through a similar channel at days 6 and 7 p.c., with the difference being the relative expression of an amiloride-sensitive component at day 7 p.c.

A recent article (Manejwala *et al.*, 1989) examined Na^+ transport in the mouse embryo. The authors conclude, on the basis of inhibitor studies, that most of the Na^+ uptake at the apical membrane occurs via a $Na^+/H+$ exchanger, and Cl^- follows passively through cell junctions. However, they do not rule out the possibility that a low-amiloride-affinity channel is responsible for the uptake of Na^+. At the highest inhibitor concentration used (50 μM), ethylisopropyl amiloride can have a greater effect on the low-affinity Na^+ channel than amiloride derivatives designed to inhibit specifically the high-amiloride-affinity epithelial Na^+ channel (Moran *et al.*, 1988). Thus, more experiments are needed to determine if the mouse and rabbit have similar apical Na^+-transport systems.

The ouabain-inhibitable Na^+ influx in the rabbit blastocyst at day 7 p.c. decreases, while paradoxically the number of ouabain binding sites remains constant (Benos *et al.*, 1985). There cannot be an uncoupling of the Na^+ pump from Na^+ influx because, in order for ouabain to bind to the

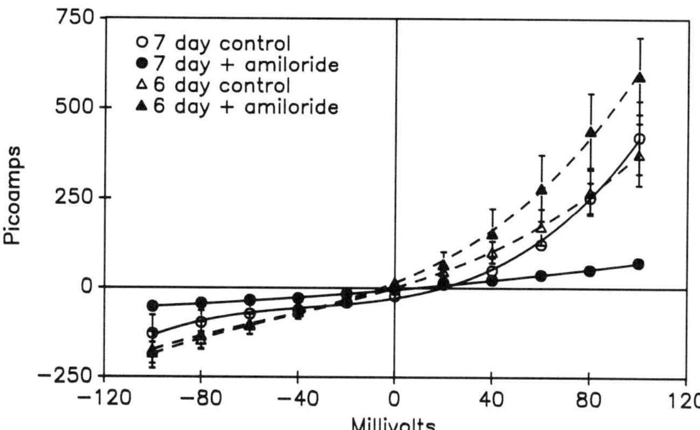

FIG. 3. Current–voltage plot of sodium currents in whole-cell patch–clamped days 6 and 7 p.c. rabbit trophectodermal cells. The current–voltage relationship in control conditions (150 m*M* sodium glutamate in both the pipette and bath) is similar at days 6 and 7 p.c. However, amiloride at 10^{-4} *M* inhibits the current only at day 7 p.c.

pump, the pump must be actively removing Na^+ from the cell and accumulating K^+ (Skou, 1975). Because tritiated ouabain binds as avidly at day 7 as at day 6 p.c. (Benos, 1981b; Robinson *et al.*, submitted), the pump-uncoupling hypothesis is not tenable. Tritiated ouabain binding studies also indicate that the pump at day 7 p.c. is basolaterally located (Robinson *et al.*, submitted). Thus, any Na^+ that crosses the apical membrane could only be actively removed from the cell at the juxtacoelic membrane. Clearly, there must be another explanation for this Na^+/K^+-ATPase-independent Na^+ influx.

The work of Cross and co-workers (Cross and Brinster, 1969, 1970; Cross, 1973, 1974) strongly suggested that bicarbonate ion was actively transported inwardly across the rabbit embryo's trophectoderm before day 7 p.c. In addition, they also found that the carbonic anhydrase inhibitor acetazolamide had no effect on the putative bicarbonate flux. However, carbonic anhydrase may not be an absolute requirement, as the spontaneous generation of bicarbonate and its subsequent transport into the blastocoel would be more than sufficient to create a blastocoel negative potential. As depicted in Fig. 4, the bicarbonate exit could be accomplished by a sodium–bicarbonate cotransport system similar to that recently described in the renal proximal tubule and kidney cell lines (Jentsch *et al.*, 1986; Alpern and Chambers, 1986). This system appears to transport three bicarbonate ions for every sodium ion, and relies on the bicarbonate electrochemical gradient. At day 7 p.c., this system may be present but not operative, thus resulting in a Na^+-dependent, transtrophectodermal positive potential. Following ouabain treatment, there would still be a pathway for Na^+ movement across the basolateral membrane. Thus, a normally inactive pathway could account for the conflicting data concerning Na^+ flux and ouabain at day 7 p.c. (Benos *et al.*, 1985).

As previously stated, Cross (1973) demonstrated that the bicarbonate ion is the major anionic species that is transported across the trophectoderm of the rabbit blastocyst, and it seems likely that transport of this ion accounts for the negative transtrophectodermal potential observed at days 5 and 6 p.c. However, these data are inferential as it is not possible to examine the influx of bicarbonate directly since it is a labile compound. A later study (Cross, 1974) examined the bicarbonate concentration in the blastocoel and found that the concentration was about 59 m*M*, which is much higher than that found in the uterus. Additionally, Cross also found that there was a net accumulation of bicarbonate over time. Based on these data, it seems likely that bicarbonate flux is coupled to an active process and responsible for the observed negative transtrophectodermal potential.

It is also evident that other ions in addition to bicarbonate are transported across at least the apical membrane. Benos and Biggers (1983)

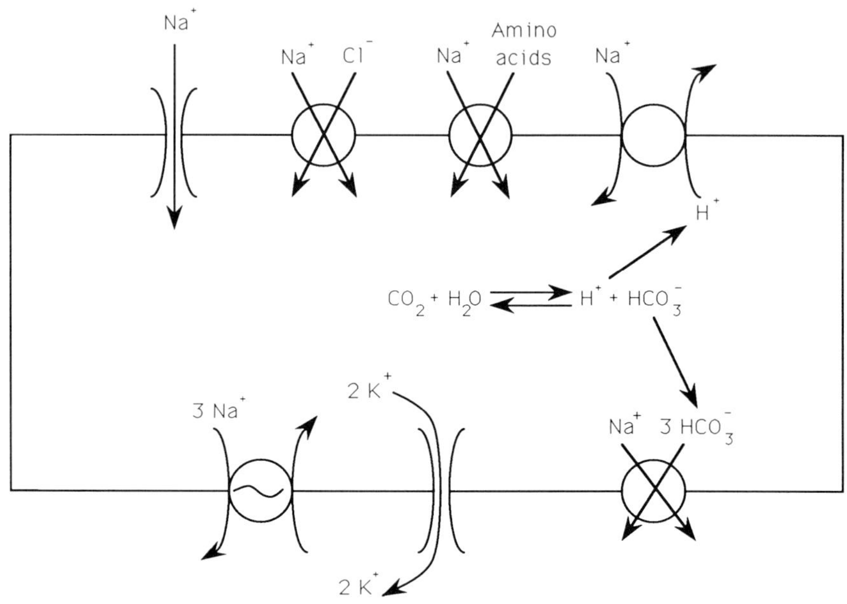

FIG. 4. Proposed model for ion transport in the rabbit trophectodermal cell. The transport systems at the utering-facing membrane are known to exist. We speculate that there is a $Na^+/HCO^-{}_3$ cotransport system at the basolateral membrane which accounts for the observed accumulation of $HCO^-{}_3$ in the blastocoel.

noted that the embryo at day 6 to 6.5 p.c. displayed a Cl^- flux that was sensitive to furosemide. The flux of Cl^- and Na^+ also seemed to be interdependent: removal of one ion affects the influx of the other. Sodium influx was also affected by furosemide: neither Cl^- nor Na^+ influx was affected by removal of bicarbonate or by treatment with the disulfonic stilbene DIDS (4,4′-diisothiocyano-2,2′-stilbenedisulfonic acid). Because potassium removal exerted no effect on Na^+ or Cl^- fluxes, the authors suggested that apical membranes of rabbit trophectoderm cells possess a Na^+/Cl^- cotransport mechanism. In contrast, the Cl − influx in the mouse blastocyst appears to move passively across the intercellular junctions (Manejwala *et al.*, 1989).

In addition to the cotransporter, data suggest other Cl^- and/or Na^+ entry pathways are present in the membrane of the day-6 p.c. rabbit embryo. Chloride influx could be inhibited by the prostaglandins PGE_1,

PGE_2, and $PGF_{2\alpha}$, while the Na^+ influx was inhibited by only PGE_1 (Benos and Biggers, 1983). The nature of these pathways has yet to be determined.

D. *Intracellular Ion Concentrations*

In all transporting epithelia, it is important to know the intracellular concentrations of different ions because the dynamics of ion entry into or exit from cellular compartments are in part dependent on the cellular concentration of the ion in question. Powers and Tupper (1977) reported that the internal K^+ and Na^+ concentrations in blastomeres of the 2-cell mouse embryo were 130 and 151 m*M*, respectively. In addition, the measured intracellular resting potential was found to be −19 mV. This Na^+ concentration is obviously much higher than that normally found in cells (10–30 m*M*; Macknight, 1980). The most likely explanation is a methodological error. Chemical methods are notorious for overestimating intracellular Na^+ concentrations because any adherent extracellular fluid has a much higher Na^+ concentration than the cellular compartment (Macknight, 1980). In the experiments of Powers and Tupper (1977), embryonic Na^+ was measured with a flame photometer after washing the embryos 5 to 6 times in a Na^+/K^+-free medium. No fluid-phase markers were employed to correct for the contribution of Na^+ from any adherent fluid. The fact that a plausible intracellular K^+ concentration was determined argues for this probability of a methodological error because extracellular fluid has a low K^+ concentration when compared to cell K^+ concentrations. This extracellular K^+ will therefore not interfere with chemical methods used for determining intracellular K^+ concentrations.

A more recent study by Lee (1987), using ion-sensitive microelectrodes, found that the intracellular Na^+ concentration was about 20 m*M*, while the K^+ concentration was 100 m*M* in the 1- to 8-cell stage mouse embryo. In addition, the resting potential was dynamic during development. As embryogenesis progressed from the 1-cell to the 8-cell stage, the cell membrane potential depolarized from −40 to −25 mV. Though this finding suggests an increase in Na^+ and/or Cl^- permeability or a decrease in K^+ permeability, the associated changes in membrane conductance have not been determined.

The thin nature (about 1 μm) of the mammalian trophectodermal cell renders it unsuitable for study with microelectrodes. Thus, other means of measuring ion content and potential are necessary. We have attempted to determine the ion concentrations in the rabbit trophectodermal cell with electron probe microanalysis, and have found reasonable values for Na^+ (22 m*M*), K^+ (138 m*M*), and Cl^- (33 m*M*) in the day-6 and day-7 p.c. rabbit embryo (D. H. Robinson, R. Rick, and D. J. Benos, unpublished observations). These values are suspect, however, because the embryos must be collapsed before freezing in order to render them amenable for cryosectioning (methods described in Rick *et al.,* 1979). Collapsing may cause rapid shifts in ion concentration that cannot be detected by this method. There are now sodium-sensitive dyes commercially available which undoubtedly will prove to be more useful, because experiments could then be done on intact embryos. Additionally, the rate of change in ion content during and after various experimental treatments could be examined.

III. FLUID TRANSPORT

The putative function of the basolateral Na^+ pump, besides maintaining cellular Na^+ and K^+ concentrations, is to transport a net quantity of ions from the uterine lumen to the blastocoel, resulting in the generation of an osmotic pressure gradient between the external environment and the blastocoel. The gradient then provides the driving force for movement of water from the uterine lumen into the blastocoel, with subsequent expansion of the embryo. Numerous studies have investigated the relationship between outer fluid osmolalities and blastocoelic ion content. An early investigation (Gamow and Daniel, 1970) noted that rabbit blastocoel fluid was hyperosmotic to the uterine fluid by anywhere from 10 to 80 m0sm. The lower value seems more probable because in 24-hr cultured day-6 p.c. rabbit embryos, the blastocoel fluid is only about 8 m0sm hypertonic to the culture medium (Borland *et al.,* 1977). Expansion of the blastocyst also displayed an energy dependence. The energy dependence of this system, when combined with the fact that expansion was also decreased by removal of external Na^+ (Gamow and Daniel, 1970), suggested that active Na^+ transport mechanisms were required for embryonic expansion to occur.

Borland *et al.* (1976) also reported that when sucrose was added to the external medium, the rate of blastocoelic volume increase was attenuated but not fully inhibited. Evidently, during exposure to sucrose, the embryo compensates for the increase in external osmolality by increasing the rate of ion transport. In support of this argument, the concentration of all the ions in the blastocoel increased following sucrose exposure (Borland *et al.,* 1976). The cellular mechanisms that cause this increase in ion transport rate are unknown. Thusfar, stimulation of ion transport in the rabbit embryo using a number of hormones (prostaglandins, or cyclic AMP, for example) has not been demonstrated (Benos and Biggers, 1983).

Mechanisms to increase fluid accumulation (and presumably ion transport) may, however, exist. Manejwala *et al.* (1986) found that treatment with cholera toxin, cAMP, or the adenylate cyclase activator forskolin caused an increase in the expansion rate of the mouse embryo. Ion transport studies in the mouse blastocyst correlated the adenylate cyclase-stimulated increase in fluid transport with increases in $^{22}Na^+$ influx (Manejwala and Schultz, 1989). In the day-6 p.c. rabbit embryo, however, Benos and Biggers (1983) found that both Na^+ and Cl^- fluxes were inhibited by cAMP.

There is an increase in the fluid transport rate in the rabbit blastocyst between days 4 and 5 p.c., followed by a nearly 2-fold increase in the water accumulation rate between days 6 and 7 p.c. (Borland *et al.,* 1976). The early increase in water transport between days 4 and 5 p.c. correlates well with the increase in Na^+ pumps (Benos, 1981b; Overstrom *et al.,* 1989; Robinson *et al.,* submitted). However, there is no increase in active Na^+ pumps between days 6 and 7 p.c. (Robinson *et al.,* submitted). This paradox may be explained by a greater efficiency in ion-coupled water transport processes. As previously stated, trophectodermal Na^+ pumps, starting at day 6 p.c., become localized to the lateral domain of the basal plasma membrane. The positioning of the pumps to this region of the basolateral membrane may result in a more efficient mechanism to transport water through junctional complexes at the intercellular cleft. This arrangement would function similarly to the Diamond and Bossert (1967) model for standing gradient osmotic water flow. However, due to the very short intercellular spaces between cells, the flow would not be isoosmotic. In fact, Borland *et al.* (1977) presented data that suggested that fluid transport is always slightly hypertonic no matter what the external osmolality.

IV. SUGAR TRANSPORT

The isolation of the inner cell mass (ICM) from the uterine environment necessitates the existence of transport systems in the trophectodermal cell that will transport metabolites from the uterine lumen to the blastocoelic compartment or to cells of the ICM directly. Glucose would appear to be the ideal candidate for a metabolic carbon source, as it is present in the uterine lumen of the rabbit at a concentration of about 1 m*M* (Lutwak-Mann, 1954, 1962). Glucose can serve not only as an energy-producing catabolic substrate, but also as a starting point for a number of anabolic reactions. It is, therefore, of interest to determine the mechanism by which glucose enters the blastocoel.

There are two fundamentally different mechanisms by which glucose can cross a cellular membrane. The first, which is found to exist at the apical border of leaky epithelia, such as renal proximal tubules and enterocytes, consists of a Na^+-dependent glucose transport system. This system, as implied, requires Na^+ as a cotransported moiety in addition to glucose. Phlorizin is a specific inhibitor, and generally this system has a high specificity for the hexose analog 4-α-methylglucoside (Esposito, 1984; Ulrich, 1979). The other sugar transport system consists of a facilitated transporter that operates independently of Na^+. This system is located in the plasma membrane of adipose, brain, and liver cells, and also on the basolateral border of the proximal tubule cells and small intestine enterocytes. A characteristic of Na^+-independent glucose transport is inhibition by cytochalasin B, dideoxyforskolin, and phloretin, a deglycosylated phlorizin analog. This particular transport system also has a high specificity for 2-deoxy-D-glucose (Kinne *et al.*, 1975; Esposito, 1984).

Within the family of Na^+-independent hexose transport systems, there exist a number of different transport proteins, all of which exhibit 55–75% peptide sequence homology with each other (Fukumoto *et al.*, 1989). Differences between brain and adipose glucose transport systems may be related to the differences in regulation: the brain system is constitutively regulated, whereas the adipose system is subject to up- or down-regulation by hormones (e.g., insulin).

In an early study on glucose utilization, Fridhandler (1961) observed that the rabbit embryo will produce CO_2 from both glucose and fructose. In addition, the generation of CO_2 from glucose could be inhibited by the addition of 2-deoxy-D-glucose. The major pathway of glucose oxidation was found to consist of the hexose monophosphate shunt pathway in the morula stage, followed by the glycolytic pathway after blastulation. However, as a caveat to this study, these experiments were performed on blastocysts that were collapsed; this treatment has been noted to decrease O_2 consumption in the rabbit blastocyst (Benos and Balaban, 1980).

Metabolic studies on the mouse embryo (Brinster, 1967) showed that CO_2 production from glucose increased 100-fold during the first 5 days of development. Clearly, there must be some mechanism in the embryo that allows for glucose uptake across the apical membrane. The relevant questions concerning this system, particularly hormonal influences during development, were not studied until Perinchief (1980) investigated hexose transport in day-5 p.c. rabbit embryos. Perinchief reported that 3-*O*-methyl glucose (a glucose analog that is generally transported by both the Na^+-dependent and -independent systems) is transported by a system that competes with 2-deoxy-D-glucose and is inhibited by cytochalasin B and phloretin. However, details of this work were not reported and transtrophectodermal fluxes were not examined. A similar system may also be present in the apical membrane of the mouse embryo (Dabich and Acey, 1982). These findings indicated that mouse blastocysts have a hexose transport system that is phloretin inhibitable. This system is Na^+ independent, and the glucose uptake on a per-embryo basis increases 4-fold between the 8-cell and blastocyst stage (Gardner and Leese, 1988). In contrast to these earlier findings, Manejwala *et al.* (1989) found that about 50% of the transport of deoxyglucose was Na^+ dependent. Further experiments are warranted to explain this discrepancy.

A recent report by Wiley and Obasaju (1989) found that phlorizin, in the absence of glucose, inhibited cavitation in the mouse embryo at 10^{-4} *M* and accelerated cavitation at 10^{-6} *M*. In addition, the authors found that phlorizin also inhibited blastomere polarization. They speculated that this effect of phlorizin was a result of an interaction with a Na^+/glucose transport protein that has the ability to transport Na^+ only when glucose is absent. However, the lack of an appropriate control (phloretin) makes these data difficult to interpret.

Transtrophectodermal transport of glucose in the rabbit embryo has only recently been addressed (Robinson *et al.*, 1990). 3-*O*-Methyl glucose traversed the trophectoderm and this movement was inhibited by phloretin, cytochalasin B, and dideoxyforskolin and was independent of Na^+. Phlorizin was without effect. These data strongly suggest the presence of a Na^+-independent system at the apical membrane. Moreover, a 55-kDa protein is detected by Western blot analysis when a rabbit blastocyst lysate is probed with a polyclonal antibody raised against either the human erythrocyte glucose transport protein or with an antibody raised against an oligopeptide homologous to the C-terminus of the rat brain glucose transport protein. Both of these glucose transporters are Na^+ independent and they share a high degree of antibody cross-reactivity. The blastocyst transport protein was localized with immunogold staining to both the apical and basolateral membranes (Fig. 5). On the basis of grain density, it appears that the apical membrane is rate-limiting for transtrophectodermal transport of hexoses. However, without direct knowledge of the kinetics of the basolateral glucose transport system, this contention is speculative. Treatment with hormones (PGE_2, $PGF_{2\alpha}$, insulin, progesterone, and cAMP) did not affect the rate of hexose transport. This lack of hormonal modulation, combined with the antibody localization studies, implies that glucose transport in rabbit blastocysts is mediated by a brain-type, Na^+-independent transport system that is constitutively expressed. The mouse embryo may also have a brain-type glucose transport protein because insulin was also found to have no effect on hexose uptake (Gardner and Leese, 1988).

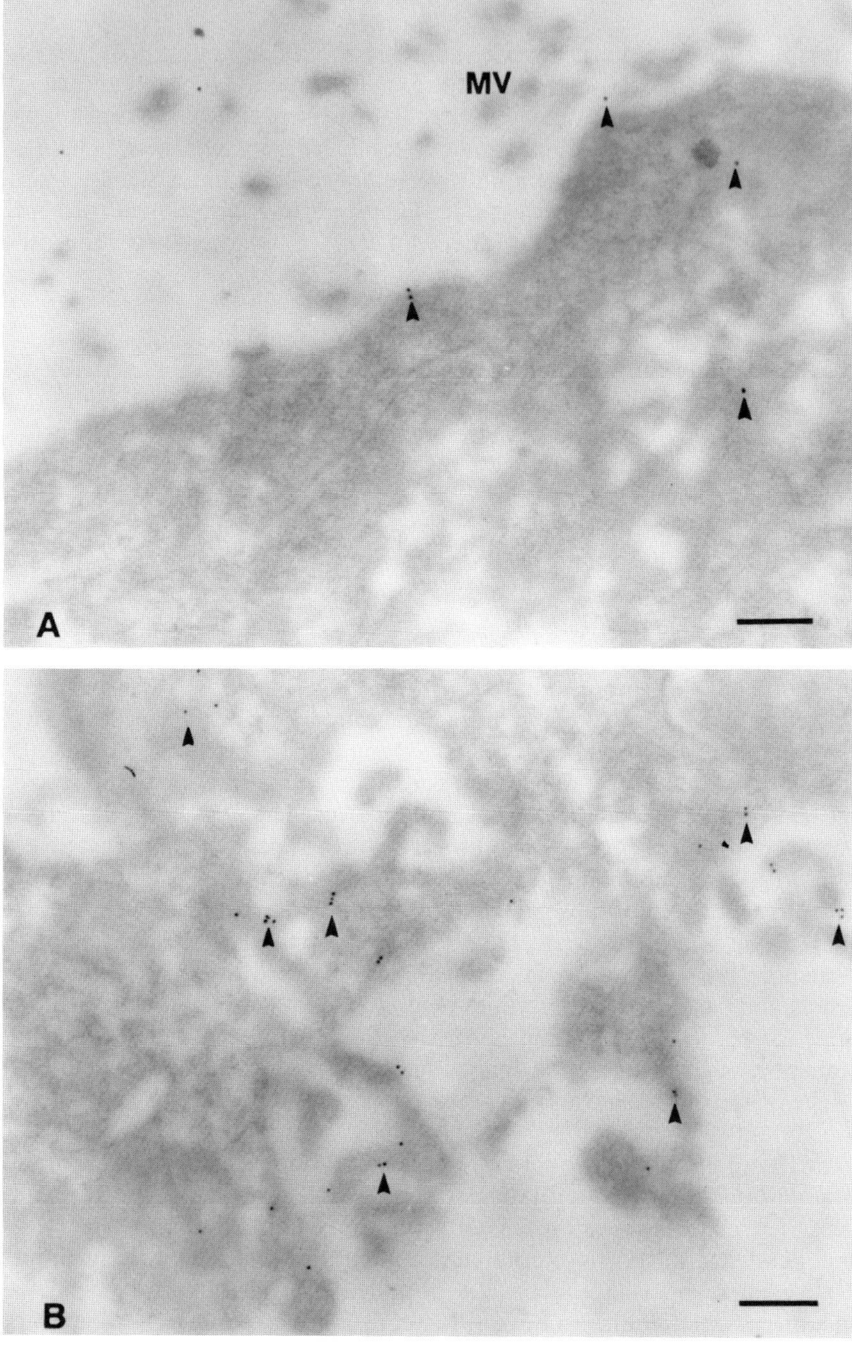
MV
A
B

It is interesting that, in Fridhandler's (1961) early study, fructose could be used by the blastocyst as a glycolytic substrate, implying that there is a separate fructose transport system. This inference arises from the data of Robinson *et al.* (1990), in which they demonstrate that glucose and fructose do not share a common transporter. The presence of a specific fructose transport system is not without precedent; a Na^+-independent fructose transport system has been observed in the rabbit small intestine (Schultz and Streeker, 1970).

V. PROTEIN TRANSPORT

Endocytotic uptake and transport of proteins by mammalian trophectodermal cells have been examined by a number of laboratories. However, results from early experiments have been conflicting, which can be explained by the use of different methodologies. With more recent studies, there does appear to be a comprehensive picture emerging.

Most of the studies investigating protein transport phenomena have utilized either biochemical (mainly gel electrophoresis) or histochemical methodologies to address the questions of protein uptake by the trophectoderm. The apparent conflict in the data results from the fact that the biochemical analyses have monitored proteins that are endogenous to the uterine environment, whereas histochemical analyses have examined movement of nonnative fluid-phase and membrane-bound markers.

In early studies that investigated the protein constituents of the blastocoel, the day-6 p.c. rabbit embryo blastocoel was discovered to contain albumin, uteroglobin (UTG; the predominant uterine secretory protein during the preimplantation period; Beier, 1970), and immunoglobin G (IgG; Beier, 1970; Hamana and Hafez, 1970; Petzoldt, 1974). These observations imply that the rabbit trophectoderm has the capacity to transcytose these elements from the uterine lumen to the blastocoel. In contrast with these data, however, Kulangara (1975, 1976) maintained that the presence of proteins within the blastocoel was an experimental artifact induced by the collection procedures commonly used. Kulangara found

FIG. 5. Micrographs demonstrating the binding of an antibody against the sodium-independent glucose transporter to the day-6 p.c. rabbit trophectodermal cell. There appear to be more transport proteins localized to the basolateral membrane (B) than to the apical membrane (A). This would allow glucose to traverse the trophectodermal cell and maintain equilibrium in the blastocoel with the glucose concentration in the uterine lumen. MV, Microvilli. Bar, 0.25 μm, (From Robinson *et al.*, 1990, with permission.)

that when the embryos were gently collected, the amount of protein measured within the blastocoel was minimal.

During the time that these biochemical studies were being performed, histochemical examinations of protein transport in mammalian embryos were also being undertaken. Schlafke and Enders (1973) and Hastings and Enders (1974a), in the rat and rabbit, respectively, found that, in the blastocyst, peroxidase could be localized to the blastocoel within 10 min of addition to the bathing medium, whereas ferritin was not able to traverse the trophectodermal cell. The observation that peroxidase was found in intracellular vesicles and not in intercellular spaces prompted the authors to hypothesize that peroxidase enters the blastocoel via transcellular mechanisms. In this case, the horseradish peroxidase was observed to accumulate on the apical membrane of the embryo, whereas ferritin did not. It is consistent with these observations to speculate that fluid-phase markers are segregated differently from membrane-bound components and, in addition, there is a transcytotic pathway that is restricted to only membrane-bound markers.

A definitive answer to the above hypothesis was provided histochemically by Fleming and Goodall (1986) in a study on mouse embryos. The authors visualized the time course of uptake and sorting of a fluid-phase marker [unconjugated horse radish peroxidase (HRP)], a nonspecific membrane-bound marker (cationic ferritin), and a specific membrane-bound marker (protein A-HRP). HRP was trapped in a secondary lysosomal compartment and was not transferred into the blastocoel. Cationic ferritin on the other hand, was localized to the blastocoel within 10 min of the initial exposure. This membrane-bound moiety was also sorted to the secondary lysosomal compartment. Evidently, a portion of the membrane-bound component could also be segregated from other membrane-bound proteins. When embryos were exposed to cationic ferritin along with protein A-HRP, these two molecules were separated in the endosomal pathway. The protein A-HRP was sorted to the lysosomes, and the cationic ferritin was preferentially transcytosed. Degradation of protein A-HRP may have precluded its localization in the blastocoel because it was visualized in this compartment in only low amounts.

Suprisingly little corroborative work has been done on the physiology of these transcytotic mechanisms until recently. Pemble and Kaye (1986) examined the uptake of ^{125}I-labeled bovine serum albumin (BSA) into the trophectodermal cell of the mouse embryo. They found that the uptake of the fluid-phase marker was low until the blastocyst stage. Other observations of BSA uptake in the mouse blastocyst included nonsaturability, indicating a fluid-phase mechanism, and also a high temperature dependence to the system, which is consistent with a cellularly mediated mechanism (Steinman *et al.*, 1983). The authors postulated that the uptake and subsequent degradation of uterine albumin in the mouse trophectoderm could provide significant quantities of amino acids to the embryo and may thus contribute to the cells' amino acid pool.

Robinson *et al.* (1989a) examined the transtrophectodermal permeabilities and temperature sensitivities of a number of biologically inert fluid-phase markers and compared these data to data obtained in similar experiments using proteins common to the rabbit uterus during the preimplantation period. Visual observations with the fluid-phase marker Lucifer Yellow demonstrated that fluid-phase markers localized in a perinuclear lysosomal compartment. There was also no apparent movement of Lucifer Yellow from this compartment during a 30-min observation period. The entry of the biologically inert markers into the blastocoel displayed temperature sensitivities characteristic of a paracellular route, most likely between leaky tight junctions. Interestingly, rabbit serum albumin and rabbit IgG both had permeability characteristics that were consistent with their entering the blastocoel by a paracellular path. Uteroglobin on the other hand, displayed a high degree of temperature sensitivity and the uptake into the blastocoel was saturable (Fig. 6). The authors speculated that this indicated a receptor-mediated transcytotic pathway. Supporting evidence for this idea came from the work of Kirchner (1976), in which he found immunoreactivity to a UTG antibody within the trophectodermal cells. Robinson *et al.* (1989a) also found that, under *in vitro* conditions,

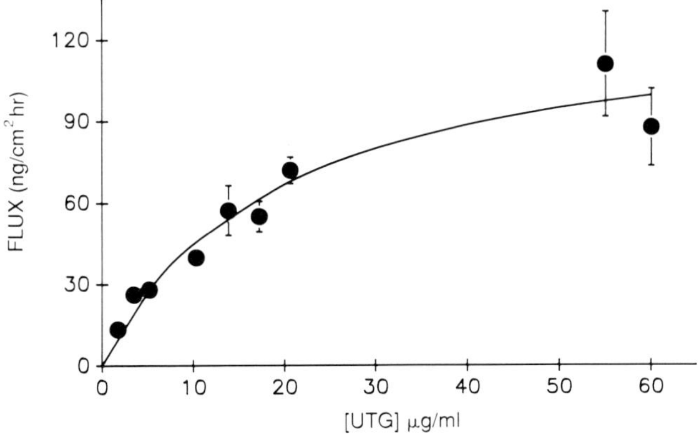

FIG. 6. Uptake of uteroglobin (UTG) as a function of concentration across the day-6 p.c. rabbit trophectodermal cell. This figure demonstrates that the uptake of UTG is saturable with a V_{max} of 132 ng cm^{-2} hr^{-1} and a K_m of 19.5 μg cm^{-2}. This indicates that there is most likely a receptor which is specific for UTG on the trophectodermal cell. (From Robinson *et al.*, 1989a, with permission.)

intact UTG did not enter the blastocoel; only fragments of the original protein were found. At this point, it is unknown if this represented lysosomal degradation prior to release into the blastocoel.

The mouse blastocyst apparently has a specific mechanism that transcytoses insulin across the trophectoderm (Heyner *et al.*, 1989). Curiously, these researchers found that insulin was only transcytosed across the trophectoderm in the area where the trophectoderm contacts the ICM, and that the cells of the ICM actively endocytose insulin. Insulin may then be modulating the growth rate of the ICM (Harvey and Kaye, 1990).

These data, when considered with the histochemical information, provide an emerging picture of the intracellular sorting events that occur in the mammalian trophectodermal cell. Constituents that remain in the bulk phase (for example, albumin and IgG in the preimplantation embryo) are endocytosed by the cell and then sorted to a lysosomal compartment. Here, they are degraded and possibly incorporated into the cell's amino acid pool. Apparently, Kulangara (1976) was correct in his assertion that proteins localized in the blastocoel were contaminants: there is no physiological or histochemical evidence for a transcytotic mechanism present for proteins in the fluid phase. Membrane-bound proteins, of which UTG is a potential candidate in the rabbit embryo, are sorted into the secondary endosomes and/or the lysosomes and then transported across the cell into the blastocoel. This is clearly the case in the postimplantation mouse embryo in which maternal IgG is found surrounding the embryo (Bernard *et al.*, 1977). Transcytosis of the undegraded maternal protein is the only reasonable explanation for this finding.

VI. AMINO ACID TRANSPORT

Of all the transport processes studied in the preimplantation embryo thusfar, mechanisms of amino acid transport appear to undergo considerable differentiation during the preimplantation process. Far more studies have been done on the mouse embryo than on the rabbit. However, generalizations between embryonic transport systems should not be made as some distinct species differences exist.

Amino acid transport itself is a rather complicated subject. There are a number of different Na^+-dependent and independent amino acid transport systems, each of which has some shared traits with other amino acid transport systems (reviewed in Christensen, 1990). In addition, researchers have discovered amino acid transport systems that are peculiar to only

TABLE II

Amino Acid Transport Systems in the Mouse and Rabbit Preimplantation Embryo

Na^+-dependent systems	
System Gly	Transports glycine
System A	Transports leucine, alanine, and other zwitterionic amino acids. Recognized MeAIB
System ASC	Transports alanine, serine, and cystine
System B^{0+}	Similar to system A, but does not recognize MeAIB
Na^+-independent systems	
System L	Transports methionine and other amino acids with branched apolar side chains. Displays a high exchange rate. Inhibited by BCH
System b^{0+}	Similar to system L, but does not recognize BCH

one cell type. Because of these complexities, our understanding of amino acid transport in the preimplantation embryo is far from complete.

In the mouse egg, Miller and Schultz (1986) found evidence for two kinetically distinct system L-type amino acid transporters (see Table II for amino acid transport characteristics). In addition, they also found evidence that the egg contains the Na^+-dependent systems ASC and A. Van Winkle *et al.* (1988a) also found evidence for system Gly, which is specific for glycine.

During the 2-cell to 8-cell stage, the mouse embryo contains system Gly (Hobbs and Kaye, 1985, 1986; Van Winkle *et al.*, 1988a). During this precompaction period of development, there may also be system A, because methylamino isobutyric acid (MeAIB), which is a system A analog, was able to inhibit glycine transport during this period (Van Winkle *et al.*, 1988a). However, it is curious that, in this study, no leucine-inhibitable glycine transport, an expected feature of system A, was noted. An explanation for this lack of effect was not provided. The Na^+-independent system L may also be present during this stage, as researchers have found that methionine uptake is Na^+ independent (Borland and Tasca, 1974, 1975). In addition, cells preloaded with methionine demonstrate an active exchange with external methionine (Kaye *et al.*, 1982). Both these features (Na^+ independence and exchange) are typical of system L.

As embryogenesis proceeds to the morula stage, system Gly appears to become the less dominant glycine transport system, and a system A-type transport becomes more dominant (Hobbs and Kaye, 1985). System L continues to be operative at the morula stage (Kaye *et al.*, 1982).

Amino acid transport in the mouse embryo changes dramatically with compaction. System Gly seems to disappear completely, and instead, a unique type of transporter becomes activated. This amino acid transporter is system A-like in that both leucine and alanine inhibit the Na^+-dependent transport of glycine (Hobbs and Kaye, 1985; Van Winkle *et al.*, 1988a). However, unlike the "classic" system A, this transporter does not recognize MeAIB (Hobbs and Kaye, 1985; Van Winkle *et al.*, 1988a), despite the fact that MeAIB is transported by the embryo (Kaye *et al.*, 1982). Van Winkle *et al.* (1988a) designated this system as B^{0+}, to denote its Na^+ dependence and its interaction with both zwitterionic and cationic amino acids.

The mouse blastocyst may also have unique Na^+-independent amino acid transport systems. Leucine and lysine share a common transporter, which is not inhibitable by 2-amino-2-norbornane-carboxylic acid (BCH). Both these findings indicate that this particular transport mechanism is not system L (Van Winkle *et al.*, 1988b). These researchers have designated this system as b^{0+}. However, system L does seem to be functional since there is a component of leucine transport which is BCH inhibitable (Van Winkle *et al.*, 1988b).

There is a paucity of data concerning amino acid transport in the rabbit embryo. Apparently, on the basis of methionine exchange experiments, the rabbit embryo transports methionine by the system L transporter (Miller and Schultz, 1983). In contrast to the mouse embryo, however, there is also a large (30–40%) portion of methionine uptake that is dependent on external Na^+ (Miller and Schultz, 1983). The Na^+-depenent portion of methionine uptake decreases from day 5 to 6 and, at day 7, methionine uptake is Na^+ independent (Benos, 1981a; Miller and Schultz, 1983). Bell *et al.* (1986) extended these earlier findings by examining glycine, aminoisobutyric acid (AIB), and leucine influxes in day-5 to day-7 p.c. rabbit blastocysts. They found, in agreement with work in the mouse blastocyst, that MeAIB influx also occurred in the rabbit embryo. However, this influx was not dependent on external Na^+, which suggests that MeAIB is not entering the trophectrodermal cell by system A. In addition, Bell *et al.* (1986) found that Na^+-dependent leucine influx was not MeAIB inhibitable, which implies that the rabbit may have a B^{0+} system similar to that found in the mouse (Van Winkle *et al.*, 1988a). In addition, leucine and AIB were found to share a common carrier. There was also clear evidence that system L was operative in the blastocyst because BCH effectively inhibited leucine influx into the tropectodermal cell. An interesting finding in this study was that, while the absolute magnitude of leucine and AIB influx does not change from day 5 to 7 p.c., the Na^+-dependent fraction decreases and the Na^+-indepenent fraction increases. This finding suggests that system L increases during blastulation and the Na^+-dependent system (B^{0+}) decreases. This observation is in contrast to the mouse, where system L decreases after cavitation.

The reason for the change in the nature of these amino acid transport systems following blastulation is unknown. The rabbit uterine fluid is rich in glycine, glutamate, and alanine, and the other amino acids are present in only small amounts (Miller and Schultz, 1987). Furthermore, the concentration of uterine glycine increases following day 3. The amino acid transport systems in the blastocyst may be more regulated, or regulated in a different manner from those found in the precompaction stages. For example, trophectoderm cellular glycine concentration decreases dramatically following blastulation (Miller and Schultz, 1987), as does the concentration of all amino acids. However, the questions of hormonal control of these systems has not yet been addressed.

VII. SUMMARY AND PROSPECTS

Our knowledge of transport systems in preimplantation mammalian embryos is still at the descriptive stage. Until we have sufficient knowledge of the transport systems present, and at which point in embryogenesis these systems arise, we will be unable to correlate adequately ion and solute transport with the developmental processes. For example, it is unknown if the development of ion channels is dependent on hormonal factors or if these transporters are expressed as a result of an internal genomic clock that controls transcription of the appropriate message. The work that has thusfar examined this question suggests that development of the amiloride-sensitive Na^+ channel is independent of steroid hormones (Nielsen *et al.*, 1987).

The question of developmental expression of epithelial transport-related proteins and its timing is an important one. For example, with the exception of the connexin proteins (Barron *et al.*, 1989), it is unknown whether proteins typically found in epithelia are expressed before or during the compaction event. Another question that may be addressed through knowledge of epithelial proteins is that of cell lineage. Researchers (Petersen *et al.*, 1986; Winkel and Petersen, 1988) have examined the question of cell lineage by peroxidase labeling cells of ICM or trophectoderm and then tracing the labeled cell's fate in other embryonic tissues. Their data strongly suggested that some cells in the trophectoderm are derived from the ICM. These data then raise the question of whether epithelially related proteins are expressed before, during, or after insertion into the trophectoderm, and are epithelially related genes expressed simultaneously, or is there a disparity in the timing of expression?

A great deal of effort in epithelial research has recently been concerned with the isolation of predominant proteins and genes that are specifically expressed in epithelial tissues. With the use of cellular and genomic probes, we will have the means to study important questions of epithelial differentiation in the preimplantation embryo.

Acknowledgments

We would like to thank Dr. Eric Overstrom for his helpful advice and criticism, Ms. Cathy Guy for her excellent secretarial assistance, and Ms. Amy Burns for the artwork. We also thank Ms. Becky Gargus for her constructive comments on the manuscript. Supported by NIH grants HD21302 and HD07069.

References

Alpern, R. J., and Chambers, M. (1986). Cell pH in the rat proximal convoluted tubule. Regulation by luminal and peritubular pH and sodium concentration. *J. Clin. Invest.* **78,** 502–510.

Barron, D. J., Valdimaisson, G., Paul, D. L., and Kidder, G. M. (1989). Connexin 32, a gap junction protein, is a persistent oogenetic product through preimplantation development of the mouse. *Dev. Genet.* **10,** 318–323.

Beier, H. M. (1970). Protein patterns of endometrial secretion in the rabbit. *In* "Ovo-Implantation: Human Gonadotropins and Prolactins" (P. O. Hubinot, F. Leroy, C. Robyn, and P. Leleuse, eds.), pp. 157–163. Karger, New York.

Bell, J. E., Begg, K. E., Sin, Y., Biggers, J. D., and Benos, D. J. (1986). Neutral amino acid influx in developing rabbit blastocysts. *Am. J. Physiol.* **251,** C285–C292.

Benos, D. J. (1981a). Developmental changes in epithelial transport characteristics of preimplantation rabbit blastocysts. *J. Physiol. (London)* **316,** 191–202.

Benos, D. J. (1981b). Ouabain binding to preimplantation rabbit blastocysts. *Dev. Biol.* **83,** 69–78.

Benos, D. J. (1982). Amiloride: A molecular probe of sodium transport in tissues and cells. *Am. J. Physiol.* **242,** C131–C145.

Benos, D. J., and Balaban, R. S. (1980). Energy requirements of the developing mammalian blastocyst for active ion transport. *Biol. Reprod.* **23,** 941–947.

Benos, D. J., and Biggers, J. D. (1983). Sodium and chloride co-transport by preimplantation rabbit blastocysts. *J. Physiol. (London)* **342,** 23–33.

Benos, D. J., Reyes, J., and Shoemaker, D. G. (1983). Amiloride fluxes across erythrocyte membranes. *Biochim. Biophys. Acta* **734,** 99–104.

Benos, D. J., Balaban, R. S. Biggers, J. D., Mills, J. W., and Overstrom, E. W. (1985). Developmental aspects of sodium-dependent transport processes of preimplantation rabbit embryos. *In* "Regulation and Development of Membrane Transport Processes" (J. S. Graves, ed.), pp. 211–235. Wiley, New York.

Bernard, O., Ripoche, M.-A., and Bennett, D. (1977). Distribution of maternal immunoglobins in the mouse uterus and embryo in the days after implantation. *J. Exp. Med.* **145,** 58–75.

Biggers, J. D., Borland, R. M., and Lechene, C. P. (1978). Ouabain-sensitive fluid accumulation and ion transport by rabbit blastocysts. *J. Physiol. (London)* **280,** 319–330.

Biggers, J. D., Bell, J. E., and Benos, D. J. (1988). Mammalian blastocyst: Transport functions in a developing epithelium. *Am. J. Physiol.* **255,** C419–C432.

Borland, R. M., and Tasca, R. J. (1974). Activation of a Na^+-dependent amino acid transport system in preimplantation mouse embryos. *Dev. Biol.* **36,** 169–182.
Borland, R. M., and Tasca, R. J. (1975). Na^+-dependent amino acid transport in preimplantation mouse embryos. *Dev. Biol.* **46,** 192–201.
Borland, R. M., Biggers, J. D., and Lechene, C. P. (1976). Kinetic aspects of rabbit blastocoel fluid accumulation: An application of electron probe microanalysis. *Dev. Biol.* **50,** 201–211.
Borland, R. M., Biggers, J. D., and Lechene, C. P. (1977). Fluid transport by rabbit preimplantation blastocysts. *J. Reprod. Fertil.* **51,** 131–135.
Bridges, R. J., Cragoe, E. J., Jr., Frizzell, R. A., and Benos, D. J. (1989). Inhibition of colonic Na^+ transport by amiloride analogues. *Am. J. Physiol.* **256,** C67–C74.
Brinster, R. L. (1967). Carbon dioxide production from glucose by the preimplantation mouse embryo. *Exp. Cell Res.* **47,** 271–277.
Christensen, H. N. (1990). Role of amino acid transport and countertransport in nutrition and metabolism. *Physiol. Rev.* **70,** 43–77.
Cross, M. H. (1973). Active sodium and chloride transport across the rabbit blastocoel wall. *Biol. Reprod.* **8,** 566–575.
Cross, M. H. (1974). Rabbit blastocoel pH. *J. Exp. Zool.* **186,** 17–22.
Cross, M. H., and Brinster, R. L. (1969). Trans membrane potential of the rabbit blastocyst trophoblast. *Exp. Cell Res.* **58,** 125–127.
Cross, M. H., and Brinster, R. L. (1970). Influence of ions, inhibitors and anoxia on transtrophectodermal potential of rabbit blastocyst. *Exp. Cell Res.* **62,** 303–309.
Dabich, D., and Acey, R. A. (1982). Transport of glucosamine (aldohexoses) by preimplantation mouse blastocysts. *Biochim. Biophys. Acta* **684,** 146–148.
Diamond, J. M., and Bossert, W. H. (1967). Standing-gradient osmotic flow: A mechanism for coupling of water and solute transport in epithelia. *J. Gen. Physiol.* **50,** 2061–2083.
Dickson, A. D. (1966). The form of the mouse blastocyst. *J. Anat.* **100,** 335–348.
DiZio, S. M., and Tasca, R. J. (1977). Sodium-dependent amino acid transport in preimplantation mouse embryos. *Dev. Biol.* **59,** 198–205.
Dubinsky, W. P., Jr., and Frizzell, R. A. (1983). A novel effect of amiloride on a H^+-dependent Na^+ transport. *Am. J. Physiol.* **245,** C157–C159.
Ducibella, T., and Anderson, E. (1975). Cell shape and membrane changes in the eight-cell mouse embryo: Pre-requisites for morphogenesis of the blastocyst. *Dev. Biol.* **47,** 45–58.
Ducibella, T., Albertini, D. F., Anderson, E., and Biggers, J. D. (1975). The preimplantation mammalian embryo: Characterization of intercellular junctions and their appearance during development. *Dev. Biol.* **45,** 231–250.
Esposito, G. (1984). Intestinal permeability of water-soluble nonelectolytes: Sugars, amino acids, peptides. *In* "Pharmacology of Intestinal Permeability" (T. Z. Czaky, ed.), pp. 567–611. Springer-Verlag, New York.
Fleming, T. P., and Goodall, H. (1986). Endocytic traffic in trophectoderm and polarized blastomeres of the mouse preimplantation embryo. *Anat. Rec.* **216,** 490–503.
Fridhandler, L. (1961). Pathways of glucose metabolism in fertilized rabbit ova at various pre-implantation stages. *Exp. Cell Res.* **22,** 303–316.
Fukumoto, H., Kayano, T., Buse, J. B., Edwards, Y., Pilch, P. F., Bell, G. I., and Seino, S. (1989). Cloning and characterization of the major insulin-responsive glucose transporter expressed in human skeletal muscle and other insulin-responsive tissues. *J. Biol. Chem.* **264,** 7776–7779.
Gamow, E., and Daniel, J. C., Jr. (1970). Fluid transport in rabbit blastocyst. *Wilhelm Roux Arch.* **164,** 261–278.

Gardiner, C. S., Grobner, M. A., and Menino, A. R., Jr. (1990). Sodium/potassium adenosine triphosphatase α-subunit and α-subunit mRNA levels in early rabbit embryos. *Biol. Reprod.* **43,** 788–794.

Gardner, D. K., and Leese, H. J. (1988). The role of glucose and pyruvate transport in regulating nutrient utilization by preimplantation mouse embryos. *Development* **104,** 423–429.

Gumbiner, B. (1987). Structure, biochemistry, and assembly of epithelial tight junctions. *Am. J. Physiol.* **253,** C749–C758.

Hamana, K., and Hafez, E. S. E. (1970). Disc electrophoretic patterns of uteroglobins and serum proteins in rabbit blastocoelic fluid. *J. Reprod. Fertil.* **21,** 555–558.

Harvey, M. B., and Kaye, P. L. (1990). Insulin increases the cell number of the inner cell mass and stimulates morphological development of mouse blastocysts *in vitro. Development* **110,** 963–967.

Hastings, R. A., and Enders, A. C. (1974a). Uptake of exogenous protein by the preimplantation rabbit. *Anat. Rec.* **179,** 311–330.

Hastings, R. A., and Enders, A. C. (1974b). Junctional complexes in the preimplantation rabbit embryo. *Anat. Rec.* **181,** 17–34.

Heyner, S., Rao, L. V., Jarett, L., and Smith, R. M. (1989). Preimplantation mouse embryos internalize maternal insulin via receptor-mediated endocytosis: Pattern of uptake and functional correlations. *Dev. Biol.* **134,** 48–58.

Hobbs, J. G., and Kaye, P. L. (1985). Glycine transport in mouse eggs and preimplantation embryos. *J. Reprod. Fertil.* **74,** 77–86.

Hobbs, J. G., and Kaye, P. L. (1986). Glycine and Na^+ transport in preimplantation mouse embryos. *J. Reprod. Fertil.* **77,** 61–66.

Jentsch, T. J., Matthes, H., Keller, S. K., and Weiderholt, M. (1986). Electrical properties of sodium bicarbonate symport in kidney epithelial cells (BSC-1). *Am. J. Physiol.* **251,** F964–F968.

Kaye, P. L., Schultz, G. A., Johnson, M. H., Pratt, H. P. M., and Church, R. B. (1982). Amino acid transport and exchange in preimplantation mouse embryos. *J. Reprod. Fertil.* **65,** 367–380.

Kinne, R., Murer, H., Kinne-Saffran, E., Thees, M., and Sachs, G. (1975). Sugar transport by renal plasma membrane vesicles. *J. Membr. Biol.* **21,** 375–395.

Kirchner, C. (1976). Uteroglobin in the rabbit I. Intracellular localization in the oviduct, uterus, and preimplantation blastocyst. *Cell Tissue Res.* **170,** 415–424.

Kulangara, A. C. (1975). Absence of maternal proteins in 5–7 day blastocyst fluid indicates limited protein passage before implantation. *J. Exp. Zool.* **193,** 101–108.

Kulangara, A. C. (1976). Quantitative studies on passage of protein into unimplanted blastocysts. *In* "The Meternofoetal Transmission of Immunoglobins" (W. A. Hemmings, ed.), pp. 313–324. Cambridge Univ. Press, New York.

Lee, S. (1987). Membrane properties in preimplantation mouse embryos. *J. In Vitro Fertil. Embryo Transplant* **4,** 331–333.

Lehtonen, E., and Bradley, R. A. (1980). Localization of cytoskeletal proteins in preimplantation mouse embryos. *J. Embryol. Exp. Morphol.* **55,** 211–225.

Lutwak-Mann, C. (1954). Some properties of the rabbit blastocyst. *J. Embryol. Exp. Morphol.* **2,** 1–13.

Lutwak-Mann, C. (1962). Glucose, lactic acid and bicarbonate in rabbit blastocyst fluid. *Nature (London)* **193,** 653–654.

Macknight, A. D. C. (1980). Comparison of analytic techniques: Chemical, isotopic and microprobe analysis. *Fed. Proc.* **39,** 2881–2887.

Macknight, A. D. C., Dibona, D. R., and Leaf, A. (1980). Sodium transport across toad urinary bladder: A model "tight" epithelium. *Physiol. Rev.* **60,** 615–715.

Magnuson, T., Demsey, A., and Stackpole, C. W. (1977). Characterization of intercellular junctions in the preimplantation mouse embryo by freeze-fracture and thin section electron microscopy. *Dev. Biol.* **61,** 252–261.

Manejwala, F., Kaji, E., and Schultz, R. M. (1986). Development of activatable adenylate cyclase in the preimplantation mouse embryo and a role for cyclic AMP in blastocoel formation. *Cell* **46,** 95–103.

Manejwala, F. M., and Schultz, R. M. (1989). Blastocoel expansion in the preimplantation mouse embryo: Stimulation of sodium uptake by cAMP and possible involvement of cAMP-dependent protein kinase. *Dev. Biol.* **136,** 560–563.

Manejwala, F. M., Cragoe, E. J., Jr., and Schultz, R. M. (1989). Blastocoel expansion in the preimplantation mouse embryo: Role of extracellular sodium and chloride and possible apical routes of their entry. *Dev. Biol.* **133,** 210–220.

Miller, J. G. O., and Schultz, G. A. (1983). Properties of amino acid transport in preimplantation rabbit embryos. *J. Exp. Zool.* **228,** 511–525.

Miller, J. G. O., and Schultz, G. A. (1986). Amino acid transport and exchange in unfertilized mouse eggs. *Gamete Res.* **15,** 1–12.

Miller, J. G. O., and Schultz, G. A. (1987). Amino acid content of preimplantation rabbit embryos and fluids of the reproductive tract. *Biol. Reprod.* **36,** 125–129.

Moran, A., Asher, C., Cragoe, E. J., Jr., and Garty, H. (1988). Conductive sodium pathway with a low affinity to amiloride in LLC-PK1 cells and other epithelia. *J. Biol. Chem.* **263,** 19586–19591.

Neilsen, L. L., Benos, D. J., and Biggers, J. D. (1987). Mineralocorticoid concentrations in unstressed female rabbits and embryonic sodium transport. *J. Reprod. Fertil.* **81,** 553–562.

O'Brien, T. G., and Prettyman, R. (1987). Phorbol esters and mitogenesis: comparison of the proliferative response of parenteral and Na^+ K^+ Cl^--cotransport-defective BALB/c 3T3 cells to 12-*O*-tetradecanoylphorbol-13-acetate. *J. Cell. Physiol.* **130,** 377–381.

Overstrom, E. J. (1987). *In vitro* assessment of blastocyst differentiation. *In* "The Mammalian Preimplantation Embryo. Regulation of Growth and Differentiation *In Vitro*" (B. D. Bavister, ed.), pp. 95–116. Plenum, New York.

Overstrom, E. J., Benos, D. J., Biggers, J. D.m and Godkin, J. D. (1984). Transtrophectodermal sodium transport during porcine blastogenesis. *Biol. Reprod.* **30(S1),** 46.

Overstrom, E. J., Benos, D. J., and Biggers, J. D. (1989). Synthesis of Na^+/K^+-ATPase by the preimplantation rabbit blastocyst. *J. Reprod. Fertil.* **85,** 283–295.

Pemble, L. B., and Kaye, P. L. (1986). Whole protein uptake and metabolism by mouse blastocysts. *J. Reprod. Fertil.* **78,** 149–157.

Perinchief, P. N. (1980). Monosaccharide transport in the preimplanted rabbit blastocyst. *Fed. Proc.* **39,** 954.

Petersen, R. A., Wu, K., and Balakier, H. (1986). Origin of the inner cell mass in mouse embryos: Cell lineage analysis by microinjection. *Dev. Biol.* **117,** 581–595.

Petzoldt, U. (1974). Micro-disc electrophoresis of soluble proteins in rabbit blastocysts. *J. Embryol. Exp. Morphol.* **31,** 479–487.

Powers, R. D., and Tupper, J. T. (1977). Developmental changes in membrane transport and permeability in the early mouse embryo. *Dev. Biol.* **56,** 306–315.

Powers, R. D., Borland, R. M., and Biggers, J. D. (1977). Amiloride-sensitive rheogenic Na^+ transport in rabbit blastocyst. *Nature (London)* **270,** 603–604.

Rick, R., Doerge, A., Gehring, K., Bauer, R., and Thurau, K. (1979). Quantitative determination of cellular electrolyte concentrations in thin freeze-dried cryosections using energy-dispersive X-ray microanalysis. *In* "Microbeam Analysis in Biology" (C. P. Lechene and R. R. Warner, eds.), pp. 517–524. Academic Press, New York.

Robinson, D. H., Kirk, K. L., and Benos, D. J. (1989a). Macromolecular transport in rabbit blastocysts: Evidence for a specific uteroglobin transport system. *Mol. Cell. Endocrinol.* **63,** 227–237.

Robinson, D. H., Bubien, J. K., Smith, P. R., and Benos, D. J. (1991). Epithelial sodium conductance in rabbit preimplantation trophectodermal cells. *Dev. Biol.* (in press).

Robinson, D. H., Smith, P. R., and Benos, D. J. (1990). Hexose transport in preimplantation rabbit blastocysts. *J. Reprod. Fertil.* **89,** 1–11.

Robinson, D. H., Mills, J. W., and Benos, D. J. (1991). [^{3}H]-Ouabain autoradiography of the preimplantation rabbit embryo. Submitted.

Rodriguez-Boulan, E., and Nelson, W. J. (1989). Morphogenesis of the polarized epithelial cell phenotype. *Science* **245,** 718–725.

Schlafke, S., and Enders, A. C. (1973). Protein uptake by rat preimplantation stages. *Anat. Rec.* **175,** 539–560.

Schultz, S. G., and Streeker, C. K. (1970). Fructose influx across the brush border of rabbit ileum. *Biochim. Biophys. Acta* **211,** 586–588.

Skou, J. C. (1975). The (Na^+-K^+) activated enzyme system and its relationship to transport of sodium and potassium ions. *Q. Rev. Biophys.* **7,** 401–434.

Soltoff, S. P., and Mandel, L. J. (1983). Amiloride directly inhibits the Na/K-ATPase activity of rabbit kidney proximal tubules. *Science* **220,** 957–958.

Sorscher, E., Accavitti, M. A., Keeton, D., Steadman, E., Frizzell, R. A., and Benos, D. J. (1988). Antibodies against purified epithelial sodium channel protein from bovine renal papilla. *Am. J. Physiol.* **255,** C835–C843.

Steinman, R. M., Mellman, I. S., Muller, W. A., and Cohn, Z. A. (1983). *J. Cell Biol.* **96,** 1–27.

Ullrich, K. J. (1979). Sugar, amino acid and Na^+ cotransport in the proximal tubule. *Annu. Rev. Physiol.* **41,** 181–195.

Van Winkle, L. J., Haghighat, N., Campione, A. L., and Gorman, J. M. (1988a). Glycine transport in mouse eggs and preimplantation conceptuses. *Biochim. Biophys. Acta* **941,** 241–256.

Van Winkle, L. J., Campione, A. L., and Gorman, J. M. (1988b). Na^+-independent transport of basis and zwitterionic amino acids in mouse blastocysts by a shared system and by processes which distinguish between these substrates. *J. Biol. Chem.* **263,** 3150–3163.

Watson, A. J., and Kidder, G. M. (1988). Immunofluorescence assessment of the timing of appearance and cellular distribution of Na/K-ATPase during mouse embryogenesis. *Dev. Biol.* **126,** 80–90.

Wiley, L. M. (1984). Cavitation in the mouse embryo: Na/K-ATPase and the origin of nascent blastocoel fluid. *Dev. Biol.* **105,** 330–342.

Wiley, L. M., and Obasaju, M. F. (1989). Effects of phlorizin and ouabain on the polarity of mouse 4-cell/16-cell stage blastomere heterokaryons. *Dev. Biol.* **133,** 375–384.

Winkel, G. K., and Petersen, R. A. (1988). Fate of the inner cell mass in mouse embryos as studied by microinjection of lineage tracers. *Dev. Biol.* **127,** 143–156.

Ziomek, C. A., and Johnson, M. H. (1980). Cell surface interaction induces polarization of mouse 8-cell blastomeres at compaction. *Cell* **21,** 935–942.

PART III

Developmental Biology of Ion and Solute Co- and Counter-Transport

CHAPTER 5

Cell Biology and Molecular Genetics of Enterocyte Differentiation

Michael W. Smith
AFRC Institute of Animal Physiology and Genetics Research, Babraham, Cambridge CB2 4AT, England

I. INTRODUCTION

A. Conflicting Requirements of Intestinal Function

The small intestine is required both to act as a barrier to penetration by pathogenic organisms and selectively absorb a wide variety of nutrients and electrolytes. This barrier function is largely maintained through a constant process of cell renewal which ensures that undamaged epithelial cells rapidly replace any enterocytes damaged by disease. The absorptive requirement is then satisfied through a process of enterocyte differentiation programmed to move to completion in a strictly limited and often variable period of time. Further interactions between barrier and absorptive requirements occur whenever immune reactions are mounted by the gut-associated immune tissue against invading microflora. Such immune responses inadvertently increase cell proliferation in intestinal crypts, further decreasing the time available for enterocytes to differentiate.

Animals require different types and amounts of nutrients throughout life, and the intestine is presented with a mixture of dietary constituents varying on a daily basis. Enterocyte differentiation of particular digestive and absorptive functions must adapt to these short- and long-term changes in nutrient availability and animal requirements. Part of this adaptive response involves control over villus structure as well as enterocyte differentiation. This in turn involves interactions taking place with mesenchymal tissue and responses being initiated to systemic and locally produced hormones. Additional organized responses involve a regional control over enterocyte expression of digestive enzymes and absorptive functions and stem cell specification of different cell types.

How the whole intestine uses different control mechanisms to maintain homeostasis in a changing environment is a complicated question only partly answered by work carried out on enterocyte differentiation. Research investigating mechanisms affecting differentiation is, however, sufficiently important to warrant attention in its own right.

B. Critical Appraisal of a Complicated Process

How enterocytes differentiate form and function can be seen from the preceding section to depend on a number of interrelated events impossible to control *in vivo* or reproduce faithfully *in vitro*. In spite of this difficulty, it is still possible to identify specific factors affecting enterocyte development under carefully controlled conditions. The purpose of the present review is to bring together work fulfilling this requirement in an attempt to construct a framework for future research. With this aim in mind, it is also worthwhile trying to identify present gaps in our knowledge of how intestinal function is organized at the cellular level.

II. MARKERS OF ENTEROCYTE DIFFERENTIATION

Differences between the structure, function, and biochemical composition of plasma membranes, as well as changes in the concentration of intracellular proteins, create a rich source of markers suitable to measure the onset, state, and progression of enterocyte differentiation. Temporal studies show differentiation of some of these markers takes place sequentially in adult as well as fetal tissue. Use of only one marker in this case gives a simplified and possibly misleading view of how the whole process operates.

A. Brush Border Cytoskeleton

The most obvious differentiation feature of an enterocyte is the brush border membrane, first discovered in the middle of the nineteenth century (Henle, 1837). Electron microscopic studies later showed villus tip microvilli to be longer than those found in the crypt (Trier, 1967). Quantitative analysis of microvillus elongation has since identified factors controlling this aspect of enterocyte differentiation (Smith and Brown, 1989).

The molecular architecture of the microvillus cytoskeleton has also been studied in detail and proteins specific for the microvillus isolated and characterized. Antibodies raised against a polypeptide–calmodulin complex, fimbrin, and villin have then been used to mark the appearance of these molecules in differentiating enterocytes (Kedinger *et al.*, 1988; Chantret *et al.*, 1988; Boller *et al.*, 1988). Cloning and sequencing villin has further allowed the distribution of specific mRNA to be determined (Boller *et al.*, 1988). The terminal web proteins fodrin and caldesmon have also been used as markers of enterocyte differentiation (Kedinger *et al.*, 1988; Younes *et al.* 1989).

B. Brush Border Hydrolases

The brush border membrane contains a number of saccharidases and peptidases involved in the terminal digestion of nutrients (Alpers, 1986). Many of these enzymes have already been used as markers of enterocyte differentiation. Estimating these enzymes biochemically in pooled cell or tissue extracts gives no idea of the state of differentiation of individual enterocytes. This information can be obtained using quantitative cytochemistry (Gutschmidt and Gossrau, 1981; Gutschmidt *et al.*, 1984). Applying this method to intact villi represents a significant advance in this field of research (James *et al.*, 1988; Tivey and Smith, 1989).

Purified hydrolases have also been used to produce polyclonal antibodies. Monoclonal antibodies have also been prepared against a number of intestinal hydrolases (Hauri *et al.*, 1985). This makes it relatively easy to detect enzymes in enterocyte brush border membranes. It is, however, still difficult to determine the amount of enzyme present by this method.

Disaccharidases and peptidases now cloned and sequenced include sucrase-isomaltase, lactase-phlorizin hydrolase, γ-glutamyl transpeptidase, aminopeptidase N, dipeptidyl peptidase IV, endopeptidase, and angiotensin-converting enzyme. mRNA probes to sucrase-isomaltase have also been used to test for a transcriptional control of enzyme expression (Sebastio *et al.*, 1986, 1987; Rousset *et al.*, 1989). Similar work using mRNA probes to different peptidases has yet to be carried out.

C. Transport Functions

Fully differentiated enterocytes transport fluid, electrolytes, and a wide variety of nutrients from the lumen to blood. Several of these transport functions have been used successfully to mark a late stage in enterocyte development.

Fluid transport across Caco-2 cells grown to confluence on an impermeable support is recognized as blister formation in the monolayer (Zweibaum and Chantret, 1989). Grown on filters, these same cells transport Na^+ electrogenically following treatment with nystatin or amphotericin B (Grasset *et al.*, 1984). Similar swelling of intercellular spaces takes place after incubating tissue *in vitro* (Cremaschi *et al.*, 1973). Seeing spaces appear beneath or between enterocytes shows only that fluid transport has taken place. Quantitative assessment of this event could be provided using methods described originally by Spring (1979).

Sugar and amino acid transport into villus-attached enterocytes can also be used as markers of enterocyte differentiation (Kinter and Wilson, 1965). Quantitative autoradiography shows amino acid and peptide transport to be confined to upper villus enterocytes (Smith, 1985b; Cheeseman, 1986). Cloning and sequencing of the Na^+-dependent sugar carrier now allows one to make antibodies to test when the carrier is first expressed during enterocyte development.

D. Membrane Receptors and Intracellular Proteins

Receptors to a number of growth factors and hormones have recently been identified and characterized in villus and crypt enterocytes. Changes in receptor abundance have also been found to be associated with the state of enterocyte differentiation [insulin receptors occur with highest frequency in fully differentiated enterocytes (Gallo-Payet and Hugon, 1984), and numbers of IGF_1, IGF_2, and EGF receptors are highest in crypt enterocytes (Gallo-Payet and Hugon, 1985; Laburthe *et al.*, 1988)].

Immunoglobulin secretory component (SC) is also less abundant in villus than crypt enterocytes and entirely absent from follicle-associated epithelial cells (Buts and Delacroix, 1985; Roy and Varvayanis, 1987). Small amounts of major histocompatibility complex (MHC) class II antigens expressed by villus and crypt enterocytes are also absent from follicle-associated M cells (Bjerke and Brandtzaeg, 1988). Immunological stimuli induce MHC class II antigen expression in crypt and villus but probably not follicle-associated M cells (Barclay and Mason, 1982). All of these antigens, together with others marking M cell basolateral membranes (Pappo, 1989), surface colonocytes (Gorr *et al.*, 1988), and carcinoembryonic antigen (CEA) or Lewis antigens (Wolf *et al.*, 1989), are potentially useful in studying mechanisms controlling enterocyte differentiation.

There are, finally, two types of intracellular fatty acid-binding protein localized preferentially in upper villus enterocytes (Shields *et al.*, 1986). These markers are now being used to investigate regulation of gene expression along the tract as well as along the villus (Gordon, 1989).

E. Choosing a Marker

Expression of any marker of enterocyte differentiation should, until proved otherwise, be considered to represent a unique event taking place with its own time course within the overall process of development. Maintaining that marker in the cell can, however, result from continual expression of a short-lived molecule or virtually no synthesis of a molecule having a long half-life. Making quantitative estimates of both the amount and turnover of the marker and relating these results to the specific mRNA content of the cell is needed in order to specify mechanisms controlling its expression.

Screening methods used to determine whether markers have or have not appeared in enterocytes can also be useful in identifying mechanisms regulating gene expression. Such studies can be carried out without making any quantitative assessment of the amount of marker present. The number of markers chosen for use depends on their ease of determination and the model chosen for study. Future work carried out with more markers in fewer systems could prove an effective way to extend knowledge in this area of research.

III. FETAL DEVELOPMENT

Studies of fetal development have already proved extremely useful in revealing mechanisms controlling tissue formation and enterocyte differentiation of structure and function. These studies also show some areas where further research is now needed.

A. *Normal Conditions*

1. Cell Proliferation and Organogenesis

Tissue-dependent changes in the development of rat fetal intestine include initial stimulation of proliferation in the endoderm (cell layers increasing from 2 to 8 between gestational days 15 and 17), cell degeneration of the outer endodermal layers by day 17, and appearance of small lumina between deeper situated endodermal cells (Mathan *et al.,* 1976). Cell proliferation then decreases, mesenchymal cells invaginate the endoderm, and villus formation occurs by day 19. Random mitoses occurring at earlier stages of development become largely restricted to the intervillus spaces at this stage of development (Hermos *et al.,* 1971). Mechanisms controlling the initial surge in cell proliferation, later cell death, and subsequent localization of mitosis remain unknown. Screening tissue sections for the presence of different growth factors and determining mRNA levels as described for transforming growth factor $\beta1$ (TGF-$\beta1$) in other tissues (Thompson *et al.,* 1989) could now help us understand how control over cell proliferation leads to onset of enterocyte differentiation. Further details of major changes taking place in enterocyte differentiation, mesenchymal invagination, and villus formation are shown for 18- and 19-day tissues in Fig. 1.

Certain important conclusions emerge from examination of these stages of development. First, it appears that differentiation of microvillus structure in outer endodermal cells does *not* depend on physical contact with mesenchymal cells or a basement membrane (Fig. 1a). Secondary specification of different cell types may, however, depend on physical contact taking place between these two tissues (Fig. 1b). Similar general changes have been shown to take place during fetal development in several other species (Klein and McKenzie, 1983).

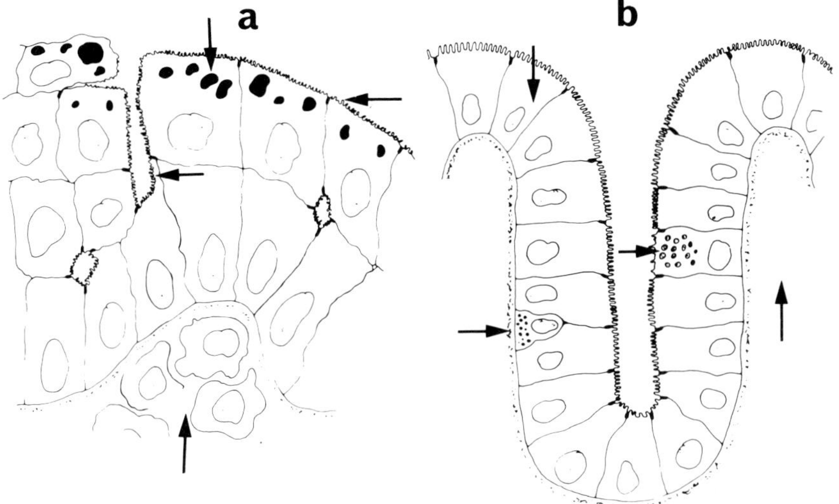

FIG. 1. Schematic diagram of duodenal mucosa taken from (a) 18- and (b) 19-day-old rat fetus. Mesenchymal cells begin to invaginate the endoderm (↑), outer cells degenerate (↓), and microvilli form (←) on day 18. Villi form (↑), the endoderm becomes a monolayer (↓), and goblet cells appear (→) on day 19. (From Mathan *et al.*, 1976.)

2. Hydrolase Expression

It is generally accepted that enterocyte differentiation takes place early in human compared with rat intestine, and that later expression of sucrase in rat intestine is specifically controlled through an intrinsically timed mechanism. These assumptions can be tested directly by comparing the normal time dependency of expression for five different hydrolases in Table I.

Alkaline phosphatase is the first hydrolase to be expressed by differentiating enterocytes. There then follows an ordered appearance of lactase, maltase, sucrase-isomaltase, and trehalase in rats and pigs. Alkaline phosphatase is also the first hydrolase to appear in human intestine, but after that there is no correlation with events taking place in rat and pig intestine. Expression of all hydrolases takes place *relatively* earlier in human compared with pig and rat intestine. The postnatal appearance of sucrase in rats probably results from the short time available to complete differentiation *in utero*. It is concluded from these comparisons that time-controlled expression of some and possibly *all* hydrolases occurs during development, but that more work is needed in order to test the generality of this statement. More recent experiments with different cytoplasmic proteins (L- and I-FABP, apoAI and AIV, and CRBP II) show a mosaic pattern of

TABLE I

First Appearance of Brush Border Hydrolases during Fetal Development

Enzyme	First recorded appearance (relative gestation time[a])		
	Rat	Pig	Human
Alkaline phosphatase	0.80	0.40	0.20
Lactase	0.85	0.40	0.50
Maltase	0.90	1.05	0.70
Sucrase-isomaltase	1.70	1.05	0.30
Trehalase	1.70	1.30	0.40

[a] Gestation times for rats (22 days), pigs (117 days), and humans (280 days) have each been given a value of one to compare relative times for enzyme appearance. Estimated times were calculated from papers by Simon-Assmann *et al.* (1982) and Henning (1981) for rat, by Kidder and Manners (1978) and D. R. Tivey (personal communication) for pigs, and by Dahlqvist and Lindberg (1966) and Semenza (1967) for humans.

gene expression occurring during late fetal development (Rubin *et al.*, 1989). It would now be interesting to extend these studies to brush border hydrolases.

B. Experimentally Manipulated Conditions

1. Xenoplastic Tissue Recombinations

The finding that chick or rat fetal intestine grafted into the coelomic cavity of a 3-day-old chick embryo could later express brush border hydrolases and that similar differentiation could be induced by mixing endodermal tissue from one species with mesenchymal cells from another (Kedinger *et al.*, 1981) has since been exploited to investigate the general ability of mesenchymal tissue to initiate enterocyte differentiation (Haffen *et al.*, 1989). Mesenchymal cells from the small intestine can be shown, in intraspecies grafting experiments, to be more effective than cells prepared from other regions of the gut in inducing differentiation of endodermal tissue. Endoderm from the small intestine is at the same time more resistant than other gut endoderm in its response to mesenchymal induction. Recent work shows, however, that intact mesenchyme from rat or chick small intestine can *inhibit* proventricular development in the chick. This effect suggests that it is the endoderm rather than the mesenchyme which initiates the interactive process (Yasugi *et al.*, 1989). Endodermal tissue also affects the differentiation and tissue redistribution of mesenchymal cells (Haffen *et al.*, 1989). Whether or not all of these effects require cell contact or interaction with a jointly formed basement membrane is questionable (see Section V,A).

2. Fetal Isografts

Enterocyte expression of sucrase activity in pieces of fetal mouse intestine transplanted under the kidney capsule of adult syngeneic animals takes place at the same time as that determined during normal postnatal development (Ferguson *et al.*, 1973). This finding was later confirmed in rats and a regional distribution of sucrase identified along the tract of fetal isografts (Kendall *et al.*, 1979; Montgomery *et al.*, 1981). Less widely publicized are the parallel findings that fetal tissue becomes committed to late expression *before* the mesenchyme begins to invaginate the endoderm, that maltase and alkaline phosphatase also develop normally, and that lactase expression is down-regulated in fetal isografts. This technique has also been used to implant newborn and 5-day-old postnatal intestine under the skin of newborn rats (Yeh and Holt, 1986). Sucrase activity in these tissues is expressed according to the age of the graft rather than the age of the animal. These results show that hormonal changes taking place in suckling rats are not needed to initiate expression of sucrase. The importance of these experiments lies in their ability to distinguish locally programmed changes in enterocyte differentiation from those arising from changes in the hormonal status or food intake of suckled animals. Further use of this model to determine the time when cell commitment to differentiate first takes place could give important information about how biological clocks operate at the cellular level.

3. Chimeric Mice

It has recently been possible to produce mouse chimeras that are recognized by a difference in H-2 antigens and membrane carbohydrates to demonstrate that adult crypts are derived from single stem cells (Ponder *et al.*, 1985; Schmidt *et al.*, 1985). Later work shows neonatal crypts to contain stem cells derived from both genotypes. Changing from a mixed to a single cell-derived crypt only becomes complete 14 days after birth (Schmidt *et al.*, 1988). Anchorage of stem cells in deepening crypts could cause this change in stem cell behavior. The clonal origin of crypts can also be studied by injecting the mutagen ethyl nitrosourea into pregnant mice to induce loss of an allele specifying *Dolichos biflorus* agglutinin (DBA) binding (Winton *et al.*, 1988). This technique also allows one to measure the kinetics of stem cell replacement throughout the animal's lifetime.

The work described above shows changes in the behavior of stem cells responsible for forming four distinct cell types in the neonatal intestine. Specification of cell type also varies during neonatal development (Smith and Jarvis, 1978). Investigation of a possible connection between these two events has yet to take place.

IV. ENTEROCYTE DIFFERENTIATION IN ADULT INTESTINE

Little mention is made of work carried out on postnatal changes taking place in rat intestine. This is because results from isograft experiments appear to reduce the importance of this work to a description of how diet affects enzyme induction and cell proliferation (Section III,B,2). Postnatal development in rats has demonstrated transcriptional control over sucrase expression (Sebastio *et al.*, 1986). Finding similar results in human fetus (Section II,B) further emphasizes the view that rats carry out a program of late fetal development postnatally.

A. *Normal Conditions*

1. Differentiation along the Intestine

It has never been difficult to measure villus structure, digestive enzymes, and absorptive function in different parts of the small intestine. From this work has appeared a general view that villus area, many hydrolase activities, and some absorptive functions decrease along the jejunoileal axis. Extrinsic factors held responsible for creating and maintaining these gradients include luminal nutrition, trophic factors in pancreaticobiliary secretions, and various hormones (Dowling, 1982). Intrinsic mechanisms operating in fetal isografts also exert regional control over villus structure (Jolma *et al.*, 1980; MacDonald and Ferguson, 1981) and the expression of different brush border enzymes (Kendall *et al.*, 1979; Montgomery *et al.*, 1981).

It has not yet been possible to assess the relative importance of extrinsic and intrinsic mechanisms regulating enterocyte differentiation along the gut. This problem has been bypassed recently by using transgenic mice to study regional control of fatty acid-binding protein expression by differentiating enterocytes (see Section IV,B,1).

2. Differentiation along the Villus

Enterocytes differentiate structure and function during migration from crypt base to villus tip. New techniques allow one to describe the kinetics of this process in individual villus-attached cells (Smith, 1985b). Recent results show the kinetics of microvillus development to be described by the simple equation $M = 0.0016C + 0.073C/R$, where M and C represent the maximal microvillus length and crypt depth, and R is the rate of enterocyte migration (Smith and Brown, 1989). A graphic representation of this relationship is shown in Fig. 2. Increasing the migration rate from 11 to 19 $\mu m\ hr^{-1}$ has very little effect on differentiation of the microvillus membrane. Reducing the migration rate from 9 to 4 $\mu m\ hr^{-1}$ causes the microvillus length to double for constant crypt depth. These differences in migration rate all fall within the range encountered under normal physiological conditions. Ideas as to how these factors operate at the cellular level remain tentative (Smith and Brown, 1989).

Similar data for hydrolase appearance can be obtained by applying quantitative cytochemistry to frozen sections of intestinal villi (Smith, 1985b). Sugar induction of α-glucosidases in mouse enterocytes has been shown, by this method, to follow the equation $I = RT$, where I is the maximal inductive effect, R is the initial rate of enzyme appearance, and T is the time over which this initial rate of synthesis can be maintained (Collins *et al.*, 1989). It is interesting to note in this work that a threefold sugar-induced increase in R is accompanied by a halving of T under conditions where the enterocyte migration rate remains constant. The cellular mechanisms responsible for this compromise remain to be identified. Quantitative cytochemistry can also be applied to intact as well as sectioned tissue (James *et al.*, 1988; Tivey and Smith, 1989). Subsequent

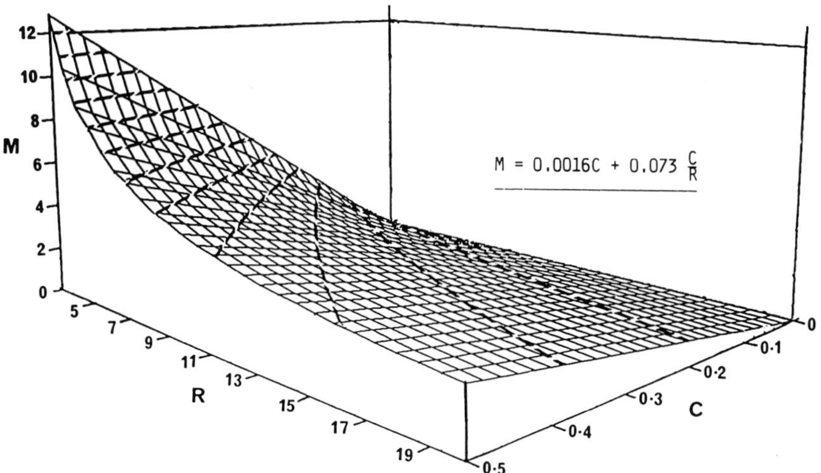

FIG. 2. Three-dimensional representation of how crypt size (C) and enterocyte migration rate (R) combine to control enterocyte differentiation of microvillus structure. An equation describing this relation for M (microvillus length) is based on a large body of experimental findings (Smith and Brown, 1989).

analysis of microdissected villi then gives measurements of total enzyme activity and villus structure as well as more accurate estimates of developmental profiles for enzyme expression.

Quantitative autoradiography of tritiated amino acid uptake by villus-attached enterocytes can also be used to measure the onset and development of absorptive function (Smith, 1985b). First appearance of amino acid transport is detected in upper villus enterocytes 15–20 hr after completion of microvillus elongation and hydrolase expression (Cremaschi *et al.*, 1986). Other transport processes appearing late in development include those responsible for sugar, peptide, and fatty acid transport (Kinter and Wilson, 1965; Haglund *et al.*, 1973; Cheeseman, 1986; Shields *et al.*, 1986). Changing potassium ion activities in mid-villus enterocytes, possibly through earlier development of the brush border membrane and synthesis of sodium pumps, could initiate gene expression of transport proteins (Cremaschi *et al.*, 1986).

3. Differentiation over Lymphoid Tissue

Enterocyte migration over lymphoid follicles takes place in a similar way to that described for villi (Smith *et al.*, 1980; Ponder *et al.*, 1985). The pattern of differentiation taking place in each case is, however, very different. Two of the most interesting differences are shown in Fig. 3.

Secretory component is entirely absent from the follicle side of common crypts producing villus and follicle-associated enterocytes. Inhibitory interactions occurring between lymphoid and epithelial tissues must be very local to produce such an effect. Lactase expression is also much reduced in follicle-associated compared with adjacent villus enterocytes. The selectivity of this inhibition is highlighted by the parallel finding that alkaline phosphatase expression is slightly *increased* in fully differentiated follicle-associated cells (Smith, 1985a). Lactase appearance in villi immediately adjacent to lymphoid follicles is also delayed compared with that found in distant villi (Fig. 3; broken line). Delayed escape from initial inhibition could account for such anomalous behavior. The biochemical cause of these effects, which bear some resemblance to other changes occurring in diseased tissue (see Section VI,A), has not been identified.

Alkaline phosphatase expression by follicle-associated enterocytes, observed *en face* on intact tissue, shows a great deal of local heterogeneity (Smith *et al.*, 1988). Some of these cells containing small amounts of enzyme have also been shown to endocytose antigens across a rudimentary brush border membrane (Owen, 1977; Owen and Bhalla, 1983). The original decision to call these enterocytes M cells (Owen, 1977), and the later suggestion that they represent an entirely new cell type (Bye *et al.*, 1984), may be difficult to justify as more becomes known about the

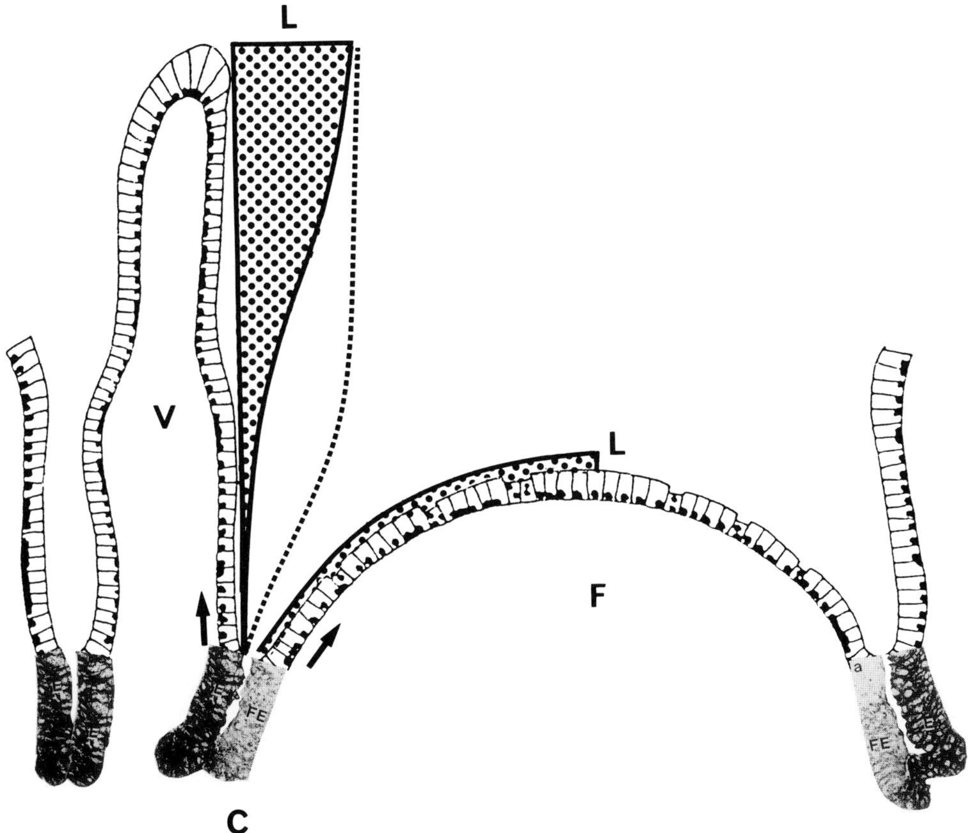

FIG. 3. Schematic reconstruction of differentiation events taking place as enterocytes migrate over a lymphoid follicle (F) or a follicle-associated villus (V). L, Developmental profile for lactase appearance (dotted area); broken line, normal developmental profile for lactase appearance in a villus that is distant from lymphoid follicles; arrows, enterocyte migration pathways from a common crypt (C). Crypts have been stained for secretory component (Owen and Pappo, 1988); lactase results come from Smith (1985a).

nature of cellular interactions taking place between microflora, enterocytes, and the gut immune system.

B. Experimentally Manipulated Conditions

Transgenic Mice

Liver and intestinal fatty acid-binding proteins (L-FABP and I-FABP) are expressed more readily in proximal than distal intestine and in villus tip rather than basal villus or crypt enterocytes (Shields *et al.*, 1986). Portions of genes encoding these two homologous proteins have been linked recently to a human growth hormone reporter gene minus its regulatory element (hGH) to investigate how genes regulate FABP synthesis. The structures of plasmids containing three different fragments of L-FABP gene are shown in Fig. 4.

pLFhGH2, a short promoter L-FABP-hGH construct, contains nucleotides −596 to +21. A long promoter, *pLFhGH4,* contains nucleotides −4000 to +21, and *pLFhGH5* contains a similar stretch of nucleotides (−4000 to −597) in reverse orientation linked to the short promoter (Sweetser *et al.,* 1988a). Two fragments of I-FABP gene have also been linked to hGH to produce short (*pIFhGH1,* −277 to +28) and long (*pIFhGH2,* −1178 to +28) promoters (Sweetser *et al.,* 1988b).

All three L-FABPhGH constructs were able to generate appropriate hGH expression along the small intestine of transgenic mice. Regional specification of hGH by the long promoter I-FABPhGH construct was also similar to that found *in vivo.* hGH expression by the short promoter I-FABPhGH was considerably less than that of the long promoter. This was most noticeable in the distal region of the small intestine. Taken together, these results show that elements regulating regional specification of FABP expression are located within the first 0.6 kb to the start side of transcription.

All of the L-FABPhGH constructs caused increased expression of hGH in crypt compared with villus enterocytes, the opposite result to that predicted from carrying out villus distribution studies for L-FABP (Shields *et al.,* 1986). Expression of hGH and I-FABP did, however, show the predicted crypt–villus distribution for both short and long promoter I-FABP constructs. These results show that elements within the first 0.3 kb of the construct direct I-FABP but not apparently L-FABP expression along the crypt–villus axis. Genes encoding for other markers can also be used provided they are effficiently transcribed and that gene products are confined to enterocytes.

V. ENTEROCYTE DIFFERENTIATION *in Vitro*

A. *Fetal Enterocytes*

Obtaining a preparation of undifferentiated crypt enterocytes, keeping them alive in culture for long periods of time, and subsequently inducing differentiation to occur at will has long been a frustrated aim of cell biologists. Cryptlike cells prepared from suckling rat intestine failed to differentiate and late fetal explants failed to maintain their differentiation

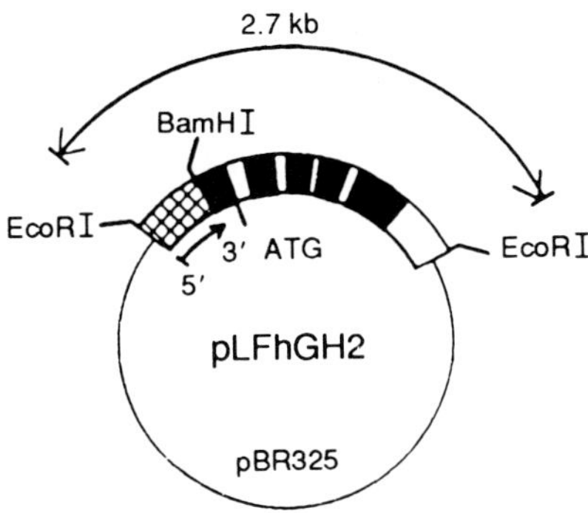

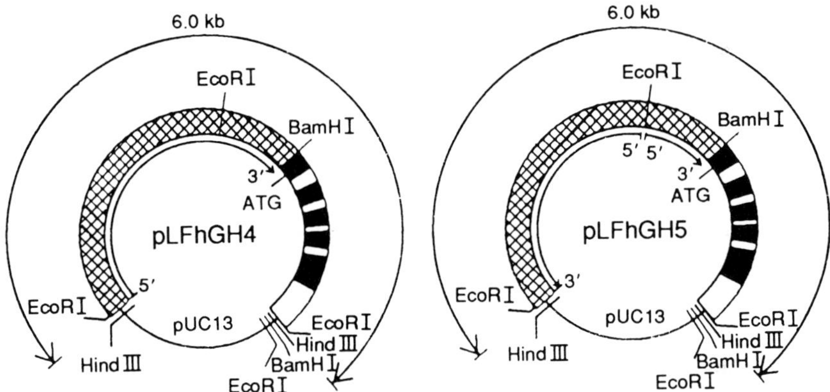

FIG. 4. Plasmid structures used to prepare fusion genes for transgenic mouse production. Thin lines denote vector sequences and cross-hatched areas denote sequences derived from the 5′ nontranscribed region of the L-FABP gene. Thick solid lines indicate exons of the hGH structural gene and open boxes the hGH introns. Outer arrows show DNA fragments isolated subsequently for injection into male mice pronuclei. Inner arrows show the 5′-3′ orientation of the promoter sequences. ATG refers to the methionine codon present in exon 1 of the hGH gene. (From Sweetser *et al.*, 1988a.)

potential in secondary culture (Quaroni, 1985). Cells separated from rat fetal intestine have since been shown to differentiate and form intestine-like organoids *in vitro* when surrounded by collagen gel, but this is still a largely uncontrollable system containing fibroblasts as well as enterocytes (Montgomery, 1986).

An important breakthrough in being able to study enterocyte differentiation *in vitro* was achieved recently by culturing endodermal cells prepared from 15-day-old fetal rat intestine on top of mesenchymal monolayers (Stallmach *et al.*, 1989). These conditions enabled endodermal cells to express three different hydrolases in the presence of dexamethasone. Mesenchymal monolayers prepared from skin and gastric mucosa did not support differentiation. This confirmed results found previously in grafting experiments (see Section III,B,1). More recent work shows rat endodermal cells to express alkaline phosphatase and lactase activities when cultured for 4 days on a substrate made from laminin or a basement membrane extract. Enzyme expression was further increased in the presence of dexamethasone (Hahn *et al.*, 1990). The importance of this work lies in the demonstration that mesenchymal cells are *not* needed to enable a committed endodermal cell to differentiate *in vitro* and that dexamethasone *can* act directly on endodermal cells to facilitate differentiation.

A final question arises as to whether it is worthwhile continuing the search for inducers of differentiation in noncommitted cells. It has, for instance, been reported that TGF-β can both inhibit proliferation and induce expression of sucrase in IEC-6 cells grown on plastic in the presence of different growth factors (Kurakowa *et al.*, 1987), but this latter effect is not reproducible (Barnard *et al.*, 1989). Getting such a system to operate reliably *in vitro* could allow one to study how proliferating cells become committed to differentiate. The past history of finding consistently negative results with this cell line is, however, worrying.

B. Carcinoma Cell Lines

Most that can be said about the advantages and disadvantages of using colon carcinoma cell lines to mimic critical aspects of normal enterocyte development has been reviewed recently by Zweibaum *et al.* (1991). Cell lines of potential interest to investigate the induction and development of differentiation include Caco-2 cells, which differentiate spontaneously when grown to confluence in culture (Pinto *et al.*, 1983), and HT-29 cells, where differentiation can be induced by changing substrates (Wice *et al.*, 1985). Both cell lines have been used to study the biosynthesis, intracellular transport, and processing of proteins in differentiated enterocytes. Changes in enzyme activities controlling glucose and glycogen metabolism also take place during differentiation of HT-29 cells (Zweibaum *et al.*, 1991). The time course describing onset of hydrolase expression in Caco-2 as well as HT-29 cells is, however, measured in days rather than hours (Pinto *et al.*, 1983; Wice *et al.*, 1985) and not all cells differentiate to the same extent (Zweibaum, 1986). Both of these restrictions make them unsuitable for measuring kinetic aspects of enterocyte differentiation.

VI. DISEASE EFFECTS ON ENTEROCYTE DIFFERENTIATION

Disease effects on intestinal function are generally assumed to result from nonspecific damage to the epithelium, leading to a reduction in villus surface area and compensatory hypertrophy of the crypts. Work with germfree animals further suggests that microflora present in the gut of apparently healthy animals can produce similar, though less pronounced, changes in intestinal structure and function (Simon and Gorbach, 1987). In this case, it is obviously desirable to define both the disease agent used and the health state of the animal before attempting to investigate the effects of disease on enterocyte development. Relating results obtained to parallel measurements of cell turnover also aids subsequent interpretation of experimental findings.

A. Brush Border Hydrolases

Infecting neonatal mice with Coxsackie B-1 or EDIM-rotavirus, or inducing GvHR in similarly aged mice, increases crypt cell proliferation and the early expression of sucrase and maltase (Hammond and Rosenberg, 1972; Collins *et al.,* 1988; Lund *et al.,* 1986a). Initiation of immune response and increasing levels of circulating corticosteroids are held to be jointly responsible for producing these effects. Previously, it was suggested that the number of mitoses taking place in crypts might determine the time needed to express sucrase during normal postnatal development (Yeh and Holt, 1986). Calculations based on GvHR reduction of cell cycle time showed no correlation with sucrase expression in mice (Lund and Smith, 1987).

Disease-induced decreases in lactase expression in human and pig intestines are greater than for sucrase–maltase under conditions where the expression of alkaline phosphatase remains virtually unaltered (Miller *et al.,* 1986; Phillips *et al.,* 1988). This correlation breaks down, however, using mice infected with the immunosuppressive parasite *Nematospiroides dubius* (Smith and Lloyd, 1989). Selective *increase* of lactase compared with alkaline phosphatase activities suggests that immune reactions taking place in normal mice might partly inhibit lactase expression. The identity of the agent(s) responsible for producing these debilitating effects on intestinal function is not known.

B. Absorptive Function

Intestinal malabsorption accompanies many disease-induced changes in brush border enzyme activities. Na^+-dependent alanine transport is, for instance, inhibited both in GvHR mice and in piglets suffering from soya bean-induced postweaning diarrhea (Lund *et al.*, 1986b; Miller *et al.*, 1986). These effects are not seen for lysine absorption or for alanine absorption measured in the absence of sodium (Lund *et al.*, 1986b; Smith *et al.*, 1985). Soya bean also fails to inhibit Na^+-dependent alanine transport in similarly aged gnotobiotic pigs (Ratcliffe *et al.*, 1989). This latter result emphasizes the role played by enteric microflora in amplifying minimal disease effects produced, in this case, by feeding soya bean.

Hormonal imbalance in the whole animal can also affect the ability of the small intestine to absorb nutrients. Recent work carried out with streptozotocin-treated diabetic rats shows increased hexose transport to be partly mediated by changes in the properties of individual enterocytes. Glucose carriers appear, from phlorizin-binding studies, to increase in diabetic rat enterocytes, and more cells bind phlorizin specifically (Fedorak *et al.*, 1989). Their membrane potential is also increased and this facilitates Na^+-dependent sugar entry across the brush border membrane (Debnam and Ebrahim, 1989). Both of these effects take place without change in enterocyte migration rate (Debnam and Ebrahim, 1990). Villi of diabetic rats are longer than control villi and this probably allows enterocytes more time to complete a normal program of development. Indirect evidence supporting this view comes from the parallel finding that enterocytes from diabetic rats also absorb valine more readily than control enterocytes (Debnam and Ebrahim, 1990).

It is of course difficult to determine which hormone(s) might be responsible for increasing villus size and sugar transport in diabetic animals. Chronic administration of glucagon into normal rats increases sugar transport and enterocyte membrane potential, but this is associated with a *decrease* in villus size and enterocyte migration rate (Thompson and Debnam, 1986; Rudo *et al.*, 1976). Further work is needed to identify all factors affecting enterocyte differentiation in the diabetic model.

VII. BARRIER FUNCTION AND ENTEROCYTE DEVELOPMENT

The barrier function of the small intestine is generally acknowledged to be maintained by exerting control over both transcellular and paracellular permeation pathways. Linking cells together near their apical surface through a series of tight junctional complexes restricts paracellular transport, while nonselective movement of material across enterocytes is restricted by the presence of a well-formed brush border membrane. The efficiency with which these barrier functions operate does, however, depend on the age of the animal and the state of enterocyte differentiation.

A. Tight Junctions

The typical arrangement of tight junctions in adult guinea pig villus and crypt enterocytes and the predicted effect these arrangements have on the relative paracellular permeabilities to electrolytes are shown in Fig. 5.

Major differences between crypt and villus structures include the number and continuity of the junctional strands, both of which increase

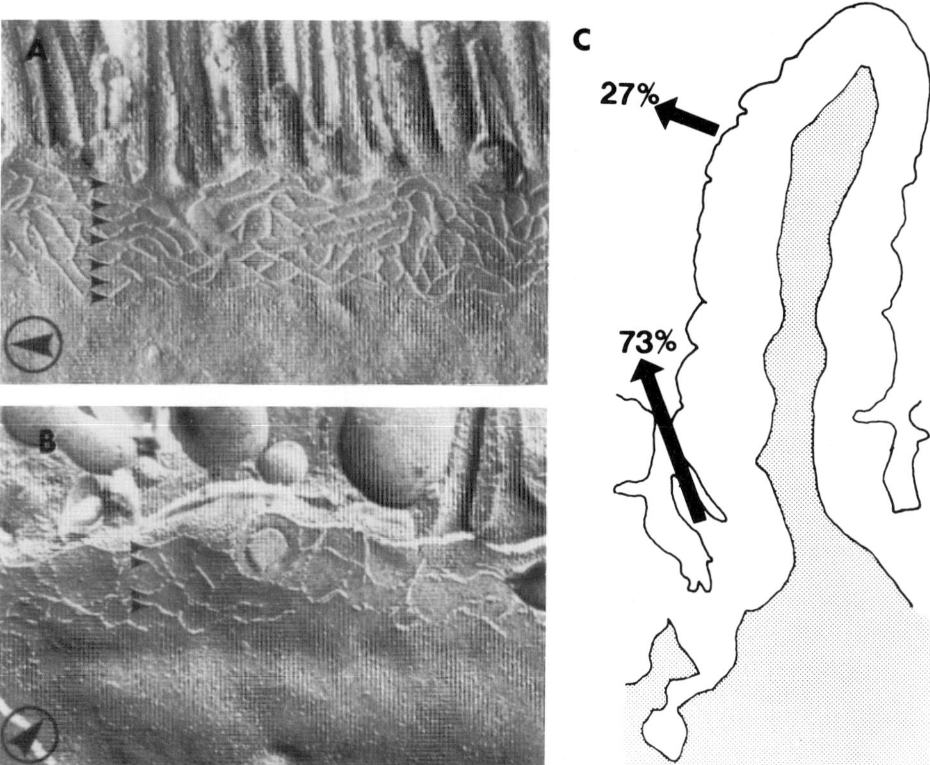

FIG. 5. Composite picture showing differences between guinea pig villus and crypt tight junctions (A and B) and the predicted effect (C) these differences have on the distribution of paracellular conductance along the crypt-villus axis. (From Marcial *et al.*, 1984.)

markedly as enterocytes differentiate. Similar findings have been reported in the stomach, large intestine, and small intestine of other mammals (for references see Mora-Galindo, 1986). The predicted effect of these differences is to change a secretory crypt epithelium into one capable of absorbing fluid (Marcial *et al.*, 1984). Tight junctions in neonatal rat intestinal villi are in many ways similar to those found between adult crypt enterocytes (Knutton *et al.*, 1978). The molecular mechanisms responsible for controlling this aspect of enterocyte development have not been studied in detail.

B. Immature Enterocytes

Enterocytes in the small intestines of all newborn mammals are able to endocytose proteins presented in ingested milk (Walker, 1979). These proteins are then either digested completely within phagolysosomes in the enterocyte or passed on to the blood circulation in a relatively undegraded form. Such endocytosis can be both selective and nonselective for homologous immunoglobulins (for references see Baintner, 1986). So-called closure of the intestine to this type of uptake is associated with the repopulation of villi with adult-type enterocytes having a well-formed terminal web. This change takes place at about the time of weaning in neonatal rats, at a time when other digestive enzymes are also changing activities in the brush border membrane (Henning, 1981). Injection of steroids can induce the precocious appearance of mature enterocytes having high sucrase activity which are also unable to endocytose proteins (Henning, 1981; Morris and Morris, 1975). Similar changes in sucrase activity also occur, however, in tissue isografts (Section III,B,2). It may be concluded from this that closure depends on an intrinsic timing mechanism rather than steroid induction. This suggestion has yet to be tested experimentally.

VIII. CONCLUDING REMARKS

Deciding what not to talk about in a review is often more revealing than what is actually written down. In the present case, for instance, it was decided to omit any theoretical discussion of cell proliferation since nobody, to my knowledge, has yet produced a *testable* hypothesis suggesting why crypt cells stop cycling and begin to differentiate. Biochemical description of how enterocytes become polarized has also been omitted on the grounds that this work says little about the way these processes adapt to changing physiological requirements. Attempts have also been made to avoid including material for old times' sake. Enterocyte differentiation is now a multidisciplinary subject attracting much interest in its own right and this leaves less opportunity to republish old findings.

Emerging techniques in cell and molecular biology are now exposing previously accepted theories to critical analysis. Some of the results appearing from this work already raise more questions than they answer. Which aspect of research is going to prove most rewarding in the future remains problematical. A more fundamental question concerns the possible illusory nature of some of the breakthroughs currently being described. Optimists and writers of grant applications will say that advances in this field of research are now so numerous as to make progress inevitable. My own opinion is that this is correct, but that any idea that this will quickly solve the secret of how differentiation takes place is premature.

References

Alpers, D. (1986). Digestion and absorption of carbohydrates and proteins. *In* "Physiology of the Gastrointestinal Tract" (L. R. Johnson, ed.), Vol. 2, pp. 1469–1488. Raven, New York.

Baintner, K. (1986). "Intestinal Absorption of Macromolecules and Immune Transmission from Mother to Young." CRC Press, Boca Raton, Florida.

Barclay, A. N., and Mason, D. W. (1982). Induction of $_{Ia}$ antigen in rat epidermal cells and gut epithelium by immunological stimuli. *J. Exp. Med.* **156,** 1665–1676.

Barnard, J. A., Beauchamp, R. D., Coffey, R. J., and Moses, H. L. (1989). Regulation of intestinal epithelial cell growth by transforming growth factor type β. *Proc. Natl. Acad. Sci. U.S.A.* **86,** 1578–1582.

Bjerke, K., and Brandtzaeg, P. (1988). Lack of relation between expression of HLA-DR and secretory component (SC) in follicle-associated epithelium of human Peyer's patches. *Clin. exp. Immunol.* **71,** 502–507.

Boller, K., Arpin, M., Pringault, E., Mangeat, P., and Reggio, H. (1988). Differential distribution of villin and villin mRNA in mouse intestinal cells. *Differentiation* **39,** 51–57.

Buts, J.-P., and Delacroix, D. (1985). Ontogenic changes in secretory component expression by villous and crypt cells of rat small intestine. *Immunology* **54,** 181–187.

Bye, W. A., Allan, C. H., and Trier, J. S. (1984). Structure, distribution and origin of M cells in Peyer's patches of mouse ileum. *Gastroenterology* **86,** 789–801.

Chantret, I., Barbat, A., Dussaulx, E., Brattain, M. G., and Zweibaum, A. (1988). Epithelial polarity, villin expression and enterocytic differentiation of cultured human colon carcinoma cells: A survey of twenty cell lines. *Cancer Res.* **48,** 1936–1942.

Cheeseman, C. I. (1986). Expression of amino acid and peptide transport systems in rat small intestine. *Am. J. Physiol.* **251,** G636–G641.

Collins, A. J., James, P. S., and Smith, M. W. (1989). Sugar-dependent selective induction of mouse jejunal disaccharidase activities. *J. Physiol. (London)* **419,** 157–167.

Collins, J., Starkey, W. G., Wallis, T. S., Clarke, G. J., Worton, K. J., Spencer, A. J., Haddon, S. J., Osborne, M. P., Candy, D. C. A., and Stephen, J. (1988). Intestinal enzyme profiles in normal and rotavirus-infected mice. *J. Pediatr. Gastroenterol. Nutr.* **7,** 264–272.

Cremaschi, D., Smith, M. W., and Wooding, F. B. P. (1973). Temperature-dependent changes in fluid transport across goldfish gallbladder. *J. Membr. Biol.* **13,** 143–164.

Cremaschi, D., James, P. S., Meyer, G., Rossetti, C., and Smith, M. W. (1986). Intracellular potassium as a possible inducer of amino acid transport across hamster jejunal enterocytes. *J. Physiol.* (*London*) **375,** 107–119.

Dahlqvist, A., and Lindberg, T. (1966). Development of the intestinal disaccharidase and alkaline phosphatase activities in the human foetus. *Clin. Sci.* **30,** 517–528.

Debnam, E. S., and Ebrahim, H. Y. (1989). Diabetes mellitus and the sodium electrochemical gradient across the brush border membrane of rat intestinal enterocytes. *J. Endocrinol.* **123,** 453–459.

Debnam, E. S., and Ebrahim, H. Y. (1990). Autoradiographic localization of Na^+-dependent L-valine uptake by the jejunum of streptozotocin-diabetic rats. *Eur. J. Clin. Invest.* **20,** 61–66.

Dowling, R. H. (1982). Small bowel adaptation and its regulation. *Scand. J. Gastroenterol.* **74,** Suppl., 53–74.

Fedorak, R. N., Gershon, M. D., and Field, M. (1989). Induction of intestinal glucose carriers in streptozotocin-treated chronically diabetic rats. *Gastroenterology* **96,** 37–44.

Ferguson, A., Gerskowitch, V. P., and Russell, R. I. (1973). Pre- and postweaning disaccharidase patterns in isografts of rat and mouse intestine. *Gastroenterology* **64,** 292–297.

Gallo-Payet, N., and Hugon, J. S. (1984). Insulin receptors in isolated adult mouse intestinal cells: Studies *in vivo* and in organ culture. *Endocrinology* (*Baltimore*) **114,** 1885–1892.

Gallo-Payet, N., and Hugon, J. S. (1985). Epidermal growth factor receptors in isolated adult mouse intestinal cells: Studies *in vivo* and in organ culture. *Endocrinology* (*Baltimore*) **116,** 194–201.

Gordon, J. I. (1989). Intestinal epithelial differentiation: New insights from chimaeric and transgenic mice. *J. Cell Biol.* **108,** 1187–1194.

Gorr, S.-U., Stieger, B., Fransen, J. A., Kedinger, M., Marxer, A., and Hauri, H.-P. (1988). A novel marker glycoprotein for the microvillus membrane of surface colonocytes of rat large intestine and its presence in small-intestinal crypt cells. *J. Cell Biol.* **106,** 1937–1946.

Grasset, E., Pinto, M., Dussaulx, E., Zweibaum, A., and Desjeux, J.-F. (1984). Epithelial properties of human colonic carcinoma cell line (Caco-2: electrical parameters) *Am. J. Physiol.* **247,** C260–C267.

Gutschmidt, S., and Gossrau, R. (1981). A quantitative histochemical study of dipeptidylpeptidase IV (DPP IV). *Histochemistry* **73,** 285–304.

Gutschmidt, S., Hoper, R., and Gossrau, R. (1984). Kinetic characterization of brush border membrane proteases in relationship to mucosal architecture by section biochemistry. *Adv. Exp. Med. Biol.* **167,** 209–218.

Haffen, K., Kedinger, M., and Simon-Assmann, P. (1989). Cell contact dependent regulation of enterocyte differentiation. *In* "Human Gastrointestinal Development" (E. Lebenthal, ed.), pp. 19–39. Raven, New York.

Haglund, U., Jodal, M., and Lundgren, O. (1973). An autoradiographic study of the intestinal absorption of palmitic and oleic acid. *Acta Physiol. Scand.* **89,** 306–317.

Hahn, U., Stallmach, A., Hahn, E. G., and Riecken, E. O. (1990). Basement membrane components are potent promoters of rat intestinal epithelial cell differentiation *in vitro.* *Gastroenterology* **98,** 322–335.

Hammond, J. B., and Rosenberg, J. L. (1972). Stimulation of small intestinal mucosal enzymes during Coxsackie virus infection in neonatal mice. *J. Lab. Clin. Med.* **79,** 814–823.

Hauri, H.-P., Sterchi, E. E., Bienz, D., Fransen, J. A. M., and Marxer, A. (1985). Expression and intracellular transport of microvillus membrane hydrolases in human intestinal epithelial cells. *J. Cell Biol.* **101,** 838–851.

Henle, J. (1837). Symbolae ad anatomia villorum intestinaticum imprimis eorum epithelii et vasorum lacteorum Berolini. M. D. Thesis, Univ. Berlin.

Henning, S. J. (1981). Postnatal development: Coordination of feeding, digestion and metabolism. *Am. J. Physiol.* **241,** G199–G214.

Hermos, J. A., Mathan, M., and Trier, J. S. (1971). DNA synthesis and proliferation by villous epithelial cells in fetal rats. *J. Cell Biol.* **50,** 255–258.

James, P. S., Smith, M. W., and Tivey, D. R. (1988). Single-villus analysis of disaccharidase expression by different regions of the mouse intestine. *J. Physiol.* (*London*) **401,** 533–545.

Jolma, V. M., Kendall, K., and Koldovsky, O. (1980). Differences in the development of jejunum and ileum as observed in fetal rat intestinal isografts. Possible implications related to villus size gradient. *Am. J. Anat.* **158,** 211–215.

Kedinger, M., Simon, P. M., Grenier, J. F., and Haffen, K. (1981). Role of epithelial–mesenchymal interactions in the ontogenesis of intestinal brush-border enzymes. *Dev. Biol.* **86,** 339–347.

Kedinger, M., Rochette-Egly, C., Simon-Assmann, P., Bouziges, F., and Haffen, K. (1988). Mesenchymal cells influence the expression of enterocytic brush border differentiation markers. *In* "Mammalian Brush Border Membrane Proteins" (M. J. Lentze and E. E. Sterchi, eds.), pp. 111–120. Thieme, New York.

Kendall, K., Jumawan, J., and Koldovsky, O. (1979). Development of jejunoileal differences of activity of lactase, sucrase and acid β-galactosidase in isografts of fetal intestine. *Biol. Neonate* **36,** 206–214.

Kidder, D. E., and Manners, M. J. (1978). Digestion of carbohydrates. *In* "Digestion in the Pig" (D. E. Kidder and M. J. Manners, eds.), pp. 96–149. Scientechnica, Bristol, England.

Kinter, W. H., and Wilson, T. H. (1965). Autoradiographic study of sugar and amino acid absorption by everted sacs of hamster intestine. *J. Cell Biol.* **25,** 19–39.

Klein, R. M., and McKenzie, J. C. (1983). The role of cell renewal in the ontogeny of the intestine. I. Cell proliferation patterns in adult, fetal and neonatal intestine. *J. Pediatr. Gastroenterol. Nutr.* **2,** 10–43.

Knutton, S., Limbrick, A. R., and Robertson, J. D. (1978). Structure of occluding junctions in ileal epithelial cells of suckling rats. *Cell Tissue Res.* **191,** 449–462.

Kurokowa, M., Lynch, K., and Podolsky, D. K. (1987). Effects of growth factors on an intestinal epithelial cell line: Transforming growth factor B inhibits proliferation and stimulates differentiation. *Biochem. Biophys. Res. Commun.* **142,** 775–782.

Laburthe, M., Rouyer-Fessard, C., and Gammeltoft, S. (1988). Receptors for insulin-like growth factors I and II in rat gastrointestinal epithelium. *Am. J. Physiol.* **254,** G457–G462.

Lund, E. K., and Smith, M. W. (1987). Effect of graft versus host reaction on cell cycle time in neonatal mouse jejunum. *Cell Tissue Kinet.* **20,** 369–378.

Lund, E. K., Bruce, M. G., Smith, M. W., and Ferguson, A. (1986a). Selective effects of graft–versus–host reaction on disaccharidase expression by mouse jejunal enterocytes. *Clin. Sci.* **71,** 189–198.

Lund, E. K., Smith, M. W., and Peacock, M. A. (1986b). Parental spleen cells accelerate the development of intestinal brush border structure and function in neonatal mice. *Comp. Biochem. Physiol. A.* **85A,** 175–181.

MacDonald, T. T., and Ferguson, A. (1981). Regulation of villus height: The role of luminal factors in determining the villus height gradient of the mouse small intestine. *In* "Mechanisms of Intestinal Adaptation" (J. W. L. Robinson, *et al.*, eds.), pp. 47–53. MTP Press, Lancaster, England.

Marcial, M. A., Carlson, S. L., and Madara, J. L. (1984). Partitioning of paracellular conductance along the ileal crypt-villus axis: A hypothesis based on structural analysis with detailed consideration of tight junction structure-function relationship. *J. Membr. Biol.* **80,** 59–70.

Mathan, M., Moxey, P. C., and Trier, J. S. (1976). Morphogenesis of fetal rat duodenal villi. *Am. J. Anat.* **146,** 73–92.

Miller, B. G., James, P. S., Smith, M. W., and Bourne, F. J. (1986). Effect of weaning in the capacity of pig intestinal villi to digest and absorb nutrients. *J. Agric. Sci.* **107,** 579–589.

Montgomery, R. K. (1986). Morphogenesis *in vitro* of dissociated fetal rat small intestinal cells upon an open surface and subsequent to collagen gel overlay. *Dev. Biol.* **117,** 64–79.

Montgomery, R. K., Sybicki, A., and Grand, R. J. (1981). Autonomous biochemical and morphological differentiation in fetal rat intestine transplanted at 17 and 20 days of gestation. *Dev. Biol.* **87,** 76–84.

Mora-Galindo, J. (1986). Maturation of tight junctions in guinea-pig cecal epithelium. *Cell Tissue Res.* **246,** 169–175.

Morris, B., and Morris, R. (1975). Globulin transmission by the gut in young rats, and the effects of cortisone acetate. *In* "Maternofoetal Transmission of Immunoglobulins" (W. A. Hemming, ed.), pp. 359–369. Elsevier, Amsterdam, The Netherlands.

Owen, R. L. (1977). Sequential uptake of horseradish peroxidase by lymphoid follicle epithelium of Peyer's patches in the normal unobstructed mouse intestine: An ultrastructural study. *Gastroenterology* **72,** 440–451.

Owen, R. L., and Bhalla, D. K. (1983). Cytochemical analysis of alkaline phosphatase, esterase activities and of lectin-binding and anionic sites in rat and mouse Peyer's patch M cells. *Am. J. Anat.* **168,** 199–212.

Owen, R. L., and Pappo, J. (1988). Absence of secretory component expression by epithelial cells overlying rabbit gut-associated tissue. *Gastroenterology* **95,** 1173–1177.

Pappo, J. (1989). Generation and characterization of monoclonal antibodies recognizing follicle epithelial M cells in rabbit gut-associated lymphoid tissues. *Cell. Immunol.* **120,** 31–41.

Phillips, A. D., Smith, M. W., and Walker-Smith, J. P. (1988). Selective alteration of brush-border hydrolases in intestinal diseases in childhood. *Clin. Sci.* **74,** 193–200.

Pinto, M., Robine-Leon, S., Appay, M. D., Kedinger, M., Triadou, N., Dussaulx, E., Lacroix, B., Simon-Assmann, P., Haffen, K., Fogh, J., and Zweibaum, A. (1983). Enterocyte-like differentiation and polarization of the human colon carcinoma cell line Caco-2 in culture. *Biol. Cell.* **47,** 323–330.

Ponder, B. A. J., Schmidt, G. H., Wilkinson, M. M., Wood, M. J., Monk, M., and Reid, A. (1985). Derivation of mouse intestinal crypts from single progenitor cells. *Nature (London)* **313,** 689–691.

Quaroni, A. (1985). Development of fetal rat intestine in organ and monolayer culture. *J. Cell Biol.* **100,** 1611–1622.

Ratcliffe, B., Smith, M. W., Miller, B. G., James, P. S., and Bourne, F. J. (1989). Effect of soya-bean protein on the ability of gnotobiotic pig intestine to digest and absorb nutrients. *J. Agric. Sci.* **112,** 123–130.

Rousset, M., Chantret, I., Darmoul, D., Trugnan, G., Sapin, C., Green, F., Swallow, D., and Zweibaum, A. (1989). Reversible forskolin-induced impairment of sucrase-isomaltase mRNA levels, biosynthesis, and transport to the brush border membrane in Caco-2 cells. *J. Cell. Physiol.* **141,** 627–635.

Roy, M. J., and Varvayanis, M. (1987). Development of dome epithelium in gut-associated lymphoid tissues: Association of IgA with M cells. *Cell Tissue Res.* **248,** 645–651.

Rubin, D. C., Ong, D. E., and Gordon, J. I. (1989). Cellular differentiation in the emerging fetal rat small intestinal epithelium: Mosaic patterns of gene expression. *Proc. Natl. Acad. Sci. U.S.A.* **86,** 1278–1282.

Rudo, N. D., Rosenberg, I. H., and Wissler, R. W. (1976). The effect of partial starvation and glucagon treatment on intestinal villus morphology and cell migration. *Proc. Soc. Exp. Biol. Med.* **152,** 277–280.

Schmidt, G. H., Wilkinson, M. M., and Ponder, B. A. J. (1985). Cell migration pathway in the intestinal epithelium: An in situ marker system using mouse aggregation chimeras. *Cell* **40,** 425–429.

Schmidt, G. H., Winton, D. J., and Ponder, B. A. J. (1988). Development of the pattern of cell renewal in the crypt-villus unit of chimaeric mouse small intestine. *Development* **103,** 785–790.

Sebastio, G., Hunziker, W., Ballabio, A., Auricchio, S., and Semenza, G. (1986). On the primary site of control in the spontaneous development of small-intestinal sucrase-isomaltase after birth. *FEBS Lett.* **208,** 460–464.

Sebastio, G., Hunziker, W., O'Neill, B., Malo, C., Ménard, D., Auricchio, S., and Semenza, G. (1987). The biosynthesis of intestinal sucrase-isomaltase in human embryo is most likely controlled at the level of transcription. *Biochem. Biophys. Res. Commun.* **149,** 830–839.

Semenza, G. (1967). Intestinal oligosaccharidases and disaccharidases. *In* "Handbook of Physiology. Sect. 6: Alimentary Canal" (C. F. Ede, ed.), Vol. 5, pp. 2543–2666. Am. Physiol. Soc., Washington, D.C.

Shields, H. M., Bates, M. L.., Bass, N. M., Best, C. J., Alpers, D. H., and Ockner, R. K. (1986). Light microscopic immunocytochemical localization of hepatic and intestinal types of fatty acid-binding proteins in rat small intestine. *J. Lipid Res.* **27,** 549–557.

Simon, G. L., and Gorbach, S. L. (1987). Intestinal flora and gastrointestinal function. *In* "Physiology of the Gastrointestinal Tract" (L. R. Johnson, ed.), pp. 1729–1747. Raven, New York.

Simon-Assmann, P. M., Kedinger, M., Grenier, J. F., and Haffen, K. (1982). Control of brush border enzymes by dexamethasone in the fetal rat intestine cultured *in vitro*. *J. Pediatr. Gastroenterol. Nutr.* **1,** 257–265.

Smith, M. W. (1985a). Selective expression of brush border hydrolases by mouse Peyer's patch and jejunal villus enterocytes. *J. Cell. Physiol.* **124,** 219–225.

Smith, M. W. (1985b). Expression of digestive and absorptive function in differentiating enterocytes. *Annu. Rev. Physiol.* **47,** 247–260.

Smith, M. W., and Brown, D. (1989). Dual control over microvillus elongation during enterocyte development. *Comp. Biochem. Physiol. A.* **93A,** 623–628.

Smith, M. W., and Jarvis, L. G. (1978). Growth and cell replacement in the newborn pig intestine. *Proc. R. Soc. London, Ser. B* **203,** 69–89.

Smith, M. W., and Lloyd, S. (1989). Intestinal infection with *Nematospiroides dubius* selectively increases lactase expression by mouse jejunal enterocytes. *Clin. Sci.* **77,** 139–144.

Smith, M. W., Jarvis, L. G., and King, I. S. (1980). Cell proliferation in follicle-associated epithelium of mouse Peyer's patch. *Am. J. Anat.* **159,** 157–166.

Smith, M. W., Miller, B. G., James, P. S., and Bourne, F. J. (1985). Effect of weaning on the structure and function of piglet small intestine. *In* "Digestive Physiology in the Pig" (A. Just, H. Jørgensen, and J. A. Fernandez, eds.), Rep. No. 580, pp. 75–78. Natl. Inst. Anim. Sci., Copenhagen, Denmark.

Smith, M. W., James, P. S., Tivey, D. R., and Brown, D. (1988). Automated histochemical analysis of cell populations in the intact follicle-associated epithelium of the mouse Peyer's patch. *Histochem. J.* **20,** 443–448.

Spring, K. R. (1979). Optical techniques for the evaluation of epithelial transport processes. *Am. J. Physiol.* **237,** F167–F174.

Stallmach, A., Hahn, U., Merker, H. J., Hahn, E. C., and Riecken, E. O. (1989). Differentiation of rat epithelial cells is induced by organotypic mesenchymal cells *in vitro. Gut* **30,** 959–970.

Sweetser, D. A., Birkenmeier, E. H., Hoppe, P. C., McKeel, D. W., and Gordon, J. I. (1988a). Mechanisms underlying generation of gradients in gene expression within the intestine: An analysis using transgenic mice containing fatty acid binding protein-human growth fusion genes. *Genes Dev.* **2,** 1318–1332.

Sweetser, D. A., Hauft, S. M., Hoppe, P. C., Birkenmeier, E. H., and Gordon, J. I. (1988b). Transgenic mice containing intestinal fatty acid-binding protein-human growth hormone fusion genes exhibit correct regional and cell-specific expression of the reporter gene in their small intestine. *Proc. Natl. Acad. Sci. U.S.A.* **85,** 9611–9615.

Thompson, C. S., and Debnam, E. S. (1986). Hyperglucagonaemia: Effects on active nutrient uptake by the rat jejunum. *J. Endrocrinol.* **111,** 37–42.

Thompson, N. L., Flanders, K. C., Smith, J. M., Ellingsworth, L. R., Roberts, A. B., and Sporn, M. B. (1989). Expression of transforming growth factor-β1 in specific cells and tissues of adult and neonatal mice. *J. Cell Biol.* **108,** 661–669.

Tivey, D. R., and Smith, M. W. (1989). Cytochemical analysis of single villus peptidase activities in pig intestine during neonatal development. *Histochem. J.* **21,** 601–608.

Trier, J. S. (1967). Structure of the mucosa of the small intestine as it relates to intestinal function. *Fed. Proc.* **26,** 1391–1404.

Walker, W. A. (1979). Gastrointestinal host defence: Importance of gut closure in control of macromolecular transport. *Ciba Found. Symp.* **70,** 201–216.

Wice, B. M., Trugnan, G., Pinto, M., Rousset, M., Chevalier, G., Dussaulx, E., Lacroix, B., and Zweibaum, A. (1985). The intracellular accumulation of UDP-*N*-acetylhexosamines is concomitant with the inability of human colon cancer cells to differentiate. *J. Biol. Chem.* **260,** 139–146.

Winton, D. J., Blount, M. A., and Ponder, B. A. J. (1988). A clonal marker induced by mutation in mouse intestinal epithelium. *Nature* (*London*) **333,** 463–466.

Wolf, B. C., Salem, R. R., Sears, H. F., Horst, D. A., Lavin, P. T., Herlyn, M., Itzkowitz, S. H., Schlom, J., and Steele, G. D. (1989). The expression of colorectal carcinoma-associated antigens in the normal colonic mucosa. An immunohistochemical analysis of regional distribution. *Am. J. Pathol.* **135,** 111–119.

Yasugi, S., Kedinger, M., Simon-Assman, P., Benziges, F., and Haffen, K. (1989). Differentiation of proventricular epithelium in xenoplastic associations with mesenchymal or fibroblastic cells. *Wilhelm Roux's Arch. Dev. Biol.* **198,** 114–117.

Yeh, K.-Y., and Holt, P. R. (1986). Ontogenic timing mechanism initiates the expression of rat intestinal sucrase activity. *Gastroenterology* **90,** 520–526.

Younes, M., Harris, A. S., and Morrow, J. S. (1989). Fodrin as a differentiation marker. Redistributions in colonic neoplasia. *Am. J. Pathol.* **135,** 1197–1212.

Zweibaum, A. (1986). Enterocytic differentiation of cultured human colon cancer cell lines: Negative modulation by D-glucose. *In* "Ion Gradient-Coupled Transport" (F. Alvarado and C. H. van Os, eds.), pp. 345–353. Elsevier, Amsterdam.

Zweibaum, A., and Chantret, I. (1989). Human colon carcinoma cell lines as *in vitro* models for the study of intestinal cell differentiation. *In* "Adaptation and Development of Gastrointestinal Function" (M. W. Smith and F. V. Sepúlveda, eds.), pp. 103–112. Manchester Univ. Press, Manchester, England.

Zweibaum, A., Laburthe, M., Grasset, E., and Louvard, D. (1991). Use of cultured cell lines in studies of intestinal cell differentiation and function. *In* "Handbook of Physiology. Absorptive and Secretory Processes of the Intestines" (M. Field and R. A. Frizzell, eds.). Am. Physiol. Soc., Washington, D.C. (in press).

CHAPTER 6

Ion Transport and Adenylyl Cyclase System in Red Blood Cells

Hyun Dju Kim
Department of Pharmacology, School of Medicine, University of Missouri, Columbia, Missouri 65212

I. INTRODUCTION

Nonnucleated mammalian as well as nucleated avian erythrocytes have long been a readily available model for the investigation of membrane transport in general, and ion transport in particular. As pointed out by Duhm (1989), an impressive number of ion transport pathways now known to exist in a variety of cells were originally identified in red cell membranes, namely, Ca^{2+}-activated K^+ channels (Gardos, 1956), Cl^-/HCO_3^- exchanger (Hamburger, 1918), ouabain inhibition of Na^+/K^+ pump (Schatzmann, 1953), Ca^{2+} pump (Schatzmann, 1966), and Na^+/K^+ cotransport (Wiley and Cooper, 1974). For more than three decades, the pump and leak hypothesis, envisioned by Tosteson and Hoffman (1960) on dimorphic sheep red cells, has been the driving force which stimulated interest in unravelling the underlying mechanism responsible for homeostasis of cell volume. Apart from metabolic energy which energizes the Na^+/K^+ pump, a different form of energy transduction which can also fuel a solute movement against its chemical potential gradient has been recognized as implicitly embodied in the pioneering discovery of Na^+ gradient-coupled glucose transport in intestinal epithelia (Crane *et al.*, 1961; Csaky, 1965). Thus, it is now well established that the transport of one ion influences transport of the other in the same direction, termed cotransport, and in the opposite direction, termed antiport, involving a common element in the membrane. Cotransport system can mediate concomitant movement of more than two ions, as evidenced by the $Na^+/K^+/Cl^-$ cotransport pathway as opposed to the Na^+/H^+ antiport in which there is no net solute movement. However, the antiporter can operate in conjunction with the Cl^-/HCO_3^- exchanger, resulting in salt and water movement.

Although both cotransport and antiport are usually minimal in most erythrocytes at their normal cell volume, a perturbation in cell volume triggers a selective and coordinate activation of these transport pathways, resulting in the restoration of cell volume. In addition, there is growing evidence that hormones and neurotransmitters regulate ion transport. In recent years, signal transduction pathways involving second messengers have been investigated in intricate detail. That agonist occupancy of β-receptors plays a profound role in the regulation of ion transport is particularly well known.

This chapter addresses the regulation of ion transport by the adenylyl cyclase system, with particular emphasis on studies with red cells used as a model system. The first section of this chapter describes recent advances in the adenylyl cyclase system in general, followed by developmental changes in adenylyl cyclase in reticulocytes and in erythrocytes. The second part discusses effects of cyclic AMP in $Na^+/K^+/Cl^-$ cotransport, Na^+/H^+ antiport, and K^+/Cl^- cotransport systems. The third section examines a novel observation pertaining to the influence of adenosine receptor agonists on KCl cotransport.

II. HORMONE-SENSITIVE ADENYLYL CYCLASE SYSTEM

The epoch-making discovery of cAMP was made more than 30 years ago by Rall *et al.* (1957) in experiments with liver phosphorylase, which is activated by epinephrine and glucagon. These authors noted the production of a heat-stable factor, which they initially termed "active factor," that stimulated the formation of liver phosphorylase.

A decade and a half later, Rodbell *et al.* (1971), while studying glucagon-activated adenylyl cyclase of liver membranes, found an indispensable requirement of guanine nucleotides. This eventually led to the discovery of the so-called guanine nucleotide-binding regulatory proteins, termed G proteins or N proteins. As it turns out, in early attempts to ascertain dual requirement of hormone and GTP for the action of adenylyl cyclase, avian erythrocytes were widely used biological sources in which key discoveries were made. Pfeuffer and Helmreich (1975) showed in pigeon erythrocyte membranes that metabolically stable analogs are far more effective in stimulating β-receptor-activated adenylyl cyclase with the rank order of decreasing potencies: GTPγs > GPP(NH)P > GPP(CH_2)P. These metabolically inert guanine nucleotides have since been immeasurably useful tools for the elucidation of G protein function. A variety of GTP-binding proteins was identified in pigeon erythrocyte membranes, including proteins with M_r of 86,000, 52,000, 42,000, and 23,000 (Pfeuffer, 1977). It is now well known that G proteins are ubiquitous, some of which have nearly identical molecular weight corresponding to the G proteins initially reported by Pfeuffer (1977). That the guanine nucleotide-binding proteins possess intrinsic GTPase activity was initially observed by Pfeuffer and Helmreich (1975). Cassel and Selinger (1976) have established the link between activation of β-receptors by catecholamines and stimulation of GTPase, which they have correctly shown to be independent from cAMP production by adenylyl cyclase in turkey erythrocyte membranes.

The notion that a guanine nucleotide-binding protein, termed G_s protein, may be the agent mediating the GTP activation of adenylyl cyclase was suggested independently and simultaneously by Pfeuffer (1977) and Ross and Gilman (1977). Pfeuffer (1977) found that myocardial adenylyl cyclase, which was rendered inactive by depletion of guanine nucleotide-binding proteins, could be restored on reconstitution with guanine nucleotide-binding proteins isolated from pigeon erythrocyte membranes. In the same *Journal of Biological Chemistry* issue in which Pfeuffer's paper was published, Ross and Gilman (1977) arrived at the same conclusion based on reconstitution experiments using a mutant S49 lymphoma cell clone that lacks detectable G proteins.

It is now firmly established that stimulatory guanine nucleotide-binding (G_s) proteins and inhibitory guanine nucleotide-binding proteins, termed G_i proteins, function as transducers linking the external signal through the receptor to the catalytic moiety of adenylyl cyclase, resulting in a stimulation and inhibition of the catalyst, respectively (Fig. 1). In analogy to G_s proteins, Katada and Ui (1979, 1980, 1981, 1982) describe the salient

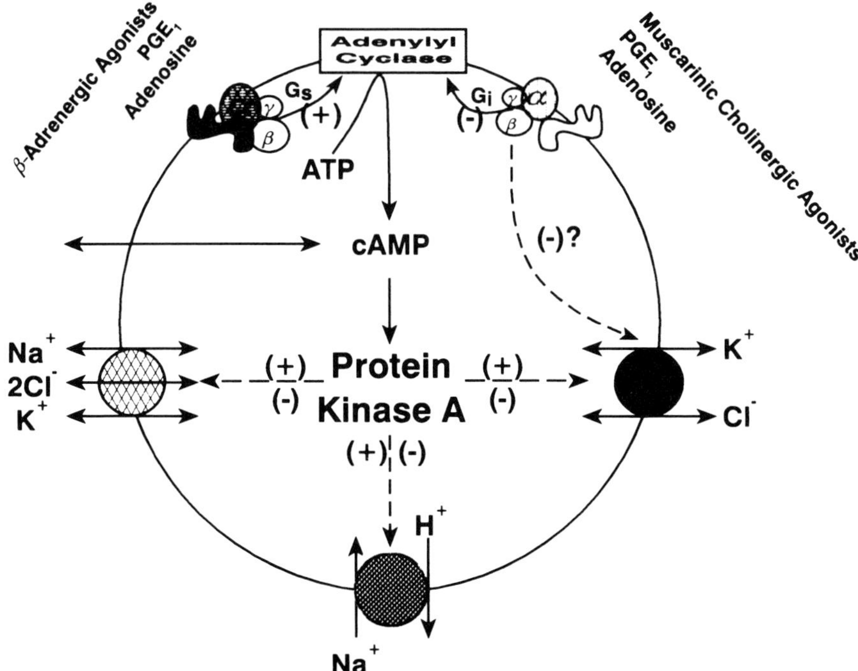

FIG. 1. Adenylyl cyclase and ion transport. + and − denote activation and inhibition of enzyme activity or transport pathways, respectively. For details see text.

feature of G_i proteins. It was found that GTP was required for adenylyl cyclase inhibition by epinephrine in cultured islet cells; the inhibitory action of epinephrine on adenylyl cyclase was markedly attenuated by pertussis toxin; and pertussis toxin catalyzed ADP ribosylation from NAD of an M_r 41,000 membrane protein, which is distinct from the cholera toxin substrate.

In addition to G_s and G_i proteins, a variety of G proteins constitutes a large G protein family (Casperson and Bourne, 1987; Gilman, 1987; Milligan, 1988; Johnson and Dhanasekaran, 1989). The guanine nucleotide-binding proteins, termed G_o proteins, whose biological function is unknown, were isolated from brain tissue (Sternweiss and Robishaw, 1984; Neer *et al.*, 1984) and G_t, known as transducin, activated the photon rhodopsin receptor linked to a cyclic GMP-specific phosphodiesterase. Pertussis toxin has not only proved to be a useful tool for elucidation of G_i protein function but also for identification and cloning of heterogeneous forms of pertussis toxin-sensitive proteins, including G_t, G_i1, G_i2, G_i3, and

G_o (Milligan, 1988). Although there exist "little" GTP-binding proteins having M_r of 20,000–25,000, a major class of G proteins that participate in signal transduction pathways is heterotrimeric in structure, composed of α, β, and γ subunits.

The α subunits of G_s proteins, which range from M_r 39,000 to 52,000 (Milligan, 1988), not only possess a high-affinity guanine nucleotide-binding site, but also stimulate the catalyst of adenylyl cyclase (Northrup *et al.*, 1983). Most of the α subunits of G_s and G_i proteins are substrates for ADP ribosylation catalyzed by cholera toxin and pertussis toxins, respectively. Moreover, the α subunits of both G_s and G_i proteins exhibit an intrinsic GTPase activity (Milligan, 1988). In addition to G proteins' being heterogeneous, α subunits of a G protein may exist in multiple forms. Katada *et al.* (1986) have identified two GTP-binding protein subunits, termed α_{41} and α_{39}, of which only α_{41} inhibits the adenylyl cyclase.

The β and γ subunits have M_r of 35,000–36,000 and 8,000–10,000, respectively, but appear to exist in tight association with each other. Although $\beta\gamma$ subunits of G_s and G_i proteins are thought to cross-react freely with both α_s and α_i subunits, more recent molecular cloning has revealed the presence of multiple genes which may encode β subunits (Fong *et al.*, 1987; Gao *et al.*, 1987).

The integration of individual subunits of G protein participating in the signal transduction pathway is complex (Katada *et al.*, 1986; Gilman, 1987; Bourne, 1988; Milligan, 1988; Johnson and Dhanasekaran, 1989). However, it may be summarized in general terms as follows:

The GDP dissociates from α subunits on interaction with agonist occupied receptors.

The α subunits, having a guanine nucleotide-binding site available, bind GTP, which results in the dissociation of the GTP α subunit from $\beta\gamma$ subunits.

GTP α subunits, arising from either stimulatory or inhibitory receptor activation, regulate the catalyst of the enzyme. The inactivation of GTP α subunits ensues as a result of the intrinsic GTPase property of α subunits which catalyzes the hydrolysis of GTP-α to GDP-α subunits. The reassociation of subunits into $\alpha\beta\gamma$ complex presumably completes the cycle.

In addition, the $\beta\gamma$ subunit, which is dissociated from G_i proteins as a consequence of the inhibitory receptor activation, may reassociate α subunits of G_s proteins. This causes a reduction of the free concentration of α subunits of G_s protein, thereby indirectly inhibiting adenylyl cyclase activity.

$\beta\gamma$ subunits may directly inhibit catalytic activity of adenylyl cyclase.

The α_{41} of G_i proteins inhibits the catalyst only when the catalyst is stimulated by α subunit of G_s.

In Table I, a partial list of ligands which activate adenylyl cyclase stimulatory and inhibitory receptors is given. In addition to β-adrenergic agonists, a variety of agents stimulate adenylyl cyclase. Even before G proteins were isolated and their function was ascertained, it was known that certain ligands listed in Table I elicited an inhibition rather than a stimulation of adenylyl cyclase in numerous cells. Adenosine was one of the first ligands found to inhibit cAMP production (Moriwaki and Fóa, 1970). While most ligands listed in Table I confer either a stimulatory or inhibitory response, a single ligand can also exert a biphasic regulation of adenylyl cyclase. Again, adenosine is a case in point. Thus, adenylyl cyclase is stimulated by low adenosine concentrations but inhibited by high adenosine concentrations in a variety of cells (Londos and Wolff, 1977; Van Calker *et al.,* 1979). The purinergic receptors for which adenosine is the most potent agonist are termed P_1 receptors, to be distinguished from those purinergic receptors, termed P_2 receptors, for which ATP is the most potent agonist (Burnstock, 1978). The P_1 receptors were initially subdivided depending on whether adenylyl cyclase is stimulated (R_a or A_2 receptors) or inhibited (R_1 or A_1 receptors) (Londos and Wolff, 1977). In light of the finding that adenosine receptor activation does not always lead to predictable changes in second messengers, the current classification of adenosine receptor subtypes is based on the rank order of potencies of several adenosine agonists.

A. Nucleated Erythrocytes

1. β-Receptors

Turkey erythrocytes have long been a widely used model of β-receptor activation of adenylyl cyclase. Using a potent photoaffinity probe, ^{125}I-labeled *P*-azidobenzylcarazolol, Sibley *et al.* (1984a) identified a protein of $M_r = 49{,}000$ as the β_1-adrenergic receptor. Although brief exposure of turkey erythrocytes to catecholamines leads to the production of cAMP, prolonged exposure to an agonist results in an attenuation of adenylyl cyclase activity by subsequent agonist stimulation. This process is referred to as "desensitization." The turkey erythrocytes exhibit not only homologous desensitization in which hormonal attenuation is limited to catecholamine stimulation, but also heterologous desensitization in which diminished responses can be elicited by a broad spectrum of activators, including guanine nucleotide, NaF, and other hormones (Stadel *et al.,* 1983a; Nambi *et al.,* 1984, 1985). Both the sequestration (Nambi *et al.,* 1984) and phosphorylation (Stadel *et al.,* 1983a; Sibley *et al.,* 1984b; Nambi *et al.,* 1985) of receptors have been postulated as underlying mechanisms responsible for the desensitization process.

TABLE I

Regulation of Adenylyl Cyclase by Receptor Activation

Ligands or Toxin	Source	References
Stimulation of Adenylyl Cyclase		
β-Adrenergic agonists	Rat reticulocytes	Montandon and Porzig (1983), Beckman and Hollenberg (1979), Kaiser *et al.* (1977)
	Turkey erythrocytes	Sibley *et al.* (1984 a,b), Stadel *et al.* (1982)
	Mouse erythrocytes	Sheppard and Burghardt (1969)
	Rat erythrocytes	Rasmussen *et al.* (1975)
	Human erythrocytes	Kaiser *et al.* (1974), Sager and Jacobsen (1985)
	Frog erythrocytes	Chuang *et al.* (1980), Stadel *et al.* (1983b)
	Baboon erythrocytes	Susanni *et al.* (1985)
	Fish erythrocytes	Mahe *et al.* (1985)
Prostaglandin E_1 and GTP	Rat reticulocytes	Yamashita *et al.* (1988)
Cholera toxin	Pigeon erythrocytes	Le Vine and Cuatrecasas (1981)
Vasoactive intestinal peptide	HT29 cells	Turner *et al.* (1986)
Adenosine	Rabbit reticulocytes	Cooper and Jagus (1982)
	Human platelet membranes	Barrington *et al.* (1989)

(continued)

TABLE I *(Continued)*

Ligands or Toxin	Source	References
Inhibition of Adenylyl Cyclase		
Epinephrine	Platelets	Jakobs *et al.* (1978), Steer and Wood (1979)
Prostaglandin E_1	Hamster fat cell	Aktories *et al.* (1979)
Muscarinic cholinergic agonists	Rabbit myocardial membranes	Jakobs *et al.* (1979)
	NG108-15	Lichtshtein *et al.* (1979)
Opioid peptide	NG108-15 cells	Blume *et al.* (1979)
Nicotinic acid	Fat cells	Aktories *et al.* (1980)
Catecholamines	NG108-15 cells	Sabol and Nirenberg (1979)
Prostaglandin E_1 and GTP	Rat reticulocytes	Yamashita *et al.* (1988)
Adenosine	Rat liver homogenate	Moriwaki and Fóa (1970)
	Fat cell ghosts	Fain *et al.* (1972)
	Brain tissue, fat cells, Ehrlich ascites	McKenzie and Bär (1973)
	Guinea pig lung	Weinryb and Michel (1974)
	Human embryonic intestinal epithelial cells	Zenser (1976)
	Rat liver membranes	Londos and Preston (1977)

2. Purification of G_s Proteins

The G_s proteins isolated from turkey erythrocytes have two putative subunits with M_r of 35,000 and 45,000 (Hanski *et al.*, 1981). The reconstitution experiments with isolated turkey erythrocyte G_s proteins and cyc$^-$S49 lymphoma cells allowed an expression of the adenylyl cyclase property characteristic of the turkey erythrocyte system.

B. Nonnucleated Erythrocytes

1. β-Receptors and Adenylyl Cyclase

While the existence of adenylyl cyclase in human erythrocyte membranes has long been known (Rodan *et al.*, 1976), the presence of β-receptors on the human erythrocyte surface has been equivocal (Jeffrey *et al.*, 1980). Contrary to the widely held view that human red cells lack β-receptors, Sager (1983) and Sager and Jacobsen (1985) demonstrated the presence of β-receptors in human erythrocytes, using an improved filtering system for binding assay. In competition binding experiments, the antagonist radioligand [^{3}H]dihydroalprenolol was displaced by agonists with the rank order of decreasing potencies isoproterenol > epinephrine > norepinephrine. These findings show that β-receptors of human erythrocytes are the β_2-adrenergic subtype, as opposed to the β_1-adrenergic receptor subtype residing on the surface of avian erythrocytes (Sibley *et al.*, 1984a). A curious finding is that both the number of β-receptors and isoproterenol-stimulated elevations of cAMP content are reported to increase upon treatment of human erythrocytes with autologous plasma (Sager and Jacobsen, 1985). Although β-receptor activation was found to cause a 4-fold increase in the intracellular cAMP content of intact human red blood cells (Kaiser *et al.*, 1974; Sager, 1982), other studies failed to corroborate this (Sheppard and Burghardt, 1969; Rasmussen *et al.*, 1975).

2. Isolation and Characterization of the Stimulatory and Inhibitory Regulatory Proteins from Human Erythrocytes

Codina *et al.* (1984a,b) have obtained 500–1000 μg each of G_s and G_i proteins from 60 units of outdated human blood. Isolated α_s subunits, which were identified by adenylyl cyclase reconstitution assay in cyc$^-$ membranes, have M_r = 42,000, whereas the α_i subunits, determined by ADP ribosylation reaction with pertussis toxin, have M_r = 40,000. The β subunit of both G_s and G_i proteins was found to be a protein of M_r = 35,000. Subsequent studies by Codina and colleagues (Iyengar *et al.*, 1987) have identified a GTP-binding protein of $M_r = 43{,}000$ which is a pertussis toxin substrate. The pertussis toxin substrate protein isolated from human erythrocytes is termed G_k to denote its capacity to couple muscarinic receptors to K^+ channels of the atrium. Raging controversy exists as to which of the α (Codina *et al.*, 1987; Cerbai *et al.*, 1988; Yatani *et al.*, 1988a) versus $\beta\gamma$ (Logothetis *et al.*, 1987) subunits is responsible for the activation of K^+ channels.

Recently, molecular cloning has provided evidence that there are three genes encoding α subunits of pertussis toxin substrates in the human genome, termed α_i-1, α_i-2, and α_i-3 (Codina *et al.*, 1988). Based on the full nucleotide sequence of the complete open reading frame of human liver α_i-3 cDNA and the animo acid sequence of proteolytic fragments of α_k, Codina *et al.* (1988) suggest that α_i-3 is α_k.

C. Comparison of Adenylyl Cyclase Activation and Inhibition in Erythrocytes

It is abundantly clear that adenylyl cyclase activation by diverse agonists has been extensively delineated in both reticulocytes and erythrocytes. By sharp contrast, little is known about agonist occupancy resulting in adenylyl cyclase inhibition of mature erythrocytes (Table I).

D. Erythrocyte Membrane Permeability to cAMP

Apart from a ligand–receptor interaction resulting in increased cAMP content, it is possible to raise intracellular cAMP substantially, since cAMP, despite its phosphorylated form, is transported across cell membranes (Brunton and Mayer, 1979; Heasley *et al.*, 1979; Holman, 1979; Brunton and Buss, 1980). Davoren and Sutherland (1963) were the first to recognize that cAMP, which was produced in response to catecholamine in pigeon cells, was transported out of cells against its concentration gradient. Subsequent studies show that cAMP efflux is a saturable and ATP-dependent process (Brunton and Mayer, 1979; Wiemer *et al.*, 1982).

Nonnucleated mammalian red cells are also permeable to cAMP (Garay, 1982; Sergeant and Kim, 1985; Kim *et al.*, 1989). Figure 2 shows cAMP uptake by pig cells which were incubated in the presence of 1 m*M* cAMP at 37°C. At the end of 1 hr incubation, cAMP was accumulated to approximately 2.7 μmol/liter packed cells.

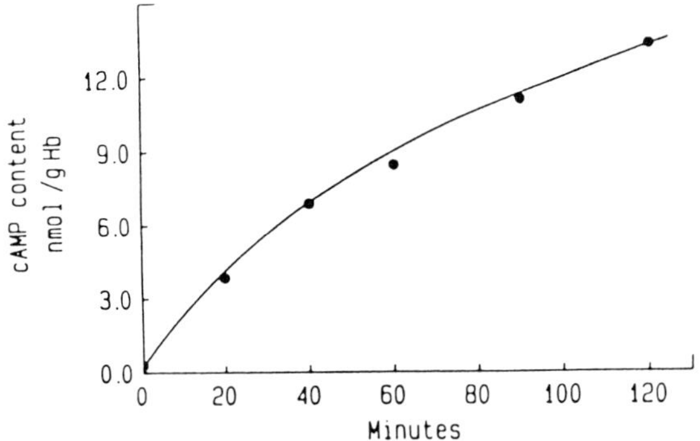

FIG. 2. cAMP uptake by pig red cells. Cells were incubated in balanced isotonic media containing 1 m*M* cAMP. At frequent intervals, cells were washed several times at 4°C and deproteinized with perchloric acid. cAMP was measured by radioimmunoassay on an aliquot of perchloric acid extract. Values represent an average of two separate experiments. (Reproduced from Kim *et al.*, 1989, with permission from the publisher.)

E. Developmental Changes in the Adenylyl Cyclase System during Reticulocyte Maturation

Our current understanding of the developmental changes in the adenylyl cyclase system comes primarily from studies with rat reticulocytes which were produced either by bleeding or by the use of hemolytic agents. As in the avian adenylyl cyclase system, the rat reticulocyte system also exhibits heterologous desensitization in that a prolonged treatment with isoproterenol results in a decrease in the subsequent activation of adenylyl cyclase by β-agonists, fluoride, and GPPNHP (Yamashita *et al.*, 1987). Moreover, regulation of the adenylyl cyclase system of rat reticulocytes is under dual control, mediated by both G_s and G_i proteins (Shane *et al.*, 1985b). As stated earlier, a single agonist such as adenosine can elicit a biphasic regulation of adenylyl cyclase depending on its concentration. A single agonist at a given concentration may also manifest a dual control of adenylyl cyclase depending on the variation of cofactors. For example, prostaglandin E_1 (PGE_1) at 10 μM causes stimulation of the adenylyl cyclase of reticulocytes if the GTP concentration is low, but an inhibition if the GTP concentration is high (Yamashita *et al.*, 1988). In addition, NaF plus $AlCl_3$ and GPPNHP were found to regulate adenylyl cyclase in a dose-dependent, biphasic manner. Treatment of rat reticulocytes with phorbol-12 myristate-13 acetate (PMA) for 1.5 hr at 30°C results in a decrease in isoproterenol stimulation, but not basal adenylyl cyclase, mimicking the phenomenon of desensitization. This suggests that protein kinase C is involved in the regulation of the stimulatory limb of adenylyl cyclase. With respect to the inhibitory pathway of adenylyl cyclase, Yamashita *et al.* (1988) examined the effects of PMA on PGE_1/GTP-regulated adenylyl cyclase activity. It was found that PMA treatment blunted the inhibition of adenylyl cyclase activity seen in the presence of a high concentration of GTP. Moreover, pertussis toxin treatment resulted in a similar reduction of the inhibitory limb of adenylyl cyclase. These findings led Yamashita *et al.* (1988) to suggest that protein kinase C also interferes with the coupling between PGE_1 receptor and G_i proteins.

Studies aimed at unraveling the mechanisms by which maturing rat reticulocytes lose catecholamine-sensitive adenylyl cyclase activity reveal complex changes. In the course of rat reticulocyte maturation, the number of β-receptors is reported to be unchanged (Shane *et al.*, 1981) or decreased by 70% (Limbird *et al.*, 1980; Larner and Ross, 1981). An 80–90% loss of adenylyl cyclase activity during reticulocyte maturation has also been observed (Larner and Ross, 1981). Larner and Ross (1981) found relatively proportional decreases in the number of β-receptors, G protein activity as assessed by reconstitution experiments with S49 cyc lymphoma cells, and the catalyst of the enzyme as determined by its intrinsic Mn^{2+} stimulation in the maturing rat reticulocyte.

In contrast, Farfel and Cohen (1984) showed that, in the human system, the catalyst of the enzyme is preferentially lost in relation to G protein activity, which remains relatively unchanged as assessed by the same reconstitution assay as used by Larner and Ross (1981). Inasmuch as blood cells obtained from patients with varying degrees of reticulocytosis were used without enriching the reticulocyte population, the magnitude of the loss of adenylyl cyclase activity or β-receptors in the course of human reticulocyte maturation is not known.

In addition to agonist–receptor interactions which regulate the catalyst of the enzyme through various G proteins, a cytosolic activator protein (RCAP) with $M_r \approx 20{,}000$ has been identified. RCAP causes an augmentation of catechol amine-sensitive adenylyl cyclase activity in reticulocyte membranes (Shane *et al.*, 1985a,b). Although the physiological role for RCAP is less clear, RCAP not only requires G_s proteins for enzyme activation but also appears to inhibit G_i function. In recent years, the characterization of this interesting cytosolic activation has not been continued (J. P. Bilezikian, personal communication).

III. EFFECTS OF cAMP ON ION CHANNELS AND TRANSPORT

Table II compiles a variety of ion movements influenced by cAMP. Perhaps the most widely known cAMP action pertains to the diarrheal disease caused by cholera toxin infection. Dramatic fluid loss, which can

TABLE II

Effects of cAMP on Ion Channels and Transport

Ion transport and channels	Stimulation	Inhibition	References
Cl^- channels	Gastrointestinal tract		Fields *et al.* (1989a,b)
Ca^{2+} transport	Human red blood cell ghosts		Weller (1978)
Na^+ influx	Brown adipocytes		Connolly *et al.* (1986)
Cl^- transport	Rat submandibular cell line		He *et al.* (1989)
	T_{84} cells		Mandel *et al.* (1986)
Ca^{2+}-activated K^+ channels	T_{84} cells		McRoberts *et al.* (1985)
HCO_3^-/Cl^- exchange		Aortic smooth muscle cells	Vigne *et al.* (1988)
$Na^+/K^+/Cl^-$ cotransport		Flounder intestine	Palfrey and Rao (1983)
		Human red blood cells	Garay (1982), Garay *et al.* (1983), Diez *et al.* (1984)
		Human fibroblasts	Owen and Prastein (1985)
		Vascular smooth muscle	Owen (1984)
	Duck red blood cells		Kregenow (1978, 1981), Kregenow *et al.* (1976), McManus and Schmidt (1978), Haas and McManus (1985)

(*continued*)

TABLE II *(Continued)*

Ion transport or channels	Stimulation	Inhibition	References
$Na^+/K^+/Cl^-$ cotransport	Turkey red blood cells		Palfrey *et al.* (1980), Ueberschär and Bakker-Grunwald (1985)
	Goose red blood cells		Palfrey and Greengard (1981)
	Pigeon red blood cells		Palfrey and Greengard (1981)
	Frog red blood cells		Rudolph and Greengard (1980)
	HT29 cells		Kim *et al.* (1988)
	Smooth muscle and endothelial cells		Smith and Smith (1987)
Na^+/H^+ exchanger		Renal brush border membranes	Kahn *et al.* (1985)
		Necturus gall bladder epithelium	Reuss and Peterson (1985)
	Trout red blood cells		Borgese *et al.* (1986), Garcia-Romeu *et al.* (1988), Borgese *et al.* (1987)
Cl^--dependent K^+ transport		Duck red blood cells	Starke and McManus (1988), Starke and McManus (1990)
	Pig red blood cells		Kim *et al.* (1989)
	Sheep red blood cells		Sohn and Kim (1991)

be counteracted by stimulation of the Na/glucose cotransport, stems from ADP ribosylation by toxin with an attendant increase in cAMP content causing activation of the Cl^- channels (Field *et al.*, 1989a,b). While the stimulatory role of cAMP on ion movements is well established, cAMP does not always result in the activation of ion channels or transport. In the following, ion transport systems in which cAMP elicits either an activation or inhibition, depending on cell type (Fig. 1), are discussed.

A. $Na^+/K^+/Cl^-$ Cotransport

Before the identity of cAMP was firmly established, Ørskov (1956) found that norepinephrine increased K^+ transport in pigeon and frog erythrocytes. More than a decade later, Riddick *et al.* (1971) confirmed and extended the initial observation of Ørskov (1956) on the norepinephrine-induced K^+ transport in duck erythrocytes. Freshly drawn duck cells were found to shrink spontaneously in isotonic salt media lacking catecholamines if the external $[K^+]$ was held below its physiological concentration. Moreover, the shrunken cells were able to restore their volume if the external $[K^+]$ was raised to a higher level and norepinephrine was supplemented (Haas and McManus, 1985). It is now well established that this recovery phase of cell volume results from net movements of salt and water by the $Na^+/K^+/Cl^-$ cotransport pathway which is stimulated as a consequence of β-receptor activation involving the second messenger cAMP as shown by several reports listed in Table II. In addition to cAMP, numerous physiological regulators are known to influence $Na^+/K^+/Cl^-$ cotransport (Haas, 1989).

In recent years, the kinetic properties and diversity of $Na^+/K^+/Cl^-$ cotransporters have been extensively reviewed (McManus and Schmidt, 1978; Kregenow, 1981; Hoffmann, 1986; Parker and Dunham, 1989; Haas, 1989). $Na^+/K^+/Cl^-$ cotransport is present in a variety of epithelial and nonepithelial cells. The salient features include the interdependency of the cotransported species, narrow substrate specificity, inhibition by loop diuretics, and insensitivity to alteration of membrane potential. The cotransport mediates bidirectional and electrically neutral movements of ions with a stoichiometry of 1 : 1 : 2 for Na^+, K^+, and Cl^-, respectively. In addition, a stoichiometry of 3 cations and 3 anions has also been reported in ferret red cells (Hall and Ellory, 1985) and squid axon (Russell, 1983). Recently, Haas (1989) cautions a the potential complication in the estimation of the stoichiometry due to the presence of Na^+/Na^+ exchange or K^+/K^+ exchange. The direction of the cotransport depends on the net driving force established by the sum of chemical gradients of all participating ions across cell membranes.

Several studies have explored loop diuretic binding in attempts to identify cotransporters. To date, use of the photoaffinity analogs of loop diuretics permitted identification of a wide variety of membrane proteins ranging from approximately 6 kDa in bovine kidney (O'Grady *et al.*, 1987) to approximately 200 kDa in shark rectal gland membranes (Haas, 1989). With the advent of the molecular cloning strategy, it is anticipated that the gene encoding the cotransporter may soon be known. In this regard, mutant cell lines with defective cotransport have already been identified (McRoberts *et al.*, 1983) and should be useful tools for the characterization of the cotransporter in analogy with the cyc^- mutant, which has been helpful in the elucidation of G protein function.

As stated, the $Na^+/K^+/Cl^-$ cotransporter plays a physiological role in the regulation of cell volume (Haas, 1989). Apart from $Na^+/K^+/Cl^-$ cotransporters, several transport systems have been implicated in the restoration of the cell volume of hypertonically shrunken cells toward the original cell volume. This process is termed regulatory volume increase (RVI). The other players in RVI include the Na^+/Cl^- cotransporter and the Cl^-/HCO_3^- exchanger coupled to the Na^+/H^+ exchanger (Hoffmann, 1986). Because Na, which enters the cells by the volume-activated transport, is extruded from cells through the Na^+/K^+ pump, the net effect is the uptake of KCl and water, causing the cells to swell (Hoffmann, 1986). As dramatic as shrinking-activated $Na^+/K^+/Cl^-$ cotransport in duck red cells is the KCl cotransport that is activated by cell swelling. In addition, shrinking or swelling turns off KCl cotransport and $Na^+/K^+/Cl^-$ cotransport, respectively. Since both cotransporters are nearly quiescent at normal cell volume, Starke and McManus (1990) define the cotransport activity at isotonic volume as the set point. The shift of the set point apparently depends on intracellular free Mg^{2+} concentration. Moreover, β-adrenergic receptor activation causes the set point to shift to a higher volume. To state it another way, $Na^+/K^+/Cl^-$ cotransport is activated by cAMP, but KCl cotransport is inhibited by cAMP in duck red cells. In this regard, it is worth noting that Kregenow *et al.* (1976) failed to find cAMP accumulation in duck cells in hypertonic media lacking catecholamine in which the cotransport was activated. Thus, this suggests that the signal transduction pathway triggered by cell shrinkage does not involve cAMP.

Nonnucleated human red cells also possess the $Na^+/K^+/Cl^-$ cotransporter (Wiley and Cooper, 1974), although ion movements mediated by the cotransport are not only highly variable in red cells of different individuals, but also represent a small component of total ion movements (Duhm and Göbel, 1984). While Canessa *et al.* (1981) have suggested that the $Na^+/K^+/Cl^-$ cotransporter normally mediates K^+ accumulation, Duhm and Göbel (1984) have postulated that a small outwardly directed net driving force for the cotransport may exist *in vivo,* causing K^+ extrusion. The human red cell cotransporter differs from the nucleated erythrocyte counterpart in that a perturbation of cell volume does not prompt a significant activation (Duhm and Göbel, 1982) and that catecholamines fail to influence cotransport (McManus and Schmidt, 1978). The lack of a catecholamine effect on human cells may stem from the variable production of cAMP content reported in human cells (Sheppard and Burghardt, 1969; Kaiser *et al.,* 1974; Rasmussen *et al.,* 1975; Sager, 1982). In any case, Garay and colleagues (1983; Garay, 1982) raised the intracellular cAMP by taking advantage of cAMP permeability, thereby circumventing the stimulatory receptor mediation altogether. Under these conditions, it was found that $Na^+/K^+/Cl^-$ cotransport in human red cells was inhibited to varying degrees by the second messenger rather than stimulated (Garay, 1982; Garay *et al.,* 1983; Diez *et al.,* 1984).

The extent to which cAMP inhibits cotransport was augmented by the addition of the phosphodiesterase inhibitor 1-methyl-3-isobutylxanthine to the flux media. In contrast to cAMP, cGMP was without effect. Garay (1982) estimated the intracellular cAMP concentrations which inhibit 50% of the cotransport activity as ranging from 3 to 150 μmol/liter of cells. By contrast, the intracellular cAMP concentration, at which the $Na^+/K^+/Cl^-$ cotransport in avian red cells is half-maximally activated, is estimated to be approximately 0.5 μmol/liter of cell water (Kregenow *et al.,* 1976; Ueberschär and Bakker-Grunwald, 1985). As noted by Garay (1982), cAMP content determined by ^{14}C-AMP uptake is likely to represent an overestimation. Intracellular cAMP, measured by radioimmunoassay in human cells after 1 hr of incubation with 1 m*M* cAMP, was ~4.2 μmol/liter of cells (Sergeant and Kim, 1985), which is in reasonable agreement with the cAMP accumulation seen in pig cells (Fig. 2). Thus, it is likely that human red cell $Na^+/K^+/Cl^-$ cotransport is inhibited by a much lower cAMP concentration than reported by Garay (1982).

While it is abundantly clear that intracellular cAMP either generated by β-receptor activation or introduced from an external source elicits a dual control of the cotransport, little is known about the underlying mechanism for cAMP action. In this regard, it is worth noting that ATP depletion results in an inhibition of $Na^+/K^+/Cl^-$ cotransport in cultured HT29 human colonic adenocarcinoma cells (Kim *et al.,* 1988) and in turkey erythrocytes (Ueberschär and Bakker-Grunwald, 1985). Palfrey and Greengard

(1981) found a close correlation between the state of phosphorylation of a membrane protein, $M_r = 180{,}000$, termed goblin, and activation of $Na^+/K^+/Cl^-$ cotransport by cyclic AMP in turkey erythrocytes. These findings imply that phosphorylation of the cotransporter itself or other membrane protein could be involved in the regulation of cotransport by cyclic AMP, but this hypothesis awaits further experimental verification.

B. Na^+/H^+ Exchange

As with the $Na^+/K^+/Cl^-$ cotransporter, the Na^+/H^+ exchanger is now known to be present in a wide variety of animal cells. The antiporter mediates transmembrane exchange of Na^+ for H^+ with a stoichiometry of 1 : 1. As a result, the antiporter, like the $Na^+/K^+/Cl^-$ cotransporter, is electrically silent. Because an inwardly directed Na^+ gradient exists across cell membranes, the antiporter ordinarily operates in the direction of Na^+ inwardly coupled to the H^+ outward mode. When the ratio of external Na^+ to internal Na^+ matches the ratio of external $[H^+]_o$ to internal $[H^+]_i$, the driving force of the antiporter dissipates and no net movements of these ions occur. Although the antiporter is bidirectional in that it can be driven in the reverse mode, the effects of pH on the inward versus outward direction of antiporter activity are remarkably asymmetrical. While raising $[H^+]_i$ results in an increase in Na^+ influx, increasing $[H^+]_o$ fails to promote significant augmentation of Na^+ efflux (Grinstein *et al.*, 1984a). These and other findings are consistent with the existence of a cytoplasmic allosteric H^+-binding site, which was first postulated by Aronson and co-workers (Aronson, 1985). The binding to this modifier site by H^+ presumably activates the antiporter.

The complete cDNA sequence encoding the human Na^+/H^+ antiporter, which is sensitive to amiloride, is now known (Sardet *et al.*, 1988, 1989, 1990). Faced with the difficulty in isolating the antiporter for which no specific antibodies or ligand binding assay was available, Pouysségur and associates (Sardet *et al.*, 1988, 1989, 1990) have ingeniously exploited the function of the antiporter. A bioassay for screening mouse fibroblast clones was devised based on cell killing by an acid load. Using the bioassay, they isolated three different clones: a mutant devoid of the antiporter (Franchi *et al.*, 1986a,b), a secondary mouse fibroblast transformant, and a variant which overexpresses the antiporter. By complementing a mouse fibroblast mutant with human genomic DNA, a 0.8-kb genomic probe was isolated which detects gene amplication of the antiporter in the variant which overexpresses the antiporter. The antiporter, whose amino acid sequence is predicted by the isolated cDNA encoding the antiporter recognized by the genomic probe, is a protein of M_r 110,000 (Sardet *et al.*, 1990).

The physiological intracellular pH of approximately 7.2, where the antiporter is nearly quiescent, is referred to as the set point by Grinstein and Rothstein (1986). One of the remarkable features of the antiporter is its ability to respond to a large number of different stimuli, including but not limited to sperm, hormones, growth factors, mitogens, and activators of protein kinases. In point of fact, no other transporter appears to be as diversely regulated as the antiporter. According to a scheme postulated by Sardet *et al.* (1988), activators of the antiporter can be grouped into two broad classes. One common pathway, which is shared by a variety of tyrosine kinase-activating growth factors such as epidermal growth factor (EGF), platelet-derived growth factor (PDGF), and fibroblast growth factor (FGF), involves transmembrane receptors that function independently from G protein activation. The other pathway, in which the activation of the antiporter may be linked to G proteins, particularly the G proteins coupled to phospholipase C, is utilized by sperm, neurohormones, neurotransmitters, and vasoactive peptides. Using newly available antibodies prepared against the antiporter, Sardet *et al.* (1990) showed convincingly that the antiporter in hamster fibroblasts and A431 human epidermoid cells is phosphorylated by a variety of growth factors, including EGF, thrombin, phorbol esters, and serum.

The antiporter has been implicated in numerous physiological roles. Since antiporter activity is progressively "turned on" as the intracellular acidity is increased, it is well suited for protecting cells against sudden acid loads. In a number of studies in which antiporter activity was measured as a function of intracellular pH, it was generally found that a variety of stimuli cause an alkaline shift of the set point by 0.2–0.3 pH units (Grinstein and Rothstein, 1986) without altering the V_{max}. To state it another way, these mitogens activate the otherwise quiescent antiporter, resulting in a cytoplasmic alkalinization which, among other things, is thought to stimulate DNA synthesis and cell division (Rozengurt, 1985). Once the new set point is reached, it is quiescent again. Apart from its role in intracellular pH homeostasis and cellular growth, the antiporter plays a crucial part in cell volume regulation. As mentioned earlier, the increase in Na^+ movement which is triggered by cell shrinkage can be mediated not only by the $Na^+/K^+/Cl^-$ cotransporter but also by the Na^+/H^+ exchanger which operates in parallel with a Cl^-/HCO_3^- exchanger. The cells which make use of the latter mechanism include lymphocytes (Grinstein *et al.*, 1983, 1985), *Amphiuma* red cells (Kregenow, 1981), dog red cells (Parker and Castranova, 1984), and trout red cells (Borgese *et al.*, 1986, 1987; Garcia-Romeu *et al.*, 1988). However, it is known that adenylyl cyclase activation does not seem to influence the Na^+/H^+ exchangers of these cells except that of trout cells (Borgese *et al.*, 1986, 1987; Garcia-Romeu *et al.*, 1988).

As in avian cells (Riddick *et al.*, 1971; McManus and Schmidt, 1978; Palfrey *et al.*, 1981), trout cells spontaneously swell to reach a new steady state in isotonic media containing high Na^+ content if a catecholamine is present. In early attempts to characterize the catecholamine-induced Na^+ uptake, Baroin *et al.* (1984) initially suggested that the Na^+ movement was mediated by the Na^+/Cl^- cotransport system. However, later studies (Borgese *et al.*, 1986, 1987; Garcia-Romeu *et al.*, 1988) rejected the existence of the Na^+/Cl^- cotransport pathway but instead demonstrated that the Na^+/H^+ exchanger and a Cl^-/HCO_3^- exchanger operating in tandem account for the catecholamine-induced cell swelling. In trout cells, there is no evidence for the $Na^+/K^+/Cl^-$ cotransport system.

An interesting feature of RVI of the trout cell is that the enlargement of the cells proceeds until it reaches a new steady state which is maintained as long as catecholamine is present. Apparently, the new steady state of cell volume is maintained due to stimulation of K^+ loss which compensates for the Na^+ gain induced by catecholamines. The K^+ extrusion, which is chloride dependent, is presumed to be mediated by a KCl cotransporter. It can be readily unmasked, if NaCl uptake is inhibited by amiloride, if β-receptors are blocked by propranolol, or if catecholamines are removed by washing. KCl cotransport, which is "switched on" or "switched off" depending on trout cell volume, represents a paradigm of a variety of transport systems that participate in the restoration of swollen cells toward their original volume. This process is called the regulatory volume decrease (RVD). Thus, the catecholamine-regulated trout cell volume is delicately balanced by a separate but simultaneous operation of RVI and RVD processes. In this regard, trout cells differ from avian cells in which the catecholamine-promoted cell enlargement does not elicit a simultaneous activation of Na^+ movement through the $Na^+/K^+/Cl^-$ cotransport and K^+ movement through the KCl cotransporter at a given cell volume (Starke and McManus, 1990).

A detailed investigation of the kinetic properties of the cAMP-dependent Na^+/H^+ antiporter activity in trout cells reveals a complex but intriguing Na^+ permeability change which Motais and associates (Garcia-Romeu *et al.*, 1988) describe as a desensitization. It was found that, in response to isoproterenol, the unidirectional Na^+ influx increases by more than 100-fold within the first 1.5 min, followed by an exponential decay of Na^+ permeability. After a 1-hr incubation, Na^+ permeability is reduced to one-fourth of the maximally stimulated rate seen within the first minutes of catecholamine exposure. Neither the intracellular pH changes nor the decrease in cAMP content can be invoked to explain the process of desensitization. However, the desensitization appears to depend on external Na^+ concentration, since Na^+ concentration below 20 m*M* fails to cause desensitization.

In addition to dog (Parker and Castranova, 1984) and *Amphiuma* (Kregenow, 1981) red cells, human (Escobales and Canessa, 1986), rabbit (Jennings *et al.*, 1986; Escobales and Rivera, 1987), and pig (Sergeant *et al.*, 1989) red cells have an amiloride-sensitive Na^+/H^+ exchanger. We reported that the Na^+/H^+ exchanger, which is stimulated on shrinking pig red cells in hypertonic media, was inhibited by cyclic nucleotides, including cAMP and cGMP (Sergeant *et al.*, 1989). However, a later study could not corroborate the earlier finding in regard to cAMP inhibition (Sohn and Kim, 1991). In our measurements of amiloride-sensitive Na^+ influx, we routinely depleted intracellular Na^+ in order to minimize the Na^+/Na^+ exchange component. Cyclic AMP or cGMP was present during the nystatin procedure, preloading, and flux period. Although the reason for this discrepancy is not clear, the earlier observation may have stemmed from the different cell volumes which were used for normalizing the Na^+ influx. In any case, cGMP is still inhibitory to the volume-activated Na^+/H^+ exchange pathway in pig cells.

In both renal brush border vesicles (Kahn *et al.*, 1985) and a cultured opossum kidney cell line (Pollock *et al.*, 1986), cAMP inhibits to varying degrees the amiloride-sensitive, but not amiloride-insensitive, Na^+ flux. Thus, the antiporter represents another example in which cAMP elicits dual control. However, the mechanism by which the antiporter is regulated by second messengers, or for that matter by a variety of growth factors, is poorly understood. That the antiporter can be phosphorylated by growth factors is already alluded to. In this regard, it is of interest to note that, as with the $Na^+/K^+/Cl^-$ cotransport, metabolic depletion causes an inhibition of the antiport (Cassel *et al.*, 1986). Repletion of ATP by subsequent metabolic manipulation restores antiporter activity. ATP depletion results in the shift of the set point toward a lower intracellular pH as opposed to the alkaline shift of the set point seen with growth factors. As a result, the antiporter in ATP-depleted cells is fully functional at intracellular pH 6.2 but shuts off at intracellular pH 6.4. Whether ATP inhibition reflects a common mechanism perhaps involving phosphorylation of the antiporter remains to be explored.

C. KCl Cotransport

The principal and common substrate of diverse transport systems in RVD is the K ion. Kregenow (1971) first showed that duck red cells, when placed in hypotonic media, initially swell but gradually shrink toward their original volume by losing KCl and osmotically obligated water. Since then, K^+ movements in relation to the RVD have been well characterized. In Fig. 3, different types of transport modes mediating K^+ movements triggered by cell swelling are compared by Lauf (1985). In *Amphiuma* red cells, cell swelling causes the extrusion of K salt as a result of the unique combined action of a K^+/H^+ exchanger coupled to a Cl^-/HCO_3^- exchanger (Cala, 1986). Lymphocytes restore their volume within a few minutes after cell swelling by activating K^+ channels in parallel with Cl^- channels (Grinstein *et al.*, 1984b). In both instances, Ca^{2+} stimulates K^+ extrusion. By contrast, Ca^{2+}, if anything, inhibits the Cl^--dependent K^+ flux which is presumably a KCl cotransport in LK sheep red cells. Since the RVD seen in *Amphiuma,* lymphocytes, and other cell types has been extensively reviewed (Kregenow, 1978; Lauf, 1985; Hoffmann, 1986; Eveloff and Warnock, 1987a; Parker and Dunham, 1989), it is not discussed further here.

KCl cotransport has been identified in a variety of cells, including Ehrlich ascites tumor cells (Thornhill and Laris, 1984, Kramhøft *et al.*, 1986), intestinal mucosa (Halm *et al.*, 1985), the basolateral membranes of *Necturus* gall bladder (Corcia and Armstrong, 1983), the thick ascending limb in nephrons (Greger and Schlatter, 1983), and the renal cortex (Eveloff and Warnock, 1987b). In addition to the avian and fish red cells already mentioned, red cells of various species have the KCl cotransport system. (For a comparative review, see Ellory *et al.*, 1985; for a comprehensive review, scc Parker and Dunham, 1989). The view that red cells possess a KCl cotransport apart from $Na^+/K^+/Cl^-$ cotransport was first suggested by reports appearing about the same time from several laboratories (Lauf and Theg, 1980; Ellory and Dunham, 1980; Dunham and Ellory, 1981). In retrospect, it is interesting that these studies employed sheep red cells of the LK phenotype as a model, for the choice of HK cells, in which the KCl transporter is now known to exist in a latent form, would have been a more difficult model for discovering this fascinating transport pathway. As with $Na^+/K^+/Cl^-$ cotransport, KCl cotransport has narrow anion specificity, with the rank order of decreasing transport: $Br^- >> Cl^- >>> NO^{3-} = I^- = SCN^-$ (Lauf, 1985). As with the $Na^+/K^+/Cl^-$ cotransport, the KCl cotransport can be inhibited by the loop diuretics bumetanide and furosemide. However, their inhibitory actions on the two transport systems are vastly different. While $Na^+/K^+/Cl^-$ cotransport can be inhibited by micromolar concentrations of these drugs, KCl cotransport is much less sensitive, requiring millimolar concentrations (Ellory *et al.*, 1982; Kaji, 1986). The rank order of inhibitory potencies of these drugs appears to be different also (Parker and Dunham, 1989).

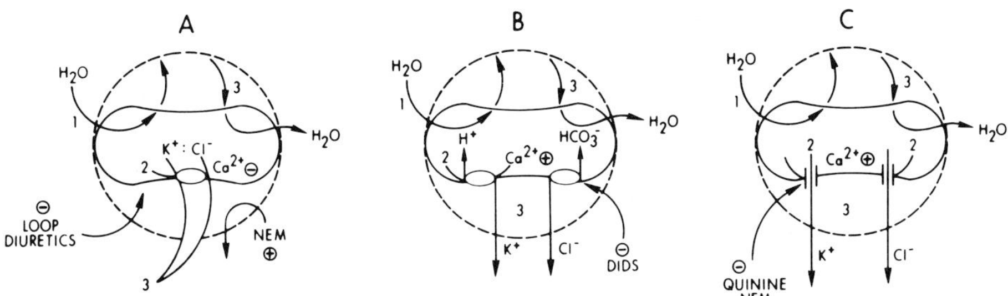

FIG. 3. K^+ flux modes regulating volume decrease by (A) $K^+ : Cl^-$ cotransport, (B) K^+/H^+ countertransport, and (C) electrogenic K^+ and Cl^- channels. Common events preceding activation of all three modes are water entry and cell swelling (1) in hyposmotic media, the unknown trigger for flux activation (2) of each mode (3), and the accompanying regulatory volume return to the original volume. ⊕, Activators; ⊖, inhibitors; NEM, *N*-ethylmaleimide; DIDS, 4,4′-diisothiocyano-2,2′-stilbene-disulfonic acid. (Reproduced from Lauf, 1985, with permission from the publisher.)

In addition to the pharmacological difference, $Na^+/K^+/Cl^-$ and KCl cotransport can be distinguished on the basis of *N*-ethylmaleimide (NEM) treatment, which causes substantial activation of KCl cotransport, whereas $Na^+/K^+/Cl^-$ cotransport is activated to a much lesser extent (Lauf *et al.*, 1984a). These observations strengthened the view that $Na^+/K^+/Cl^-$ cotransport and KCl cotransport may be a reflection of separate molecular entities rather than alternative modes of expression of the same transporter.

In red cells, the accompaniment of Cl^- with K^+ movement has not been rigorously demonstrated due to powerful Cl^-/HCO_3^- exchange activity which cannot be completely inhibited. However, coupling between K^+ and Cl^- has been implied in elegant experiments by Brugnara *et al.* (1989), who examined effects of the Cl^- gradient across human red cell membranes on K^+ movement at constant membrane potential. These authors were able to establish a Cl^- gradient under conditions in which the membrane potential was "clamped" by using NO_3 or SCN, which are not transported by the cotransport, but are transported by the anion exchange pathway. The outwardly directed Cl^- gradient, in which the Cl^- assumes the role of the driver ion, was found to drive net uphill K^+ efflux against a K^+ activity gradient. Since it has not been possible to assess net Cl^- flux movement through the KCl cotransporter in red cells, in practice, the Cl^--dependent K^+ flux is determined as the difference between K^+ flux measured in Cl^- media and in NO_3^- media.

Because sheep red cells lack $Na^+/K^+/Cl^-$ cotransport as well as Ca^{2+}-activated K^+ channels, these cells represent a relatively simple but valuable model which permits investigations aimed at unravelling the kinetic properties of cotransport in detail. In coupled KCl cotransport, net flux will be zero when the product of $[K^+]_o$ and $[Cl^-]_o$ equals the product of $[K^+]_i$ and $[Cl^+]_i$. In LK sheep cells, the external $[K^+]$ at which zero net flux occurs was found to be 15 m*M*, corresponding to the $[K^+]_o/[K^+]_i$ ratio of 0.7 that is interestingly close to the Donnan ratio of chloride ions (Lauf, 1983a). Thus, the direction of cotransport, resulting in net K^+ influx or efflux, depends on whether the external $[K^+]$ is above or below 15 m*M*. Treatment of LK cells with NEM, which stimulates K^+ movement, did not affect the external $[K^+]$ at which zero net K^+ flux occurs (Lauf, 1983a).

As with $Na^+/K^+/Cl^-$ cotransport and Na^+/H^+ exchange pathways, the depletion of ATP by starving cells elicits complex responses of KCl cotransport. Upon metabolic depletion, K^+ influx, but not K^+ efflux, measured either in isotonic or hypotonic media, was increased rather than decreased in LK sheep cells (Lauf, 1983b, 1984; Logue *et al.*, 1983). These findings prompted Parker and Dunham (1989) to raise the possibility of asymmetry of the KCl cotransport. A far more discernible inhibition of both influx and efflux was seen in NEM-treated cells after ATP depletion (Lauf, 1983b, 1984; Logue *et al.*, 1983). NEM stimulated KCl cotransport which is inhibited by metabolic depletion can be recovered by subsequent metabolic rejuvenation. Metabolic regulation was also examined in intact human red cells (Lauf *et al.*, 1985) and in resealed ghosts (Dunham and Logue, 1986; Sachs, 1988). In human cells, metabolic effects were likewise reversible in that metabolic depletion and subsequent repletion resulted in an inhibition and restoration, respectively, of NEM-stimulated, but not basal, cotransport (Lauf, 1985). In resealed human ghosts, the finding that the cotransport was dependent on MgATP, but not nonhydrolyzable ATP analog, lends support to the notion that ATP, rather than other metabolites formed by metabolic manipulation, is a likely agent regulating the KCl cotransport and that the KCl cotransporter may undergo phosphorylation (Sachs, 1988). However, since metabolic depletion produces profound changes in a plethora of metabolites, the presumed link between ATP and passive ion flux should be viewed with caution.

Like sheep cells, pig red cells lack the $Na^+/K^+/Cl^-$ cotransporter and Ca^{2+}-activated K^+ channels, but possess a KCl cotransporter which is volume sensitive (Kim *et al.*, 1989). Figure 4 shows the effects of cAMP loading on the K^+ efflux of pig cells in hypotonic media (Kim *et al.*, 1989). It is clear that ouabain-resistant, volume-activated, Cl^--dependent K^+ efflux can be stimulated by cAMP loading. It is worth noting that both K^+ influx and K^+ efflux do not immediately respond to cell swelling but instead require a lag period of approximately 5 min before initiating a volume response. Figure 5 shows cAMP activation of KCl cotransport as a

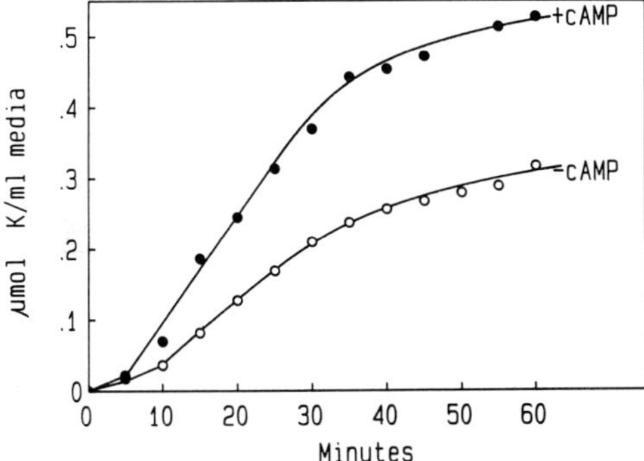

FIG. 4. Effects of cAMP on K^+ efflux from pig cells suspended in hypotonic media. K^+ efflux media was measured in 240 mosM balanced salt solution buffered with 10 m*M* sodium phosphate (pH 7.4). cAMP loading was carried out in Cl^- and NO_3^- media as in Fig. 2. Cl^--dependent K^+ efflux was deduced as the difference between K^+ efflux measured from Cl^- and NO_3^- loaded cells. (Reproduced from Kim *et al.*, 1989, with permission from the publisher.)

function of cell volume. Unlike human cells, which can be swollen up to 60% of the original cell volume, pig cells are more rigid in the sense that they reach the critical hemolytic volume at approximately 20% swelling. The enlargement of cells causes a progressive increase in KCl cotransport which is further potentiated by cAMP loading. The KCl cotransport, which is barely detectable in pig cells at normal volume (Lauf *et al.*,

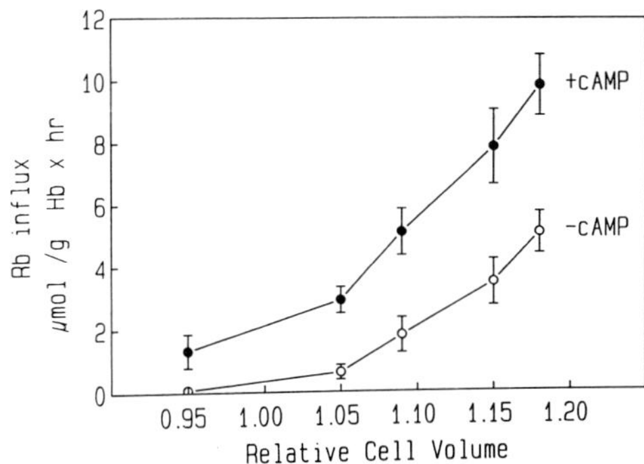

FIG. 5. Effects of cell volume and cAMP on Cl^--dependent Rb^+ influx. Flux media osmolarity was varied from 320 to 240 mosM by adjusting either NaCl or $NaNO_3$; RbCl or $RbNO_3$ was held constant at 10 m*M*. Influx was measured in the presence of 0.1 m*M* ouabain either with or without 1 m*M* cAMP. Values represent means ± SE from seven determinations of control and four determinations of cAMP-loaded cells. (Reproduced from Kim *et al.*, 1989, with permission from the publisher.)

1984b), is also responsive to cAMP (Kim *et al.*, 1989) in analogy to NEM activation of a latent KCl cotransport seen in HK cells. In Table III, effects of cAMP analogs on the volume-activated KCl cotransport are summarized. Cyclic AMP or its analogs are stimulatory, whereas cGMP is without effect. Adenosine 5′-monophosphate (AMP), as opposed to cAMP, was found to be inhibitory to KCl cotransport. Although the inhibitory effect of AMP puzzled us initially, it now seems feasible that AMP may act through purinergic receptors, as discussed in Section IV.

In the course of reticulocyte maturation, cotransport activity diminishes (Lauf, 1983c; Lauf *et al.*, 1984b; Lauf and Bauer, 1987). In the case of the mature HK phenotype, cotransport activity has practically disappeared, but it can be "unmasked" by hypotonic swelling or treatment with NEM just as stated (Fujise and Lauf, 1987). The maturational changes in ion transport have also been examined in naturally occurring reticulocytes which were isolated from a 7-day-old piglet (Lauf *et al.*, 1984b). It was found that the Na^+/K^+ pump and the Cl^--dependent K^+ flux were 25 to 50-fold higher in reticulocytes than in mature red cells. NEM activated the KCl cotransport measured in isotonic media in reticulocytes by 2-fold and

TABLE III

Activation of a Cl-Dependent Rb Influx of Pig Red Cells by Various Nucleotides[a,b]

Incubation conditions	n	Cl^--dependent Rb^+ influx (μmol g Hb^{-1} hr^{-1})
Control	3	6.28 ± 1.34
CAMP	4	9.56 ± 1.34
DBcAMP	4	10.68 ± 1.27
8-BrcAMP	4	9.92 ± 1.24
cGMP	5	6.89 ± 0.81
AMP	4	4.40 ± 1.47

[a] Reproduced from Kim *et al.* (1989), with permission of the publisher.

[b] Values are means ± SE from n determinations. Pig cells were preincubated in isotonic media containing 1mM test nucleotide for 1 hr at 37°C. Cells were washed once and suspended in hypotonic media of 240 mosM containing 10 mM containing 10 mM Rb^+ with the principal anion being either Cl^- or NO_3^- for 1 hr. The difference between Rb^+ flux measured in Cl^- and NO_3^- media was taken to represent Cl^--dependent Rb^+ influx. DBcAMP, Dibutyryl 3′,5′-cyclic adenosine monophosophate; 8-BrcAMP, 8-bromoadenosine 3″,5′-cyclic monophosphate; cGMP, 3′,5′-cyclic guanosine monophosphate.

in mature cells by 13-fold. In *in vitro* tissue culture conditions, where reticulocytes were permitted to mature, both the pump and cotransport activity were drastically diminished. Since younger red cells have higher KCl cotransport compared to older cells (Hall and Ellory, 1986; Kim *et al.*, 1989), dismantling of the cotransporter apparently continues not only throughout erythroid differentiation but also during cellular aging. Figure 6 compares fetal pig red cells and postnatal reticulocytes isolated from 7-day-old piglets with respect to volume-activated KCl cotransport as a function of external [Rb^+]. It is evident that fetal red cells respond to cAMP as do mature red cells taken from adult animals. However, unexpectedly, cAMP loading fails to activate Rb^+ flux in reticulocytes. It seems plausible that the reticulocyte KCl^+ cotransport, which is an order of magnitude larger than the red cell counterpart, may be operating in full capacity under volume-activated conditions so that further activation by the second messenger is not possible. In any case, it seems abundantly clear that the evolvement of ion transport systems in the course of cellular development is far more complicated than previously thought. The dismantling of the KCl cotransport system now appears to have another maturational feature in that the sensitivity to second messenger regulation appears to be incorporated during the last stage of transition from the reticulocyte to the erythrocyte stage. In this regard, it is of considerable interest to delineate whether the avian system, in which cAMP inhibits the KCl cotransport pathway, would have a similar maturational feature.

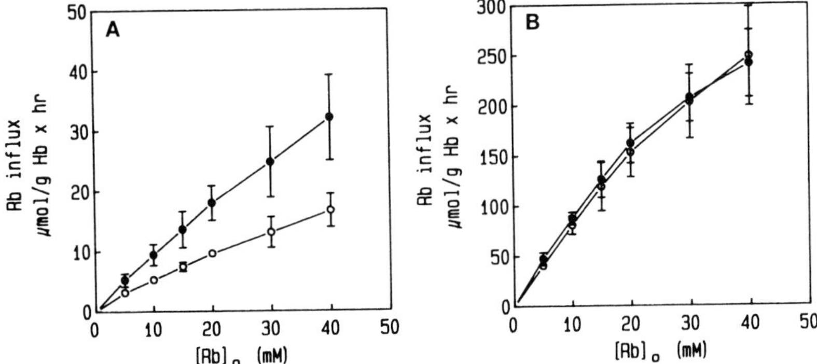

FIG. 6. Effects of cAMP on Cl^--dependent Rb^+ influx in (A) fetal red blood cells and (B) reticulocytes. Cells were first loaded with cAMP in isotonic media. To activate the Rb^+ influx, cells were then suspended in 240 mosM media containing various concentrations of RbCl or $RbNO_3$. Values represent means ± SE from three determinations.

IV. ADENOSINE INFLUENCE ON ION MOVEMENTS

It is now well known that adenosine elicits diverse responses of ion movements in a variety of cells, as summarized in Table IV. Adenosine or other P_1 receptor agonists have been shown to increase K^+ conductance or K^+ channel activity in cardiac muscle, neurons, and *Xenopus* oocytes. In contrast, Ca^{2+} channel activities and Ca^{2+} uptake are depressed by other adenosine or adenosine receptor agonists in neurons, cardiac muscle, and brain cortical synaptosomes. In epithelia, adenosine produces yet another response in that Cl^- secretion is stimulated. While these paradigms clearly illustrate widespread but complex adenosine influences on ion movements, red blood cells, in which many transport pathways have been characterized, conspicuously have not been investigated with respect to adenosine-mediated ion transport.

A recent finding, in which adenosine used as an energy source for pig red cells influenced the KCl cotransport, prompted us to investigate the possible role of adenosine receptors in red cells. The use of adenosine stemmed from an unusual metabolic feature of pig red cells, which occurs during the postnatal period and during the transition from the reticulocyte to the erythrocyte stage when glucose transporters are discarded, resulting in nonglycolytic cells. An *in vivo* energy source has been identified as inosine (Watts *et al.*, 1979; Kim *et al.*, 1980; Jarvis *et al.*, 1980; Zeidler *et al.*, 1985; Young *et al.*, 1986). Although glucose cannot be used, pig cells have a broad affinity for a variety of substrates, including adenosine, ribose, dihydroxyacetone, and glyceraldehyde (Kim, 1983). In keeping with a biphasic regulation of adenylyl cyclase activity by adenosine (Londos and Wolff, 1977), we found that the KCl cotransport in pig red cells was activated by low adenosine concentrations (H. D. Kim, unpublished results), but inhibited by high adenosine concentrations (Sohn and Kim, 1991). While we have not yet investigated the mechanism by which adenosine at low concentrations elicits activation of KCl cotransport, we have delineated certain aspects of the inhibitory effects of adenosine at high concentrations on KCl cotransport. In addition to adenosine, it was found that KCl cotransport in pig cells was inhibited by adenosine receptor agonists as shown in Fig. 7 (Sohn and Kim, 1991). The rank order of decreasing inhibitory potencies was CHA > 2Cl-Ado > NECA, which is consistent with the activation of A_1 type receptors.

As shown in Table I, there is a paucity of information regarding adenosine receptors in erythrocytes. Cooper and Jagus (1982) have identified the presence of adenosine receptors in bleeding induced rat reticulocytes in which NECA was found to stimulate the adenylyl cyclase activity with a K_a of 70 nM. 3-Isobutyl-1-methylxanthine was found to competitively

TABLE IV

Regulation of Ionic Movements by Adenosine or Its Agonists

Ion movement	Cell type	Adenosine agonists[a]	Receptor type	Adenosine antagonists[b]	Guanine nucleotide-binding protein[c]	Cyclic nucleotide	References
K^+ conductance ↑	Atrial cardiac preparations	PIA, adenosine	—	—	PT inhibitory		Böhm *et al.* (1986)
K^+ channel activity ↑	Isolated atrial cells	Adenosine	—	Theophylline	PT inhibitory	cAMP not required	Kurachi *et al.* (1986)
K^+ conductance ↑	Neurons in striated and hippocampal cells	2-CA, adenosine	—	—	PT inhibitory GTP required	cAMP not required	Trussel and Jackson (1985, 1987)
K^+ current ↑ (outward)	*Xenopus* ooctyes	Adenosine, NECA, PIA	R_a receptor subtype	Theophylline	—	Intracellular cAMP levels ↑ cAMP mimics the response	Lotan *et al.* (1985)
Voltage-dependent Ca^{2+} conductance ↓	Mouse sensory neurons	*R*(−)PIA > *S*(+)PIA>CHA > 2-CA ≫ 1-methylisoguanine > adenosine	A novel receptor?	Theophylline?	—	cAMP mimics the response	MacDonald *et al.* (1986)
Ca^{2+} current (ICa^{2+}) and action potential duration ↓	Dorsal root ganglion neurons	2-CA	A_1	IBMX	—	8-Bromo-cAMP mimics the response	Dolphin *et al.* (1986)
K^+-evoked Ca^{2+} uptake ↓	Brain cortical synaptosomes	CHA, PIA, 2-CA, NECA, adenosine dideoxyadenosine	A_2	Theophylline	—	—	Wu *et al.* (1982)
Ca^{2+} uptake ↓	Atrial muscle	Adenosine	—	—	—	—	Schrader *et al.* (1975)

(continued)

TABLE IV (*Continued*)

Ion movement	Cell type	Adenosine agonists[a]	Receptor type	Adenosine antagonists[b]	Guanine nucleotide-binding protein[c]	Cyclic nucleotide	References
Action potential ↓							
Intracellular free Ca^{2+} ↓	Aortic vascular smooth muscle	Adenosine	—	—	—	—	Kai *et al.* (1987)
Ca^{2+} spikes ↓	Hippocampal neurons	Adenosine	—	—	—	—	Procter and Dunwiddie (1983)
Cl^- secretion ↑	Mucosal tracheal epithelium	2-CA	A_1?	8-Phenyltheophylline	—	Intracellular cAMP levels ↑	Pratt *et al.* (1986)
Cl^- secretion ↑	Basolateral descending colon	NECA, 2-CA, CHA, adenosine	R_a subtype	Theophylline	—	Intracellular cAMP ↑	Grasl and Turnheim (1984)
Cl^- secretion ↑	Mucosal ileum	NECA, 2-CA, PIA, adenosine	—	8-Phenyltheophylline	—	Intracellular cAMP ↑	Dobbins *et al.* (1984)
cAMP-stimulated and swelling activated KCl cotransport ↓	Red blood cells, sheep and pig	CHA, 2-CA, NECA, adenosine	A_1	—	—	No change in cAMP levels	Sohn and Kim (1991)

[a] 2-CA, 2-Chloroadenosine; CHA, N^6-cyclohexyladenosine; *R*(−)PIA, LPIA or N^6-(L-phenylisoporopyl) adenosine or N^6-(*R*)-1-methyl-2-phenythyl)adenosine; *S*(+)PIA, DPIA (the *S* isomer of LPIA); NECA, 5′-*N*-ethylcarboxaminoadenosine.

[b] IBMX, 3-Isobutyl-1-methylxanthine.

[c] PT, Pertussis toxin.

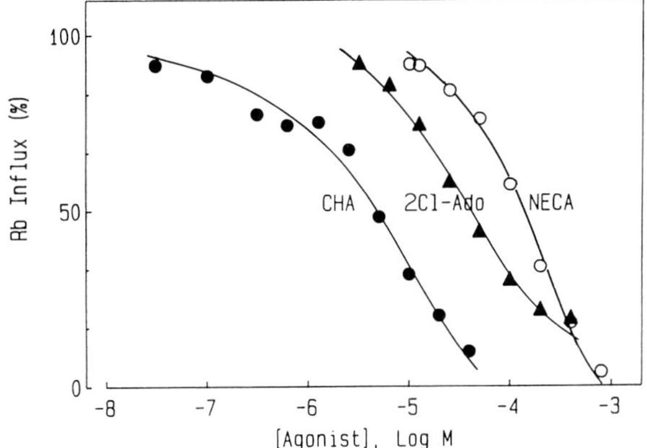

FIG. 7. Effects of various adenosine receptor agonists on the Cl^--dependent Rb^+ influx. Cells were pretreated with various concentrations of adenosine receptor agonists for 1 hr in isotonic balanced salt solution. The cells were then suddenly exposed to 240 mosM flux media containing 30 mM RbCl in $RbNO_3$ to activate KCl cotransport. A typical dose–response curve for each adenosine receptor agonist from different cell preparations is plotted. The Rb^+ flux of control cells for each of adenosine receptor agonists (CHA, 2Cl-Ado, NECA) is as follows: 15.7 μmol/g Hb × hr (CHA); 23.25 μmol/g Hb × hr (2Cl-Ado); 14.04 μmol/g Hb × hr (NECA). CHA, N^6-Cyclohexyladenosine; 2Cl-Ado, 2-chloroadenosine; NECA, 5′-N-ethylcarboxaminoadenosine. (From Sohn and Kim, 1991, with permission from the publisher.)

antagonize the NECA activation of adenylyl cyclase with a K_i of 1.1 μM. Thus, the reticulocyte adenosine receptors are likely to be of the A_2 type. Both turkey (Newman and Levitzki, 1982) and human (Kumar *et al.*, 1978) erythrocyte membranes are also known to bind to adenosine, although the type of receptors residing on the surface of these cells is not known. That red cells may possess purinergic receptors was also suggested by Parker *et al.* (1977; Parker and Snow, 1972), who found a prompt increase in the permeability of both Na^+ and K^+ by exogenously added ATP. While these findings taken together are suggestive of the presence of purinergic receptors in red cells, it is difficult to reconcile with high EC_{50} values found for adenosine agonists inhibiting KCl cotransport. Consequently, we cannot entirely rule out the possibility that adenosine receptor agonists act directly on the KCl cotransporter. Thus, the existence and classification of purinergic receptors on red cells await further experimental verification with ligand binding assays.

The volume-activated KCl cotransport that is potentiated by cAMP was also inhibited in the presence of CHA. Neither the basal level of cAMP nor the cAMP content after cAMP loading was altered by treatment with CHA (Sohn and Kim, 1991). Thus, adenosine agonists appear to elicit their effects on KCl cotransport without involving cAMP. This is not entirely surprising in that there are numerous precedents for this. For example, adenosine activates K^+ channels in neurons (Trussell and Jackson, 1987) and isolated atrial cells (Kurachi *et al.*, 1986) without involving cAMP. The muscarinic receptor-coupled K^+ channels activated by acetylcholine are also regulated without involving the cyclic nucleotide (Pfaffinger *et al.*, 1985), although G_i subunit mediation of the signal remains controversial. As stated earlier, both α subunits (Codina *et al.*, 1987; Kirsch *et al.*, 1988; Yatani *et al.*, 1988a,b; Cerbai *et al.*, 1988) and $\beta\gamma$ subunits (Logothetis *et al.*, 1987) have also been reported to be responsible for activating muscarinic receptor-coupled K^+ channels. Interestingly, the G protein subunit termed α_κ, which was purified from human red blood cell membranes, has been used in several reconstitution experiments on K^+ channels (Codina *et al.*, 1987; Cerbai *et al.*, 1988; Yatani *et al.*, 1988a,b). Although the α_κ function in red blood cells is poorly understood, these experiments demonstrate the cross-reactivity of G proteins from different cells.

Based on the possible regulation of ion transport in red cells by the adenosine receptor, which is presumably linked to G proteins, it is tempting to speculate about the signal transduction mechanism for volume-activated KCl cotransport. By analogy to muscarinic receptor-coupled K^+ channels, it can be envisioned that a change in cell volume signals, through a putative volume sensor, for the dissociation of G proteins into subunits. According to this working hypothesis, activated G protein subunits somehow stimulate ion transporters. Perhaps, the lag time seen before the onset of activation of KCl cotransport in swollen pig red cells reflects the period in which the volume signal is processed at the level of G proteins. We are currently investigating the possible involvement of G proteins in relation to volume-activated ion transport.

V. CONCLUSIONS

In this chapter, we have examined the influence of cAMP on ion transport in red cells. There is compelling evidence that, like avian red cells, nonnucleated mammalian red cells are also capable of responding to cAMP introduced exogenously. However, to date, the link between receptor activation and ion transport has not been demonstrated in mammalian red cells.

It is now well established that cAMP exerts a dual regulation of the $Na^+/K^+/Cl^-$ cotransporter, the KCl cotransporter, and the Na^+/H^+ exchanger, depending on cell type. Needless to say, the intriguing central issue to be unravelled is the mechanism by which cAMP influences ion transport. Cyclic AMP action is presumably mediated by cAMP-dependent protein kinase, which phosphorylates membrane proteins. A specific membrane protein undergoing phosphorylation may mediate the cAMP effect as, for example, postulated by Palfrey and Greengard (1981) in their description of goblin. In addition, Sardet *et al.* (1990) showed that the Na^+/H^+ antiporter with M_r = 110,000 is phosphorylated in hamster fibroblasts by mitogenic activation and in A431 human epidermoid cells by a variety of stimulatory agents. Moreover, in light of the finding that metabolic depletion of ATP causes an inhibition of cotransport, it is anticipated that the cotransporters likewise can be phosphorylated. In any case, identification and eventual isolation of transporters would be an essential step toward the elucidation of second messenger regulation of ion transport. However, as pointed out by Parker and Dunham (1989), the use of red cells has limitations in that the number of copies of many transporters in red cell membranes is low.

On the other hand, red cells represent a promising model in which to delineate the volume-activated signal transduction pathway. In analogy with agonist occupancy resulting in receptor activation, it seems reasonable to view the perturbation of cell volume as triggering the signal transduction pathway which regulates ion transport systems. Indeed, intracellular cAMP content is now known to increase in S49 mouse lymphoma cells during hypotonic swelling (Watson, 1990), in hearts with elevated aortic pressure (Watson *et al.*, 1989), and in distended lungs after partial pneumonectomy (Russo *et al.*, 1989). Thus, it is tempting to speculate that a perturbation of cell volume leads to activation of G proteins which, in turn, influences ion transport. Alternatively, Starke and McManus (1990) envision intracellular free Mg^{2+} ion concentration as playing a role in volume-activated ion transport. In view of experimental manipulations which permit reconstitution of the ion transport system in ghosts, it seems feasible to test these hypotheses with red blood cells.

Acknowledgments

The author wishes to express his gratitude to Ms. Judy Richey for her unstinting dedication and assistance in the preparation of this manuscript. The technical assistance of Ms. Jane Burnett is also gratefully acknowledged. In addition, the author wishes to thank Dr. John Turner for stimulating discussions during the preparation of this manuscript. Supported in part by grant NIH DK33456.

References

Aktories, K., Schultz, G., and Jakobs, K. H. (1979). Inhibition of hamster fat cell adenylate cyclase by prostaglandin E_1 and epinephrine: Requirement for GTP and sodium ions. *FEBS Lett.* **107,** 100–104.

Aktories, K., Jakobs, K. H., and Schultz, G. (1980). Nicotinic acid inhibits adipocyte adenylate cyclase in a hormone-like manner. *FEBS Lett.* **115,** 11–14.

Aronson, P. S. (1985). Kinetic properties of the plasma membrane Na^+-H^+ exchanger. *Annu. Rev. Physiol.* **47,** 545–560.

Baroin, A., Garcia-Romeu, F., Lamarre, T., and Motais, R. (1984). Hormone-induced co-transport with specific pharmacological properties in erythrocytes of rainbow trout, *Salmo gairdneri. J. Physiol.* (*London*) **350,** 137–157.

Barrington, W. W., Jacobson, K. A., Hutchison, A. J., Williams, M., and Stiles, G. L. (1989). Identification of the A_2 adenosine receptor binding subunit by photoaffinity crosslinking. *Proc. Natl. Acad. Sci. U.S.A.* **86,** 6572–6576.

Beckman, B. S., and Hollenberg, M. D. (1979). Beta-adrenergic receptors and adenylate cyclase activity in rat reticulocytes and mature erythrocytes. *Biochem. Pharmacol.* **28,** 239–248.

Blume, A. J., Lichtshtein, D., and Boone, G. (1979). Coupling of opiate receptors to adenylate cyclase: Requirement for Na^+ and GTP. *Proc. Natl. Acad. Sci. U.S.A.* **76,** 5626–5630.

Böhm, M., Brückner, R., Neumann, J., Schmitz, W., Scholz, H., and Starbatty, J. (1986). Role of guanine nucleotide-binding protein in the regulation by adenosine of cardiac potassium conductance and force of contraction. Evaluation with pertussis toxin. *Naunyn-Schmiedebergs Arch. Pharmakol.* **332,** 403–405.

Borgese, F., Garcia-Romeu, F., and Motais, R. (1986). Catecholamine-induced transport systems in trout erythrocyte. Na^+/H^+ countertransport or NaCl cotransport? *J. Gen. Physiol.* **87,** 551–566.

Borgese, F., Garcia-Romeu, F., and Motais, R. (1987). Control of cell volume and ion transport by β-adrenergic catecholamines in erythrocytes of rainbow trout, *Salmo gairdneri. J. Physiol.* (*London*) **382,** 123–144.

Bourne, H. R. (1988). Summary: Signals past, present, and future. *Cold Spring Harbor Symp. Quant. Biol.* **53,** 1019–1031.

Brugnara, C., Ha, T. V., and Tosteson, D. C. (1989). Role of chloride in potassium transport through a K-Cl cotransport system in human red blood cells. *Am. J. Physiol.* **256,** C994–C1003.

Brunton, L. L., and Buss, J. E. (1980). Export of cyclic AMP by mammalian reticulocytes. *J. Cyclic Nucleotide Res.* **6,** 369–377.

Brunton, L. L., and Mayer, S. E. (1979). Extrusion of cyclic AMP from pigeon erythrocytes. *J. Biol. Chem.* **254,** 9714–9720.

Burnstock, G. (1978). A basis for distinguishing two types of purinergic receptor. *In* "Cell Membrane Receptors for Drugs and Hormones: A Multidisciplinary Approach" (R. W. Straub and L. Bolis, eds.), pp. 107–118. Raven, New York.

Cala, P. M. (1986). Volume-sensitive ion fluxes in *Amphiuma* red blood cells: General principles governing Na-H and K-H exchange transport and Cl-HCO_3 exchange coupling. *Curr. Top. Membr. Transp.* **27,** 193–218.

Canessa, M., Bize, J., Solomon, H., Andragna, N., Tosteson, D. C., Dagher, G., Garay, R., and Meyer, P. (1981). Na countertransport and cotransport in human red cells: Function, dysfunction, and genes in essential hypertension. *Clin. Exp. Hypertens.* **3,** 783–795.

Casperson, G. F., and Bourne, H. R. (1987). Biochemical and molecular genetic analysis of hormone-sensitive adenylyl cyclase. *Annu. Rev. Pharmacol. Toxicol.* **27,** 371–384.

Cassel, D., and Selinger, Z. (1976). Catecholamine-stimulated GTPase activity in turkey erythrocyte membranes. *Biochim. Biophys. Acta* **452,** 538–551.

Cassel, D., Katz, M., and Rotman, M. (1986). Depletion of cellular ATP inhibits Na^+/H^+ antiport in cultured human cells. Modulation of the regulatory effect of intracellular protons on the antiporter activity. *J. Biol. Chem.* **261,** 5460–5466.

Cerbai, E., Klockner, U., and Isenberg, G. (1988). The α subunit of the GTP binding protein activates muscarinic potassium channels of the atrium. *Science* **240,** 1782–1783.

Chuang, D.-M., Kinnier, W. J., Farber, L., and Costa, E. (1980). A biochemical study of receptor internalization during β-adrenergic receptor desensitization in frog erythrocytes. *Mol. Pharmacol.* **18,** 348–355.

Codina, J., Hildebrandt, J. D., Sekura, R. D., Birnbaumer, M., Bryan, J., Manclark, C. R., Iyengar, R., and Birnbaumer, L. (1984a). N_s and N_i, the stimulatory and inhibitory regulatory components of adenylyl cyclases. Purification of the human erythrocyte proteins without the use of activating regulatory ligands. *J. Biol. Chem.* **259,** 5871–5883.

Codina, J., Rosenthal, W., Hildebrandt, J. D., Sekura, R. D., and Birnbaumer, L. (1984b). Updated protocols and comments on the purification without use of activating ligands of the coupling proteins N_5 and N_i of the hormone sensitive adenylyl cyclase. *J. Recept. Res.* **4,** 411–442.

Codina, J., Yatani, A., Grenet, D., Brown, A. M., and Birnbaumer, L. (1987). The α subunit of the GTP binding protein G_k opens atrial potassium channels. *Science* **236,** 442–445.

Codina, J., Olate, J., Abramowitz, J., Mattera, R., Cook, R. G., and Birnbaumer, L. (1988). α_i-3 cDNA encodes the α subunit of G_k, the stimulatory G protein of receptor-regulated K^+ channels. *J. Biol. Chem.* **263,** 6746–6750.

Connolly, E., Nånberg, E., and Nedergaard, J. (1986). Norepinephrine-induced Na^+ influx in brown adipocytes is cyclic AMP-mediated. *J. Biol. Chem.* **261,** 14377–14385.

Cooper, D. M. F., and Jagus, R. (1982). Impaired adenylate cyclase activity of phenylhydrazine-induced reticulocytes. *J. Biol. Chem.* **257,** 4684–4687.

Corcia, A., and Armstrong, W. McD. (1983). KCl cotransport: A mechanism for basolateral chloride exit in *Necturus* gallbladder. *J. Membr. Biol.* **76,** 173–182.

Crane, R. K., Miller, D., and Bihler, I. (1961). The restrictions on possible mechanisms of intestinal active transport of sugars. *Membr. Transp. Metab., Proc. Symp., Prague, 1960* pp. 439–449.

Csaky, T. Z. (1965). Transport through biological membranes. *Annu. Rev. Physiol.* **27,** 415–450.

Davoren, P. R., and Sutherland, E. W. (1963). The effect of L-epinephrine and other agents on the synthesis and release of adenosine 3′,5′-phosphate by whole pigeon erythrocytes. *J. Biol. Chem.* **238,** 3009–3015.

Diez, J., Braquet, P., Nazaret, C., Hannaert, P., Verna, R., and Garay, R. (1984). The effects of cyclic nucleotides and eicosanoids on Na^+ and K^+ transport systems in human red cells and mouse macrophages. *Adv. Cyclic Nucleotide Protein Phosphorylation Res.* **17,** 621–630.

Dobbins, J. W., Laurenson, J. P., and Forrest, J. N., Jr. (1984). Adenosine and adenosine analogues stimulate adenosine cyclic 3′,5′-monophosphate-dependent chloride secretion in the mammalian ileum. *J. Clin. Invest.* **74,** 929–935.

Dolphin, A. C., Forda, S. R., and Scott, R. H. (1986). Calcium-dependent currents in cultured rat dorsal root ganglion neurones are inhibited by an adenosine analogue. *J. Physiol.* (*London*) **373,** 47–61.

Duhm, J. (1989). Possible role of plasma lipids in alterations of cell membrane sodium transport in essential hypertension: Pathogenic implications. *Atheroscler. Rev.* **19,** 247–269.

Duhm, J., and Göbel, B. O. (1982). Sodium-lithium exchange and sodium-potassium cotransport in human erythrocytes. Part 1: Evaluation of a simple uptake test to assess the activity of the two transport systems. *Hypertension* **4,** 468–476.

Duhm, J., and Göbel, B. O. (1984). Role of furosemide-sensitive Na^+/K^+ transport system in determining the steady-state Na^+ and K^+ content and volume of human erythrocytes *in vitro* and *in vivo*. *J. Membr. Biol.* **77,** 243–254.

Dunham, P. B., and Ellory, J. C. (1981). Passive potassium transport in low potasium sheep red cells: Dependence upon cell volume and chloride. *J. Physiol. (London)* **318,** 511–530.

Dunham, P. B., and Logue, P. J. (1986). Potassium-chloride cotransport in resealed human red cell ghosts. *Am. J. Physiol.* **250,** C578–C583.

Ellory, J. C., and Dunham, P. B. (1980). Volume dependent passive potassium transport in LK sheep red cells. *In* "Membrane Transport in Erythrocytes" (I. V. Lassen, H. H. Ussing, and J. O. Wieth, eds.), Alfred Benzon Symp., Vol. 14, pp. 409–427. Munksgaard, Copenhagen.

Ellory, J. C., Dunham, P. B., Logue, P. J., and Stewart, G. W. (1982). Anion-dependent cation transport in erythrocytes. *Philos. Trans. R. Soc. London, Ser. B* **299,** 483–495.

Ellory, J. C., Hall, A. C., and Stewart, G. W. (1985). Volume-sensitive cation fluxes in mammalian red cells. *Mol. Physiol.* **8,** 235–246.

Escobales, N., and Canessa, M. (1986). Amiloride-sensitive Na^+ transport in human red cells: Evidence for a Na/H exchange system. *J. Membr. Biol.* **90,** 21–28.

Escobales, N., and Rivera, A. (1987). Na^+ for H^+ exchange in rabbit erythrocytes. *J. Cell. Physiol.* **132,** 73–80.

Eveloff, J. L., and Warnock, D. G. (1987a). Activation of ion transport systems during cell volume regulation. *Am. J. Physiol.* **252,** F1–F10.

Eveloff, J., and Warnock, D. G. (1987b). K-Cl transport systems in rabbit renal basolateral membrane vesicles. *Am. J. Physiol.* **252,** F883–F889.

Fain, J. N., Pointer, R. H., and Ward, W. F. (1972). Effects of adenosine nucleosides on adenylate cyclase, phosphodiesterase, cyclic adenosine monophosphate accumulation, and lipolysis in fat cells. *J. Biol. Chem.* **247,** 6866–6872.

Farfel, Z., and Cohen, Z. (1984). Adenylate cyclase in the maturing human reticulocyte: Selective loss of the catalytic unit, but not of the receptor-cyclase coupling protein. *Eur. J. Clin. Invest.* **14,** 79–82.

Field, M., Rao, M. C., and Chang, E. B. (1989a). Intestinal electrolyte transport and diarrheal disease, Part I. *N. Engl. J. Med.* **321,** 800–806.

Field, M., Rao, M. C., and Chang, E. B. (1989b). Intestinal electrolyte transport and diarrheal disease, Part II. *N. Engl. J. Med.* **321,** 879–883.

Fong, H. K. W., Amatruda, T. T., III, Birren, B. W., and Simon, M. I. (1987). Distinct forms of the β subunit of GTP-binding regulatory proteins identified by molecular cloning. *Proc. Natl. Acad. Sci. U.S.A.* **84,** 3792–3796.

Franchi, A., Perucca-Lostanlen, D., and Pouysségur, J. (1986a). Functional expression of a human Na^+/H^+ antiporter gene transfected into antiporter-deficient mouse L cells. *Proc. Natl. Acad. Sci. U.S.A.* **83,** 9388–9392.

Franchi, A., Cragoe, E., Jr., and Pouysségur, J. (1986b). Isolation and properties of fibroblast mutants overexpressing an altered Na^+/H^+ antiporter. *J. Biol. Chem.* **261,** 14614–14620.

Fujise, H., and Lauf, P. K. (1987). Swelling, NEM, and A23187 Cl^--dependent K^+ transport in high-K^+ sheep red cells. *Am. J. Physiol.* **252,** C197–C204.

Gao, B., Gilman, A. G., and Robishaw, J. D. (1987). A second form of the β subunit of signal-transducing G proteins. *Proc. Natl. Acad. Sci. U.S.A.* **84,** 6122–6125.
Garay, R. P. (1982). Inhibition of the Na^+/K^+ cotransport system by cyclic AMP and intracellular Ca^{2+} in human red cells. *Biochim. Biophys. Acta* **688,** 786–792.
Garay, R., Nazaret, C., Diez, J., Dagher, G., Hannaert, P., and Braquet, P. (1983). The effect of cyclic nucleotides and icosanoids on Na^+ and K^+ transport in human red cells. *Biomed. Biochim. Acta* **42,** S53–S57.
Garcia-Romeu, F., Motais, R., and Borgese, F. (1988). Desensitization by external Na of the cyclic AMP-dependent Na^+/H^+ antiporter in trout red blood cells. *J. Gen. Physiol.* **91,** 529–548.
Gardos, G. (1956). The permeability of human erythrocytes to potassium. *Acta Physiol. Acad. Sci. Hung.* **10,** 185–189.
Gilman, A. G. (1987). G proteins: Transducers of receptor-generated signals. *Annu. Rev. Biochem.* **56,** 615–649.
Grasl, M., and Turnheim, K. (1984). Stimulation of electrolyte secretion in rabbit colon by adenosine. *J. Physiol. (London)* **346,** 93–110.
Greger, R., and Schlatter, E. (1983). Properties of the basolateral membrane of the cortical thick ascending limb of Henle's loop of rabbit kidney. *Pfluegers Arch.* **396,** 325–334.
Grinstein, S., and Rothstein, A. (1986). Mechanisms of regulation of the Na^+/H^+ exchanger. *J. Membr. Biol.* **90,** 1–12.
Grinstein, S., Clarke, C. A., and Rothstein, A. (1983). Activation of Na^+/H^+ exchange in lymphocytes by osmotically induced volume changes and by cytoplasmic acidification. *J. Gen. Physiol.* **82,** 619–638.
Grinstein, S., Goetz, J. D., and Rothstein, A. (1984a). $^{22}Na^+$ fluxes in thymic lymphocytes. II. Amiloride-sensitive Na^+/H^+ exchange pathway: Reversibility of transport and asymmetry of the modifier site. *J. Gen. Physiol.* **84,** 585–600.
Grinstein, S., Rothstein, A., Sarkadi, B., and Gelfand, E. W. (1984b). Responses of lymphocytes to anisotonic media: Volume-regulating behavior. *Am. J. Physiol.* **246,** C204–C215.
Grinstein, S., Cohen, S., Goetz, J. D., and Rothstein, A. (1985). Osmotic and phorbol ester-induced activation of Na^+/H^+ exchange: Possible role of protein phosphorylation in lymphocyte volume regulation. *J. Cell Biol.* **101,** 269–276.
Haas, M. (1989). Properties and diversity of (Na-K-Cl) cotransporters. *Annu. Rev. Physiol.* **51,** 443–457.
Haas, M., and McManus, T. J. (1985). Effect of norepinephrine on swelling-induced potassium transport in duck red cells. Evidence against a volume-regulatory decrease under physiological conditions. *J. Gen. Physiol.* **85,** 649–667.
Hall, A. C., and Ellory, J. C. (1985). Measurement and stoichiometry of bumetanide-sensitive (2Na:1K:3Cl) cotransport in ferret red cells. *J. Membr. Biol.* **85,** 205–213.
Hall, A. C., and Ellory, J. C. (1986). Evidence for the presence of volume-sensitive KCl transport in 'young' human red cells. *Biochim. Biophys. Acta* **858,** 317–320.
Halm, D. R., Krasny, E. J., Jr., and Frizzell, R. A. (1985). Electrophysiology of flounder intestinal mucosa. I. Conductance properties of the cellular and paracellular pathways. *J. Gen. Physiol.* **85,** 843–864.
Hamburger, H. J. (1918). Anionenwanderungen im serum und blut unter dem einfluss von CO_2, säure and alkali. *Biochem. Z.* **86,** 309–327.
Hanski, E., Sternweis, P. C., Northup, J. K., Dromerick, A. W., and Gilman, A. G. (1981). The regulatory component of adenylate cyclase. *J. Biol. Chem.* **256,** 12911–12919.
He, X., Ship, J., Wu, X., Brown, A. M., and Wellner, R. B. (1989). β-Adrenergic control of

cell volume and chloride transport in an established rat submandibular cell line. *J. Cell. Physiol.* **138,** 527–535.

Heasley, L. E., Azari, J., and Brunton, L. L. (1979). Export of cyclic AMP from avian red cells. Independence from major membrane transporters and specific inhibition by prostaglandin A_1. *Mol. Pharmacol.* **27,** 60–66.

Hoffmann, E. K. (1986). Anion transport systems in the plasma membrane of vertebrate cells. *Biochim. Biophys. Acta* **864,** 1–31.

Holman, G. D. (1979). Infinite *cis* influx of cyclic AMP into human erythrocyte ghosts. *Biochim. Biophys. Acta* **553,** 489–494.

Iyengar, R., Rich, K. A., Herberg, J. T., Dagoberto, G., Mumby, S., and Codina, J. (1987). Identification of a new GTP-binding protein. A. M_r-43,000 substrate for pertussis toxin. *J. Biol. Chem.* **262,** 9239–9245.

Jakobs, K. H., Saur, W., and Schultz, G. (1978). Inhibition of platelet adenylate cyclase by epinephrine requires GTP. *FEBS Lett.* **85,** 167–170.

Jakobs, K. H., Aktories, K., and Schultz, G. (1979). GTP-dependent inhibition of cardiac adenylate cyclase by muscarinic cholinergic agonists. *Naunyn-Schmiedeberg's Arch. Pharmacol.* **310,** 113–119.

Jarvis, S. M., Young, J. D., Ansay, M., Archibald, A. L., Harkness, R. A., and Simmonds, R. J. (1980). Is inosine the physiological energy source of pig erythrocytes? *Biochim. Biophys. Acta* **597,** 183–188.

Jeffrey, D. R., Charlton, R. R., and Venter, J. C. (1980). Reconstitution of turkey erythrocyte β-adrenergic receptors into human erythrocyte acceptor membranes. Demonstration of guanine nucleotide regulation of agonist affinity. *J. Biol. Chem.* **255,** 5015–5018.

Jennings, M. L., Douglas, S. M., and McAndrew, P. E. (1986). Amiloride-sensitive sodium-hydrogen exchange in osmotically shrunken rabbit red blood cells. *Am. J. Physiol.* **251,** C32–C40.

Johnson, G. L., and Dhanasekaran, N. (1989). The G-protein family and their interaction with receptors. *Endocr. Rev.* **10,** 317–330.

Kahn, A. M., Dolson, G. M., Hise, M. K., Bennett, S. C., and Weinman, E. J. (1985). Parathyroid hormone and dibutyryl cAMP inhibit Na^+/H^+ exchange in renal brush border vesicles. *Am. J. Physiol.* **248,** F212–F218.

Kai, H., Kanaide, H., Matsumoto, T., Shogakiuchi, Y., and Nakamura, M. (1987). Adenosine decreases intracellular free calcium concentrations in cultured vascular smooth muscle cells from rat aorta. *FEBS Lett.* **212,** 119–122.

Kaiser, G., Quiring, K., Gauger, D., Palm, D., Becker, H., and Schoeppe, W. (1974). Occurrence of adenyl cyclase activity in human erythrocytes. *Blood* **29,** 115–122.

Kaiser, G., Palm, D., Quiring, K., and Gauger, D. (1977). The adrenergic β-receptor system of the premature erythrocyte: Indication for adrenergic control of the erythron? *Pharmacol. Res. Commun.* **9,** 93–103.

Kaji, D. (1986). Volume-sensitive K transport in human erythrocytes. *J. Gen. Physiol.* **88,** 719–738.

Katada, T., and Ui, M. (1979). Islet-activating protein. Enhanced insulin secretion and cyclic AMP accumulation in pancretic islets due to activation of native calcium ionophores. *J. Biol. Chem.* **254,** 469–479.

Katada, T., and Ui, M. (1980). Slow interaction of islet-activating protein with pancreatic islets during primary culture to cause reversal of α-adrenergic inhibition of insulin secretion. *J. Biol. Chem.* **255,** 9580–9588.

Katada, T., and Ui, M. (1981). Islet-activating protein. A modifier of receptor-mediated regulation of rat islet adenylate cyclase. *J. Biol. Chem.* **256,** 8310–8317.

Katada, T., and Ui, M. (1982). ADP ribosylation of the specific membrane protein of C6 cells by islet-activating protein associated with modification of adenylate cyclase activity. *J. Biol. Chem.* **257,** 7210–7216.

Katada, T., Oinuma, M., and Ui, M. (1986). Mechanisms for inhibition of the catalytic activity of adenylate cyclase by the guanine nucleotide-binding proteins serving as the substrate of islet-activating protein, pertussis toxin. *J. Biol. Chem.* **261,** 5215–5221.

Kim, H. D. (1983). Postnatal changes in energy metabolism of mammalian red blood cells. *In* "Red Blood Cells of Domestic Mammals" (N. S. Agari and P. G. Board, eds.), pp. 339–355. Elsevier, Amsterdam.

Kim, H. D., Watts, R. P., Luthra, M. G., Schwalbe, C. R., Conner, R. T., and Brendel, K. (1980). A symbiotic relationship of energy metabolism between a 'nonglycolytic' mammalian erythrocyte and the liver. *Biochim. Biophys. Acta* **589,** 256–263.

Kim, H. D., Tsai, Y.-S., Franklin, C. C., and Turner, J. T. (1988). Characterization of $Na^+/K^+/Cl^-$ cotransport in cultured HT29 human colonic adenocarcinoma cells. *Biochim. Biophys. Acta* **946,** 397–404.

Kim, H. D., Sergeant, S., Forte, L. R., Sohn, D. H., and Im, J. H. (1989). Activation of a Cl-dependent K flux by cAMP in pig red cells. *Am. J. Physiol.* **256,** C772–C778.

Kirsch, G. E., Yatani, A., Codina, J., Birnbaumer, L., and Brown, A. M. (1988). α-Subunit of G_k activates atrial K^+ channels of chick, rat, and guinea pig. *Am. J. Physiol.* **254,** H1200–H1205.

Kramhøft, B., Lambert, I. H., Hoffman, E. K., and Jørgenson, F. (1986). Activation of Cl-dependent K transport in Ehrlich ascites tumor cells. *Am. J. Physiol.* **251,** C369–C379.

Kregenow, F. M. (1971). The response of duck erythrocytes to nonhemolytic hypotonic media. Evidence for a volume-controlling mechanism. *J. Gen. Physiol.* **58,** 372–395.

Kregenow, F. M. (1978). An assessment of the co-transport hypothesis as it applies to the norepinephrine and hypertonic responses. *In* "Osmotic and Volume Regulation" (C. B. Jorgensen and E. Skadhauge, eds.), Alfred Benzon Symp., Vol. 11, pp. 379–391. Munksgaard, Copenhagen.

Kregenow, F. M. (1981). Osmoregulatory salt transporting mechanisms: Control of cell volume in anisotonic media. *Annu. Rev. Physiol.* **43,** 493–505.

Kregenow, F. M., Robbie, D. E., and Orloff, J. (1976). Effect of norepinephrine and hypertonicity on K influx and cyclic AMP in duck erythrocytes. *Am. J. Physiol.* **231,** 306–312.

Kumar, R., Yuh, K.-C., and Tao, M. (1978). Human erythrocyte proteins associated with adenosine 3′,5′-cyclic monophosphate action. *Enzyme* **23,** 73–83.

Kurachi, Y., Nakajima, T., and Sugimoto, T. (1986). On the mechanism of activation of muscarinic K^+ channels by adenosine in isolated atrial cells: Involvement of GTP-binding proteins. *Pfluegers Arch.* **407,** 264–274.

Larner, A. C., and Ross, E. M. (1981). Alteration in the protein components of catecholamine-sensitive adenylate cyclase during maturation of rat reticulocytes. *J. Biol. Chem.* **256,** 9551–9557.

Lauf, P. K. (1983a). Thiol-dependent passive K/Cl transport in sheep red cells: I. Dependence on chloride and external K^+ [Rb^+] ions. *J. Membr. Biol.* **73,** 237–246.

Lauf, P. K. (1983b). Thiol-dependent passive K^+-Cl^- transport in sheep red blood cells. V. Dependence on metabolism. *Am. J. Physiol.* **245,** C445–C448.

Lauf, P. K. (1983c). Thiol-dependent passive K/Cl transport in sheep red cells: II. Loss of Cl^- and N-ethylmaleimide sensitivity in maturing high K^+ cells. *J. Membr. Biol.* **73,** 247–256.

Lauf, P. K. (1984). Thiol-dependent passive K^+-Cl^- transport in sheep red blood cells: VI. Functional heterogeneity and immunologic identity with volume-stimulated K^+(Rb^+) flux. *J. Membr. Biol.* **82,** 167–178.

Lauf, P. K. (1985). K^+:Cl^- cotransport: Sulfhydryls, divalent cations, and the mechanism of volume activation in a red cell. *J. Membr. Biol.* **88,** 1–13.

Lauf, P. K., and Bauer, J. (1987). Direct evidence for chloride-dependent volume reduction in monocytic sheep reticulocytes. *Biochem. Biophys. Res. Commun.* **144,** 849–855.

Lauf, P. K., and Theg, B. E. (1980). A chloride dependent K^+ flux induced by *N*-ethylmaleimide in genetically low K^+ sheep and goat erythrocytes. *Biochem. Biophys. Res. Commun.* **92,** 1422–1428.

Lauf, P. K., Adragna, N. C., and Garay, R. P. (1984a). Activation by *N*-ethylmaleimide of a latent K^+-Cl^- flux in human red blood cells. *Am. J. Physiol.* **246,** C385–C390.

Lauf, P. K., Zeidler, R. B., and Kim, H. D. (1984b). Pig reticulocytes. V. development of Rb^+ influx during *in vitro* maturation. *J. Cell. Physiol.* **121,** 284–290.

Lauf, P. K., Perkins, C. M., and Adragna, N. C. (1985). Cell volume and metabolic dependence of NEM-activated K^+-Cl^- flux in human red blood cells. *Am. J. Physiol.* **249,** C124–C128.

Le Vine, H., III, and Cuatrecasas, P. (1981). Activation of pigeon erythrocyte adenylate cyclase by cholera toxin. Partial purification of an essential macromolecular factor from horse erythrocyte cytosol. *Biochim. Biophys. Acta* **672,** 248–261.

Lichtshtein, D., Boone, G., and Blume, A. (1979). Muscarinic receptor regulation of NG108-15 adenylate cyclase: Requirement for Na^+ and GTP. *J. Cyclic Nucleotide Res.* **5,** 367–375.

Limbird, L. E., Gill, D. M., Stadel, J. M., Hickey, A. R., and Lefkowitz, R. J. (1980). Loss of β-adrenergic receptor-guanine nucleotide regulatory protein interactions accompanies decline in catecholamine responsiveness of adenylate cyclase in maturing rat erythrocytes. *J. Biol. Chem.* **255,** 1854–1861.

Logothetis, D. E., Kurachi, Y., Galper, J., Neer, E. J., and Clapham, D. E. (1987). The $\beta\gamma$ subunits of GTP-binding proteins activate the mucarinic K^+ channel in heart. *Nature (London)* **325,** 321–326.

Logue, P. J., Anderson, C., Kanik, C., Farguharson, B., and Dunham, P. (1983). Passive potassium transport in LK sheep red cells. *J. Gen. Physiol.* **81,** 861–885.

Londos, C., and Preston, M. S. (1977). Regulation by glucagon and divalent cations of inhibition of hepatic adenylate cyclase by adenosine. *J. Biol. Chem.* **252,** 5951–5956.

Londos, C., and Wolff, J. (1977). Two distinct adenosine-sensitive sites on adenylate cyclase. *Proc. Natl. Acad. Sci. U.S.A.* **74,** 5482–5486.

Lotan, I., Dascal, N., Oron, Y., Cohen, S., and Lass, Y. (1985). Adenosine-induced K^+ current in *Xenopus* oocyte and the role of adenosine 3′,5′-monophosphate. *Mol. Pharmacol.* **28,** 170–177.

MacDonald, R. L., Skerritt, J. H., and Werz, M. A. (1986). Adenosine agonists reduce voltage-dependent calcium conductance of mouse sensory neurones in cell culture. *J. Physiol. (London)* **370,** 75–90.

Mahe, Y., Garcia-Romeu, G., and Motais, R. (1985). Inhibition by amiloride of both adenylate cyclase activity and the Na^+/H^+ antiporter in fish erythrocytes. *Eur. J. Pharmacol.* **116,** 199–206.

Mandel, K. G., Dharmsathaphorn, K., and McRoberts, J. A. (1986). Characterization of a cyclic AMP-activated Cl^- transport pathway in the apical membrane of a human colonic epithelial cell line. *J. Biol. Chem.* **261,** 704–712.

McKenzie, S. G., and Bär, H. P. (1973). On the mechanism of adenyl cyclase inhibition by adenosine. *Can. J. Physiol. Pharmacol.* **51,** 190–196.

McManus, T. J., and Schmidt, W. F., III (1978). Ion and co-ion transport in avian red cells. *In* "Membrane Transport Processes" (J. F. Hoffman, ed.), Vol. 1, pp. 79–106. Raven, New York.

McRoberts, J. A., Tran, C. T., and Saier, M. H. (1983). Characteristics of low potassium-resistant mutants of the Madin-Darby canine kidney cell line with defects in NaCl/KCl symport. *J. Biol. Chem.* **258,** 12320–12326.

McRoberts, J. A., Beuerlein, G., and Dharmsathaphorn, K. (1985). Cyclic AMP and Ca^{2+}-activated K^+ transport in a human colonic epithelial cell line. *J. Biol. Chem.* **260,** 14163–14172.

Milligan, G. (1988). Techniques used in the identification and analysis of function of pertussis toxin-sensitive guanine nucleotide binding proteins. *Biochem. J.* **255,** 1–13.

Montandon, J.-B., and Porzig, H. (1983). *In vitro* maturation of the rat reticulocyte beta-adrenoceptor adenylate cyclase system. *Biomed. Biochim. Acta* **42,** S197–S201.

Moriwaki, K., and Fóa, P. P. (1970). Inhibition of rat liver adenyl cyclase by adenosine and adenine nucleotides. *Experientia* **26,** 22.

Nambi, P., Sibley, D. R., Stadel, J. M., Michel, T., Peters, J. R., and Lefkowitz, R. J. (1984). Cell-free desensitization of catecholamine-sensitive adenylate cyclase. Agonist- and cAMP-promoted alterations in turkey erythrocyte β-adrenergic receptors. *J. Biol. Chem.* **259,** 4629–4633.

Nambi, P., Peters, J. R., Sibley, D. R., and Lefkowitz, R. J. (1985). Desensitization of the turkey erythrocyte β-adrenergic receptor in a cell-free system. *J. Biol. Chem.* **260,** 2165–2171.

Neer, E. J., Lok, J. M., and Wolf, L. G. (1984). Purification and properties of the inhibitory guanine nucleotide regulatory unit of brain adenylate cyclase. *J. Biol. Chem.* **259,** 14222–14229.

Newman, M., and Levitzki, A. (1982). Characteristics of high-affinity [^{3}H]adenosine binding to rat brain synaptosomes and turkey erythrocyte membranes. *Biochim. Biophys. Acta* **685,** 129–136.

Northup, J. K., Smigel, M. D., Sternweis, P. C., and Gilman, A. G. (1983). The subunits of the stimulatory regulatory component of adenylate cyclase. Resolution of the activated 45,000-Dalton (α) subunit. *J. Biol. Chem.* **258,** 11369–11376.

Ørskov, S. L. (1956). Experiments on the influence of adrenaline and noradrenaline on the potassium absorption of red blood cells from pigeons and frogs. *Abstr. Commun., Int. Physiol. Congr., Brussels* **20,** 694.

O'Grady, S. M., Palfrey, H. C., and Field, M. (1987). Characteristics and functions of Na-K-Cl cotransport in epithelial tissues. *Am. J. Physiol.* **253,** C177–C192.

Owen, N. E. (1984). Regulation of Na/K/Cl cotransport in vascular smooth muscle cells. *Biochem. Biophys. Res. Commun.* **125,** 500–508.

Owen, N. E., and Prastein, M. L. (1985). Na/K/Cl cotransport in cultured human fibroblasts. *J. Biol. Chem.* **260,** 1445–1451.

Palfrey, H. C., and Greengard, P. (1981). Hormone-sensitive ion transport systems in erythrocytes as models for epithelial ion pathways. *Ann. N.Y. Acad. Sci.* **372,** 291–308.

Palfrey, H. C., and Rao, M. C. (1983). Na/K/Cl co-transport and its regulation. *J. Exp. Biol.* **106,** 43–54.

Palfrey, H. C., Feit, P. W., and Greengard, P. (1980a). cAMP-stimulated cation cotransport in avian erythrocytes: Inhibition by "loop" diuretics. *Am. J. Physiol.* **238,** C139–C148.

Palfrey, H. C., Greengard, P., and Feit, P. W. (1980b). Specific inhibition by "loop" diuretics of an anion-dependent $Na^+ + K^+$ cotransport system in avian erythrocytes. *Ann. N.Y. Acad. Sci.* **341,** 134–138.

Parker, J. C., and Castranova, V. (1984). Volume-responsive sodium and proton movements in dog red blood cells. *J. Gen. Physiol.* **84,** 379–401.

Parker, J. C., and Dunham, P. B. (1989). Passive cation transport. *In* "Red Blood Cell

Membranes: Structure, Function, Clinical Implications" (P. Agre and J. C. Parker, eds.), Vol. 11, pp. 507–561. Dekker, New York.

Parker, J. C., and Snow, R. L. (1972). Influence of external ATP on permeability and metabolism of dog red blood cells. *Am. J. Physiol.* **223,** 888–893.

Parker, J. C., Castranova, V., and Goldinger, J. M. (1977). Dog red blood cells: Na and K diffusion potentials with extracellular ATP. *J. Gen. Physiol.* **69,** 417–430.

Pfaffinger, P. J., Martin, J. M., Hunter, D. D., Nathanson, N. M., and Hille, B. (1985). GTP-binding proteins couple cardiac muscarinic receptors to a K channel. *Nature (London)* **317,** 536–538.

Pfeuffer, T. (1977). GTP-binding proteins in membranes and the control of adenylate cyclase activity. *J. Biol. Chem.* **252,** 7224–7234.

Pfeuffer, T., and Helmrich, E. J. M. (1975). Activation of pigeon erythrocyte membrane adenylate cyclase by guanylnucleotide analogues and separation of a nucleotide binding protein. *J. Biol. Chem.* **250,** 867–876.

Pollock, A. S., Warnock, D. G., and Strewler, G. J. (1986). Parathyroid hormone inhibition of Na^+-H^+ antiporter activity in a cultured renal cell line. *Am. J. Physiol.* **250,** F217–F225.

Pratt, A. D., Clancy, G., and Welsh, M. J. (1986). Mucosal adenosine stimulates chloride secretion in canine tracheal epithelium. *Am. J. Physiol.* **251,** C167–C174.

Proctor, W. R., and Dunwiddie, T. V. (1983). Adenosine inhibits calcium spikes in hippocampal pyramidal neurons *in vitro*. *Neurosci. Lett.* **35,** 197–201.

Rall, T. W., Sutherland, E. W., and Berthet, J. (1957). The relationship of epinephrine and glucagon to liver phosphorylase. IV. Effect of epinephrine and glucagon on the reactivation of phsophorylase in liver homogenates. *J. Biol. Chem.* **224,** 463–475.

Rasmussen, H. R., Lake, W., and Allen, J. E. (1975). The effect of catecholamines and prostaglandins upon human and rat erythrocytes. *Biochim. Biophys. Acta* **411,** 63–73.

Reuss, L., and Peterson, K. (1985). Cyclic AMP inhibits Na^+/H^+ exchange at the apical membrane of *Necturus* gallbladder epithelium. *J. Gen. Physiol.* **85,** 409–429.

Riddick, D. H., Kregenow, F. M., and Orloff, J. (1971). The effect of norepinephrine and dibutyryl cyclic adenosine monphosphate on cation transport in duck erythrocytes. *J. Gen. Physiol.* **57,** 752–766.

Rodan, S. B., Rodan, G. A., and Sha'afi, R. I. (1976). Demonstration of adenylate cyclase activity in human red blood cell ghosts. *Biochim. Biophys. Acta* **428,** 509–515.

Rodbell, M., Birnbaumer, L., Pohl, S. L., and Krans, H. M. J. (1971). The glucagon-sensitive adenyl cyclase system in plasma membranes of rat liver. *J. Biol. Chem.* **246,** 1877–1882.

Ross, E. M., and Gilman, A. G. (1977). Resolution of some components of adenylate cyclase necessary for catalytic activity. *J. Biol. Chem.* **252,** 6966–6969.

Rozengurt, E. (1985). The mitogenic response of cultured 3T3 cells: Integration of early signals and synergistic effects in a unified framework. *In* "Molecular Aspects of Cellular Regulation. Vol. 4: Molecular Mechanisms of Transmembrane Signalling" (P. Cohen and M. D. Houslay, eds.), pp. 429–452. Elsevier, Amsterdam.

Rudolph, S. A., and Greengard, P. (1980). Effects of catecholamines and prostaglandin E_1 on cyclic AMP, cation fluxes, and protein phosphorylation in the frog erythrocyte. *J. Biol. Chem.* **255,** 8534–8540.

Russell, J. M. (1983). Cation-coupled chloride influx in squid axon. Role of potassium and stoichiometry of the transport process. *J. Gen. Physiol.* **81,** 909–925.

Russo, L. A., Rannels, S. R., Laslow, K. S., and Rannels, D. E. (1989). Stretch-related changes in lung cAMP after partial pneumonectomy. *Am. J. Physiol.* **257,** E261–E268.

Sabol, S. L., and Nirenberg, M. (1979). Regulation of adenylate cyclase of neuroblastoma × glioma hybrid cells by α-adrenergic receptors. I. Inhibition of adenylate cyclase mediated by α receptors. *J. Biol. Chem.* **254,** 1913–1920.

Sachs, J. R. (1988). Volume-sensitive K influx in human red cell ghosts. *J. Gen. Physiol.* **92,** 685–711.

Sager, G. (1982). Receptor binding sites for beta-adrenergic ligands on human erythrocytes. *Biochem. Pharmacol.* **31,** 99–104.

Sager, G. (1983). β-2 Adrenergic receptors on intact human erythrocytes. *Biochem. Pharmacol.* **32,** 1946–1949.

Sager, G., and Jacobsen, S. (1985). Effect of plasma on human erythrocyte beta-adrenergic receptors. *Biochem. Pharmacol.* **34,** 3767–3771.

Sardet, C., Franchi, A., and Pouysségur, J. (1988). Molecular cloning of the growth-factor-activatable human Na^+/H^+ antiporter. *Cold Spring Harbor Symp. Quant. Biol.* **53,** 1011–1018.

Sardet, C., Franchi, A., and Pouysségur, J. (1989). Molecular cloning, primary structure, and expression of the human growth factor-activatable Na^+/H^+ antiporter. *Cell* **56,** 271–280.

Sardet, C., Counillon, L., Franchi, A., and Pouysségur, J. (1990). Growth factors induce phosphorylation of the Na^+/H^+ antiporter, a glycoprotein of 100 kD. *Science* **247,** 723–726.

Schatzmann, H.-J. (1953). Herzglycoside als hemmstoffe fur den aktiven kalium- und natriumtransport durch die erythrocytenmembran. *Helv. Physiol. Pharmacol. Acta* **11,** 346–354.

Schatzmann, H.-J. (1966). ATP-dependent Ca^{++}-extrusion from human red cells. *Experientia* **22,** 282–289.

Schrader, J., Rubio, R., and Berne, R. M. (1975). Inhibition of slow action potentials of guinea pig atrial muscle by adenosine: A possible effect on Ca^{2+} influx. *J. Mol. Cell. Cardiol.* **7,** 427–433.

Sergeant, S., and Kim, H. D. (1985). Inhibition of 3-*O*-methylglucose transport in human erythrocytes by forskolin. *J. Biol. Chem.* **260,** 14677–14682.

Sergeant, S., Sohn, D. H., and Kim, H. D. (1989). Volume-activated Na/H exchange activity in fetal and adult pig red cells: Inhibition by cyclic AMP. *J. Membr. Biol.* **109,** 209–220.

Shane, E., Gammon, D. E., and Bilezikian, J. P. (1981). A cellular activator of catecholamine-sensitive adenylate cyclase in rat reticulocytes and erythrocytes: Changes during reticulocyte development and effects on the β receptor. *Arch. Biochem. Biophys.* **208,** 418–425.

Shane, E., Yeh, M., Feigin, A. S., Owens, J. M., and Bilezikian, J. P. (1985a). Reticulocyte cytosol activator protein: Effects on the stimulatory and inhibitory regulatory proteins of adenylate cyclase. *Endocrinology* (*Baltimore*) **117,** 264–270.

Shane, E., Yeh, M., Feigin, A. S., Owens, J. M., and Bilezikian, J. P. (1985b). Cytosol activator protein from rat reticulocytes requires the stimulatory guanine nucleotide-binding protein for its actions on adenylate cyclase. *Endocrinology* (*Baltimore*) **117,** 255–263.

Sheppard, H., and Burghardt, C. (1969). Adenyl cyclase in non-nucleated erythrocytes of several mammalian species. *Biochem. Pharmacol.* **18,** 2576–2578.

Sibley, D. R., Peters, J. R., Nambi, P., Caron, M. G., and Lefkowitz, R. J. (1984a). Photoaffinity labeling of turkey erythrocyte beta-adrenergic receptors: Degradation of the M_r = 49,000 protein explains apparent heterogeneity. *Biochem. Biophys. Res. Commun.* **119,** 458–464.

Sibley, D. R., Peters, J. R., Nambi, P., Caron, M. G., and Lefkowitz, R. J. (1984b). Desensitization of turkey erythrocyte adenylate cyclase. β-Adrenergic receptor phosphorylation is correlated with attenuation of adenylate cyclase activity. *J. Biol. Chem.* **259,** 9742–9749.

Smith, J. B., and Smith, L. (1987). $Na^+/K^+/Cl^-$ cotransport in cultured vascular smooth muscle cells: Stimulation by angiotensin II and calcium ionophores, inhibition by cyclic AMP and calmodulin antagonists. *J. Membr. Biol.* **99,** 51–63.

Sohn, D. H., and Kim, H. D. (1991). Effects of adenosine receptor agonists on volume-activated ion transport in pig red cells, *J. Cell Physiol.* **146,** 318–324.

Stadel, J. M., Nambi, P., Lavin, T. N., Heald, S. L., Caron, M. G., and Lefkowitz, R. J. (1982). Catecholamine-induced desensitization of turkey erythrocyte adenylate cyclase. Structural alterations in the β-adrenergic receptor revealed by photoaffinity labeling. *J. Biol. Chem.* **257,** 9242–9245.

Stadel, J. M., Nambi, P., Shorr, R. G. L., Sawyer, D. F., Caron, M. G., and Lefkowitz, R. J. (1983a). Catecholamine-induced desensitization of turkey erythrocyte adenylate cyclase is associated with phosphorylation of the β-adrenergic receptor. *Proc. Natl. Acad. Sci. U.S.A.* **80,** 3173–3177.

Stadel, J. M., Strulovici, B., Nambi, P., Lavin, T. N., Briggs, M. M., Caron, M. G., and Lefkowitz, R. J. (1983b). Desensitization of the β-adrenergic receptor of frog erythrocytes. Recovery and characterization of the down-regulated receptors in sequestered vesicles. *J. Biol. Chem.* **258,** 3032–3038.

Starke, L. C., and McManus, T. J. (1988). Control of the volume-regulatory set point in duck red cells. *J. Gen. Physiol.* **92,** 42a. (Abstr.)

Starke, L. C., and McManus, T. J. (1990). Intracellular free magnesium determines the volume regulatory set point in duck red cells. *FASEB J.* **4,** A818.

Steer, M. L., and Wood, A. (1979). Regulation of human platelet adenylate cyclase by epinephrine, prostaglandin E_1, and guanine nucleotides. Evidence for separate guanine nucleotide sites mediating stimulation and inhibition. *J. Biol. Chem.* **254,** 10791–10797.

Sternweis, P. C., and Robishaw, J. D. (1984). Isolation of two proteins with high affinity for guanine nucleotides from membranes of bovine brain. *J. Biol. Chem.* **259,** 13806–13813.

Susanni, E. E. T., Ross, F. P., Scriven, D. R. L., and Rosendorff, C. (1985). Baboon erythrocyte ghosts contain β-adrenergic receptors. *Am. J. Physiol.* **249,** C15–C19.

Thornhill, W. B., and Laris, P. C. (1984). KCl loss and cell shrinkage in the Ehrlich ascites tumor cell induced by hypotonic media, 2-deoxyglucose and propranolol. *Biochim. Biophys. Acta* **773,** 207–218.

Tosteson, D. C., and Hoffman, J. F. (1960). Regulation of cell volume by activation transport in high and low potassium sheep red cells. *J. Gen. Physiol.* **44,** 169–194.

Trussell, L. O., and Jackson, M. B. (1985). Adenosine-activated potassium conductance in cultured striatal neurons. *Proc. Natl. Acad. Sci. U.S.A.* **82,** 4857–4861.

Trussell, L. O., and Jackson, M. B. (1987). Dependence of an adenosine-activated potassium current on a GTP-binding protein in mammalian central neurons. *J. Neurosci.* **7,** 3306–3316.

Turner, J. T., Jones, S. B., and Bylund, D. B. (1986). A fragment of vasoactive intestinal peptide, VIP(10-28), is an antagonist of VIP in the colon carcinoma cell line, HT29. *Peptides* **7,** 849–854.

Ueberschär, S., and Bakker-Grunwald, T. (1985). Effects of ATP and cyclic AMP on the (Na^+ + K^+ + $2Cl^-$)-cotransport system in turkey erythrocytes. *Biochim. Biophys. Acta* **818,** 260–266.

Van Calker, D., Muller, D. M., and Hamprecht, B. (1979). Adenosine regulates via two different types of receptors, the accumulation of cyclic AMP in cultured brain cells. *J. Neurochem.* **33,** 999–1005.

Vigne, P., Breittmayer, J.-P., Frelin, C., and Lazdunski, M. (1988). Dual control of the intracellular pH in aortic smooth muscle cells by a cAMP-sensitive HCO_3^-/Cl^- antiporter and a protein kinase C-sensitive Na^+/H^+ antiporter. *J. Biol. Chem.* **263,** 18023–18029.

Watson, P. A., Haneda, T., and Morgan, H. E. (1989). Effect of higher aortic pressure on ribosome formation and cAMP content in rat heart. *Am. J. Physiol.* **256,** C1257–C1261.

Watson, P. A., (1990). Direct stimulation of adenylate cyclase by mechanical forces in S49 mouse lymphona cells during hyposmotic swelling. *J. Biol. Chem.* **265,** 6569–6575.

Watts, R. P., Brendel, K., Luthra, M. G., and Kim, H. D. (1979). Inosine from liver as a possible energy source for pig red blood cells. *Life Sci.* **25,** 1577–1582.

Weinryb, I., and Michel, I. M. (1974). Potent magnesium-dependent inhibition of adenylate cyclase activity from guinea pig lung by adenosine and other 9-substituted adenines. *Biochim. Biophys. Acta* **334,** 218–225.

Weller, M. (1978). The effect of cyclic nucleotides and protein phosphorylation on the permeability of human erythrocyte ghosts to certain cations. *Mol. Cell. Biochem.* **20,** 95–102.

Wiemer, G., Hellwich, U., Dietz, W. J., Hellwich, M., and Palm, D. (1982). Energy-dependent extrusion of cyclic 3′,5′-adenosine-monophosphate. A drug-sensitive regulatory mechanism for the intracellular nucleotide concentration in rat erythrocytes. *Naunyn-Schmiedeberg's Arch. Pharmacol.* **321,** 239–246.

Wiley, J. S., and Cooper, R. A. (1974). A furosemide-sensitive cotransport of sodium plus potassium in the human red cell. *J. Clin. Invest.* **53,** 745–755.

Wu, P. H., Phillis, J. W., and Thierry, D. L. (1982). Adenosine receptor agonists inhibit K^+-evoked Ca^{2+} uptake by rat brain cortical synaptosomes. *J. Neurochem.* **39,** 700–708.

Yamashita, A., Kurokawa, T., Dantur, T., Yanagiuchi, H., and Ishibashi, S. (1987). Characterization of heterologous desensitization of rat reticulocyte adenylate cyclase system. *J. Pharmacobio.-Dyn.* **10,** 250–254.

Yamashita, A., Kurokawa, T., Une, Y., and Ishibashi, S. (1988). Phorbol ester regulates stimulatory and inhibitory pathways of the hormone-sensitive adenylate cyclase system in rat reticulocytes. *Eur. J. Pharmacol.* **151,** 167–175.

Yatani, A., Hamm, H., Codina, J., Mazzoni, M. R., Birnbaumer, L., and Brown, A. M. (1988a). A monoclonal antibody to the α subunit of G_k blocks muscarinic activation of atrial K^+ channels. *Science* **241,** 828–831.

Yatani, A., Mattera, R., Codina, J., Graf, R., Okabe, K., Padrell, E., Iyengar, R., Brown, A. M., and Birnbaumer, L. (1988b). The G protein-gated atrial K^+ channel is stimulated by three distinct $G_i\alpha$-subunits. *Nature (London)* **336,** 680–682.

Young, J. D., Jarvis, S. M., Clanachan, A. S., Henderson, J. F., and Paterson, A. R. P. (1986). Nitrobenzylthionosine: An *in vivo* inhibitor of pig erythrocyte energy metabolism. *Am. J. Physiol.* **251,** C90–C94.

Zeidler, R. B., Metzler, M. H., Moran, J. B., and Kim, H. D. (1985). The liver is an organ site for the release of inosine metabolized by non-glycolytic pig erythrocytes. *Biochim. Biophys. Acta* **838,** 321–328.

Zenser, T. V. (1976). Inhibition of cholera toxin-stimulated intestinal epithelial cell adenylate cyclase by adenosine analogs (39342). *Proc. Soc. Exp. Biol. Med.* **152,** 126–129.

PART IV

Ion Channel Development

CHAPTER 7

Development, Maintenance, and Modulation of Voltage-Dependent Sodium Channel Topography in Nerve Cells

Kimon J. Angelides and Eun-hye Joe

Departments of Molecular Physiology and Biophysics, and Neuroscience, Baylor College of Medicine, Houston, Texas 77030

I. INTRODUCTION

Neurons are highly specialized cells whose morphology, with multiple branching dendrites and axons, is tailored for the receipt and transmission of information. These highly polarized cells are characterized by a cell surface organization where proteins are segregated and maintained in discrete functional domains. Although in other polarized cells, such as epithelial cells, it is thought that the different plasma membrane domains are constructed as a result of a targeted transport of proteins to particular destinations, in neurons it is not known how specific cell surface domains are created.

A neuron's electrical fingerprint arises from how voltage-sensitive ion channels and receptors are placed and maintained in specific regions of the cell surface. In myelinated nerve, clustering of voltage-dependent sodium channels (NaChs) to nodes of Ranvier creates sites of large inward sodium current and enables conduction to jump from node to node, thereby facilitating rapid conduction down the axon. The segregation of NaChs to nodes of Ranvier is a classic and striking example of how specific cell surface components are localized and segregated to compartments and even within local domains of the axon. Because the NaCh lends such unique excitability characteristics to the cell membrane, the distribution of this channel is clearly critical in determining some of the differential electrical properties of one part versus another part of the neuron. Often, alteration of this precise distribution leads to the functional changes which are seen in neurological disorders such as multiple sclerosis.

It is the object of this review to examine some of the mechanisms by which voltage-dependent NaChs are distributed and maintained on the nerve cell surface. In particular, we focus on the development of the node of Ranvier, where NaChs are clustered, and attempt to develop a hypothesis based on current information in order to stimulate further experimentation.

A. Organization of the Axon Membrane

1. Molecular Heterogeneity of the Myelinated Axon

In the peripheral nervous system (PNS), each internodal segment on the axon is ensheathed by a single motile Schwann cell, while in the central nervous system (CNS) an oligodendrocyte has the ability to ensheathe several axons. Placement of nodes in both the CNS and PNS is very precise, with little variation in the ~1.4 mm node-to-node distance.

Cytochemical probes have been useful in describing the architecture of myelinated axons and have shown that the axolemma is structurally heterogeneous. (Waxman, 1981, 1987). Combined electrophysiological and cytochemical studies have also shown that the distribution of NaChs on the axon is nonuniform (Black *et al.*, 1990). Specifically, gating current measurements (Conti *et al.*, 1976), intramembrane particle (IMP) distribution (Rosenbluth, 1976, 1981), and equilibrium binding of [^{3}H]saxitoxin (STX) (Ritchie and Rang, 1983; Ritchie and Rogart, 1979), a specific NaCh marker, suggest that NaChs are sequestered at nodes with densities of between 1000 and 5000/μm^2. The inexcitability of the internodal membrane is presumably due to the low density of NaChs. Differences between nodal and internodal axolemmas are also seen by cytochemical markers. Electron-dense metal staining (Quick and Waxman, 1977; Foster *et al.*, 1980), various IMP specializations (Kristol *et al.*, 1978), and a subaxolemmal electron-dense undercoating are all characteristic of nodes (Fig. 1). In *Sternarchus* axons, some nodes are excitable and others inexcitable; only those regions of the axon which are excitable show high IMP density and preferential staining by ferric ion-ferrocyanide (Fe-FeCN) (Quick and Waxman, 1977). In addition, dense staining and clusters of IMPs are localized at spinal cord motoneuron initial segments, an area thought to be rich in NaChs and important for initiation of the action potential. Although each of these morphological features clearly shows that the axon membrane is heterogeneous, there is little direct evidence to link these structures to NaChs. Antibodies raised against the purified eel and muscle NaCh proteins do indeed show staining at nodes in *Electrophorus electricus* and peripheral nerve (Ellisman and Levinson, 1982; Haimovich *et al.*, 1984) that appears to correlate with the IMPs and heavy metal stains.

This organization is not static and can change, however. When axons are demyelinated, conduction switches from the saltatory mode to continuous conduction along the axon, suggesting that NaChs redistribute and confer electrical excitability (Bostock and Sears, 1978) to the internodal membrane. These changes are not immediate because cytochemical data suggest that NaChs and associated structures remain temporarily clustered at nodes (Foster *et al.*, 1980; Coria *et al.*, 1984), but as demyelination proceeds, stain gradually appears in the demyelinated internodal membrane (Foster *et al.*, 1980; Coria *et al.*, 1984), paralleling the change to continuous conduction in this region. As some of these fibers recover from demyelination, prior to remyelination, again the electrical properties change to a saltatory-like conduction suggestive of NaCh clustering on the axon surface. Thus, all data suggest that, even though segregated and maintained in distinct regions of the axon, NaCh distribution is dynamic where the membrane can reorganize and imply that the axon has sufficient plasticity. However, the intriguing question of the origin and placement of these NaChs remains unanswered. There are several, as yet unexplained,

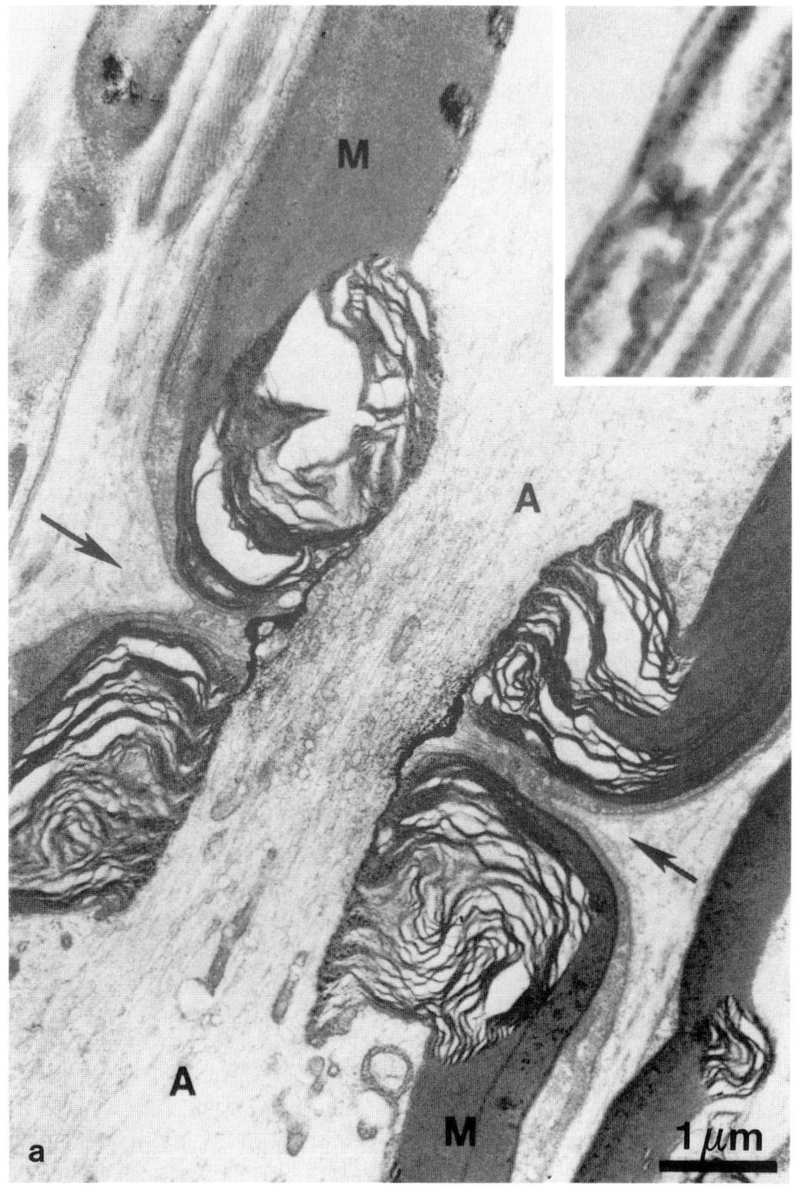
M
A
A
M
a
1 μm

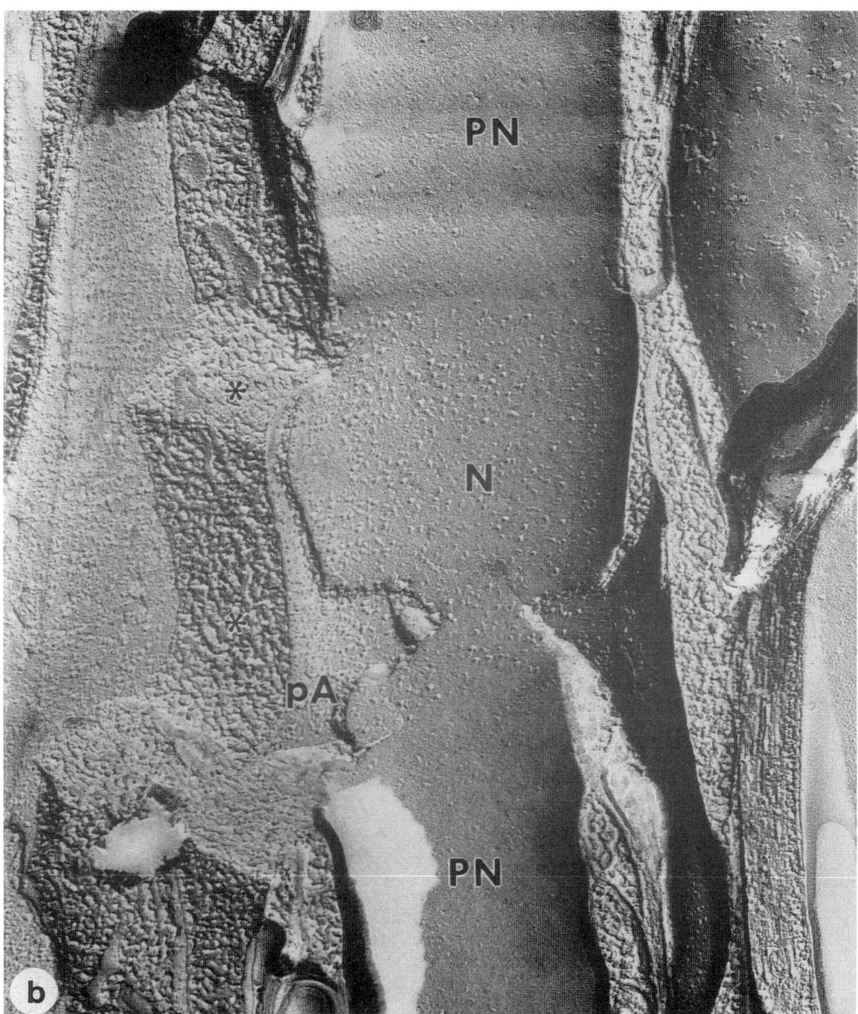

FIG. 1. Cytochemical heterogeneity of the axon membrane. (a) Electron micrograph of a node of Ranvier (arrow) in rat sciatic nerve. The tissue was stained with ferric ion-ferrocyanide, which results in a dense deposit of electron-opaque stain located immediately subjacent to the nodal axolemma. Other parts of the axon are not stained. (b) Freeze-fracture electron micrograph of nodal region from rat optic nerve. The E-face of nodal membrane exhibits a high density of intramembranous particles, while the E-face of paranodal axolemma displays relatively few particles and has a scalloped appearance because of the association with underlying terminal paranodal loops. A, Axon; M, myelin; N, node; pA, perioxonal; PN, paranode. (From Quick and Waxman, 1977.)

possibilities on how the heterogeneity of the axon arises. NaChs could redistribute via lateral diffusion from nodal sites along the axolemma (Foster *et al.,* 1980; Coria *et al.,* 1984; Shrager, 1988; Waxman, 1987), insert locally into the demyelinated zone from an intracellular axonal pool, or be delivered to the axon by the surrounding glial cells (Gray and Ritchie, 1985).

2. Development of Axonal Membrane Heterogeneity

How does this striking differentiation of the axon membrane develop and become established?

Studies of NaCh biosynthesis in cultured cortical neurons (Schmidt and Catterall, 1986) and in retinal ganglion cells (Wollner *et al.,* 1988) and the acquisition of neurotoxin binding (Baumgold *et al.,* 1983) on cultured nerve cells suggest that NaChs are expressed early in development and reach cell surface densities that are comparable to those of fully differentiated tissue within 3 weeks postpartum (Schmidt and Catterall, 1986). The distribution of NaChs during axonal development is generally assumed to be uniform, but this view seems to reflect mainly a lack of experimental information. In the few instances in which the matter has been tested, the data suggest otherwise. A morphometric study of IMP distribution, correlated to [^{3}H]STX binding in growing axons, suggested that if IMPs represent NaChs, then patches of NaChs are inserted proximally in the cell soma and laterally diffuse through the plasmalemma to distal locations where they may be trapped at sites such as the initial segment and developing node (Strichartz *et al.,* 1984).

Cytochemical and electrophysiological studies on developing and demyelinating nerve have provided some additional insights into how the heterogeneity of the axon membrane develops. During development or when axons are demyelinated, it appears that, in some fibers, "nodal" (phinodes) areas are formed prior to remyelination which are sites of high Na^+ current and, presumably, NaCh density (Smith *et al.,* 1982a). In addition, these NaCh clusters are of such high density that normal saltatory conduction can be supported prior to myelination (Bostock and Sears, 1978; Smith *et al.,* 1982a; Rasminsky and Sears, 1972). Modeling studies suggest that impulse conduction could be sustained in axons focally devoid of myelin if NaChs remained in clusters (Rasminsky and Sears, 1972), or if a demyelinated axon membrane reorganized such that the density of NaChs was sufficient to sustain impulse conduction. The sequence of events through which patching of NaChs proceeds is not known, although cytochemical evidence in regenerating, demyelinating, and remyelinating fibers suggest that it is existing NaChs that are recruited to these sites.

How NaCh hotspots could be formed in the absence of ensheathment is not clear. It seems that the neuron would be required to coordinate the expression, assembly, and targeting of NaChs and/or channel-associated structures to specific axonal sites. However, even for developing neurons, the cellular details that lead to the establishment and subsequent maintenance of restricted membrane domains are not well understood. As discussed below, NaChs may associate and segregate on the basis of specializations inherent to the channel protein, axon membrane, and/or associated cytoskeletal structures, and/or because node formation involves close apposition and interaction of Schwann cell and axon, cellular adhesion signals could also play early roles in specifying the clustering and localization of NaChs at nodes (Reiger *et al.*, 1986).

B. *Implications of Altered Sodium Channel Distribution in Disease States*

One severe consequence of altered NaCh distribution is seen in demyelinating diseases such as multiple sclerosis. When myelin is disrupted, a reduction in the action potential velocity is observed which frequently results in conduction block as the action potential arrives at a demyelinated segment (Waxman, 1977). A major question is whether the conduction block is due to a low density of NaChs or an alteration in the distribution of NaChs at nodes of Ranvier.

In both PNS and CNS axons that have been demyelinated, Fe-FeCN staining is distributed diffusely over the length of demyelinated axon (Foster *et al.*, 1980), which could arise by lateral diffusion of preexisting NaChs that were previously concentrated at nodes. Apparently, this low density of NaChs is insufficient to support conduction, although in some fibers, staining is confined to a restricted region of axolemma at the beginning of a demyelinated segment (Coria *et al.*, 1984). During remyelination, nodal-like stained areas are formed at contact zones between preremyelinating Schwann cells, sites which could be alternative sites for impulse generation in demyelinated fibers (Smith *et al.*, 1982a).

Studies with monoclonal antibodies generated to eel electroplax membranes appear to label nodes of Ranvier in normal fibers and show staining throughout demyelinated regions of peripheral nerve (Meiri *et al.*, 1984). Although the antigen to which the monoclonal is directed has not been determined, these results support the view that some axolemmal NaChs may redistribute following demyelination. If the changing axonal topography is related to NaChs, this may explain the restoration of impulse conduction in some demyelinated fibers.

II. MOLECULAR PROPERTIES OF VOLTAGE-DEPENDENT SODIUM CHANNELS

While many studies have utilized IMPs and various electron-dense stains to define the topography of NaChs over the nerve cell surface, recent advances in the isolation of the NaCh protein, its characterization, and immunocytochemical analysis of its distribution have provided considerably more detail on how NaChs are distributed and maintained on the nerve cell surface.

A. Molecular Composition

NaChs have been purified from electric eel, rat brain, and rat skeletal muscle. In preparations of rat brain (Hartshorne and Catterall, 1984; Elmer *et al.,* 1985; Barhanin *et al.,* 1985), rat muscle (Kraner *et al.,* 1985), and eel electroplax (Miller *et al.,* 1983) NaChs the major peptide is a highly glycosylated (29% by mass) protein of M_r 260 kDa (α-peptide) (Fig. 2). In some laboratories (Hartshorne and Catterall, 1984), the rat brain NaCh contains additional subunits of 33 kDa (β_2) and 36 kDa (β_1), where β_2 is linked to the 260-kDa protein by disulfide bonds. Although several roles have been suggested for the smaller subunits, including the assembly, targeting, and maintenance of NaCh distribution (Schmidt and Catterall, 1986; Wollner *et al.,* 1988), reconstitution of ion flux with the 260-kDa protein (Rosenberg *et al.,* 1984), binding of cytoskeletal proteins (Srinivasan *et al.,* 1988), expression of functional channels in *Xenopus* oocytes after injection of mRNA encoding only the 260-kDa protein (Suzuki *et al.,* 1988; Stuhmer *et al.,* 1987) or in a somatic cell line (Scheuer *et al.,* 1990) indicate that the 260-kDa component contains most, if not all, of the structural and functional information for ion conduction, targeting, and association with the cytoskeleton.

Analysis of the primary sequence derived from cDNA has revealed several interesting features (Noda *et al.,* 1986). All NaCh cDNA sequences encode proteins containing four internal repeats (domains I–IV) of 300 to 400 amino acids, with an approximately 50% identity or conserved amino acid sequence. The hydropathy profile of each domain shows that each domain contains five hydrophobic segments (S1, S2, S3, S5, and S6) at equivalent positions and one positively charged segment (S4) between segment S3 and S5. One of the unusual features is the presence of four to eight arginine or lysine residues at every third position in S4. From mutagenesis experiments this S4 segment appears to act as the voltage sensor (Stuhmer *et al.,* 1989) where the positive charges respond to changes in transmembrane voltage and initiate channel opening. Recent experiments have implicated the cytoplasmic loop between III and IV as an important component of the inactivation or closing of the channel.

Structural predictions from the primary sequence have suggested that the transmembrane part of the channel is formed by the symmetric array of the four homologous transmembrane domains (Fig. 2). Experimental measurements with antibodies directed against the N- and C-termini suggest that these segments lie on the cytoplasmic side of the membrane. As a result, a ~200 amino acid segment between domains I and II is predicted to be hydrophilic and cytoplasmic; this has been borne out experimentally since this segment can be phosphorylated at five separate sites by the cAMP-dependent protein kinase (Rossie *et al.*, 1987).

In addition to posttranslational modifications in the cytoplasmic loops, several of the extracellular segments bear consensus sequences for N-glycosylation. The carbohydrate structure of the channel has been partially determined and is composed of *N*-acetyl-β-D-glucosaminyl oligomers with highly negative charges contributed by the terminal polysialic acid residues (James and Agnew, 1987). This N-linked protein glycosylation is obviously important because tunicamycin inhibits the expression of NaChs at the cell surface (Waechter *et al.*, 1983). On the other hand, when core glycosylation is allowed and only subsequent processing and sialylation are inhibited, sodium channels are synthesized, assembled, and inserted into the cell surface with normal physiological properties (Schmidt and Catterall, 1986). Although the extensive processing and terminal sialylation of oligosaccharide chains during maturation of the α subunit may not be essential functionally (Wollner *et al.*, 1988), it is possible that these modifications may have a role in specifying the distribution of NaChs along the axon.

B. Existence of Multiple Sodium Channel Subtypes

From cloning studies in brain it became apparent that three distinct NaCh genes exist (Noda *et al.*, 1986). The three genes encode channels which are highly homologous if not identical in their transmembrane domains and differ primarily in their cytoplasmically located segments. This is true, for example, in RI, whose cytoplasmic loop between domains I and II is ~200 amino acids long, while RII and RIII NaChs have loops which are less than 100 amino acids with little or no homology in this segment among each of the channels. Despite these structural differences, expression of RII and RIII NaChs in *Xenopus* oocytes is functionally identical.

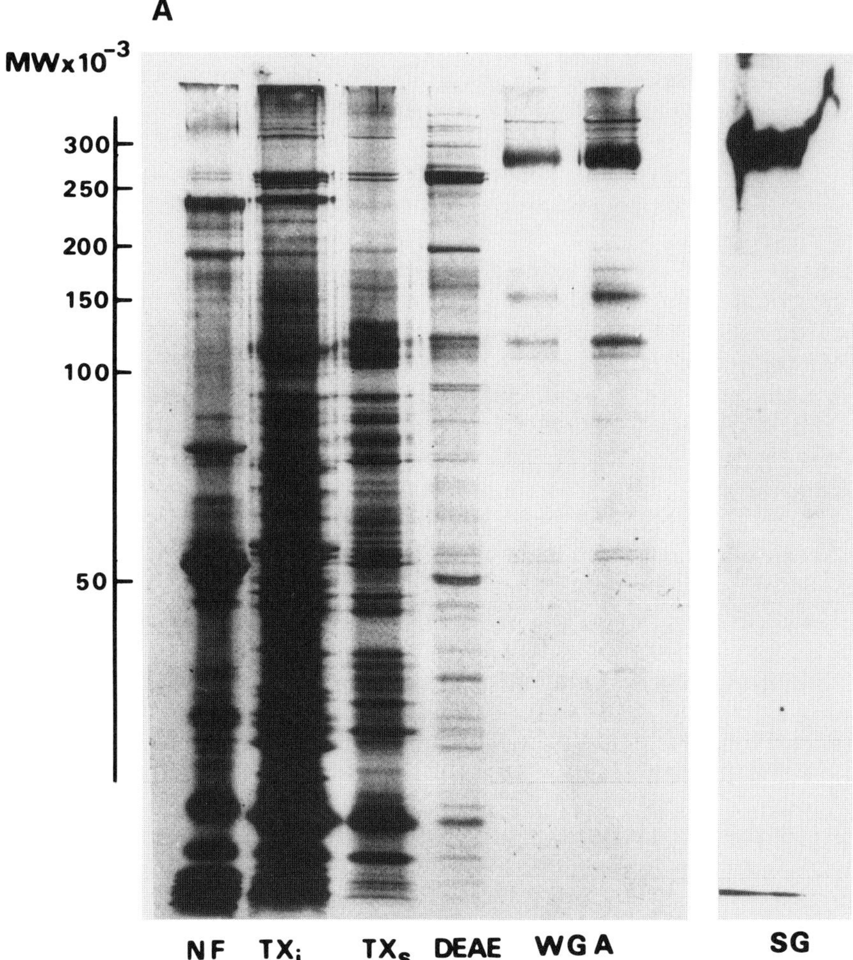

FIG. 2. Molecular properties of the NaCh protein. (A) Biochemical purification of the NaCh protein from rat brain. After solubilization, the protein is purified by sequential chromotograpy and ion exchange, hydroxylapatite, wheat germ agglutinin (WGA), and, finally, through sucrose gradient sedimentation. The purified protein has an apparent M_r of 260 kDa. (From Elmer *et al.*, 1985.) (B) Proposed transmembrane topology of the 260-kDa NaCh protein. (From Catterall, 1988.)

C. Biosynthesis and Intracellular Processing

NaCh biosynthesis has been examined in primary cultures of rat brain neurons (Wollner *et al.*, 1988; Waechter *et al.*, 1983; Schmidt *et al.*, 1985). Although a 203-kDa protein is synthesized in the presence of tunicamycin,

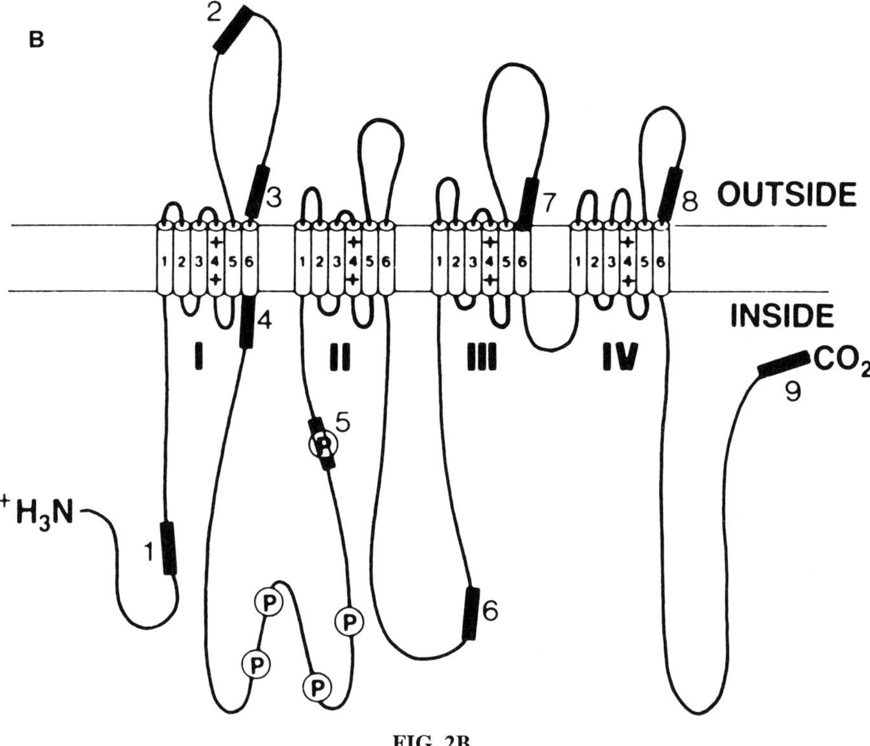

FIG. 2B

which blocks the initial stage of glycosylation, STX binding at the cell surface is not detected, and indicates that N-glycosylation is essential for trafficking of the NaCh α subunit to the plasma membrane. This 203-kDa mass correlates well with the reported size of the core protein of the α peptide after enzymatic deglycosylation and with the molecular mass determined from the cDNA analysis.

In pulse–chase experiments when glycosylation is allowed to proceed normally, several intermediate forms of the protein are identified. These include two components of 224 kDa, one of which contains sufficient *N*-acetylglucosamine to be retained by wheat germ agglutinin, and an additional discrete form of 249 kDa. The latter is then processed, possibly through the addition of further sialic acid residues, to its final size of 260 kDa. The wheat germ binding polypeptide is presumably processed in the medial Golgi, taking about 30 min, and polysialic acid residues are added; the product is then terminally glycosylated in the trans Golgi to form mature channels.

Nearly 70% of this newly synthesized glycosylated channel protein has been reported to equilibrate with an intracellular, membrane-associated pool that is not covalently associated with either β_1 or β_2 (Schmidt *et al.*, 1985). Just prior to insertion in the plasma membrane, maturation of the channel may be accompanied by the covalent attachment of fatty acid acyl chains to the core peptide in the form of palmityl and sulfonyl groups.

Several factors appear to influence the density of NaChs expressed on the cell surface. For example, factors that tend to increase cytoplasmic free calcium result in a down-regulation of NaChs, while factors that increase cAMP up-regulate the cell surface density of NaChs. It is not known whether the level of control is transcriptional, translational, or exit from the Golgi. Increases in mRNA levels corresponding to NaCh II occur in response to nerve growth factor (NGF), which is mediated primarily through cAMP-responsive elements at the transcriptional level (Rudy *et al.*, 1982; Sherman *et al.*, 1985).

III. LOCALIZATION AND MAINTENANCE OF VOLTAGE-DEPENDENT SODIUM CHANNELS

A. Distribution of Sodium Channels in Mature Nerve

While differences between nodal and internodal axolemma are revealed by cytochemical means, it is not clear whether these stains represent NaChs. Only recently have probes specific for the NaCh been developed to determine its distribution and dynamics on the neuronal membrane.

Because the NaCh has a particularly rich pharmacology and can be selectively inhibited by several naturally occurring toxins, such as tetrodotoxin (TTX), STX, and polypeptide scorpion toxins, biologically active probes have been prepared that retain their high affinity and can be used to determine NaCh distribution and dynamics on living nerve (Angelides and Nutter, 1983; Darbon and Angelides, 1984; Angelides, 1986).

When these probes are used to examine NaCh distribution on living neurons, digital fluorescent images show that NaChs are nonuniformly distributed over the cell surface and are localized to morphologically distinct regions (Angelides *et al.*, 1988). In spinal cord neurons, for example, both fluorescent and radiolabeled TTX and scorpion toxin specifically stain the axon hillock intensely, with punctate and patchy fluorescence confined to this region, while staining in the cell body is sparse and diffuse (Fig. 3). Several foci of high channel density are also seen on the axon shaft. In myelinated nerve, nodes are very heavily labeled, as expected from electrophysiological and biochemical estimates. The distribution of

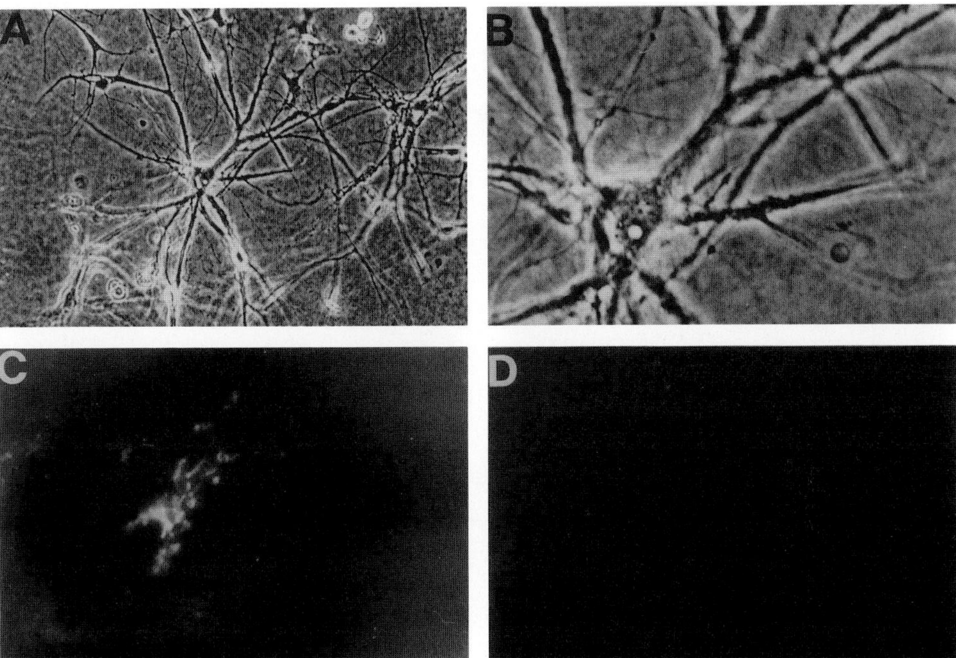

FIG. 3. Distribution of NaChs in cultured rat spinal cord neurons as shown using a fluorescent scorpion toxin derivative, tetramethylrhodamine toxin V from *Leiurus quinquestriatus*. (A and B) phase micrographs; (C) fluorescence of neurons labeled with fluorescent scorpion toxin; (D) nonspecific labeling after depolarization of the cells with 135 mM K^+, which promotes dissociation of the label.

NaChs on cultured neurons is not a general feature of most membrane receptors because voltage-dependent Ca^{2+} channels are localized in high density on cell bodies and dendrites (Jones *et al.*, 1989), and GABA receptors are maintained in discrete patches on the cell body (Velazquez *et al.*, 1989) in regions where NaChs are diffusely and sparsely distributed.

Although these specific probes have been useful for revealing the distribution of NaChs at the light microscopic level, their dissociation has inhibited further characterization at the ultrastructural level. For such studies, antibodies specific for the NaCh have been developed and utilized. In retinal ganglion cells, NaChs recognized by a polyclonal antiserum against α subunits, are localized at the axon hillock, initial segment, and premyelinated segments (Fig. 4a) (Wollner and Catterall, 1986). Myelinated axons, on the other hand, display a nonuniform distribution of NaChs, as expected from electrophysiological measurements. In myelinated nerve, the first clear identification of NaChs at the immuno-ultrastruc-

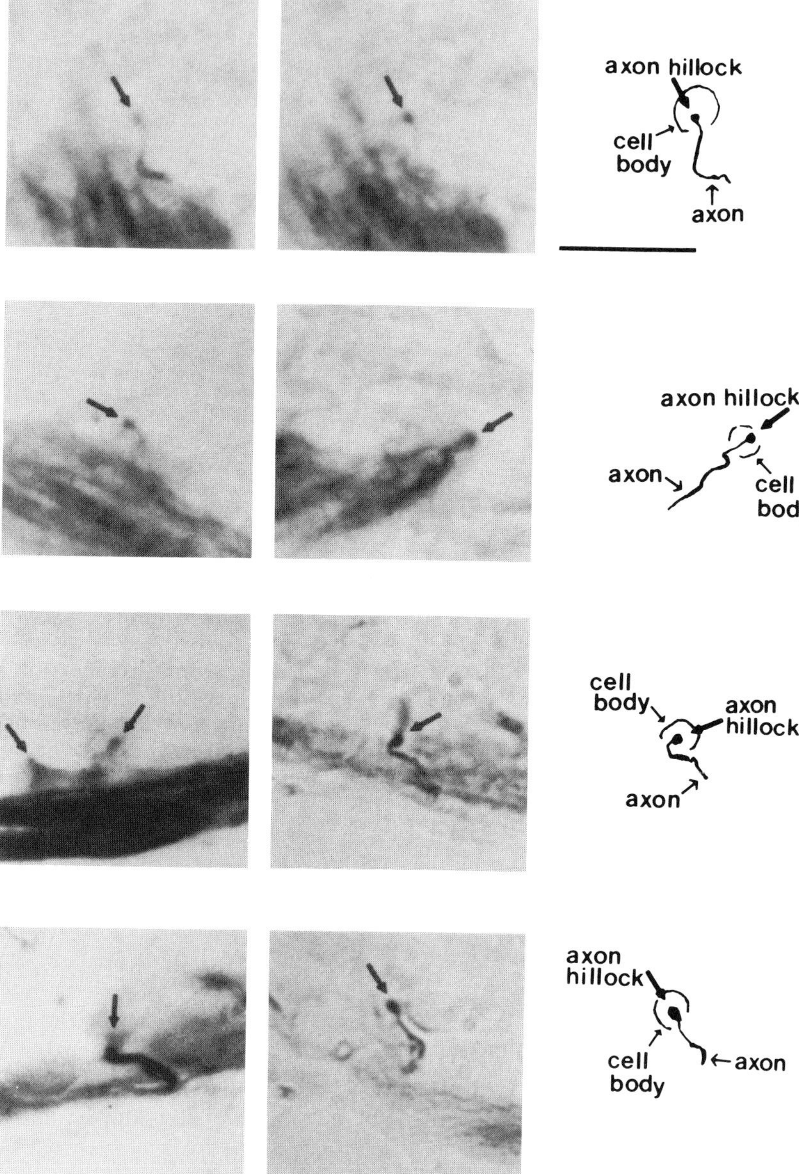

FIG. 4. Immunocytochemical localization of NaChs by NaCh-specific antibodies. Staining of NaChs is shown (a) in frog retinal ganglion cells by a polyclonal antibody raised against rat brain NaChs (Wollner and Catterall, 1986); (b) in nodes of Ranvier from dorsal columns of eel spinal cord (Ellisman and Levinson, 1982); (c) in rat optic nerve with specific polyclonal antibodies, panels 1 and 2, immune, and panel 3, preimmune (Black *et al.*, 1989); and (d) at the node of Ranvier and at astrocytic processes which abut the node (Black *et al.*, 1989). pn, Paranode; as, astrocyte.

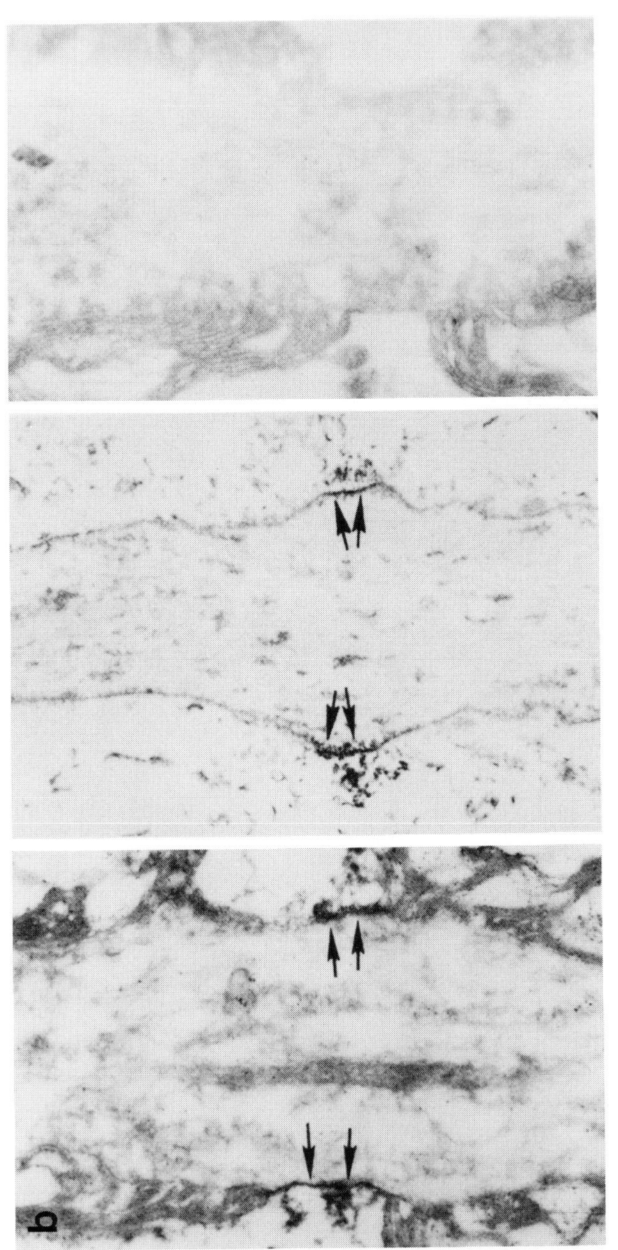
b

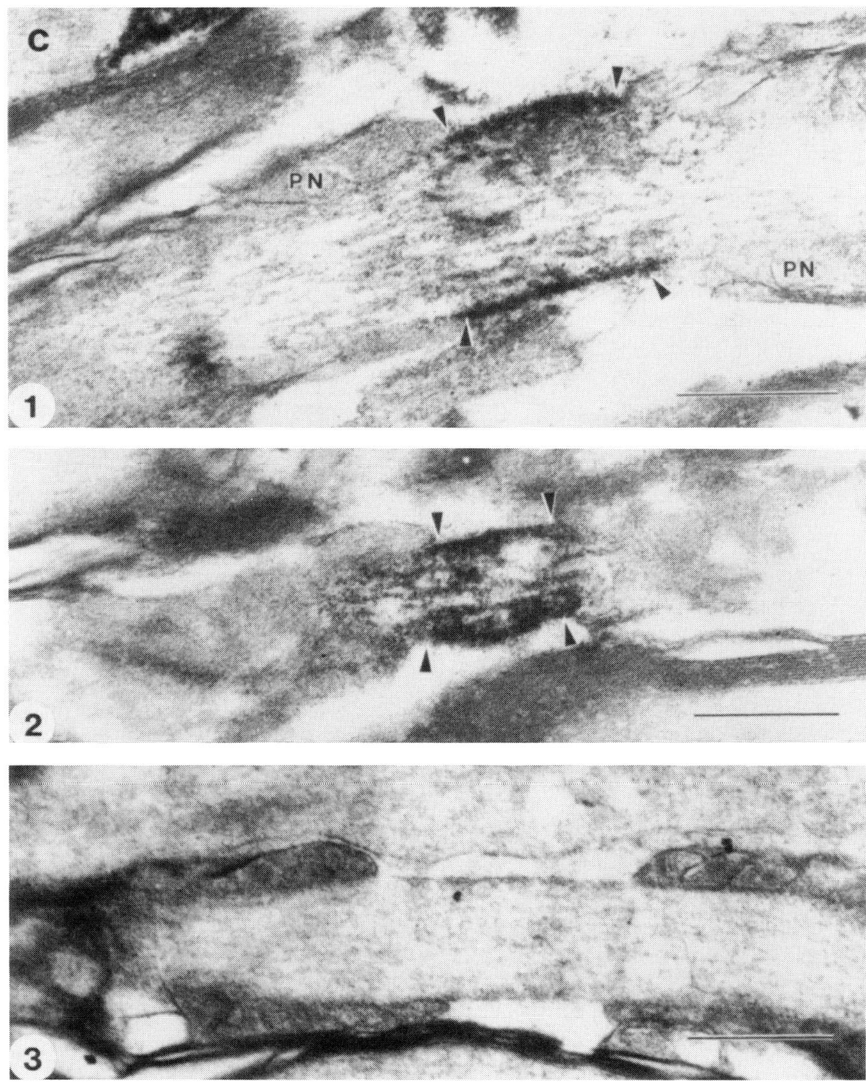

FIG. 4C

tural level was by Ellisman and Levinson (1982), who stained *Electrophorus electricus* nodes of Ranvier using antibodies raised against the purified eel channel protein (Fig. 4b). Unfortunately, these antibodies could not be

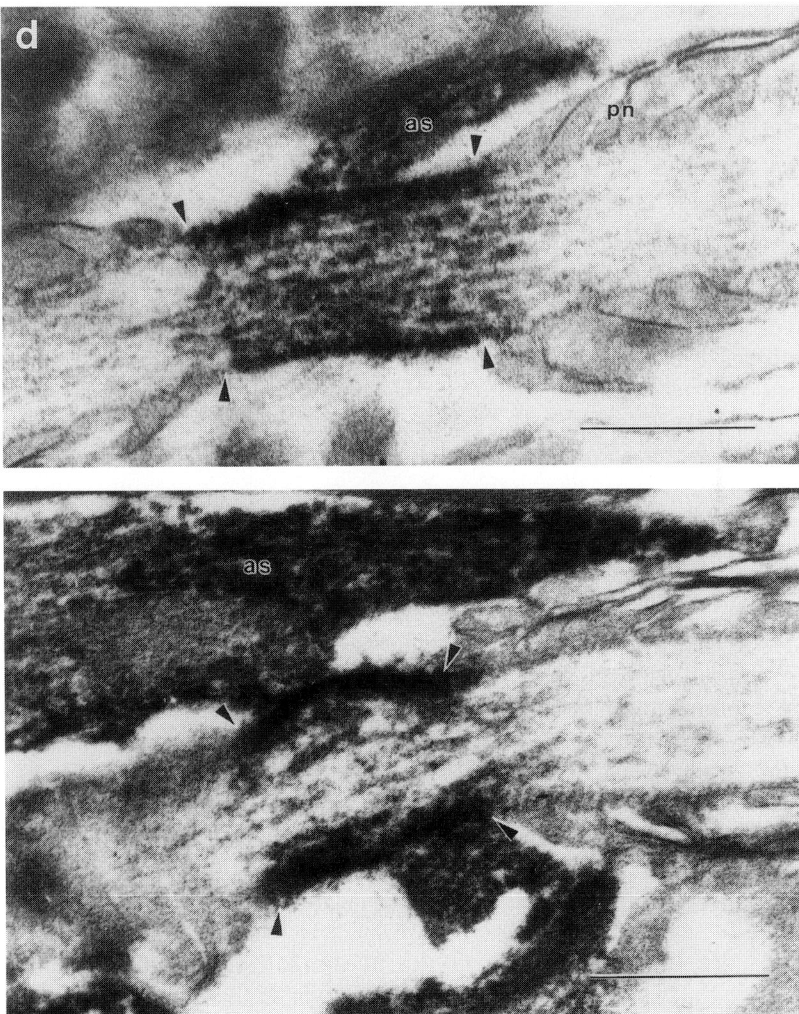

FIG. 4D

used on mammalian fibers because of antigenic differences between the eel and mammalian NaChs. Recently, the first immuno-ultrastructural localization of NaChs in mammalian fibers has been described with antibodies generated against purified NaChs from rat brain (Black *et al.*, 1989). Rat optic nerve and rat sciatic nerve, incubated with polyclonal and monoclonal antibodies directed against the rat brain 260-kDa α-peptide, show dense immunoreactivity that is specifically associated with the axon membrane at the node of Ranvier (Fig. 4c). The density of NaChs is consistent with the electrophysiological measurements which report densities of 3,000–6,000 NaChs/μm^2. In contrast, internodal axon membrane does not stain with the antibody, despite its accessibility to a variety of other antibodies, such as those directed against neurofilaments, and indicates that the density in the internodal axolemma is 100-fold lower. In CNS fibers, some perinodal astrocyte processes that are associated with and abut nodes are immunoreactive while the terminal perinodal oligodendroglia loops or glia limitans are not stained (Fig. 4d). The latter observations are consistent with the staining pattern and electrophysiological characteristics of type I and type II astrocytes in cultures (Barres *et al.,* 1989) that suggest type II astrocytes, which express primarily the "neuronal" form, are associated with nodal specializations (Ffrench-Constant and Raff, 1986).

The localization and distribution of NaChs are somewhat complicated by the fact that there are at least three distinct NaCh subtypes which are found in brain (Noda *et al.,* 1986). Where is each of these subtypes found? Using subtype-specific antibodies generated from peptides corresponding to the RI and RII subtypes, Westenbroek *et al.* (1989) have shown that these subtypes are differentially localized in central neurons (Figs. 5 and 6). Previous biochemical work showed that the majority of NaChs in whole brain, cerebral cortex, hippocampus (97%), cerebellum (84%), and midbrain are the RII NaChs subtype, while the remainder of the subtypes are of the RI and other unidentified subtypes (Gordon *et al.,* 1987).

The subcellular localization of NaCh subtypes has been investigated by immunocytochemical analysis with subtype-specific peptide antibodies (Westenbroek *et al.,* 1989; Gordon *et al.,* 1987). The RI-specific peptide antibodies preferentially label, but relatively weakly, the area where cell bodies are segregated, such as the pyramidal and dentate granule cell layers in the hippocampus, Purkinje cell layer in the cerebellum, and gray matter in the spinal cord. In contrast, RII-specific peptides antibodies strongly stain unmyelinated mossy fibers in the hippocampus and the molecular layer of the cerebellum. These results are consistent with the evidence that NaChs are localized in higher density in axons and axon hillock than in the adjacent cell body (Barrett and Crill, 1980; Schwartzkroin, 1977). Further dissection of the labeling pattern with these subtype-specific antibodies show that, within pyramidal cells of CA3, the RI and RII NaCh subtypes are differentially localized in cell bodies and axons, respectively. Cell bodies primarily contain RI NaChs, whereas

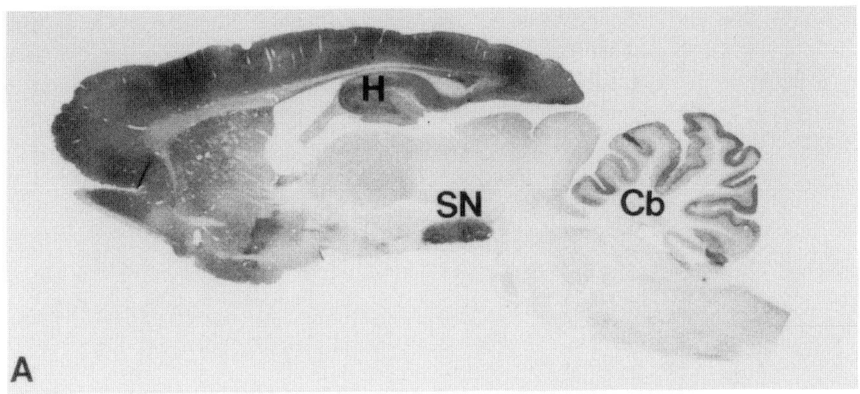

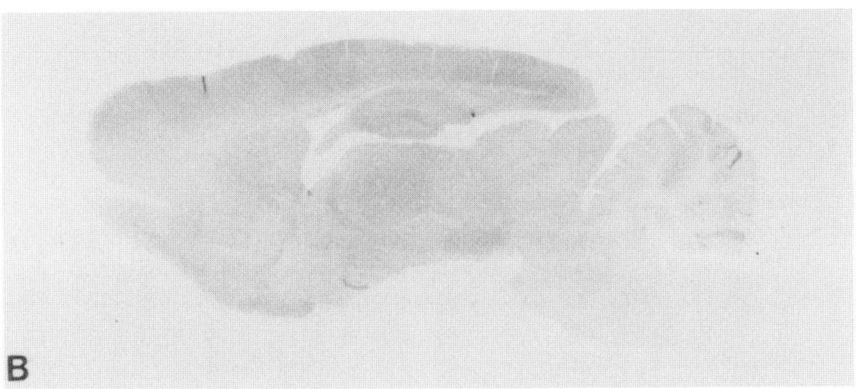

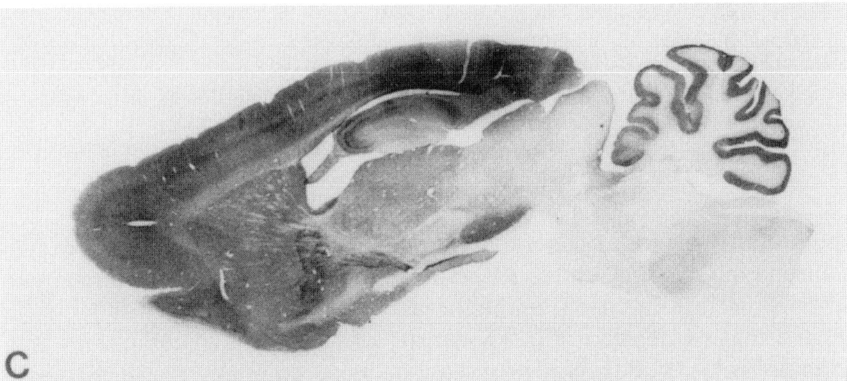

FIG. 5. Overall localization of NaCh subtypes RI and RII in rat brain using polyclonal anti-peptide antibodies. (A) RII subtypes, (B) RI subtypes, and (C) RII subtypes; H, hippocampus; SN, substantial nigron; Cb, cerebellum using a different SPII antibody. (From Westenbroek *et al.*, 1989.)

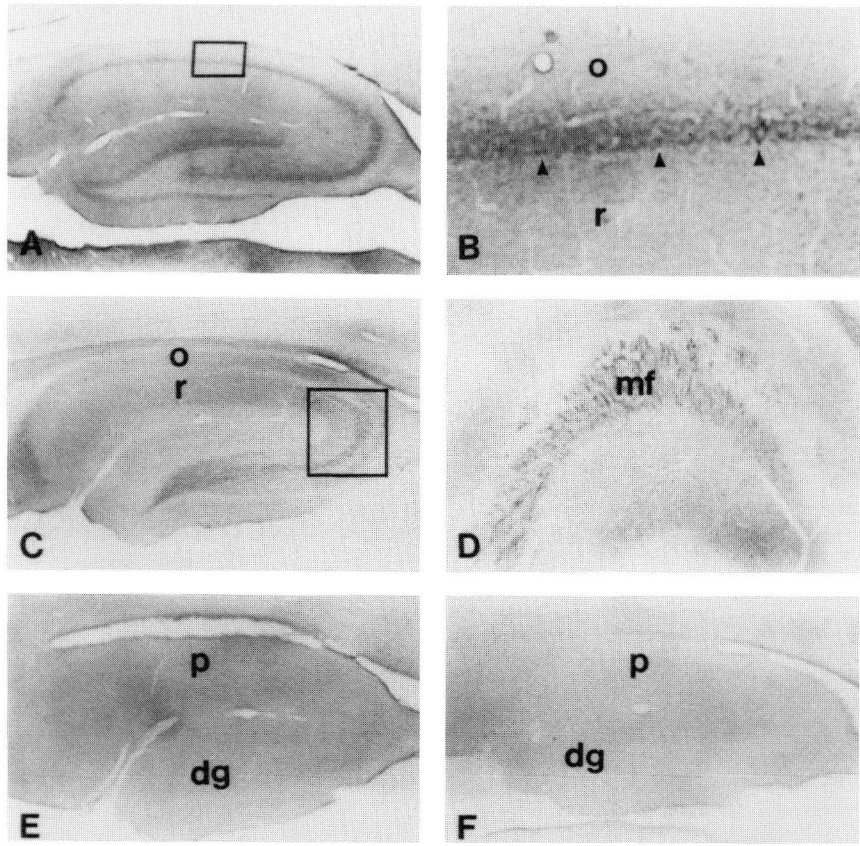

FIG. 6. Localization of NaCh RI and RII subtypes in rat hippocampus. (A and B) CA1 subfield, (C and D) CA3 subfield. B and D are enlargements of boxes in A and C, respectively. (E and F) Dendrite gyrus. mf, Mossy fibers; p, stratum pyramidale; dg, dentate gyrus; o, stratum oriens; r, stratum radiatum. (From Westenbroek *et al.*, 1989.)

their axons, the mossy fibers, are immunoreactive for RII. The pyramidal neurons in CA1–CA3 have RI-positive somata, whereas the surrounding fiber layers are intensely immunoreactive for RII. The results have suggested that expression of a low density of RI NaChs in cell bodies and a high density of RII NaChs in axons is a common characteristic of projection neurons in different brain regions and also emphasizes that, in addition to the localization of NaChs at discrete sites on the neuron, there is differential expression and targeting of NaCh subtypes on a single neuron.

An intriguing idea is that, in addition to specific targeting, the differential localizations of the RI and RII NaChs may have specific functions in the cell body and axon (Westenbroek *et al.*, 1989). NaChs in axons show rapid activation and complete inactivation and RII NaChs expressed in oocytes have similar functional properties. In contrast, NaChs in cell bodies are not quickly inactivated and the sustained depolarization may be important in controlling the frequency of firing and duration of action potentials. It has been proposed that the RI NaCh subtype is responsible for generation of sustained Na^+ current.

Electrophysiological experiments have established that the threshold for action potential generation is lower at the axon initial segment compared with the adjacent cell body in spinal cord motoneurons, cerebellar Purkinje cells, and hippocampal pyramidal cells (Barrett and Crill, 1980; Schwartzkroin, 1977; Coombs *et al.*, 1955; Eccles *et al.*, 1966). In each of these neurons, as well as in spinal cord neurons in cell culture and retinal ganglion cells *in vivo,* it has been found that a relatively low density of NaChs in the cell body (primarily the RI) compared with the adjacent axon and initial segment regions (RII) (Wollner and Catterall, 1986; Black *et al.*, 1989; Westenbroek *et al.*, 1989). The high density of NaChs at the initial segment is likely to be the basis for low threshold for action potential generation at this site, a critical feature of signal processing at the cellular level in central neurons.

How might this differential density of NaChs be established and maintained? It is possible, and there is some evidence to support this idea, that the relative number of NaChs in these two membrane domains is controlled in part by the differential expression of the genes encoding RI and RII NaCh subtypes, which are then preferentially targeted by intracellular protein transport systems to cell bodies and axons, respectively. Once present in the appropriate membrane compartment, these two different gene products may be immobilized by interaction with the cytoskeleton to maintain their differential localization (see below) (Srinivasan *et al.*, 1988). Differential localization of the RI and RII NaCh subtypes may also serve to provide specific functional specializations to the axon and the cell body. This differential localization obviously contributes to the nonuniform distribution of electrical excitability in neuron.

B. Lateral Mobility of Sodium Channels

Although immunocytochemistry and imaging provide a useful view of the distribution of NaChs, how is this topography controlled and maintained during neuronal development?

Because biological membranes can be considered to be two-dimensional fluids, where the cell membrane is in a fluid state at physiological temperatures, membrane proteins dispersed in this fluid lipid bilayer are mobile within the plane of the membrane. The mobility of membrane proteins is one mechanism by which cells can organize their plasma membranes during differentiation and important physiological events (Jacobson *et al.*, 1987). Selected examples are the diffusion-mediated aggregation of some peptide hormone receptors into microclusters prior to endocytosis and the metabolically driven capping of cell surface receptors. Application of the principles of membrane fluidity has led to molecular models for certain steps in hormonal regulation, photoreception, electron transfer, and many other membrane processes (Oliver and Berlin, 1982; Elson and Schlessinger, 1979). However, there are also numerous aspects of cell function which apparently require long-range solid-like membrane organization, e.g., the maintenance of polarity in epithelia and neurons.

The lateral mobility of NaChs in various regions of the neuron has been measured and has shed some light on those constraints that may confine NaChs to each of these regions (Angelides *et al.*, 1988). Both statistical analysis of the differential distribution of cytochemical stains and IMPs (Strichartz *et al.*, 1984; Small and Pfenninger, 1984; Small *et al.*, 1984) and fluorescence photobleach recovery (FPR) using NaCh-specific antibodies and neurotoxins have been performed (Angelides, 1986; Angelides *et al.*, 1988). From the static distribution of IMPs (which are correlated with NaChs) lateral diffusion rates on the order of 10^{-7} cm^2/sec have been reported (Small *et al.*, 1984). While this rate is 10-fold faster than the rates of two-dimensional lipid diffusion through a bilayer, these data have been interpreted to reflect perikaryal insertion of NaChs and lateral movement toward the tip of the growing axon. There is, however, some uncertainty whether these IMPs actually represent voltage-dependent NaChs.

Using FPR and channel-specific neurotoxins, NaCh mobility has been directly measured on living cells under a variety of experimental conditions to assess the role and magnitude of intracellular and/or extracellular constraints to NaCh diffusion and to elucidate the mechanisms by which NaCh topography is controlled (Angelides *et al.*, 1988).

Fluorescence photobleaching is an equilibrium perturbation method where the signal from fluorescently labeled NaChs on the neuron is photochemically depleted in a small region on the cell surface by bleaching with a laser beam (Axelrod *et al.*, 1976a; Axelrod, 1983). This creates an area which is now devoid of fluorescence. If fluorescently labeled NaChs in surrounding areas are able to move on the cell surface into the bleached region, the fluorescence will increase with time as the fluorescencing molecules are delivered back into the bleached region. The rate of this recovery is related to the diffusion or lateral mobility of the receptors on the cell surface. Frequently, the fluorescence fails to recover to the same intensity observed before bleaching and can be attributed to a fraction of molecules which are immobile on the time scale of the experiment. For photobleaching, a small area of the cell surface (about 1–5 μm) is chosen and briefly exposed (10–200 msec) to a pulse of laser light (5 mW) exciting only into the fluorophore's absorption, thereby photochemically bleaching only the fluorophore in that region. The diffusion coefficient is determined from the time dependence of the fluorescence recovery.

Because digital images of fluorescently labeled NaChs using neurotoxin probes on live neurons show that NaChs are segregated in high density at the hillock compared to a diffuse pattern on the cell body (Angelides *et al.*, 1988), the first question is whether the distribution is related to the mobility of NaChs in these regions. Photobleaching of NaChs in different regions of the neuron shows that NaChs on the cell body are freely mobile, with diffusion coefficients of 10^{-9} cm^2/sec, while >80% of the labeled NaChs located at the hillock and presynaptic terminal are immobile on the time scale of the measurement (<10^{-12} cm^2/sec) (Fig. 7). The small mobile fraction had diffusion coefficients in the range 10^{-10}–10^{-11} cm^2/sec. Is this differential mobility related to differences in the geometry between each of these regions? This is unlikely because the mobilities for tetramethylrhodamine (TmRhd)-phosphatidylethanolamine as a probe of lipid diffusion (10^{-9} cm^2/sec) and FITC–succinyl-concanavalin A as a probe for glycoproteins (10^{-11} cm^2/sec) are the same over all parts of the neuron. The rates measured for glycoproteins also indicate that the restricted distribution and immobility of NaChs are not general characteristics of neuronal cell proteins, or due to differences in membrane fluidity in the hillock created by the partitioning of specialized phospholipids, or the result of regional domains created by aggregation of glycoproteins.

On the cell body the mobility of NaChs is close to that of the phospholipids and indicates that NaCh movement is limited only by the viscosity of the membrane lipid. Similar diffusion coefficients have been measured for acetylcholine receptors in embryonic muscle (Poo, 1982) using nonfluorescence methods, for rhodopsin in amphibian rod outer segment disks (Poo and Cone, 1984), for surface antigens such as H-2Ld (Edidin and Zuniga, 1984) and the antigen restricted to the posterior tail region of guinea pig sperm (PT-1) (Myles *et al.*, 1984), and for plasma membrane proteins in which the underlying cytoskeleton may be modified or is separated from the membrane (Tank *et al.*, 1982).

In contrast to the cell body, the axon hillock has a patchy NaCh distribution, which is restricted within the boundaries of the axon hillock (see Fig. 3). At the hillock the different rates of diffusion of NaChs and FITC-

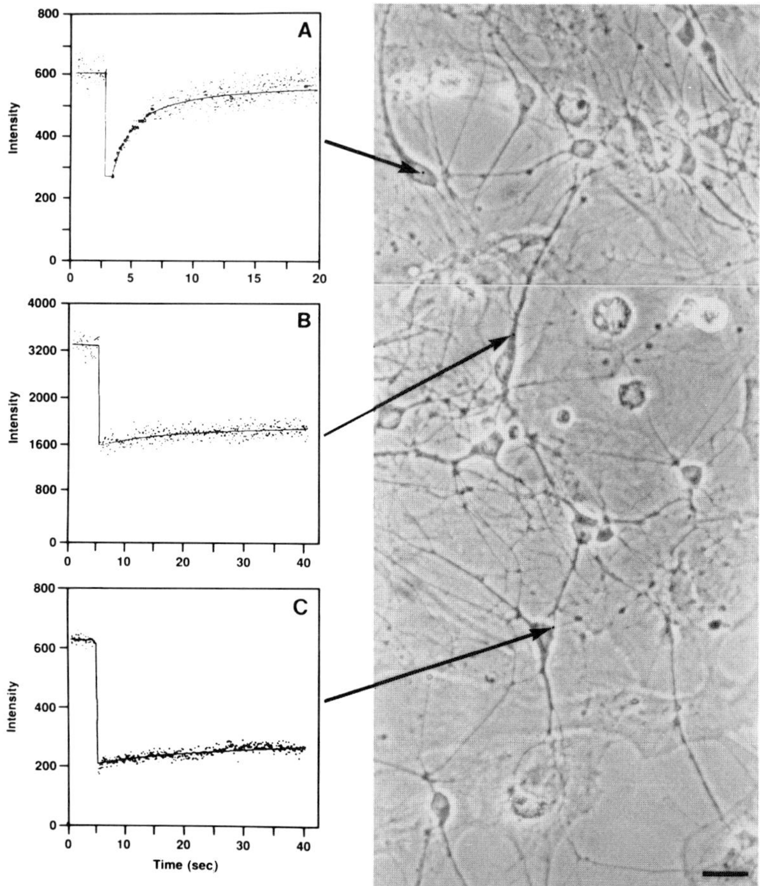

FIG. 7. Lateral mobility of NaChs on spinal cord neurons. Cells were labeled with fluorescent scorpion toxins and FPR measurements were performed. Photobleaching of cell body (A), axon hillock (B), and at the neuritic terminal (C) is shown.

glycoproteins indicate that a substantial fraction of the NaChs in this region experience stronger retarding interactions than do most glycoproteins. Hence, the forces that retard protein diffusion on this part of the plasma membrane must have some measure of molecular specificity. The forces could arise from several kinds of interactions within the plasma membrane or between the membrane and the cytoskeleton: (1) NaChs might interact directly with the specialized cytoskeleton present in this region of the cell. An example of this kind of interaction occurs in erythrocytes, where the diffusion of band 3 is retarded due to interaction with spectrin via the linking protein ankyrin (Nigg and Cherry, 1980); (2) NaChs might interact with other membrane proteins which in turn interact with the cytoskeleton. This seems to occur with sterol dextrans inserted into cell membranes and with histocompatibility antigens from which the cytoplasmic domain has been deleted (Edidin and Zuniga, 1984; Wolf *et al.*, 1980); (3) the channel might interact sterically with other proteins, which are present at high concentrations in the plasma membrane (Wier and Edidin, 1986). This kind of interaction typically seems to have only a weak effect on membrane protein diffusion but may have been observed for IMPs in regenerating axon membranes; (4) in addition, the channel might be preferentially concentrated into membrane lipid domains with very high viscosity.

Measurements of the distribution and mobility of other channels by FPR suggest that the distribution and mobility of NaChs have a great deal of molecular specificity, because γ-aminobutyric acid receptors (Velazquez *et al.*, 1989) and voltage-sensitive Ca^{2+} channels (Jones *et al.*, 1989) are concentrated and immobilized on cell bodies and dendrites, respectively, of the same neurons, regions where NaChs are diffusely distributed and mobile or absent.

It seems likely that in these regions where channel mobility is reduced the forces that retard mobility also contribute to the regional localization of channels. However, if NaChs are mobile on the cell body, how is the high density at the hillock maintained? One possibility is that the dense cytoplasmic architecture about the soma–hillock annulus (Ellisman and Porter, 1980) might impose a specific barrier to the diffusion of NaChs while allowing other proteins and lipids to diffuse freely between these regions. The existence of a selective barrier in the hillock is suggested because the fluorescence of labeled NaChs in the hillock can be depleted after several bleaches (Angelides *et al.*, 1988), while the signal from phospholipids cannot. Electron micrographs provide some basis for this observation because a dense cytoskeletal matrix, a submembranous electron-dense region, and small IMPs on the P-face of the membrane are found at the soma–hillock junction (Waxman, 1981). An example of a selective barrier to diffusion is the PT-1 antigen on sperm cells, which is confined to but diffuses freely within the 200 μm boundaries of the posterior tail region (Myles *et al.*, 1984).

This type of restriction of NaCh lateral mobility between compartments on the neuron could arise from direct association or from a physical barrier established by the underlying cytoskeleton (Koppel *et al.*, 1981), much like a corral that would trap NaChs. Within these 1–2 μm boundaries, NaChs would indeed appear laterally immobile by FPR yet, if not directly attached, would be rotationally mobile. Most NaChs at the hillock, however, are rotationally immobile and suggest that perhaps, in addition to corralling, most NaChs may be linked directly to components of the cytoskeleton. In the presence of colchicine and cytochalasin B, to disrupt microtubules and actin filaments, respectively, no change in NaCh lateral mobility was observed, suggesting that these specific elements are not directly involved in restricting channel mobility. However, when the plasma membrane is physically detached from the underlying cytoskeleton, NaChs located at the hillock become freely mobile (10^{-9} cm^2/sec). The results provided some evidence that attachments or barriers formed by the cytoskeleton could immobilize or confine NaChs and regulate their distribution. Thus, two modes could be employed to localize and maintain NaChs at high concentrations at hillocks: an attachment to the underlying cytoskeleton and/or sequestration in a confined region by a fencelike mechanism.

Restriction of channel mobility at neuritic terminals might employ some of the same elements at the axon hillock, although cell–cell contacts and interactions with extracellular components specific to these regions may also contribute. The extensive glycosylation of the channel protein, which is required to maintain functional NaChs in neuroblastoma cells, could also be involved in confining channels to specific cellular regions.

In addition to these factors, NaCh mobility can be modulated by the phosphorylation state of the channel as well as through alterations in the phosphorylation of the cytoskeleton. It is known that the affinity of interaction between proteins such as ankyrin and band 4.1 and the anion exchanger can be decreased with phosphorylation and may be one means to regulate NaCh distribution (Cianci *et al.*, 1988; Harris *et al.*, 1986).

C. Control of Sodium Channel Mobility in Nerve: What Regulates Channel Mobility and Distribution?

1. Role of the Cytoskeleton

A number of studies have indicated that the regional distribution and restricted mobilities of plasma membrane proteins are not due to low plasma membrane fluidity but have their origin in cytoskeletal structures associated with the plasma membrane. Immobilization is definitely used by some cells to maintain concentration gradients. Junctional acetylcholine receptors (AChRs) and NaChs on muscle fibers have mobilities that are at least several thousand times less than expected for an unrestrained solitary protein in a fluid lipid bilayer (Axelrod *et al.*, 1976b). For example, the lateral diffusion rate of NaChs is $1–3 \times 10^{-8}$ cm^2/sec in reconstituted lipid bilayers, and at the neuromuscular junction it is $<10^{-12}$

cm^2/sec. Surface blebs on nerve hillocks lead to an enhanced molecular diffusibility at the hillock, indicative that lateral constraints for the NaCh can be released by separation of the plasma membrane from the cytoskeleton (Tank *et al.*, 1982). When lipids are extracted from muscle cells by detergents, AChR aggregates in embryonic muscle cells remain associated with the cell, indicating that they are attached to constituents that are soluble neither in water nor in detergent (Stya and Axelrod, 1983).

Cytoskeletal restraints on mobility and the specific elements responsible for this restraint have been suspected for some time. Indeed, for the red cell membrane, a good case can be made for restraint on the diffusion of fluorescein-labeled band 3. If the spectrin network is genetically depleted or its connections to membrane components weakened, the diffusion coefficient of band 3 increases from about 4×10^{-11} cm^2/sec to about 2×10^{-9} cm^2/sec, close to the value measured for membrane proteins embedded in lipid bilayers which are devoid of peripheral structures (Sheetz *et al.*, 1980).

In nerve cells, electron microscopy and immunofluorescence show that initial segments and nodes are characterized by a dense subaxolemmal staining and by packing of microtubules, neurofilaments, and microfilaments (Ellisman and Porter, 1980). Based purely on morphological grounds, it is tempting to speculate that these features contribute to the restricted channel mobility seen in these regions and serve to maintain the steep concentration of channels in these regions. Indeed, in other cells, biochemical, electron microscopic, and immunofluorescence evidence has shown direct links between the cell surface and the cytoskeleton that are mediated by proteins such as vinculin and ankyrin. Recent biochemical evidence indicates that, in mature nerve, the cytoskeletal protein ankyrin associates with NaChs and links NaChs to the underlying cytoskeleton, and that this linkage could help maintain axolemmal membrane heterogeneity and control NaCh mobility (Srinivasan *et al.*, 1988).

Preparations of rat brain NaCh protein often appear to be "contaminated" with two bands around M_r 260 and 265 kDa which resemble the α and β components of spectrin. The comigration and diffuse staining of the NaCh protein in most of the gel systems mask these bands. Indeed, these two bands are identified as spectrin throughout the purification of the NaCh, and an additional immunoreactive ankyrin band can also be unmasked by immunoblotting (Fig. 8). Although immunoblotting does not prove that these components associate with NaChs, the [^{3}H]STX-binding activity that is associated with the channel can be precipitated from solution via anti-ankyrin antibodies or with spectrin linked to Sepharose 6 MB (Fig. 9) only in the presence of increasing concentrations of added erythrocyte ankyrin. Further experiments by direct binding of [^{125}I]-labeled an-

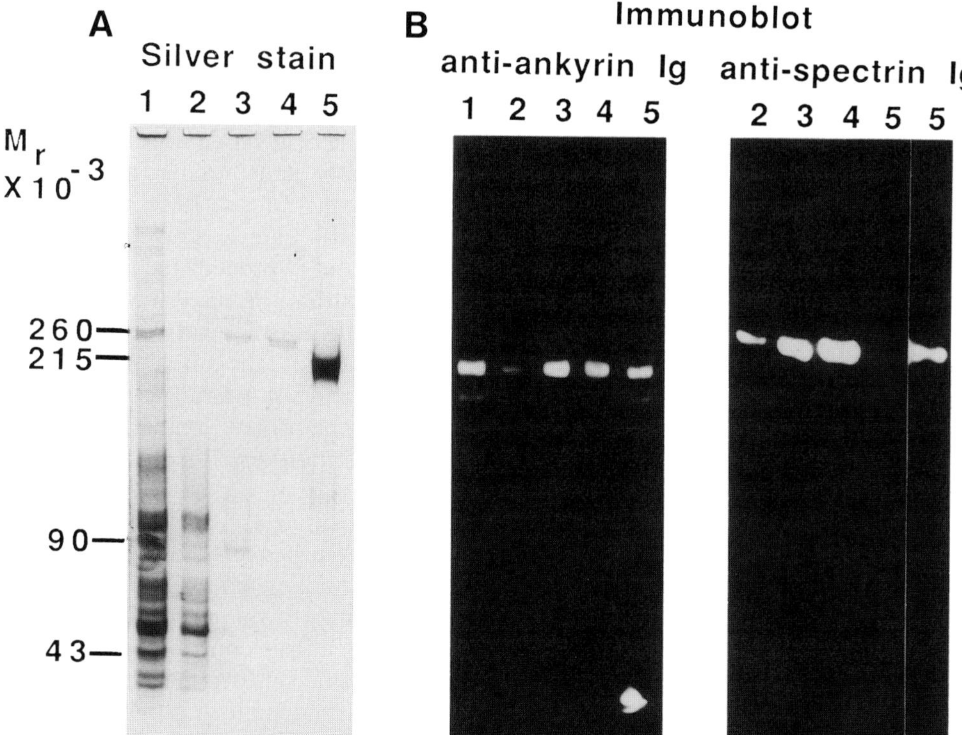

FIG. 8. Analytical SDS-PAGE of rat brain NaCh purification (A) and immunoblots of spectrin and ankyrin in this preparation (B). Lane 1, rat brain membranes showing spectrin doublet at 260–265 kDa; lane 2, Triton X-100 extract; lane 3, peak fraction from DEAE-Sephadex; lane 4, hydroxylapatite peak fraction; lane 5, WGA-Sepharose. Samples were electrophoresed in three sets of Fairbank's 3.5–17.5% exponential gradient polyacrylamide gels. Immunoblots were probed with anti-ankyrin or anti-spectrin antibodies. (From Srinivasan *et al.*, 1988.)

kyrin to the reconstituted NaCh show that ankyrin can bind directly to the channel with a K_d of 20 nM and a maximal binding capacity of 1 mole of ankyrin per mole of NaCh, and this binding depends on the cytoplasmic domain of the NaCh (Fig. 10). The purified 43-kDa cytoplasmic domain of the erythrocyte anion transporter, band 3, which also binds this form of ankyrin, also competes for the NaCh site on ankyrin with a K_i of 150 nM (Fig. 10), suggesting structural or conformational similarities between these cytoplasmic regions. Importantly, the interaction of ankyrin with NaChs is selective since neither the neuronal γ-aminobutyric acid receptor nor the dihydropyridine-sensitive Ca^{2+} channel immunoprecipitate with or bind brain ankyrin. Both of these receptors also have different distributions on the neuron (Velazquez *et al.*, 1989; Jones *et al.*, 1989).

The idea that in mature nerve NaChs are restrained by cytoskeletal linkages is consistent with recent immunofluorescence and electron microscope studies in rat sciatic nerve by Kordeli *et al.* (1990) which showed that a specialized form of red blood cell ankyrin is localized at the nodes of Ranvier (Fig. 11), while the brain ankyrin isoform is uniformly distributed along the axon. A working hypothesis based on these observations is that NaChs are immobilized to specific domains on the neuronal cell surface through linkages with ankyrin and spectrin isoforms that are assembled in these regions. Immunoelectron microscope studies of mature neurons support the idea of cytoskeletal heterogeneity because $\alpha\gamma$-spectrin is localized to axons and $\alpha\beta$-spectrin is localized to cell bodies (Riederer *et al.*, 1986; Nelson and Lazarides, 1984). These findings may be pertinent not only to the process of developmental segregation but also to the maintenance of this segregation in the mature neuron.

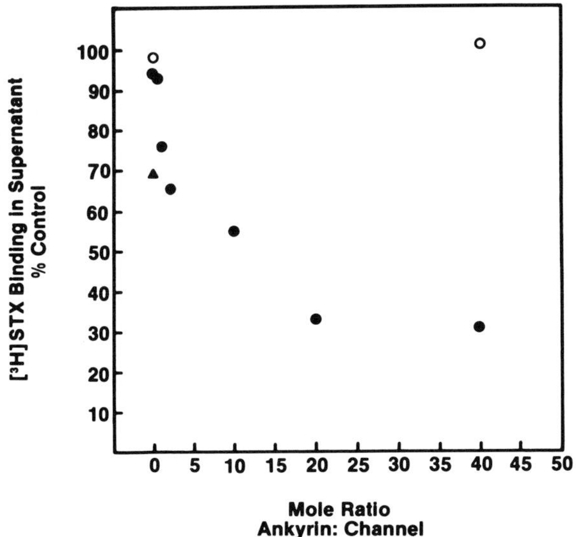

FIG. 9. Association of ankyrin with the NaCh. Depletion of [^{3}H]STX-binding activity from supernatants of purified NaCh in increasing concentrations of added ankyrin (●) by anti-ankyrin antibodies and by nonimmune IgG (○, controls) is shown. (▲), Partially pure NaCh sample from the hydroxylapatite column-depletion of [^{3}H]STX-binding activity by anti-ankyrin antibodies without added ankyrin. Ankyrin plus anti-ankyrin antibodies does not deplete [^{3}H]flunitrazepam-labeled γ-aminobutyric acid or [^{3}H]PN 200-labeled dihydropyridine receptors. (From Srinivasan *et al.*, 1988.)

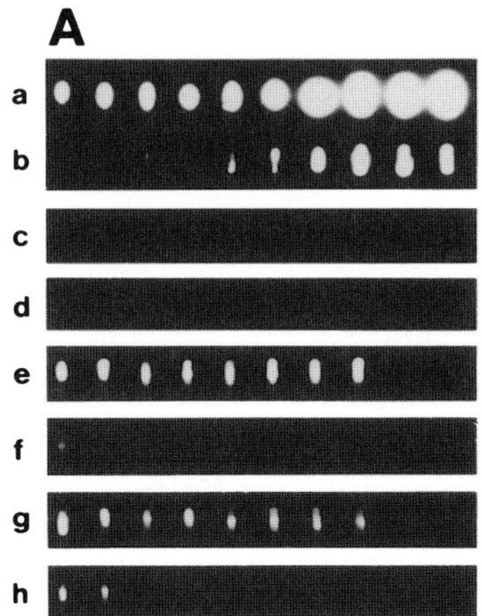

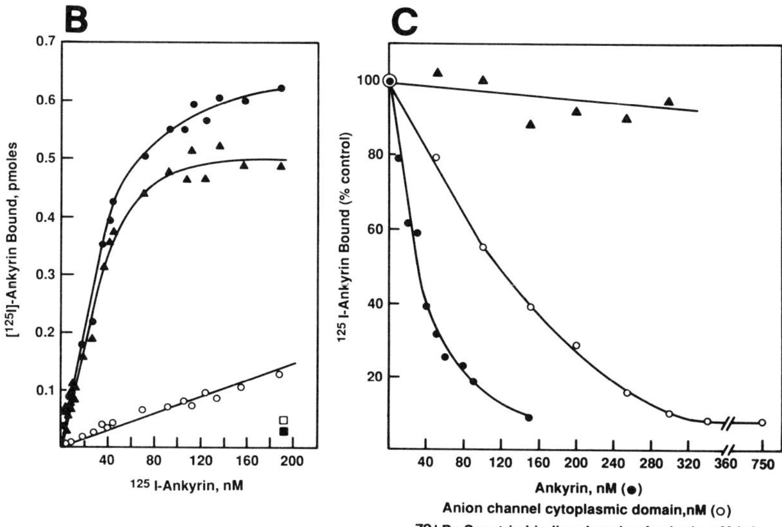

FIG. 10. Binding of ^{125}I-labeled ankyrin to the reconstituted NaCh. Purified NaChs were reconstituted into phosphatidylcholine/phosphatidylethanolamine vesicles which were then absorbed onto nitrocellulose paper. Varying concentrations of ^{125}I-labeled ankyrin were added to wells of a dot-blot apparatus in the absence or presence of 250 n*M* unlabeled ankyrin. Parallel controls were performed with vesicles that contained no protein. The nitrocellulose blots were autoradiographed, and the paper in the wells then excised and counted in a gamma counter. (A) Autoradiograms of ^{125}I-labeled ankyrin binding to NaCh. Lane a, increasing concentrations of ^{125}I-labeled ankyrin (total binding); lane b, binding in 250 n*M* unlabeled ankyrin (nonspecific binding); lane c, the binding to vesicles without

protein; lane d, the binding to reconstituted NaChs whose cytoplasmic domain has been removed after digestion with 0.02% trypsin. Lanes e–h show displacement of 10 nM ^{125}I-labeled ankyrin (lane e) with increasing concentrations of unlabeled ankyrin (lane f), the 72-kDa spectrin-binding fragment of ankyrin (lane g), and the 43-kDa cytoplasmic fragment of band 3 (lane h). (B) Quantitative data from the autoradiograms in A. Total (●), nonspecific (○), and specific (▲) binding of ^{125}I-labeled ankyrin to the NaCh is shown. Also shown is the binding to vesicles without protein (■) and to reconstituted channels with the cytoplasmic domain removed (□). (C) Quantitative data on displacement from A, lanes e–h, showing competition by unlabeled ankyrin (●), the 72-kDa spectrin-binding domain of ankyrin (▲), and the 43-kDa cytoplasmic fragment of band 3 (○). (From Srinivasan *et al.*, 1988.)

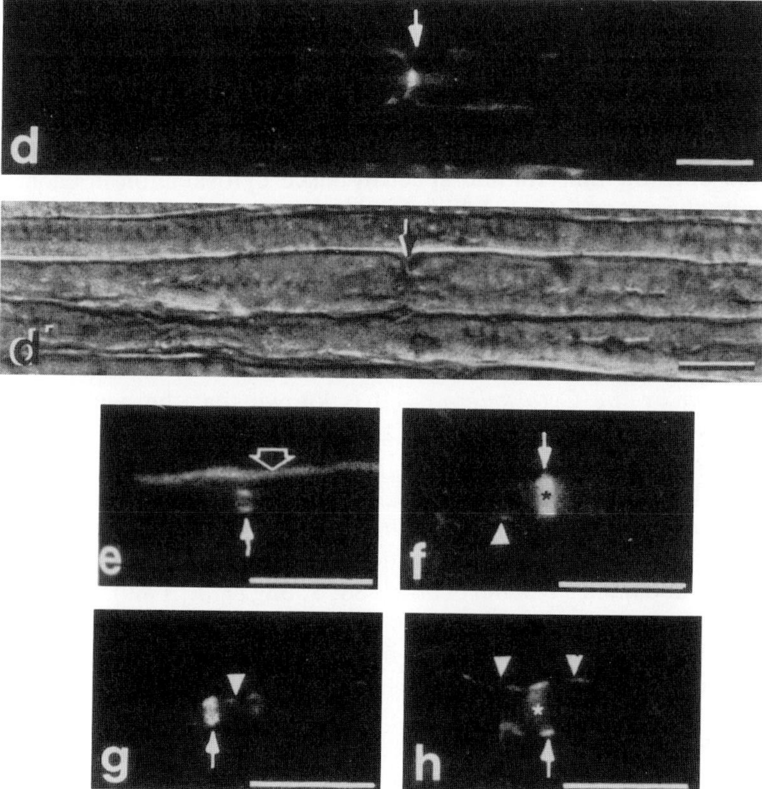

FIG. 11. Immunofluorescence localization of the erythrocyte form of ankyrin in rat peripheral nerve. The antibody stains nodes of Ranvier (arrow) but not the internodal axolemma. (From Kordeli *et al.*, 1990.)

Although there is evidence that ankyrin binds to NaChs *in vitro* and is colocalized at nodes of Ranvier, is NaCh mobility controlled by these associations? In the absence of experiments where these proteins have been expressed in mammalian cells, this idea has been tested by preparing heterokaryons of purified NaChs in liposomes and normal and spherocytic erythrocytes, the latter lacking major components of the cytoskeleton (spectrin, actin, and ankyrin). In reconstituted vesicles alone, all NaChs are very mobile ($D = 10^{-9}$ cm^2/sec). After fusion of the vesicles with red blood cells, ~80% of the NaChs are immobile ($D \leq 10^{-12}$ cm^2/sec), presumably through their binding and association with the endogenous spectrin/ankyrin network present in red blood cells. In control experiments with spherocytic erythrocytes lacking ankyrin, nearly all (85%) NaChs retain free mobility ($D = 10^{-9}$ cm^2/sec), as in liposomes.

Experiments similar to those described for the AChR have also been performed to investigate whether the lateral mobility correlates with the detergent extractability of NaChs. Differential extraction of labeled NaChs by gentle detergent treatment leaves only those NaChs that are assembled onto the ankyrin and spectrin-based cytoskeleton in each region. Thus, the detergent insolubility has been correlated with the lateral immobility of NaChs and has shown that those regions on the axon surface where NaChs are immobile correspond to regions of low detergent extractability. Thus, at least *in vitro,* the association or confinement of NaChs with ankyrin and spectrin may control channel mobility.

Modulation by the Cytoskeleton. It is known that the neuronal cytoskeleton is not a static structure but a dynamic one which can be modulated. As the neuron grows, develops, and extends its processes, the cell surface components are also changing. The NaCh, ankyrin, and spectrin can be phosphorylated *in vivo* and *in vitro,* and *in vitro* phosphorylation of ankyrin in red blood cells appears to modulate the association of ankyrin with spectrin and band 3 (Cianci *et al.*, 1988; Soong *et al.*, 1989). Similarly, phosphorylation of ankyrin and NaChs decreases their affinity for interaction by 100-fold. This is one potential means to alter NaCh distribution in growing neurons.

2. Role of Extracellular Elements

Although interaction with the subplasmalemmal cytoskeleton is a primary means whereby NaCh topography is maintained, NaChs could also be immobilized by virtue of their association with extracellular matrix components and/or adhesion molecules. In the myelinated nerve, these extracellular matrix components are contributed largely by the ensheathing/myelinating Schwann cell or oligodendrocytes. To what degree is NaCh mobility modulated by glial cell contact or by interaction with extracellular matrix components? There are several reports which have shown that cell–cell contact can modulate membrane protein distribution and mobility (Wolf *et al.,* 1980; Chow and Poo, 1982). Chow and Poo (1982) found that, in embryonic muscle cells, cell surface lectin receptors underwent rapid redistribution (even at sites remote from the contact site) which was dependent on the degree of cell contact. In addition, there is some evidence that the glycocalyx itself can restrain lateral mobility. Stearoylated dextrans, for example, diffuse at rates similar to plasma membrane glycoproteins (Wolf *et al.,* 1980), suggesting a coupling of the stearoylated dextran to some external immobilizing structure, since no direct interaction with subplasmalemmal cytoskeleton is possible.

Although not yet known for NaChs, in some cells there is evidence to suggest that development and interaction of membrane proteins with the basal lamina and lectinlike molecules serve to restrict plasma membrane mobility (Wolf *et al.,* 1980). It is possible, like acetylcholinesterase and AChR, where the basal lamina appears very early in the differentiation of the synapse and sequesters acetylcholinesterase at the synaptic cleft, that the attachment of basal lamina to the axon and its interaction with NaChs is sufficient to "seed" the first NaCh aggregates before Schwann cells arrive or before the cytoskeleton is assembled (Fraser and Poo, 1982). The NaCh is a very heavily glycosylated protein whose carbohydrate is not required for function but may play a structural role in its cellular organization. Because the basal lamina may contain specific lectin-like molecules that interact with NaChs, such interactions could be specific for a variety of lectins specific for polysialic acid or *N*-acetyl-β-D-glucosaminyl oligomers which are structural features of the NaCh (Miller *et al.,* 1983; James and Agnew, 1987). When NaChs are enzymatically deglycosylated with neuraminidase to remove the polysialic acid residues and/or with endoglycosidase F to remove the N-linked carbohydrate moities, the lateral and rotational mobilities show modest increases.

IV. DIFFERENTIATION OF THE AXON MEMBRANE: LOCALIZATION OF SODIUM CHANNELS IN DEVELOPING NERVE

A. *Sodium Channel Topography in Developing Nerve*

Major questions are How and when does the striking differentiation of the axon membrane develop and How are NaChs targeted to specific regions of the neuron? Although there is no information using active or live preparations, several studies indicate that the segregation and clustering of

NaChs may occur as a relatively early event in development (Waxman, 1981; Strichartz *et al.*, 1984; Wiley-Livingston and Ellisman, 1980). Particle patches in freeze-fracture, deposition of Fe-FeCN stain, and clusters of Na^+ current and nonuniform conduction (Bostock and Sears, 1978; Rasminsky and Sears, 1972; Smith *et al.*, 1982b; Black *et al.*, 1982) have been observed in developing axons even before the elaboration of the myelin sheath. If myelination is delayed, axolemmal maturation continues, with the establishment of a regular nodal geometry, increased axolemmal undercoating, and changes in the conduction properties characteristic of myelinated fibers (Black *et al.*, 1986).

Physiological studies indicate that even in spinal cord motoneurons early in development a nonuniform distribution of electrical excitability can be found which, as maturation proceeds, develops with a lower threshold for generation of the action potential. The morphological correlates of the developing high density are also seen early on in development, with intensely stained Fe-FeCN hillocks, a high density of 500 Å IMPs at the developing initial segment, and intense fluorescence of labeled NaChs at the initial segment of developing spinal cord cells in culture.

B. Mechanisms for Development of Sodium Channel Localization

1. Surface Transport and Diffusion Trapping

How then are channels targeted to the axon and how do such channels segregate in the axonal membrane? In growing axons, freeze-fracture studies suggest that plasmalemmal IMPs, which may be related to NaChs and other axolemmal proteins, are inserted proximally into the plasma membrane. These IMP clusters reach distal locations by lateral diffusion at a rate for iso-concentration systems of 0.5 to 1.8×10^{-7} cm^2/sec (Small *et al.*, 1984; Small and Pfenninger, 1984), and the rate of movement appears only limited by the addition of new membrane at the growing tip of the axon. If these rates do indeed apply to NaChs, the shifting distributions of IMPs that characterize the growing neuron give insight into some of the first cellular processes that establish polarity and maturation of the axon. The question then becomes, How do NaChs become deposited along the axon and how are they sequestered at specific sites along the axon? One idea is that these laterally diffusing NaChs become trapped at sites where intra- or extracellular boundaries are created. Alternatively, as discussed below, NaChs could be recruited and assembled onto a preexisting and assembled cytoskeleton or their assembly coordinated with the developing cytoskeleton.

Some direct measurements of the distribution and mobility of NaChs in developing spinal cord and dorsal root ganglion neurons have been made with antibody and fluorescent neurotoxin probes. Using neurons developing in culture and direct markers for NaChs, in young spinal cord neuron cultures (3–7 days) a gradient of NaChs that originates from the cell body and decreases dramatically toward the growth cone is observed. Examination of the lateral mobility of NaChs in these cultures shows that the lateral mobility rate increases from the cell body to the growth cone, reaching a value of 10^{-7} cm^2/sec (which is 10-fold faster than most lipids in cell membranes) at the growth cone. Careful analysis of the data, consideration of this unusually rapid rate of two-dimensional diffusion, and further experimentation suggest that this could be due to membrane flow toward the tip of the advancing growth cone and may be consistent with a mechanism in which NaChs in developing axons reach their distal location by lateral diffusion through the plasma membrane. Within 2 weeks in culture, however, NaCh clusters appear, accumulate, and immobilize at the hillock and at several locations along the axon. How do these clusters form?

There are several possible ways in which NaChs could segregate in the developing nerve: (1) NaChs form physical associations, aggregate, and form archipelagos in the membrane of developing neurons and/or are excluded by inhomogeneities in membrane lipid composition created by the segregation of certain lipids which impede lateral and rotational diffusion in these parts of the cell; (2) NaChs interact with extracellular matrix or adhesion molecules expressed by ensheathing cells; and (3) NaChs interact with the developing cytoskeleton, e.g., ankyrin and spectrin, either through physical associations or through a diffusional barrier established by the underlying meshwork. There is experimental evidence for each of these mechanisms.

There are several observations that demonstrate that NaChs do aggregate, which could lead to the formation of NaCh patches on sensory neurons during the early stages of axonal growth. First, IMPs seen on the axons of developing neuron have sizes of ~1000 Å (Small *et al.*, 1984), which is larger than the 80 Å particle diameter of a single channel (Elmer *et al.*, 1985). Direct measurement of NaCh rotational diffusion shows that the cluster size is ~2000 Å, which is still compatible with the rapid translational diffusion rates (10^{-7} cm^2/sec). Second, electron microscopy of colloidal gold-decorated NaChs shows that NaChs both in reconstituted vesicles and in developing neurons aggregate, the extent depending on the lipid composition. Given that NaChs aggregation *in vitro* can be induced in part by the properties of the bilayer, it is possible that in cells a high protein concentration in the membrane would lead to NaCh clustering. There are examples of such protein archipelagos in cells which include the light-harvesting complexes in photosynthetic bacteria, the Ca^{2+}-ATPase in sarcoplasmic reticulum (Squier *et al.*, 1988), and the unliganded epidermal growth factor receptor (Zidovetski *et al.,* 1986), all of which reach high concentrations in their membranes and aggregate. Alternatively, NaChs could be excluded from membrane domains by inhomogeneities in lipid distribution in the cell. Evidence for the segregation of membrane proteins by exclusion from regions with specialized lipids has been shown by protein lateral diffusion measurements in lipid mixtures consisting of alternating domains of fluid and solid lipid (Owicki and McConnell, 1980) and from lateral diffusion measurements of inner mitochondrial membrane proteins [ADP/ATP translocator (Muller *et al.,* 1984) and cytochrome oxidase (Sowers and Hackenbrock, 1985)], where the protein/lipid ratio is high and where there are specialized lipids.

2. Role of Extracellular Elements in Development of Sodium Channel Distribution

Obviously, when the myelin membrane is laid down the electrical properties of the axonal membrane are profoundly changed. Does the Schwann cell and/or extracellular matrix have any role in segregating and/or maintaining NaChs on the developing axon surface? The extent to which development and maintenance of the nodal differentiation and NaCh segregation are dependent on sheath cell contacts remains unresolved, although two possibilities have been proposed. Rosenbluth suggested that glial cell contact is necessary to form nodes (Rosenbluth, 1981). Ellisman, on the other hand, suggests NaChs and nodal specializations develop independently of sheath cell contact (Wiley and Ellisman, 1980; Wiley-Livingston and Ellisman, 1980); he obscrvcd small patches of IMP and nodelike accumulations on segments of the unensheathed axons of dystrophic mouse spinal roots. Physiological studies are consistent with both these hypotheses, because in remyelinating and developing fibers saltatory-like conduction is observed which implies that hotspots of NaChs may precede myelination (Smith *et al.,* 1982b). On the other hand, when the axon–Schwann cell relationship of myelinated fibers is disrupted, the intense Fe-FeCN staining at nodes is lost (Foster *et al.,* 1980), and, if related to NaChs, suggests that continuing axon–glial interactions are required for the maintenance of high NaCh density at nodes.

Basal lamina expression is an early event in axonal growth, Schwann cell proliferation, and myelinogenesis (Kleitman *et al.,* 1988), and like clustering AChRs at developing synapses (Salpeter and Podleski, 1982), basal lamina could sequester NaChs in the axon surface. At early stages, as Schwann cells associate with axons, some IMPs resembling those seen at nodes of Ranvier accumulate (Rosenbluth, 1976, 1981). The question of whether these axolemmal particles aggregate or form prior to ensheathment or afterward, however, has not been resolved (Rosenbluth, 1976;

Ellisman, 1979). The second point is that the paracrystalline pattern characteristic of the axon–glial junction appears at the region of the myelin loop closest to the presumptive node and then develops later at loops further removed from the nodal region. Thus, there could be some axolemmal component that serves as an instructional molecule in early steps of axon–glial recognition. Recent reports indicate that adhesion molecules may help define the periodicity of nodes prior to myelination (Reiger *et al.*, 1986). Specifically, neural cell adhesion molecule (NCAM) and cytotactin are found on the axon surface and node, respectively, during development. Moreover, the continued presence of these adhesive molecules in a restricted area could help to maintain the development of the paranodal region.

The maintenance of NaCh polarization and segregation could also be dependent on a fence mechanism that is created by the close apposition of glial cell membranes (although different from the tight junction of epithelial cells), or on some other restriction mechanisms, perhaps complex aggregation or active exclusion from the internodal domain. Even high aggregation numbers would have a low effect on the measured value for the diffusion coefficient, although aggregation of the NaCh protein might be a significant factor in forming patches if "filter"-like fences existed in the bilayer. Such structures, perhaps incomplete cytoskeleton or junctions formed between Schwann cells processes and axon membrane, might permit passage of single molecules (as would be the case for mobile NaChs or other molecules) but restrict the diffusion of aggregates. However, some indirect evidence suggests that filters are operational even in confluent cells without Schwann cell contact.

3. Role of the Developing Cytoskeleton

As a third possibility, NaChs channels can be sequestered at specific sites by interactions with the developing cytoskeleton either through physical association or by a diffusional barrier created by the underlying meshwork (Srinivasan *et al.*, 1988; Angelides *et al.*, 1988; Koppel *et al.*, 1981). In mature nerve there is good evidence that NaChs associate and colocalize with ankyrin and spectrin. However, ankyrin binding cannot account for the mobile class of NaChs seen at hillocks; these NaChs are mobile yet confined within the hillock. Although aggregation or modulation by the lipid environment could account for the initial clustering of NaChs, the segregation and restricted lateral mobility at the hillock and at foci along the axon of NaChs that is observed as development proceeds must involve additional factors. This probably occurs, as previously suggested

(Srinivasan *et al.*, 1988), through interactions with the developing cytoskeleton via physical association, by a diffusional barrier created by the underlying meshwork (Koppel *et al.*, 1981), or perhaps by active exclusion from the internodal domain. In developing nerve, results from several laboratories lead to the idea that some NaChs must be confined within the hillock by a domain or barrier that is established by the underlying cytoskeletal meshwork but without actually binding directly to the cytoskeleton. Formation and lateral immobilization of the large NaCh clusters that arise later in development originate in principle by diffusion-mediated trapping, where smaller mobile NaCh aggregations are trapped within the boundaries of a "filter"-like fence that has developed in the membrane. Structures formed, for example, by an incompletely assembled cytoskeleton might permit passage of single molecules but restrict the free exchange of NaCh aggregates between developing node and internode.

Although lateral diffusion measurements do not discriminate whether intra- or extracellular elements provide this barrier, NaCh lateral immobility coincides with the assembly of the spectrin and ankyrin network at day 7 even without Schwann cells. Although the distribution of these cytoskeletal proteins is uniform along the axon, these observations are consistent with the hypothesis that at early stages the axon itself provides a selective barrier to diffusion between compartments. Confinement within an underlying corral-type structure would result in restricted lateral diffusion, but free rotational mobility within this corral would still be possible. Rotational diffusion measurements in both developing and mature neurons have revealed that NaChs are laterally confined but rotationally mobile in developing nerve and become both rotationally and translationally immobile as development proceeds. Thus, it appears that interactions with the developing cytoskeleton serve to stabilize the distribution and mobility of Na^+ channels in restricted regions of the neuron as development and ensheathment proceed.

Finally, in developing axons, an interesting proposal for the localization of NaChs at nodes has been proposed (Ritchie, 1988; Gray and Ritchie, 1985). Based primarily on the finding that NaChs are expressed in Schwann cell membranes (Chiu *et al.*, 1984; Shrager *et al.*, 1985) and by consideration of the difficulties faced by a neuron in maintaining an adequate density of channels at nodes or in response to injury, it has been suggested that NaChs can be delivered to nodes by their ensheathing cell. Although this hypothesis is highly speculative, there is some morphological and biochemical support for it. There is evidence that glial cells express NaChs which have electrophysiological and pharmacological properties similar to those of NaChs in neurons (Barres *et al.*, 1990; Shrager *et al.*, 1985). Furthermore, at the node of Ranvier, both the Schwann cell in the

PNS and astrocytes in the CNS direct processes toward the axon that closely abut the nodal axolemma, the two cell membranes approaching each other much more closely than cell membranes normally do. These selected processes also show NaCh immunoreactivity (Black *et al.*, 1989). Third, Schwann cell are very active and process and turnover NaChs rapidly (Ritchie, 1988).

4. Role of the Cytoskeleton in Establishment of Restricted Membrane–Cytoskeletal Domains in Neurons

In this last section, we wish to explore the possible mechanisms by which NaChs and specific cytoskeleton elements assemble at the node. The major question is whether NaChs assemble onto a preexisting cytoskeleton or whether NaChs aggregate and are subsequently stabilized by the cytoskeleton at nodes. The first of course, requires that the necessary components, ankyrin and spectrin, are sorted and targeted to appropriate axonal domains prior to NaChs, while the latter depends on interaction with Schwann cells through the elaboration of certain components of the extracellular matrix and/or adhesion molecules.

Although the development of patterned plasma membranes is a feature demonstrated by many cell types, only in neurons is this so striking. The organization of the node of Ranvier at distinct places along the axon is a classic example of plasmalemma specialization, and its biogenesis is an important event in the terminal differentiation of myelinated neurons. Although the mechanisms that regulate NaCh distribution during myelination are not well understood, the structural asymmetry of the neuronal membrane suggests that the cytoskeleton could play an important role. In mature neurons, for example, not only is there clear heterogeneity in the distribution of membrane proteins but there is also a striking asymmetry in the molecular composition and distribution of the neuronal cytoskeleton. It is known that neurofilaments are localized to axons, while MAP-2 is localized predominantly in the perikarya and dendrites. Spectrin has been found to be asymmetrically distributed in neurons, where the $\alpha\beta$ isoform is distributed in the cell soma and the dendrites and the $\alpha\gamma$ form is concentrated along axons and presynaptic terminals (Riederer *et al.*, 1986). It is also known that a specific erythrocyte isoform of ankyrin colocalizes with NaChs at mature nodes. The differences in the cytoplasmic membrane distribution of $\alpha\beta$- and $\alpha\gamma$-spectrin in neurons may be pertinent not only to the assembly of NaChs at nodes but also to the maintenance of this distribution in the mature neuron. The observation that there is an asymmetry in the molecular composition and distribution of the neuronal membrane skeleton and NaChs raises a question regarding the developmental sequences involved in their assembly during neuronal morphogenesis and myelination.

Morphogenesis of the plasma membrane–cytoskeleton has been investigated mostly in red blood cells (Lazarides, 1987) and recently in polarized epithelial cells (Nelson and Veshnock, 1987). Although the heterogeneity of ankyrin and spectrin isoforms in these cells is not as dramatic or complicated as in neurons, it seems that, at least in red blood cells, membrane transport proteins are expressed after establishment of the spectrin- and ankyrin-based cytoskeleton (Lazarides, 1987), while in epithelial cells the distribution of ankyrin to basolateral domains occurs after sorting of their membrane attachment site (e.g., Na^{+}/K^{+}-ATPase) to this domain (Nelson and Veshnock, 1987). In red blood cells, membrane proteins associate with a preassembled cytoskeleton, while in epithelial cells external signals appear to be required to form a template for ankyrin/spectrin assembly at the plasma membrane.

In myelinated axons, how do restricted membrane–cytoskeletal domains such as the node of Ranvier develop? Specifically, does the differentiating neuron direct NaChs onto a preassembled cytoskeleton and immobilize them at nodes of Ranvier or are other extracellular elements contributed by the ensheathing Schwann cell responsible for the initial events of node development? In light of the information on the distribution, mobility, and close association of the NaCh with ankyrin, it seems appropriate that the organization of NaChs on the axon surface is somehow linked to the differentiation of the neuronal cytoskeleton.

Recent work in our laboratory has explored this question by examining the "steady-state" distribution of NaChs and cytoskeleton during myelination in cell culture. Sensory neurons in combination with Schwann cells can be arrested at specific developmental stages of myelination and probed using immuno-light and electron microscopy with specific antibodies to ankyrin, spectrin, NaChs, and adhesion molecules.

In this system, which appears to reproduce myelination *in vivo,* Schwann cells contact axons and engulf only those axons of appropriate caliber. As myelination is initiated, extracellular matrix components are synthesized and specific adhesion molecules such as L1 and myelin-associated glycoprotein (MAG) are expressed by Schwann cells. With the elaboration of basal laminal components by Schwann cells and spreading of Schwann cell processes along axons, L1 is expressed with the formation of primitive paranodal structures. Final spiralling of the myelin sheath and expression of P_0 completes the formation of compact myelin. Specifically, the expression, distribution, and localization of NaChs, ankyrin, and spectrin isoforms, have been examined during the first week of neuronal growth, after 1–2 weeks, where Schwann cells proliferate and migrate, 3–5 weeks, where extensive axonal contact by Schwann cell processes is observed and prior to ensheathment, and after the initiation of myelination (both prior to basal lamina formation and formation of compact myelin 8–10 weeks in culture).

To summarize, the following initial observations have been made: (1) Up to 2 weeks in culture, NaChs and ankyrin remain homogeneously distributed on sensory neuron axons, although in the very early stages (1 week) there is a gradient of NaChs decreasing from the cell body toward the growth cone. In weeks 2–4, small punctate NaCh clusters can be discerned on some axons. (2) When Schwann cell processes contact and extend along sensory neuron axons (5 weeks in culture), spectrin and ankyrin remain homogeneously distributed on most axons, but NaChs are very highly clustered. On some axons where Schwann cells processes have spread considerable distances along the axon, foci of NaChs are observed at developing adhesion sites, while NaCh density at the developing internodal region is lower. Even in those axons where NaCh staining is localized, the pattern of erythrocyte ankyrin staining is still homogeneous. In many axons, the ankyrin reactivity is contained in vesicle-like structures confined to the cytoplasm. This intra-axonal staining is particularly prominent in those organelle-rich varicosities identified by Hollenbeck and Bray (1987). Through-focus series of these varicosities indicated that the immunoreactive material is of a granular nature, consistent with the presence of closely packed organelles, such as those seen in light and electron micrographs of neuronal varicosities, and that contain α-spectrin and actin (Hollenbeck and Bray, 1987). It is possible that these vesicles may contain ankyrin en route to their sites of assembly. (3) As basal lamina and myelin formation is initiated with ascorbic acid, only until the expression of detectable levels of P_o (compact myelin) do regularly spaced NaCh patches occur on the axon at the developing nodes and NaCh density at the internodal membrane disappear. Only after elaboration of compact myelin and the myelin expression of P_o does the homogeneous pattern of ankyrin change to one where ankyrin is colocalized with NaChs. Based on these preliminary observations and consistent with freeze-fracture data, it seems that nodes and NaCh clustering are probably structurally determined well after Schwann cell contact and as myelination proceeds. NaCh–ankyrin complexes appear to be formed at terminal stages of myelination. This is unlike red blood cells, where membrane proteins appear to be recruited and immobilized to the cell membrane only after assembly of the cytoskeleton. It appears that the cytoskeletal elements important in maintaining NaChs localized to the node are expressed and assembled after clustering and trapping of NaChs and occur after Schwann cell contact. The present view implies that (1) Schwann cells are necessary for the formation of NaCh clusters and nodes and there is no targeting mechanism programmed by the neuron to deliver NaChs to predetermined sites on the axon; (2) NaCh clusters probably serve as a template for the assembly of a cytoskeleton-specific ankyrin isoform; and (3) assembly of ankyrin with the cortical spectrin cytoskeleton stabilizes and immobilizes NaChs at these well-formed nodes. Thus, depending on the extent of glial cell contact, NaCh distribution can be uniform or patched. Such a hypothesis, i.e., varying degrees of axon–glial contact, could account for the differences in IMP distribution reported by Rosenbluth (1981) and Ellisman (Wiley and Ellisman, 1980) using the same murine mutant.

V. CONCLUSIONS AND PERSPECTIVES

It is clear that the axon membrane exhibits a high degree of regional specialization and that the segregation of voltage-dependent NaChs to specific sites on the axon is tightly controlled. The development and maintenance of this precise distribution appear to arise through the coordinated expression and targeting of NaChs to the axon, the assembly of the membrane-based cytoskeleton, and differentiation of axon–glial interactions during the development of myelinated fibers. Recent work has begun to clarify the rules that regulate axon–glial interactions and the assembly of membrane–cytoskeletal domains in neurons.

Although we have begun to understand some of the elements important for the organization of NaChs on axon surfaces, a number of important questions remain, including how NaCh subtypes are routed to their specific destinations, the mechanisms of this transport (i.e., vesicular or by insertion and diffusion), the events in the assembly of the axonal cytoskeleton and stabilization of NaCh distribution, and the role of glial cells in the segregation of NaChs and the development of nodal specializations.

Finally, there is an important need to understand the molecular events that occur in pathological conditions such as demyelination in multiple sclerosis, the ways axons respond and reorganize following injury, and how myelination and conduction can be restored.

Acknowledgments

Work in the authors' laboratory has been supported by the National Multiple Sclerosis Society, the National Institutes of Health, and by a Career Development Award from the National Institutes of Health. The authors also wish to acknowledge the contributions by members in the authors' laboratory and those laboratories that contributed figures and micrographs.

References

Angelides, K. J. (1986). Fluorescent and photoactivable fluorescent derivatives of tetrodotoxin to probe the sodium channel of excitable membranes. *Biochemistry* **20,** 4107.

Angelides, K. J., and Nutter, T. J. (1983). Preparation and characterization of fluorescent scorpion toxins from *Leiurus quinquestriatus* as probes of the sodium channel of excitable cells. *J. Biol. Chem.* **258,** 11948.

Angelides, K. J., Elmer, L. W., Loftus, D., and Elson, E. L. (1988). Distribution and lateral mobility of voltage-dependent sodium channels in neurons. *J. Cell Biol.* **106,** 1911.

Axelrod, D. (1983). Lateral motion of membrane proteins and biological function. *J. Membr. Biol.* **75,** 1.

Axelrod, D. A., Koppel, D. E., Schlessinger, J., Elson, E. L., and Webb, W. W. (1976a). Mobility measurements of fluorescence photobleaching recovery kinetics. *Biophys. J.* **16,** 1055.

Axelrod, P., Ravdin, R., Koppel, D. E., Schlessinger, J., Webb, W. W., Elson, E. L., and Podleski, T. (1976b). Lateral motion of fluorescently labeled acetylcholine receptors in membranes of developing muscle cells. *Proc. Natl. Acad. Sci. U.S.A.* **73,** 4594.

Barhanin, J., Meiri, M., Romey, G., Pauron, D., and Lazdunski, M. (1985). A monoclonal immunotoxin acting on the Na^+ channel with properties similar to those of a scorpion toxin. *Proc. Natl. Acad. Sci. U.S.A.* **82,** 1842.

Barres, B. A., Chun, L. L. Y., and Corey, D. P. (1990). Glial and neuronal forms of the voltage-dependent sodium channel: Characteristics and cell-type distribution. *Neuron* **2,** 1375.

Barrett, J. N., and Crill, W. E. (1980). Voltage-clamp of motoneuron somata: Properties of the fast inward current. *J. Physiol. (London)* **304,** 231.

Baumgold, J., Zimmerman, I., and Bambrick, L. (1983). Appearance of [^{3}H]saxitoxin binding sites in developing rat brain. *Brain Res.* **285,** 405.

Black, J. A., Foster, R. E., and Waxman, S. G. (1982). Rat optic nerve: Freezer-fracture studies during development of myelinated axon. *Brain Res.* **250,** 1.

Black, J. A., Waxman, S. G., Sims, T. J., and Gilmore, S. A. (1986). Effects of delayed myelination by oligodendrocytes and Schwann cells on the macromolecular structure of axonal membrane in rat spinal cord. *J. Neurocytol.* **15,** 745.

Black, J. A., Friedman, B., Waxman, S. G., Elmer, L. W., and Angelides, K. J. (1989). Immunoultrastructural localization of sodium channels at nodes of Ranvier and perinodal astrocytes in rat optic nerve. *Proc. R. Soc. London, Ser. B* **238,** 39.

Black, J. A., Kocsis, J. D., and Waxman, S. G. (1990). Ion channel organization of the myelinated fiber. *Trends Neurosci.* **13,** 48–54.

Bostock, H., and Sears, T. A. (1978). The internodal axon membrane: Electrical excitability and continuous conduction in segmental demyelination. *J. Physiol. (London)* **280,** 273.

Catterall, W. A. (1988). Structure and function of voltage-sensitive ion channels. *Science* **242,** 50.

Chiu, S. Y., Schrager, P., and Ritchie, J. M. (1984). Neuronal-type Na^+ and K^+ channels in rabbit cultured Schwann cells. *Nature (London)* **311,** 156.

Chow, I., and Poo, M. M. (1982). Redistribution of cell surface receptors induced by cell–cell contraction. *J. Cell Biol.* **95,** 510.

Cianci, C. D., Giorgi, M., and Morrow, J. S. (1988). Phosphorylation of ankyrin down-regulates its cooperative interaction with spectrin and protein 3. *J. Cell. Biochem.* **37,** 301.

Conti, F., Hille, B., Neumcke, B., Nonner, W., and Stampfli, R. (1976). Conductance of the sodium channel in myelinated nerve fibers with modified sodium inactivation. *J. Physiol. (London)* **262,** 729.

Coombs, J. S., Eccles, J. C., and Fatt, P. (1955). The electrical properties of the motoneuron membrane. *J. Physiol. (London)* **130,** 291.
Coria, R., Berciano, M. J., Berciano, J., and Lafarga, M. (1984). Axon membrane remodeling in the lead-induced demyelinating neuropathy of the rat. *Brain Res.* **291,** 369.
Darbon, H., and Angelides, K. J. (1984). Structural mapping of the voltage-dependent sodium channel. *J. Biol. Chem.* **259,** 6074.
Eccles, J. C., Llinas, R., and Sasaki, K. (1966). Intracellularly recorded responses of the cerebellar Purkinje cells. *Exp. Brain Res.* **1,** 161.
Edidin, M., and Zuniga, M. (1984). Lateral diffusion of wild-type and mutant Ld antigens in L cells. *J. Cell Biol.* **99,** 2333.
Ellisman, M. H. (1979). Molecular specializations of the axon membrane at nodes of Ranvier are not dependent on myelination. *J. Neurocytol.* **8,** 719.
Ellisman, M. H., and Levinson, S. R. (1982). Immunocytochemical localization of sodium channel distributors in the excitable membranes of *Electrophorus electricus*. *Proc. Natl. Acad. Sci. U.S.A.* **79,** 6707.
Ellisman, M. H., and Porter, K. R. (1980). Microtabecular structure of the axoplasmic matrix: Visualization of cross-linking structures and their distribution. *J. Cell Biol.* **87,** 464.
Elmer, C. W., O'Brien, B., Nutter, T. J., and Angelides, K. J. (1985). Physicochemical characterization of the α-peptide of the sodium channel from rat brain. *Biochemistry* **24,** 8128.
Elson, E. L., and Schlessinger, J. (1979). Long-range motions on cell surfaces. *Neurosci. Study Program, 4th* p. 691.
Ffrench-Constant, L., and Raff, M. C. (1986). The oligodendrocyte-type-2 astrocyte cell lineage is specialized for myelination. *Nature (London)* **323,** 335.
Foster, R. E., Whalen, C. C., and Waxman, S. G. (1980). Reorganization of the axon membrane in demyelinated peripheral nerve fibers: Morphological evidence. *Science* **210,** 661.
Fraser, S. E., and Poo, M. M. (1982). Development and modulation of a patterned membrane topography based on the acetylcholine receptor. *Curr. Top. Dev. Biol.* **17,** 77.
Gordon, D., Merrick, D., Auld, U., Dunn, R., Goldin, A. L., Davidson, N., and Catterall, W. A. (1987). Tissue-specific expression of the R_1 and R_{11} sodium channel subtypes. *Proc. Natl. Acad. Sci. U.S.A.* **84,** 8682.
Gray, P. T. A., and Ritchie, J. M. (1985). Ion channels in Schwann and glial cells. *Trends Neurosci.* **8,** 411.
Haimovich, B., Bonilla, E., Casadei, J., and Barchi, R. (1984). Immunocytochemical localization of the mammalian voltage-dependent sodium channel using polyclonal antibodies against the purified protein. *J. Neurosci.* **4,** 2259.
Harris, A. S., Anderson, J. P., Yurchenco, P. D., Green, L. A., Ainger, K. J., and Morrow, J. S. (1986). Mechanisms of cytoskeletal regulation. *J. Cell. Biochem.* **30,** 51.
Hartshorne, R. P. and Catterall, W. A. (1984). The sodium channel from rat brain. *J. Biol. Chem.* **259,** 1667.
Hollenbeck, P. J., and Bray, D. (1987). Rapidly transported organelles containing membrane and cytoskeletal components: Their relation to axonal growth. *J. Cell Biol.* **105,** 2827.
Jacobson, K., Ishihara, A., and Inman, R. (1987). Lateral diffusion of proteins in membranes. *Annu. Rev. Physiol.* **49,** 163.
James, W. M., and Agnew, W. S. (1987). Multiple oligosaccharide chains in the voltage-sensitive Na channel from *Electrophorus electricus:* Evidence for α-2,8-linked polysialic acid. *Biochem. Biophys. Res. Commun.* **148,** 817.

Jones, O. T., Kunze, D. L., and Angelides, K. J. (1989). Localization and mobility of Ω-conotoxin-sensitive Ca^{2+} channels in hippocampal CA1 neurons. *Science* **244,** 1189.

Kleitman, N., Wood, P., Johnson, M. I., and Bunge, R. P. (1988). Schwann cell surfaces but not extracellular matrix organized by Schwann cells support neurite outgrowth from embryonic rat retina. *J. Neurosci.* **8,** 653.

Koppel, D. E., Sheetz, M. P., and Schindler, M. (1981). Matrix control of protein diffusion in biological membranes. *Proc. Natl. Acad. Sci. U.S.A.* **78,** 3576.

Kordeli, E., Davis, J., Trapp, B. D., and Bennett, V. (1990). An isoform of ankyrin is localized at nodes of Ranvier in myelinated axons of central and peripheral nerves. *J. Cell Biol.* **110,** 1341.

Kraner, S. D., Tanaka, J. C., and Barchi, R. L. (1985). Purification and functional reconstitution of the voltage-sensitive sodium channel from rabbit T-tubular membranes. *J. Biol. Chem.* **260,** 6341.

Kristol, C., Sandri, C., and Akert, K. (1978). Intramembranous particles at the node of Ranvier of the cat spinal cord: A morphometric study. *Brain Res.* **142,** 391.

Lazarides, E. (1987). From genes to structural morphogenesis: The genesis and epigenesis of a red blood cell. *Cell* **51,** 345.

Meiri, H., Zeitoun, I., Grunhagen, H. H., Lev-Ram, V., Eshhar, Z., and Schlessinger, J. (1984). Monoclonal antibodies associated with sodium channel block nerve impulse and stain nodes of Ranvier. *Brain Res.* **310,** 168.

Miller, J. A., Agnew, W. S., and Levinson, S. R. (1983). Principal glycopeptide of the tetrodotoxin/saxitoxin binding protein from *Electrophorus electricus:* Isolation and partial chemical and physical characterization. *Biochemistry* **22,** 462.

Muller, M., Krebs, J. J. R., Cherry, R. J., and Kawato, S. (1984). Rotational diffusion of the ADP/ATP translocator in the inner membrane of mitochondria and in proteoliposomes. *J. Biol. Chem.* **259,** 3037.

Myles, D. G., Primakoff, P., and Koppel, D. E. (1984). A localized surface protein of guinea pig sperm exhibits free diffusion in its domain. *J. Cell Biol.* **98,** 1905.

Nelson, W. J., and Lazarides, E. (1984). The patterns of expression of two ankyrin isoforms demonstrate distinct steps in the assembly of the membrane skeleton in neuronal morphogenesis. *Cell* **39,** 309.

Nelson, W. J., and Veshnock, P. J. (1987). Modulation of fodrin (membrane skeleton) stability by cell–cell contact in Madin–Darby canine kidney epithelial cells. *J. Cell Biol.* **104,** 1527.

Nigg, E. A., and Cherry, R. J. (1980). Anchorage of a band 3 population at the erythrocyte membrane surface: Protein rotational diffusion measurements. *Proc. Natl. Acad. Sci. U.S.A.* **77,** 4702.

Noda, M., Ikeda, T., Kayano, T., Suzuki, M., Takeshima, H., Kurasaki, M., Takahashi, H., and Numa, S. (1986). Existence of distinct sodium channel messenger RNAs in rat brain. *Nature (London)* **320,** 188.

Oliver, J. M., and Berlin, R. D. (1982). Mechanisms that regulate the structural and functional architecture of cell surfaces. *Int. Rev. Cytol.* **74,** 55.

Owicki, J. C., and McConnell, H. M. (1980). Lateral diffusion in homogeneous membranes. *Biophys. J.* **30,** 383.

Poo, M. M. (1982). Rapid lateral diffusion of functional Ach receptors in embryonic muscle cell membrane. *Nature (London)* **295,** 332.

Poo, M. M., and Cone, R. A. (1984). Lateral diffusion of rhodopsin in the photoreceptor membrane. *Nature (London)* **247,** 438.

Quick, D. C., and Waxman, S. G. (1977). Specific staining of the axon membrane at nodes of Ranvier with ferric ion and ferrocyanide. *J. Neurol. Sci.* **31,** 1.

Rasminsky, M., and Sears, T. A. (1972). Internodal conduction in undissected demyelinated nerve fibers. *J. Physiol. (London)* **227,** 323.

Reiger, F., Daniloff, J. K., Pincon-Raymond, M., Cresin, K. C., Grumet, M., and Edelman, G. (1986). Neuronal cell adhesion molecules and cytotactin are colocalized at the node of Ranvier. *J. Cell Biol.* **103,** 379.

Riederer, B. M., Zagon, I. S., and Goodman, S. R. (1986). Brain spectrin (240-235) and brain spectrin (240/235E): Two distinct spectrin subtypes with different locations within mammalian neural cells. *J. Cell Biol.* **102,** 2088.

Ritchie, J. M. (1988). Sodium-channel turnover in rabbit cultured Schwann cells. *Proc. R. Soc. London, Ser. B* **233,** 423.

Ritchie, J. M., and Rang, H. P. (1983). Extraneuronal saxitoxin binding sites in rabbit myelinated nerve. *Proc. Natl. Acad. Sci. U.S.A.* **80,** 2803.

Ritchie, J. M., and Rogart, R. B. (1979). Density of sodium channels in mammalian myelinated nerve fibers and nature of the axonal membrane under the myelin sheath. *Proc. Natl. Acad. Sci. U.S.A.* **74,** 211.

Rosenberg, R. L., Tomiko, S. A., and Agnew, W. S. (1984). Single-channel properties of the reconstituted voltage-regulated Na channel isolated from the electroplax of *Electrophorus electricus*. *Proc. Natl. Acad. Sci. U.S.A.* **81,** 5594.

Rosenbluth, J. (1976). Intramembraneous particle distribution at the node of Ranvier and adjacent axolemma in myelinated axons of the frog brain. *J. Neurocytol.* **5,** 709.

Rosenbluth, J. (1981). Intramembranous particle distribution at the node of Ranvier and adjacent axolemma in myelinating axons of the frog brain. *Adv. Neurol.* **31,** 391.

Rossie, S., Gordon, D., and Catterall, W. A. (1987). Identification of an intracellular domain of the sodium channel having multiple cAMP-dependent phosphorylation sites. *J. Biol. Chem.* **262,** 17530.

Rudy, B., Kirschenbaum, B., and Greene, L. (1982). Nerve growth factor-induced increase in saxitoxin binding to rat PC12 pheochromocytoma cells. *J. Neurosci.* **2,** 1405.

Salpeter, M. M., and Podleski, T. (1982). *In* "Muscle Development" (R. Dearson, ed.), p. 481. Cold Spring Harbor Press, Cold Spring Harbor, New York.

Scheuer, T., Auld, V. J., Boyd, S., Offord, J., Dunn, R., and Catterall, W. A. (1990). Functional properties of rat brain sodium channels expressed in a somatic cell line. *Science* **247,** 854.

Schmidt, J. W., and Catterall, W. A. (1986). Biosynthesis and processing of the α subunit of the voltage-sensitive sodium channel in rat brain neurons. *Cell* **46,** 437.

Schmidt, J., Rossie, S., and Catterall, W. A. (1985). A large intracellular pool of inactive Na channel α subunits in developing rat brain. *Proc. Natl. Acad. Sci. U.S.A.* **82,** 4847.

Schwartzkroin, P. A. (1977). Further characteristics of hippocampal Ca1 cells *in vitro*. *Brain Res.* **128,** 53.

Schwartzkroin, P. A. (1977). Further characteristics of hippocampal cells *in vitro*. *Brain Res.* **128,** 53.

Sheetz, M. P., Schindler, M., and Koppel, D. E. (1980). Lateral mobility of integral membrane proteins is increased in spherocytic membranes. *Nature (London)* **285,** 510.

Sherman, S. J., Chrivia, J., and Catterall, W. A. (1985). Fluorescently labelled Na^+ channels are localized and immobilized to synapses of innervated muscle fibres. *J. Neurosci.* **5,** 1570.

Shrager, P. (1988). Ionic channels and signal conduction in single remyelinating frog nerve fibers. *J. Physiol. (London)* **404,** 695.

Shrager, P., Chiu, S. Y., and Ritchie, J. M. (1985). Voltage-dependent sodium and potassium channels in mammalian cultured Schwann cells. *Proc. Natl. Acad. Sci. U.S.A.* **82,** 948.

Small, R. K., and Pfenninger, K. H. (1984). Components of the plasma membrane of growing axon. I. Size and distribution of intramembrane particles. *J. Cell Biol.* **98,** 1434.

Small, R. K., Blank, M., Ghez, R., and Pfenninger, K. H. (1984). Components of the plasma membrane of growing axons. II. Diffusion of membrane protein complexes. *J. Cell Biol.* **98,** 1434.

Smith, K., Bostock, H., and Hall, S. M. (1982a). Saltatory conduction precedes remyelination in axons demyelinated with lysophosphatidyl choline. *J. Neurol. Sci.* **54,** 13.

Smith, K. J., Bostock, H., and Hall, S. M. (1982b). Node formation precedes remyelination. *Trends Neurosci.* **5,** 196.

Soong, C. J., Lu, P. W., and Tao, M. (1989). Analysis of band 3 cytoplasmic domain phosphorylation and association with ankyrin. *Arch. Biochem. Biophys.* **254,** 509.

Sowers, A. E., and Hackenbrock, C. R. (1985). Variation in protein lateral diffusion coefficients is related to variation in protein concentration found in mitochondrial inner membranes. *Biochim. Biophys. Acta* **821,** 85.

Squier, T. C., Hughes, S. E., and Thomas, D. D. (1988). Lipid fluidity directly modulates the overall protein rotational mobility of the Ca^{2+}-ATPase in sarcoplasmic membranes. *J. Biol. Chem.* **263,** 9162.

Srinivasan, Y., Elmer, L. W., Davis, J. Q., Bennett, V., and Angelides, K. J. (1988). Ankyrin and spectrin associate with voltage-dependent sodium channels in brain. *Nature (London)* **333,** 177.

Strichartz, G. R., Small, R. K., and Pfenninger, K. H. (1984). Components of the plasma membrane of growing axons. III. Saxitoxin binding to sodium channels. *J. Cell Biol.* **98,** 1444.

Stuhmer, W., Methfessel, C., Sakmann, B., Noda, M., and Numa, S. (1987). Patch clamp characterization of sodium channels expressed from rat brain cDNA. *Eur. Biophys. J.* **14,** 131.

Stuhmer, W., Conti, F., Suzuki, H., Wang, X. D., Noda, M., Yahagi, N., Kubo, H., and Numa, S. (1989). Structural parts involved in activation and inactivation of the sodium channel. *Nature (London)* **339,** 597.

Stya, M., and Axelrod, D. (1983). Mobility and detergent extractability of acetylcholine receptors on cultured rat myotubes, A. Correlation. *J. Cell Biol.* **97,** 49.

Suzuki, H., Beckh, S., Kubo, H., Yahagi, N., Ishida, H., Kayano, T., Noda, M., and Numa, S. (1988). Functional expression of cloned cDNA encoding sodium channel III. *FEBS Lett.* **228,** 195.

Tank, D. W., Wu, E. S., and Webb, W. W. (1982). Enhanced molecular diffusibility in muscle membrane blebs: Release of lateral constraints. *J. Cell Biol.* **92,** 707.

Velazquez, J., Thompson, C. L., Barnes, E. M., and Angelides, K. J. (1989). Distribution and lateral mobility of GABA/benzodiazepine receptors on nerve cells. *J. Neurosci.* **9,** 2163.

Waechter, C. J., Schmidt, J. W., and Catterall, W. A. (1983). Glycosylation is required for maintenance of functional sodium channels in neuroblastoma cells. *J. Biol. Chem.* **258,** 5117.

Waxman, S. G. (1977). Conduction in myelinated, unmyelinated, and demyelinated fibers. *Arch. Neurol.* **34,** 585.

Waxman, S. G. (1981). Cytochemical heterogeneity of the axon membrane. *Tends Neurosci.* **4,** 7.

Waxman, S. G. (1987). Molecular neurobiology of the myelinated nerve fiber: Ion channel distributions and their implications for demyelinating diseases. *In* "Molecular Neurobiology and Psychiatry" (E. R. Kandel, ed.), p. 7. Raven, New York.

Westenbroek, R. E., Merrick, D. K., and Catterall, W. A. (1989). Differential subcellular localization of the R_I and R_{II} Na^+ channel subtypes in central neurons. *Neuron* **3,** 695.

Wier, M. L., and Edidin, M. (1986). Effects of cell density and extracellular matrix on the lateral diffusion of major histocompatibility antigens in cultured fibroblasts. *J. Cell Biol.* **103,** 215.

Wiley, C. A., and Ellisman, M. H. (1980). Rows of dimeric-particles within the axolemma and juxtaposed particles within glia, incorporated into a new model for the paranodal glial axonal junction at the node of Ranvier. *J. Cell Biol.* **84,** 261.

Wiley-Livingston, C. A., and Ellisman, M. H. (1980). Development of axonal membrane specialization defines nodes of Ranvier and precedes Schwann cell myelin elaboration. *Dev. Biol.* **79,** 334.

Wolf, E. E., Henkart, P., and Webb, W. W. (1980). Diffusion, patching, and capping of stearoylated dextrans on 3T3 cell plasma membranes. *Biochemistry* **19,** 3893.

Wollner, D. A., and Catterall, W. A. (1986). Localization of sodium channels in axon hillocks and initial segments of retinal ganglion cells. *Proc. Natl. Acad. Sci. U.S.A.* **83,** 8424.

Wollner, D. A., Scheinman, R., and Catterall, W. A. (1988). Sodium channel expression and assembly during development of retinal ganglion cells. *Neuron* **1,** 727.

Zidovetski, R., Yarden, Y., Schlessinger, Y., and Jovin, T. M. (1986). Microaggregation of hormone-occupied epidermal growth factor receptors on plasma membrane. *EMBO J.* **5,** 247.

CHAPTER 8

Biogenesis of the Mouse Muscle Nicotinic Acetylcholine Receptor

Paul Blount and John Paul Merlie

Departments of Molecular Biology and Pharmacology, Washington University School of Medicine, St. Louis, Missouri 63110

I. INTRODUCTION

The muscle nicotinic acetylcholine receptor (AChR) is a member of a family of molecules also including the neuronal nicotinic acetylcholine (Boulter *et al.*, 1986; Goldman *et al.*, 1987), $GABA_A$ (Schofield *et al.*, 1987; Pritchett *et al.*, 1989), glutamate (Hollmann *et al.*, 1989), and glycine (Grenningloh *et al.*, 1987) receptors. Each of these molecules is composed of several subunits assembled to form a specific receptor/ion channel complex. In addition to obvious questions concerning the mechanisms of assembly of multimeric membrane proteins, this family of ligand-gated channels has the complexity that alternate subunits may be substituted either during development or in a cell-type-specific manner. For example, switching from γ to ε subunit expression conferes a change in AChR channel properties during muscle development (Mishina *et al.*, 1986). Although differential assembly of subunits is perhaps a major mechanism for the expression of receptor subtypes in the central nervous system, studies of the specificity and regulation of assembly in this family of molecules are still in their infancy. The domains responsible for the specific protein–protein interactions that must occur for assembly and the mechanisms that assure that only correctly and completely assembled receptors are expressed on the cell surface have yet to be defined. One goal of our laboratory has been to understand the processing and assembly of this family of molecules. One approach we have taken is to transfect expression vectors encoding subunits of the mouse muscle AChR into a fibroblast cell line that does not normally express any of the AChR subunits. By expressing individual and combinations of subunits in this manner, the processing and functional properties of unassembled or partially assembled complexes have been easily monitored. In addition, by expressing mutated subunits generated by site-directed mutagenesis, we have begun to define amino acids and proposed structural elements that are required for normal processing and assembly. This chapter reviews our work using a fibroblast expression system to study the processing and assembly of AChR subunits in the context of other studies on the structure and function of this molecule.

II. SUBUNIT STRUCTURE AND PROCESSING

A. Some Common Features

The muscle-type AChR was the first well-characterized ligand-gated channel, and remains the prototype for all receptors in this family. Structurally, the AChR is composed of four different but highly homologous subunits (Noda *et al.*, 1983) in a stoichiometry of $\alpha_2\beta\gamma\delta$ (Karlin, 1980) (Fig. 1A) or $\alpha_2\beta\varepsilon\delta$ (Mishina *et al.*, 1986). The α subunit contains all the binding domains for high-affinity binding of α-bungarotoxin (BTX) (Merlie and Lindstrom, 1983; Blount and Merlie, 1988) and some binding domains for agonists and other competitive antagonists (Kao *et al.*, 1984; Dennis *et al.*, 1988; Langenbuch-Cachat *et al.*, 1988; Blount and Merlie, 1989). The two identical α subunits are not juxtaposed (Wise *et al.*, 1981; Zingsheim *et al.*, 1982; Kistler *et al.*, 1982; Bon *et al.*, 1984; Mitra *et al.*, 1989) and, therefore, must be associated with each of the remaining subunits when assembled in the rosette structure of the pentameric complex. Analysis of predicted protein sequences derived from cDNA clones for each of the five subunits of the muscle AChR from several species (Kubo *et al.*, 1985), and

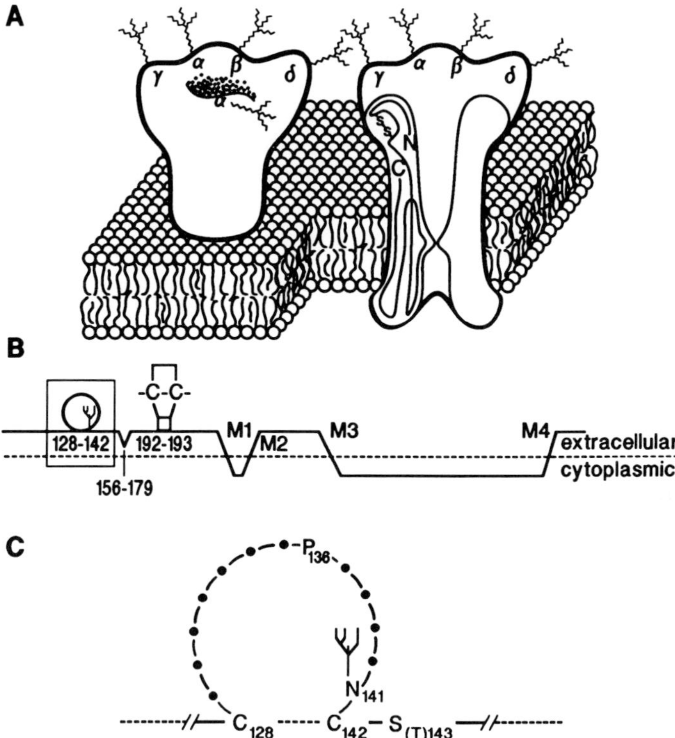

FIG. 1. Structure of the muscle nicotinic AChR and its α subunit. A, AChR in the lipid bilayer. The receptor is a pentameric structure composed of two α subunits and one β, γ (or ε), and δ subunit. In B, some structural features of the α subunit are shown. These include the four highly conserved membrane-spanning domains labeled M1 to M4 and a putative loop structure formed by a disulfide bridge, depicted here as a circle, between the amino acids at positions 128 to 142. The two cysteines located at positions 192 and 193 are found only in the α subunit of muscle and neuronal AChR subunits. An epitope between amino acid positions 156 and 179 has been mapped to the cytoplasmic side of the plasma membrane by antibody binding to *Torpedo* vesicles (Criado *et al.*, 1985; Pedersen and Cohen, 1990a) and is depicted here as a "V" dipping from the extracellular surface. The putative loop structure boxed in B is shown in greater detail in C. The proline at position 136 in the AChR α subunit and the glycosylation site just prior to the second cysteine are highly conserved among the glycine, $GABA_A$, and ACh receptors.

subunits cloned thus far for the neuronal nicotinic ACh (Boulter *et al.*, 1986; Goldman *et al.*, 1987), $GABA_A$ (Schofield *et al.*, 1987; Pritchett *et al.*, 1989), and glycine (Grenningloh *et al.*, 1987) receptors indicates several shared structural features in this family of proteins, suggesting that they all were derived from a common ancestral gene. Four putative transmembrane regions, M1 → M4, and a putative loop structure formed by a disulfide bridge in the NH_2-terminal extracellular domain are all conserved (see Fig. 1B). Within the loop structure itself, there are several conserved features, including a glycosylation site just prior to the second cysteine and a proline approximately midway between the two cysteines (see Fig. 1C). Although the loop structure is highly conserved, its functional role has remained obscure.

B. Conformational Maturation

Pulse-chase labeling experiments with the BC3H-1 muscle cell line demonstrated that high-affinity BTX binding was not an intrinsic property of the α subunit polypeptide chain; newly synthesized α subunit acquired BTX-binding sites in a time-dependent manner referred to as conformational maturation (Merlie and Lindstrom, 1983). A fibroblast clone that expressed the α subunit in the absence of other subunits provided a simplified model system for study of the acquisition of BTX binding (Blount and Merlie, 1988). The results from this cell line confirmed that high-affinity BTX binding was formed independently of the expression of other subunits (Fig. 2). Because of the time course of acquisition of BTX binding and the observation that it did not occur in a cell-free translation system, it was suggested that a covalent modification such as disulfide bond formation was required for the conformational change leading to the high-affinity BTX-binding form of the α subunit (α_{Tx}) (Merlie and Lindstrom, 1983; Merlie *et al.*, 1983). The formation of the loop structure by disulfide bridging of the cysteines at position 128 and 142, or a second disulfide bridge of adjacent cysteines at positions 192 and 193 in the α subunit (Kao and Karlin, 1986) (Fig. 1) seemed likely candidates.

The critical nature of the loop structure formed by residues 128 and 142 was suggested by some of the earliest studies of AChR formed in a *Xenopus* oocyte expression system. When the α, β, γ, and δ or the α, β, ε, and δ AChR subunit mRNAs were expressed in this way, carbamylcholine-inhibitable BTX-binding sites and ACh-gated channels were detected on the oocyte's surface (Mishina *et al.*, 1984, 1986). This expression system was also used to study the properties of mutated α in combination with wild-type β, γ, and δ *Torpedo* AChR subunits (Mishina *et al.*, 1985), demonstrating that mutation of the cysteine at either position 192 or 193 to serine had only a small effect on expression of carbamylcholine-inhibitable BTX-binding sites at the cell surface. However, when the cysteine at position 128 or 142 was mutated to serine, the appearance of AChR detected either as high-affinity BTX-binding sites or as ACh-activatable ion

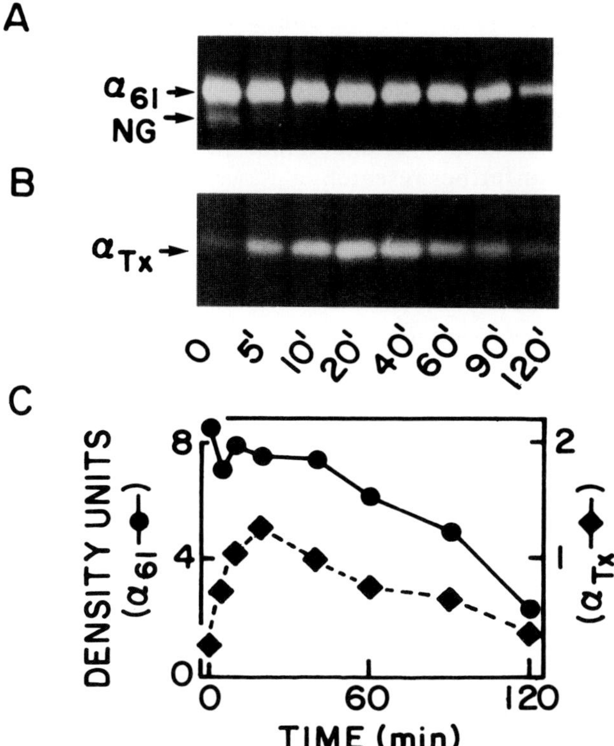

FIG. 2. Kinetics of acquisition of α-bungarotoxin binding of α subunit in transfected fibroblasts. The transfected cell line, Q-α5, was pulse-labeled with [^{35}S]methionine for 5 min and chased for the times indicated. Equal amounts of [^{35}S]methionine-labeled protein were added to each immunoprecipitation. A and B show 5- and 9-day exposures of immunoprecipitated α subunit using mAb61 (α_{61}) (Tzartos *et al.*, 1981) and toxin–antitoxin (α_{Tx}) (Merlie and Sebbane, 1981; Merlie and Lindstrom, 1983), respectively. The initials NG in A indicate the nonglycosylated form of the subunit. Appropriate exposures of α_{61} (●) and toxin–antitoxin (◆) were anlayzed by quantitative densitometry and the results are shown in C. "Density units" are relative units that have been corrected for exposure times and the efficiency of each immunoprecipitation. (From Blount and Merlie, 1988, with permission.)

channels on the surface of the injected oocytes was completely abolished. Because only cell surface AChRs were assayed, these results were inconclusive in determining if the cysteines at 128 and 142 were necessary for formation of high-affinity BTX-binding sites on the α subunit, or, alternatively, if the mutations resulted in inhibition of assembly and/or translocation of AChR to the cell surface. Our more recent data indicate that, in fibroblasts expressing α subunits with mutations of the cysteines at positions 128 and 142 (but not at positions 192 and 193), the α subunit never acquired any high-affinity BTX-binding sites (Blount and Merlie, 1991). Therefore, we suggest that formation of a disulfide bridge between cysteines 128 and 142 is required for normal processing and folding of newly synthesized subunits. Of course, the process of folding such multidomain membrane proteins is not yet understood in detail, and will provide an interesting area for further research.

III. ASSEMBLY

A. *The Dynamics of Normal Assembly*

Many of the initial studies on assembly of AChR subunits were performed in the mouse muscle-like cell line BC3H-1. In experiments with tunicamycin, a drug that inhibits N-linked glycosylation, the formation of high-affinity BTX-binding sites and assembly of α subunit with other subunits were blocked (Merlie *et al.,* 1982). This finding suggested that correct processing and conformational maturation of the α subunit were requirements for efficient assembly. However, since a large fraction of the normally glycosylated α subunit in BC3H-1 cells failed to acquire BTX binding, it was concluded that glycosylation, although necessary, was not sufficient for the acquisition of high-affinity BTX binding. Combined, these findings resulted in the hypothesis that the primary translation product of the α subunit (α_0) was efficiently and cotranslationally glycosylated; a relatively small percentage of the glycosylated subunit was modified by a conformational change (we now believe to be dependent on the disulfide bridging of cysteines 128 and 142) to a high-affinity BTX-binding form (α_{Tx}), and the subunit was subsequently assembled into the pentameric AChR. Conformational maturation and assembly of subunits occur within the endoplasmic reticulum (Smith *et al.,* 1987); the assembled complex is then transported to the cell surface.

$$\alpha_0 \rightarrow \alpha_{Tx} \rightarrow \alpha_2\beta\gamma\delta \rightarrow \text{Surface AChR}$$

The time dependence of the formation of α_{Tx} seemed sufficient to explain the relatively slow assembly observed. However, because experimental intervention in the normal process of α subunit conformational maturation was not possible, the evidence that α_{Tx} was required for subunit assembly relied on kinetic studies. And indeed, these predictions were recently challenged in experiments with tunicamycin-treated *Xenopus* oocytes injected with messenger RNA of the *Torpedo* AChR α, β, γ, and δ subunits; subunits did indeed assemble but were retained in an intracellular compartment (Sumikawa and Miledi, 1989). What, then, are the structural requirements for assembly? How does the cell know not to express improperly processed but assembled subunits on the surface? Before we address these questions, a more basic question must be answered: Why is AChR assembly apparently dependent on glycosylation in BC3H-1 cells, but not in oocytes? As is discussed in the next section, the fibroblast expression system may provided a tool for the resolution of these questions.

B. The Specificity and Regulation of Subunit Assembly

The BC3H-1 cell line and the *Xenopus* oocyte expression system have both been useful in the study of the processing and assembly of the AChR. Each of these systems has advantages and limitations. The BC3H-1 cell line has the advantage that large populations of cells producing receptor at normal metabolic rates can be obtained; therefore, binding and pulse-chase labeling experiments are performed easily. However, this system is not amenable to manipulation of the type or ratio of expressed subunits. Moreover, because assembly intermediates containing more than one but less than five subunits have not been observed in the BC3H-1 cell line, investigations of normal subunit interactions in this system have been limited by their complexity. The oocyte expression system has allowed for the study of mutated subunits (Mishina *et al.,* 1985) and mixtures of subunits in different combination (Kurosaki *et al.,* 1987), but because the time course for expression of AChR in injected oocytes is slow (Kobayashi and Aoshima, 1986) and proteins are synthesized in relatively small amounts, pulse-chase labeling and detailed ligand-binding studies are impractical. In contrast, a fibroblast expression system is amenable to both the manipulation and selective expression of subunits, as well as pulse-chase labeling and ligand-binding studies. We have isolated transfected fibroblast cell lines stably coexpressing α together with either β, γ, or δ subunit (Blount and Merlie, 1989). Coimmunoprecipitation using subunit-specific antibodies demonstrated that the α and γ and the α and δ subunits associated efficiently into $\alpha\gamma$ and $\alpha\delta$ complexes, but cell lines coexpressing the α and β subunits assembled only inefficiently (Fig. 3). Thus, fibroblast cell lines expressing pairs of subunits present an opportunity to study subunit interactions and to access the functional properties of partially assembled AChRs.

One approach used to infer the stoichiometry of free and associated subunits is velocity centrifugation. Sucrose gradient analysis previously demonstrated the presence of at least two forms of high-affinity BTX-binding sites solubilized with Triton X-100 from the BC3H-1 cell line

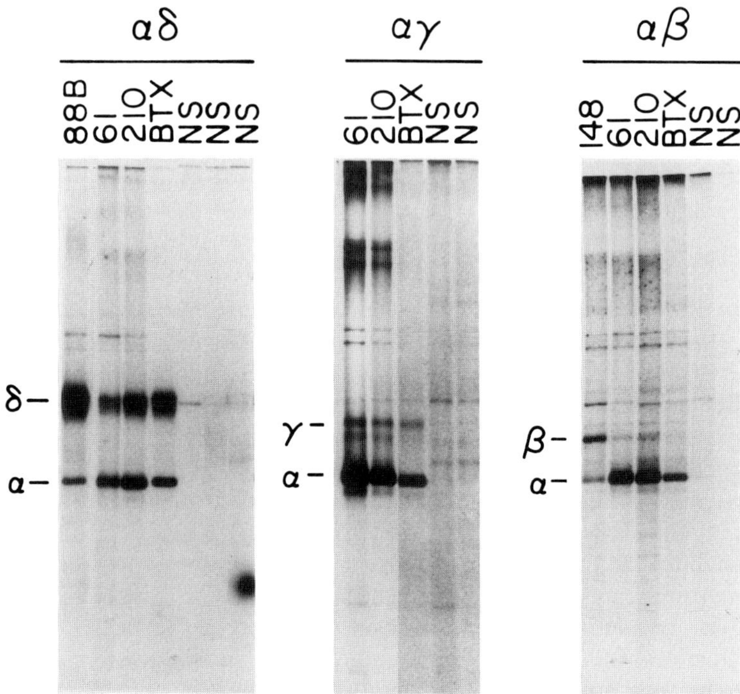

FIG. 3. Association of the δ, γ and β subunits with the α subunit in stably transfected fibroblasts. Transfected QT-6 fibroblasts stably coexpressing the α and δ (αδ), α and γ (αγ), and the α and β (αβ) subunits of the mouse AChR were pulse-labeled for 1 hr with [^{35}S]methionine; immunoprecipitations were performed and assayed by SDS-PAGE. The α-subunit-specific monoclonal antibodies mAb61 (Tzartos *et al.*, 1981) and mAb210 (Ratnam *et al.*, 1986), and a polyclonal antibody against BTX for precipitating prebound toxin (Merlie and Sebbane, 1981; Merlie and Lindstrom, 1983) were all used to specifically immunoprecipitate the α subunit. The δ-subunit-specific monoclonal antibody mAb88B (Froehner *et al.*, 1983) was used to specifically immunoprecipitate the δ subunit of the AChR, and the β-subunit-specific monoclonal antibody mAb148 (Gullick and Lindstrom, 1983) was used to immunoprecipitate β subunit. The amount of subunit labeled in these cell lines in a 1-hr pulse is roughly equivalent to or greater than the amount previously characterized for the BC3H-1 cell line (Merlie *et al.*, 1982). The first NS lane in each panel indicates a nonspecific control in which only the second antibody against the rat monoclonal antibodies and Staph A was added. The second NS lane in each panel indicates precipitation with only the anti-toxin antibody and Staph A. The additional NS lane in the αδ panel is an analogous nonspecific control for the mouse monoclonal antibody mAb88B. High-molecular-weight bands as seen in the αγ panel were observed inconsistently and remain uncharacterized. All subunits comigrated with the subunits observed in the BC3H-1 cell line. (From Blount and Merlie, 1989, with permission.)

(Merlie and Lindstrom, 1983; Carlin *et al.*, 1986). The predominant species with a sedimentation coefficient of 9.5S was derived from pentameric receptor complexes found both intracellularly and at the cell surface. Another species with a sedimentation coefficient of approximately 5S was found exclusively in an intracellular pool. Pulse–chase labeling experiments demonstrated that the newly synthesized 5S species appeared soon after labeling and had the characteristics of a precursor to the fully assembled 9.5S AChR. Moreover, fibroblasts transfected with and expressing only the α subunit produce this 5S BTX-binding form of the α subunit (Blount and Merlie, 1988). Therefore, all data are consistent with the 5S species being the unassembled, probably monomeric, α subunit. Sucrose gradient fractionation of the $\alpha\gamma$ and $\alpha\delta$ subunit complexes demonstrated that discrete and novel assembly intermediates were formed (Blount *et al.*, 1990). Both the $\alpha\gamma$ and $\alpha\delta$ complexes labeled with ^{125}I-labeled BTX had sedimentation coefficients of approximately 6.3S. Assuming the 5S α subunit is monomeric, 6.3S is consistent with a dimer of heterologous $\alpha\gamma$ or $\alpha\delta$ subunit complexes. An additional peak at 8.5S was observed for $\alpha\gamma$ complexes, suggesting that γ may have more than one binding site for the α subunit, leading to the formation of trimers or tetramers. This is a property expected only of the lone subunit between the two α subunits in the pentameric AChR (see Fig. 1); therefore, these data are consistent with models where γ is the lone subunit between the α subunit pair (Karlin *et al.*, 1983; but see Kubalek *et al.*, 1987). Thus, the association of the α and γ, and the α and δ subunits in fibroblasts appears to be specific, leading to the formation of partially assembled AChRs not previously observed by other methods.

Judging from the lack of detectable unassembled AChR subunits expressed on the cell surface (Merlie and Lindstrom, 1983; Carlin *et al.*, 1986), it seems likely that some aspect(s) of assembly or transport must be regulated. Expression systems now allow us to monitor the requirements for surface expression of the subunits of the AChR. Several studies have indicated that expression of the α, β, γ, and δ, or the α, β, ε and δ subunits are necessary for the efficient expression of AChR at the oocyte cell surface (Mishina *et al.*, 1984, 1986; Kurosaki *et al.*, 1987). Similarly, although transfected fibroblasts expressing the *Torpedo* AChR α, β, γ, and δ subunits express functional AChR on their cell surface (Claudio *et al.*, 1987), clonal cell lines expressing the unassembled mouse AChR α subunit, or complexes of $\alpha\gamma$ or $\alpha\delta$ subunits, expressed less than 0.5% of their BTX-binding sites on the cell surface (Blount and Merlie, 1988; Blount *et al.*, 1990). Lectin binding and pharmacological studies suggest that complex oligosaccharides, normally present on the cell surface AChR, are absent from $\alpha\gamma$ and $\alpha\delta$ complexes (Blount *et al.*, 1990). Since the modifications resulting in complex oligosaccharides are performed in the Golgi apparatus, and since γ and δ are the only subunits thought to contain complex oligosaccharides (Nomoto *et al.*, 1986), a likely interpretation is that assembly intermediates are formed in and confined to a pre-Golgi compartment and that only fully formed pentameric AChR is transported through the Golgi to the cell surface. This is consistent with the finding that α subunit maturation and AChR assembly occur in the endoplasmic reticulum (ER) of BC3H-1 cells (Smith *et al.*, 1987). Thus, similar to what has been observed for other multisubunit membrane proteins (Carlin and Merlie, 1986; Klausner, 1989), one mode of regulation to assure that only properly assembled receptors are expressed on the cell surface appears to be retention of incompletely assembled receptors in the ER. Recently, a consensus sequence of the amino acids KDEL at the carboxyl end of a protein has been found to be sufficient for retention in the ER (for review, see Pelham, 1989). Thus, the binding of unassembled subunits, but not assembled receptors, to such a KDEL-containing protein is one proposed mechanism assuring that incompletely assembled multimeric proteins are retained intracellularly.

The remarkable specificity and efficiency of $\alpha\delta$ subunit assembly in fibroblasts has provided us with an assay to determine the ability of mutant α subunits (discussed above) to assemble with other subunits. Mutations that inhibited the glycosylation or the normal processing of the α_0 into the α_{Tx} subunit did not inhibit association of the α and δ subunits (Blount and Merlie, 1991). We found this result surprising given the time-dependent and sequential nature of subunit maturation and assembly. Therefore, we set out to determine if association of mutant α subunit with the wild-type δ subunit was "normal." In fibroblasts coexpressing the wild-type or mutant α subunit with the wild-type δ subunit, pulse-chase labeling with $[^{35}S]$methionine was used to determine the rate of assembly of newly synthesized subunits and the rates of degradation of subunits versus assembled complexes (Fig. 4). These experiments suggested that normal subunit assembly in fibroblasts had a time dependency similar to that seen in the BC3H-1 cell line. The correctly assembled $\alpha\delta$ complex had a very long half-life for degradation (>13 hr) when compared to the unassembled α or δ subunit (about 2 hr). In contrast, mutated α subunits that were not glycosylated or did not acquire the α_{Tx} conformation showed little if any time dependency in the formation of $\alpha\delta$ complexes, and both the $\alpha\delta$ complexes and unassembled subunits were rapidly degraded ($t_{1/2}$s from 20 min for the nonglycosylated subunit to 2 hr for the glycosylated α_0). In the model proposed in Fig. 4, correctly processed and assembled subunits are retained and accumulate in the endoplasmic reticulum, while unassembled subunits and assembled complexes of incorrectly processed subunits are

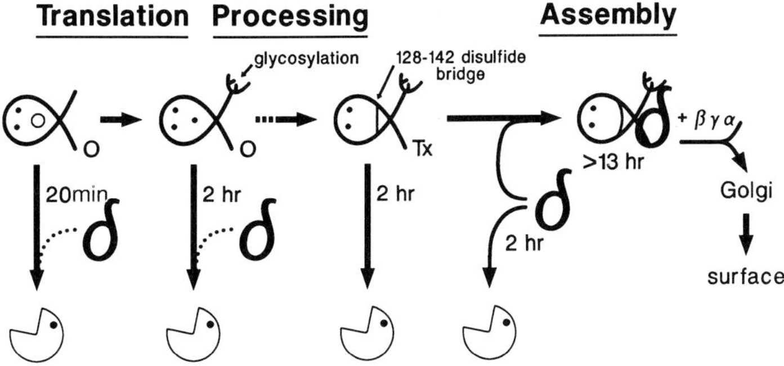

FIG. 4. A simple model indicating the kinetics of degradation for different forms of the α subunit. Evidence suggests that unassembled and partially assembled subunits are confined to a pre-Golgi compartment (Blount *et al.*, 1990) and that subunit processing and assembly normally occur in the endoplasmic reticulum (Smith *et al.*, 1987). The times shown are $t_{1/2}$s for degradation of the unassembled and assembled complexes. Normally, the α subunit is cotranslationally glycosylated, acquires high-affinity BTX binding, is assembled with other subunits, and is transported out of the endoplasmic reticulum. Mutational analysis of the α subunit expressed in fibroblasts and treatment of BC3H-1 cells with tunicamycin have shown that the nonglycosylated subunit has a $t_{1/2}$ for degradation of 20 min. Both the unassembled and assembled nonglycosylated α subunits have this half-life for degradation. Similarly, mutant α subunits that do not acquire high-affinity BTX binding have a 2-hr half-life for degradation in transfected fibroblasts regardless of whether they associate with the δ subunit or not. Processed but unassembled α and unassembled δ subunits have a 2-hr half-life for degradation in transfected fibroblasts. In contrast, the correctly assembled $\alpha\delta$ complex has a half-life for degradation of greater than 13 hr in this cell line. Therefore, correctly processed and assembled subunit complexes are recognized by the cell in a manner different from incorrectly processed and assembled subunit complexes or unassembled subunits.

degraded. In cells expressing four subunits (α, β, γ, and δ subunits, as has been observed in BC3H-1), assembly proceeds until completion of a pentameric AChR, which will then be transported to the Golgi and, ultimately, to the plasma membrane. This model predicts that nonglycosylated subunits should assemble, but that the assembled complexes are short-lived. Since this conclusion contradicts our previous interpretation of the effects of tunicamycin in BC3H-1 cells (see above), we repeated the BC3H-1 experiment, but assayed for assembly by coimmunoprecipitation of subunits from cells pulse-labeled with [^{35}S]methionine and chased for only very short times. Consistent with the mutational analysis, the very short labeling protocol showed that nonglycosylated subunits do assemble in BC3H-1, but the subunit complexes are short lived, being degraded with a half-life of 20 min (Blount and Merlie, 1991). Our recent experiments also provide an explanation for the apparent differences observed in tunicamycin-treated BC3H-1 cells and *Xenopus* oocytes expressing

AChR. Assembly of nonglycosylated subunits occurred but was not previously observed in the BC3H-1 cell line because the sensitivity of the assay was too low to observe accumulation of the rapidly degraded products. In contrast, oocytes have a slow metabolism, resulting in a slow rate of AChR expression (Kobayashi and Aoshima, 1986) and, most likely, a slow rate of degradation of all unassembled and assembled subunit species. Therefore, the accumulation of even relatively rapidly degraded subunits may be more easily observed in the oocyte expression system. The model proposed in Fig. 4 leaves several points unresolved. Why is normal assembly time dependent? Are the subunit–subunit interactions of conformationally immature subunits the same as conformationally mature subunits? Can immature subunits that are assembled into subunit complexes subsequently undergo conformational maturation? Although these and other questions remain unanswered, one conclusion that seems certain is that correctly processed and assembled subunit complexes are recognized by the cell transport and degradation machinery in a manner different from incorrectly processed and assembled subunit complexes or unassembled subunits. These differences must therefore be important aspects of the editing or regulation process that result in the expression of only normal and functional AChRs on the cell surface.

C. The Functional Properties of Assembly Intermediates

In an attempt to determine the role of the α subunit in ligand binding, a series of experiments were performed on membranes isolated from a fibroblast cell line expressing the unassembled α subunit (Blount and Merlie, 1988). These experiments demonstrated that the α subunit contained all of the domains for high-affinity BTX binding, but had low or undetectable affinity for small ligands, including the competitive antagonist *d*-tubocurarine and the agonist carbamylcholine. In order to determine whether other subunits were necessary for normal ligand binding, membranes isolated from fibroblasts stably expressing pairs of subunits (wild-type α with one other subunit) were assayed for binding of small ligands (Blount and Merlie, 1989). These studies demonstrated that the $\alpha\gamma$ and $\alpha\delta$ complexes formed distinctly different high-affinity binding sites for the competitive antagonist *d*-tubocurarine that, together, completely accounted for the two nonequivalent antagonist-binding sites in native AChR (Table I). The $\alpha\delta$ complex and native AChR had similar affinities for the agonist carbamylcholine. In contrast, although the $\alpha\gamma$ complex contains the higher affinity competitive antagonist-binding site, it had an affinity for carbamylcholine an order of magnitude less than that of the $\alpha\delta$ complex or

TABLE I

Quantitative Analysis of Relative Affinities for Agonist and Competitive Antagonist to Membranes Prepared from the Cell Types Producing Various Subunits of the AChR

	IC_{50} (M)[a]	
Cell Type[b]	*D*-Tubocurarine	Carbamylcholine
α	2.3×10^{-4}	$>10^{-2}$
$\alpha\beta$	2.2×10^{-4}	$>10^{-2}$
$\alpha\gamma$	1.9×10^{-8}	1.3×10^{-5}
$\alpha\delta$	2.8×10^{-7}	1.0×10^{-6}
AChR (BC3H-1)	$2.7 \times 10^{-8}/2.6 \times 10^{-7}$	1.7×10^{-6}

[a] Calculated affinities assuming a two-site model for receptor binding of *d*-tubocurarine to AChR and a one-site model for all other calculations. From data published in Blount and Merlie (1989).

[b] The cell types indicate QT-6 fibroblasts stably transfected with and expressing the α, $\alpha\beta$, $\alpha\gamma$, or $\alpha\delta$ subunits of the AChR. Data from the AChR produced in the BC3H-1 cell line are also presented for comparison.

of the AChR. These data supported a model of two nonequivalent binding sites within the AChR and implied that the molecular basis for this nonequivalence was the association of the α subunit with the γ or δ subunit. These findings, combined with binding properties of the native AChR expressed in membranes isolated from BC3H-1 cells (Sine and Taylor, 1980, 1981; Blount and Merlie, 1989), led to a model for the binding of competitive antagonists and agonists to the AChR (Fig. 5). Two nonequivalent sites located on an α subunit combined with either the γ or δ subunit are proposed to exist for competitive antagonists. The contribution of γ and δ subunits to the antagonist-binding sites is also suggested by affinity-labeling experiments (Langenbuch-Cachat *et al.*, 1988; Pedersen and Cohen, 1990b). In our model, the two binding sites have different affinities for competitive antagonist, with $\alpha\gamma$ having the higher of the two. There is little or no cooperativity between the two competitive antagonist-binding sites. Analogous to the situation for antagonists, agonist binding is also dependent on the association of α with the γ or δ subunits, giving rise to two nonequivalent agonist-binding sites as previously proposed (Sine and Taylor, 1980). Cooperativity in agonist binding has been proposed by several laboratories (Weber and Changeux, 1974; Neubig and Cohen, 1979; Sine and Taylor, 1980). Therefore, our data suggest that the lower affinity agonist-binding site ($\alpha\gamma$) must be allosterically regulated by binding of agonist to the higher affinity ($\alpha\delta$) site. The reverse order of affinities of agonist and antagonist for the $\alpha\gamma$ and $\alpha\delta$ sites is not inconsistent with binding data from the AChR, but would have been difficult to deduce from

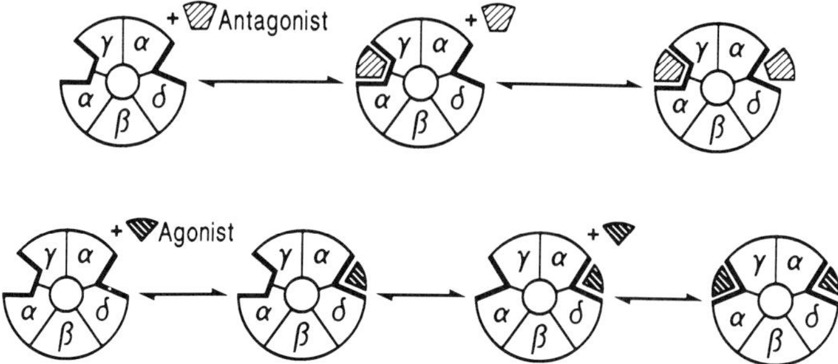

FIG. 5. A plausible model for the binding of competitive antagonists and agonists to the native AChR. For competitive antagonists, two nonequivalent sites located on an α subunit combined with either the γ or δ subunit exist. Evidence suggests that the γ and δ subunits are close to or, as depicted here, contribute to the antagonist binding site (Langenbuch-Cachat *et al.*, 1988; Pedersen and Cohen, 1990b). The two binding sites have different affinities for antagonist, with $\alpha\gamma$ having the higher of the two. There is little or no cooperativity between the two competitive antagonist-binding sites. Analogous to the situation for antagonists, agonist binding is also dependent on the association of α with the γ or δ subunits. Two nonequivalent agonist-binding sites are proposed to exist. To account for cooperativity in agonist binding (Weber and Changeux, 1974; Neubig and Cohen, 1979; Sine and Taylor, 1980), the lower affinity agonist-binding site ($\alpha\gamma$) is shown to be allosterically regulated by binding of agonist to the higher affinity ($\alpha\delta$) site. The placement of the subunits relative to each other is consistent with our data on the functional homology of γ and δ, with intuitive ideas about protein symmetry, and with the sedimentation coefficients obtained for the $\alpha\gamma$ complexes, but may not be the only interpretation. (From Blount and Merlie, 1989, with permission.)

native receptor-binding data alone. Thus, by studying the binding properties of assembly intermediates we have obtained a better understanding of the molecular basis of ligand binding to the intact AChR.

IV. CONCLUSIONS

We have used the expression of individual subunits and combinations of subunits in fibroblasts to study the processing and assembly of the AChR. Mutational analysis suggests that the acquisition of high-affinity BTX binding is due to the formation of a disulfide bridge between two cysteines in the α subunit. Because this processing step is time-dependent and relatively inefficient, it remains a possible site of regulation for AChR expression. The assembly of α_{Tx} with other subunits and transport to the cell surface of the AChR appear to be specific and highly regulated. Discrete assembly intermediates are formed when the α and γ or the α and δ subunits are coexpressed. These assembly intermediates are formed in the endoplasmic reticulum and are apparently confined there until a complete AChR is formed. Mutational analysis suggests that incompletely processed subunits can assemble, but the incorrect assembly intermediates are rapidly degraded. The structural domains responsible for the association of subunits and the signals that allow for the cell to differentiate unassembled and incorrectly assembled subunits from correctly assembled subunits remain to be investigated. Thus, the fibroblast expression system has allowed for the study of normal processing and assembly of the AChR. By manipulating this system, we have gained insight into the mechanism that determines that only properly assembled, functional AChR is expressed on the cell surface.

Selective expression of subunits in fibroblasts has contributed to our understanding of the functional role of subunits in the ligand-binding properties of the intact AChR. The assembly intermediates formed by association of the α and γ and the α and δ subunits account for the two different binding sites previously observed in the AChR (Neubig and Cohen, 1979; Sine and Taylor, 1980, 1981), thereby suggesting a model for the molecular basis of the nonequivalence of the two ligand-binding sites.

Acknowledgments

PB was supported by National Research Service Award 2 T32 GM 07805. This work was also supported by funds from the Senator Jacob Javits Center of Excellence in the Neurosciences and research grants from the National Institutes of Health and the Muscular Dystrophy Associations of America.

References

Blount, P., and Merlie, J. P. (1988). Native folding of an acetylcholine receptor α subunit expressed in the absence of other subunits. *J. Biol. Chem.* **263,** 1072–1080.

Blount, P., and Merlie, J. P. (1989). Molecular basis of the two nonequivalent ligand binding sites of the muscle nicotinic acetylcholine receptor. *Neuron* **3,** 349–357.

Blount, P., and Merlie, J. P. (1990). Mutational analysis of muscle nicotinic aceylcholine receptor subunit assembly. *J. Cell Biol.* **111,** 2613–2622.

Blount, P., Smith, M. M., and Merlie, J. P. (1990). Assembly intermediates of the mouse muscle nicotinic acetylcholine receptor in stably transfected fibroblasts. *J. Cell. Biol.* **111,** 2601–2611.

Bon, F., Lebrun, E., Gomel, J., Rapenbusch, R. V., Cartaud, J., Popot, J.-L., and Changeux, J.-P. (1984). Image analysis of the heavy form of the acetylcholine receptor from *Torpedo marmorata. J. Mol. Biol.* **176,** 205–237.

Boulter, J., Evans, K., Goldman, D., Martin, G., Treco, D., Heinemann, S., and Patric, J. (1986). Isolation of a cDNA clone coding for a possible neuronal nicotinic acetylcholine receptor α-subunit. *Nature (London)* **319,** 368–374.

Carlin, B. E., and Merlie, J. P. (1986). Assembly of multisubunit proteins. *In* "Protein Comparmentalization" (A. Strauss, I. Boime, and G. Kreil, eds.), pp. 71–86. Springer Verlag, New York.

Carlin, B. E., Lawrence, J. C., Lindstrom, J. M., and Merlie, J. P. (1986). An acetylcholine receptor precursor α subunit that binds α-bungarotoxin but not *d*-tubocurarine. *Proc. Natl. Acad. Sci. U.S.A.* **83,** 498–502.

Claudio, T., Green, W. N., Hartman, D. S., Hayden, D., Paulson, H. L., Sigworth, F. J., Sine, S. M., and Swedlund, A. (1987). Genetic reconstitution of functional acetylcholine receptor channels in mouse fibroblasts. *Science* **238,** 1688–1694.

Criado, M., Hochschwender, S., Sarin, V., Fox, J. L., and Lindstrom, J. (1985). Evidence for unpredicted transmembrane domains in acetylcholine receptor subunits. *Proc. Natl. Acad. Sci. U.S.A.* **82,** 2004–2008.

Dennis, M., Giraudat, J., Kotzyba-Hibert, F., Goeldner, M., Hirth, C., Chang, J.-Y., Lazure, C., Chretien, M., and Changeux, J.-P. (1988). Amino acids of the *Torpedo marmorata* acetylcholine receptor α subunit labeled by a photoaffinity ligand for acetylcholine binding site. *Biochemistry* **27,** 2346–2357.

Froehner, S. C., Douville, K., Klink, S., and Clup, W. J. (1983). Monoclonal antibodies to cytoplasmic domains of the acetylcholine receptor. *J. Biol. Chem.* **258,** 7112–7120.

Goldman, D., Deneris, E., Luyten, W., Kochhar, A., Patrick, J., and Heinemann, S. (1987). Members of a nicotinic acetylcholine receptor gene family are expressed in different regions of the mammalian central nervous system. *Cell* **48,** 965–973.

Green, W. N., Ross, A. F., and Claudio, T. (1991). cAMP stimulation of acetylcholine receptor expression is mediated through posttranslational mechanisms. *Proc. Natl. Acad. Sci. U.S.A.* **88,** 854–858.

Grenningloh, G., Reinitz, A., Schmitt, B., Methfessel, C., Zensen, M., Beyreuther, K., Gundelfinger, E. D., and Betz, H. (1987). The strychnine-binding subunit of the glycine receptor shows homology with nicotinic acetylcholine receptors. *Nature (London)* **328,** 215–220.

Gu, Y., Camacho, P., Gardner, P. D., and Hall, Z. W. (1991). Identification of two amino acid residues in the ε subunit that promote mammalian muscle acetylcholine receptor assembly in COS cells. *Neuron* **6,** 879–887.

Gullick, W. J., and Lindstrom, J. M. (1983). Mapping the binding of monoclonal antibodies to the acetylcholine receptor. *Biochemistry* **22,** 3312–3320.

Hollmann, M., O'Shea-Greenfield, A., Rogers, S. W., and Heinemann, S. (1989). Cloning by functional expression of a member of the glutamate receptor family. *Nature (London)* **342,** 643–648.

Kao, P. N., and Karlin, A. (1986). Acetylcholine receptor binding site contains a disulfide crosslink between adjacent half-cystinyl residues. *J. Biol. Chem.* **261,** 8085–8088.

Kao, P. N., Dwork, A. J., Kaldany, R.-R. J., Silver, M. S., Wideman, J., Stein, S., and Karlin, A. (1984). Idtification of the α subunit half-cystine specifically labeled by an affinity reagent for the acetylcholine receptor binding site. *J. Biol. Chem.* **259,** 11662–11665.

Karlin, A. (1980). Molecular properties of nicotinic acetylcholine receptors. *In* "Cell Surface and Neuronal Function" (C. W. Cotman, ed.), pp. 191–260. Elsevier/North-Holland, New York.

Karlin, A., Holtzman, E., Yodh, N., Lobel, P., Wall, J., and Hainfeld, J. (1983). The arrangement of the subunits of the actylcholine receptor of *Torpedo californica*. *J. Biol. Chem.* **258,** 6678–6681.

Kistler, J., Stroud, R. M., Klymkowsky, M. W., Lalancett, R. A., and Fairclough, R. H. (1982). Structure and function of an acetylcholine receptor. *Biophys. J.* **37,** 371–383.

Klausner, R. D. (1989). Architectural editing: Determining the fate of newly synthesized membrane proteins. *New Biol.* **1,** 3–8.

Kobayashi, S., and Aoshima, H. (1986). Time course of the induction of acetylcholine receptors in *Xenopus* oocytes injected with mRNA from *Electrophorus electricus* electroplax. *Dev. Brain Res.* **24,** 211–216.

Kubalek, E., Ralston, S., Lindstrom, J., and Unwin, N. (1987). Location of subunits within the acetylcholine receptor by electron image analysis of tubular crystals from *Torpedo marmorata*. *J. Cell Biol.* **105,** 9–18.

Kubo, T., Noda, M., Takai, T., Tanabe, T., Kayano, T., Shimizu, S., Tanaka, K., Takahashi, H., Hirose, T., Inayama, S., Kikuno, R., Miyata, T., and Numa, S. (1985). Primary structure of δ subunit precursor of calf muscle acetylcholine receptor deduced from cDNA sequence. *Eur. J. Biochem.* **149,** 5–13.

Kurosaki, T., Fukuda, K., Konno, T., Mori, Y., Tanaka, K., Mishina, M., and Numa, S. (1987). Functional properties of nicotinic acetylcholine receptor subunits expressed in various combinations. *FEBS Lett.* **214,** 253–258.

Langenbuch-Cachat, J., Bon, C., Mulle, C., Goeldner, M., Hirth, C., and Changeux, J.-P. (1988). Photoaffinity labeling of the acetylcholine binding sites on the nicotinic receptor by an aryldiazonium derivative. *Biochemistry* **27,** 2337–2345.

Merlie, J. P., and Lindstrom, J. (1983). Assembly *in vivo* of mouse muscle acetylcholine receptor: Identification of an α subunit species that may be an assembly intermediate. *Cell* **34,** 747–757.

Merlie, J. P., and Sebbane, R. (1981). Acetylcholine receptor subunits transit a precursor pool before acquiring α-bungarotoxin binding activity. *J. Biol. Chem.* **256,** 3605–3608.

Merlie, J. P., Sebbane, R., Tzartos, S., and Lindstrom, J. (1982). Inhibition of glycosylation with tunicamycin blocks assembly of newly synthesized acetylcholine receptor subunits in muscle cells. *J. Biol. Chem.* **257,** 2694–2701.

Merlie, J. P., Sebbane, R., Gardner, S., Olson, E., and Lindstrom, J. (1983). The regulation of acetylcholine receptor expression in mammalian muscle. *Cold Spring Harbor Symp. Quant. Biol.* **48,** 135–146.

Mishina, M., Kurosaki, T., Tobimatsu, T., Morimoto, Y., Noda, M., and Numa, S. (1984). Expression of functional acetylcholine from cloned cDNAs. *Nature (London)* **307,** 604–608.

Mishina, M., Tobimatsu, T., Imoto, K., Tanaka, K., Fujita, Y., Fukuda, K., Kurasaki, M., Takahashi, H., Morimoto, Y., Hirose, T., Inayama, S., Takahashi, T., Kuno, M., and Numa, S. (1985). Location of functional regions of acetylcholine receptor α subunit by site-directed mutagenesis. *Nature (London)* **313,** 364–369.

Mishina, M., Takai, T., Imoto, K., Noda, M., Takahashi, T., Numa, S., Methfessel, C., and Sakmann, B. (1986). Molecular distinction between fetal and adult forms of muscle acetylcholine receptor. *Nature (London)* **321,** 406–411.

Mitra, A. K., McCarthy, M. P., and Stroud, R. M. (1989). Three dimensional structure of the nicotinic acetylcholine receptor and localization of the major associated 43-kD cytoskeletal protein, determined at 22 Å by low dose electron microscopy and X-ray diffraction to 12.5 Å. *J. Cell Biol.* **109,** 755–774.

Neubig, R. R., and Cohen, J. B. (1979). Equilibrium binding of [^{3}H]tubocurarine and [^{3}H]acetylcholine by *Torpedo* postsynaptic membranes: Stoichiometry and ligand interactions. *Biochemistry* **18,** 5464–5475.

Noda, M., Takahashi, H., Tanabe, T., Toyosato, M., Kikyotani, S., Furutani, Y., Hirose, T., Takashima, H., Inayama, S., Miyata, T., and Numa, S. (1983). Structural homology of *Torpedo californica* acetylcholine receptor subunits. *Nature (London)* **302,** 528–532.

Nomoto, H., Takahashi, N., Nagaki, Y., Endo, S., Arata, Y., and Hayashi, K. (1986). Carbohydrate structures of acetylcholine receptor from *Torpedo californica* and distribution of oligosaccharides among the subunits. *Eur. J. Biochem.* **257,** 233–242.

Pedersen, S. E., and Cohen, J. B. (1990a). Identification of a cytoplasmic region of the *Torpedo* nicotinic acetylcholine receptor α subunit by epitope mapping. *J. Biol. Chem.* **265,** 569–581.

Pedersen, S. E., and Cohen, J. B. (1990b). *d*-Tubocurarine binding sites are located at α-γ and

α-δ subunit interfaces of the nicotinic acetylcholine receptor. *Proc. Natl. Acad. Sci. U.S.A.* **87,** 2785–2789.

Pelham, H. R. B. (1989). Heat shock and the sorting of luminal ER proteins. *EMBO J.* **8,** 3171–3176.

Pritchett, D. B., Sontheimer, H., Shivers, B. D., Ymer, S., Kettenmann, H., Schofield, P. R., and Seeburg, P. H. (1989). Importance of a novel $GABA_A$ receptor subunit for benzodiazepine pharmacology. *Nature (London)* **338,** 582–585.

Ratnam, M., Sargent, P. B., Sarin, V., Fox, J. L., Le Nguyen, D., River, J., Criado, M., and Lindstrom, J. (1986). Location of antigenic determinants on primary sequences of subunits of nicotinic acetylcholine receptor by peptide mapping. *Biochemistry* **25,** 2621–2632.

Ross, A. F., Green, W. N., Hartman, D. S., and Claudio, T. (1991). Efficiency of acetylcholine receptor subunit assembly and its regulation by cAMP. *J. Cell Biol.* **113,** 623–636.

Schofield, P. R., Darlison, M. G., Fujita, N., Burt, D. R., Stephenson, F. A., Rodriguez, H., Rhee, L. M., Ramachandren, J., Reale, V., Glencorse, T. A., Seeburg, P. H., and Barnard, E. A. (1987). Sequence and functional expression of the $GABA_A$ receptor shows a ligand-gated receptor super-family. *Nature (London)* **328,** 221–22.

Sine, S. M., and Taylor, P. (1980). The relationship between agonist occupation and the permeability response of the cholinergic receptor revealed by bound cobra α-toxin. *J. Biol. Chem.* **255,** 10144–10156.

Sine, S. M., and Taylor, P. (1981). Relationship between reversible antagonist occupancy and the functional capacity of the acetylcholine receptor. *J. Biol. Chem.* **255,** 10144–10156.

Smith, M. M., Lindstrom, J., and Merlie, J. P. (1987). Formation of the α bungarotoxin binding site and assembly of the nicotinic acetylcholine receptor subunits occur in the endoplasmic reticulum. *J. Biol. Chem.* **262,** 4367–4376.

Sumikawa, K., and Miledi, R. (1989). Assembly of *N*-glycosylation of all ACh receptor subunits are required for their efficient insertion into plasma membranes. *Mol. Brain Res.* **5,** 183–192.

Tzartos, S. J., Rand, D. E., Einarson, B. L., and Lindstrom, J. M. (1981). Mapping of surface structures of electrophorous acetylcholine receptor using monoclonal antibodies. *J. Biol. Chem.* **256,** 8635–8645.

Weber, M., and Changeux, J.-P. (1974). Binding of *Naji nigrocollis* [^{3}H]α-toxin to membrane fragments from *Electrophorus* and *Torpedo* electric organs. II. Effect of cholinergic agonists and antagonists on the binding of the tritiated α-neurotoxin. *Mol. Pharmacol.* **10,** 15–34.

Wise, D. S., Wall, J., and Karlin, A. (1981). Relative locations of the β and δ chains of the actylcholine receptor determined by electron microscopy of isolated receptor trimer. *J. Biol. Chem.* **256,** 12624–12627.

Yu, X-M., and Hall, Z. W. (1991). Extracellular domains mediating ε subunit interactions of muscle acetylcholine receptor. *Nature (London)* **352,** 64–67.

Zingsheim, H. P., Barrantes, F. F., Frank, J., Hanicke, W., and Neugebauer, D.-C. (1982). Direct structural localization of two toxin-recognition sites on an ACh receptor protein *Nature (London)* **299,** 81–84.

Note Added in Proof

Four important papers on the potential regulation of (Green *et al.*, 1991; Ross *et al.*, 1991), and protein domains required for (Yu and Hall, 1991; Gu *et al.*, 1991), AChR assembly were published subsequent to the submission of this manuscript.

CHAPTER 9

Functional Properties of Voltage-Dependent Calcium Channels

Edwin W. McCleskey and Jean E. Schroeder
Department of Cell Biology and Physiology, Washington University School of Medicine, St. Louis, Missouri 63110

I. INTRODUCTION

Voltage-dependent calcium channels (Ca^{2+} channels) are present on all excitable cells and are central to all biological processes which occur on a time scale of seconds or less, e.g., sensations, motions, higher brain functions, and the release of many hormones. They are molecular pores on the surface membrane that open in response to depolarization of the membrane voltage and selectively allow Ca^{2+} to enter the cell. The entering Ca^{2+} can perform a variety of tasks. The first, temporally, is to provide additional positive charge to depolarize the cell membrane further. Thus, after sensing an electrical depolarization, Ca^{2+} channels act to amplify it to still more positive voltages. When the accumulating Ca^{2+} reaches a sufficient concentration, it can activate ion channels such as Ca^{2+}-activated K^+ channels that allow positive charge out the cell and thereby repolarize the membrane. Thus, Ca^{2+} channels serve as elements that can sense, amplify, and terminate electrical signals.

But the most unique and crucial role of Ca^{2+} channels is to translate the electrical signal on the surface membrane into a chemical signal within the cytoplasm. The entering Ca^{2+} serves as the cytoplasmic trigger for movement, secretion, or a variety of enzymatic processes by binding to different Ca^{2+}-sensitive proteins such as troponin or calmodulin. The creation of a Ca^{2+} flux is the only known way by which a surface membrane electrical signal can get a job started within a cell.

Calcium channels were discovered in 1958 by Fatt and Ginsborg when they explored the ionic basis of a Na^+-independent action potential in crab muscle. They showed that the action potential was dependent on the concentration of Ca^{2+}, that Sr^{2+} or Ba^{2+} could substitute for Ca^{2+}, that Mn^{2+} was a blocker, and that Mg^{2+} was impermeant and a weak blocker at best. Ca^{2+}, Sr^{2+}, and Ba^{2+} remain as the only unequivocally permeant divalent ions of Ca^{2+} channels, and a wide variety of multivalent ions are blockers.

The broad significance of Ca^{2+} channels was appreciated as Hagiwara and collaborators, starting in the mid 1960s, performed a 20-year comparative survey of cells throughout different organs and phyla. Hagiwara's body of work serves as the foundation for the Ca^{2+} channel field and has been summarized in a monograph (Hagiwara, 1983). Among his discoveries was the essential role of the Ca^{2+} channel in the prolonged action potential of the heart.

This chapter describes the functional properties of Ca^{2+} channels (selectivity, activation, inactivation, and facilitation) and discusses the issues regarding multiple types of Ca^{2+} channels. There is no attempt to be exhaustive in the literature survey and no systematic coverage of the pharmacology and modulation of Ca^{2+} channels, which would justify a separate review. Other recent reviews covering Ca^{2+} channels include those by Hess (1990), Bean (1989a), and Tsien *et al.* (1987). The earlier work in the field is covered in Hagiwara and Byerly (1981). In addition, in *Trends in Neurosciences,* two special issues of review articles have been published on Ca^{2+} channels (October, 1988) and on Ca^{2+}-sensitive metabolic events (October, 1989).

II. Ca^{2+} SELECTIVITY

A. Permeability versus Selectivity

The term "Ca^{2+} channel" is sometimes used loosely to refer to any channel that lets Ca^{2+} pass. Here, we use the term strictly to refer to those channels which select Ca^{2+} and reject other ions. Many ion channels, such as the NMDA (*N*-methyl-D-aspartate) receptor and the acetycholine receptor, are Ca^{2+}-permeable, but will not be called Ca^{2+} channels since they readily pass other physiological ions such as Na^+ and K^+. The Ca^{2+} channel prefers Ca^{2+} over Na^+ at a ratio of 1000 : 1, the highest known selectivity ratio for an ion channel (Lee and Tsien, 1982; Hess *et al.*, 1986). Compare this to the 10 : 1 selectivity ratio for Na^+ over K^+ in voltage-dependent Na^+ channels (Hille, 1984). The high selectivity of Ca^{2+} channels allows exclusive passage of a minority ion and can be considered their defining feature.

B. The Role of a High-Affinity Ca^{2+} Binding Site

The extraordinary selectivity for Ca^{2+} over Na^+ occurs even though the naked ions are identical in size (2 Å diameter) and are both positively charged. It is unlikely that rejection from a simple molecular sieve, as proposed in other channels, could confer such high selectivity between such similar ions. The discovery that the diameter of the pore of the Ca^{2+} channel is nearly three times that of Ca^{2+} and Na^+ ions seems to exclude sieving as a mechanism for selectivity (McCleskey and Almers, 1985).

Is Ca^{2+} selected through its affinity for an intrapore site? Working on barnacle muscle, Hagiwara's laboratory provided early evidence for this mechanism. Normally, the current carried by Ba^{2+} through the barnacle Ca^{2+} channels is larger than that carried by the same concentration of Ca^{2+}. However, in the presence of a blocking ion (cobalt) at a concentration causing only incomplete blockade, Ca^{2+} gave a larger current than Ba^{2+} (Hagiwara *et al.*, 1974). Hagiwara *et al.* proposed that translocation through the pore involved binding to a site for which Ca^{2+} had greater affinity than Ba^{2+}; blocking ions could bind to the site without permeating. Ca^{2+} would compete more successfully with the blockers than would Ba^{2+}, thereby causing the reversal of current amplitude in the presence of cobalt. Further evidence for a transport site within the pore is that Ca^{2+} flux saturates when Ca^{2+} concentration increases above 10 m*M* (Hagiwara and Takahashi, 1967).

The modern era of the study of Ca^{2+} channel selectivity began when Kostyuk and Krishtal, working in Kiev, recorded Ca^{2+} channel activity in the presence of ethylenediaminetetraacetic acid (EDTA) (to chelate completely all extracellular Ca^{2+}) (Kostyuk and Krishtal, 1977). Figure 1 shows results from skeletal muscle Ca^{2+} channels that are similar to the Kiev work on molluskan neurons. The five traces show currents through a cell's Ca^{2+} channels under five different ionic conditions which vary in Na^{+} concentration (indicated on the left) or Ca^{2+} concentration (indicated in negative log units below). When there is 10 m*M* Ca^{2+} present (left pair of traces), addition or subtraction of 32 m*M* Na^{+} is virtually without effect, i.e., Na^{+} is invisible to the Ca^{2+} channel when Ca^{2+} is present. When the Ca^{2+} concentration is dropped to about 100 μM in the middle traces, the Ca^{2+} current is largely lost, as expected. However, when the Ca^{2+} concen-

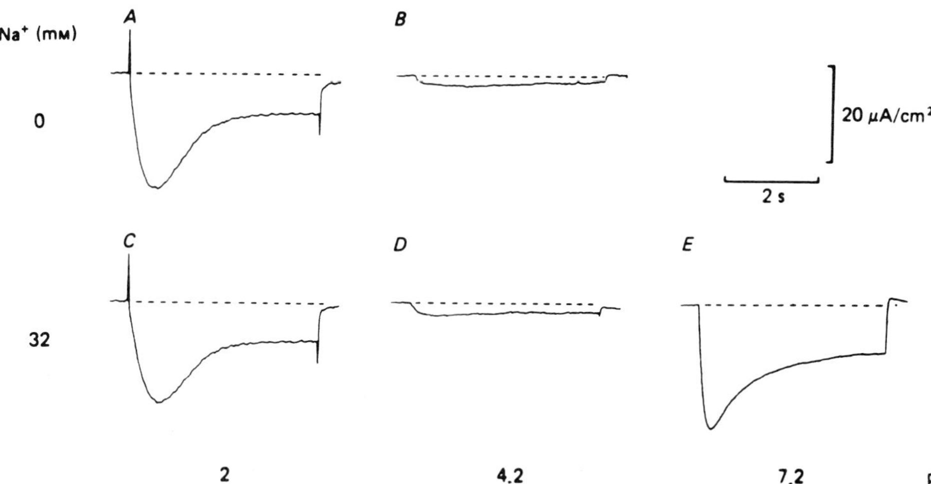

FIG. 1. Current through Ca^{2+} channels of a skeletal muscle cell under five different ionic conditions. Na^{+} concentration is indicated on the left and Ca^{2+} concentration is indicated below in negative log units. Na^{+} does not pass through the Ca^{2+} channel when there is high Ca^{2+} (left traces). When Ca^{2+} concentration drops below 1 μM, Na^{+} becomes permeant. From Almers *et al.* (1984).

tration is dropped to submicromolar concentrations, a novel current appears. This new current is carried by Na^+, and a variety of kinetic and pharmagological experiments demonstrate that it passes through Ca^{2+} channels (Kostyuk *et al.*, 1983; Almers *et al.* 1984).

The observation that Na^+ and other monovalent ions could readily pass through Ca^{2+} channels under extreme conditions was largely ignored until it was shown to involve a high-affinity Ca^{2+}-binding site on the Ca^{2+} channel. In the absence of Ca^{2+}, many monovalent ions can pass through the channel, but their flux is blocked when Ca^{2+} from the extracellular side binds to a site which has a dissociation constant of about 1 μM (Kostyuk *et al.*, 1983; Almers *et al.*, 1984). Thus, the selectivity of the Ca^{2+} channel is conferred on it by Ca^{2+} itself. Extracellular Ca^{2+} acts at a site of such high affinity that it would be permanently saturated under physiological conditions. This curious observation became problematic if one proposed that the site was within the pore of the channel. The maximum off-rate from such a high-affinity site is about 1000/sec. This is 100 times less than the measured flux rate through single Ca^{2+} channels and would be a lower flux than any ion channel. Such a pore would be selective for Ca^{2+}, but it would be too sticky to give the observed flux rates.

Evidence that the site is indeed within the pore came from measurements of the voltage dependence of the Ca^{2+} blockade. If a blocking ion must penetrate within the pore, the electric field across the pore will affect access and exit from the site. An inside-positive voltage will impede access of a blocking ion from the extracellular medium and speed exit of the blocker into the intracellular medium. Fukushima and Hagiwara (1985) documented this with whole-cell measurements of Ca^{2+} channel currents in a cell line, and Lansman *et al.* (1986) showed it at the single-channel level on cardiac Ca^{2+} channels. The single-channel measurements described blocking events by individual Ca^{2+} ions, the first and only measurement of the unitary interaction of a channel with its physiologically preferred ion.

C. Models of Permeation and Selectivity

A solution to the sticky pore problem was proposed by Hess and Tsien (1984) and Almers and McCleskey (1984). Observations on Ca^{2+} channel permeation and selectivity were quantitatively fit by a model in which the pore contained a pair of high-affinity Ca^{2+}-binding sites rather than just a single site. Ions are required to hop sequentially from one site to the other and multiple occupancy is disfavored because of electrostatic repulsion between similarly charged ions within the same pore. Thus, the binding of a single Ca^{2+} is stable and acts to guard the channel against foreign ions. The binding of a second Ca^{2+} ion effectively lessens the affinity of the pore for Ca^{2+} because of the powerful repulsion between pairs of divalent ions. Occupancy of the second site thereby promotes the exit of a Ca^{2+} ion; Ca^{2+} flux is created while leaving the pore singly occupied and still guarded against the entry of foreign ions. Quantitatively, the degree of repulsion which led to a fit of the data corresponded to a distance between Ca^{2+}-binding sites of 10 Å (assuming a dielectric constant of 20 within the pore). This is nearly identical to the distance between neighboring Ca^{2+} sites on the calmodulin molecule.

Figure 2 provides a cartoon view of the behavior of such a pore when exposed to three very different Ca^{2+} concentrations. Below 1 μM Ca^{2+}, monovalent ions readily pass through the channel since it is unoccupied by Ca^{2+}. Monovalent flux is blocked above 1 μM Ca^{2+}. Significant Ca^{2+} flux does not occur until Ca^{2+} rises to the millimolar range and the pore can become occupied by two Ca^{2+} ions. Though the model proposes only one kind of binding site in the pore, the pore shows two affinities for Ca^{2+}. When singly occupied, the pore has high affinity for Ca^{2+}, as measured by

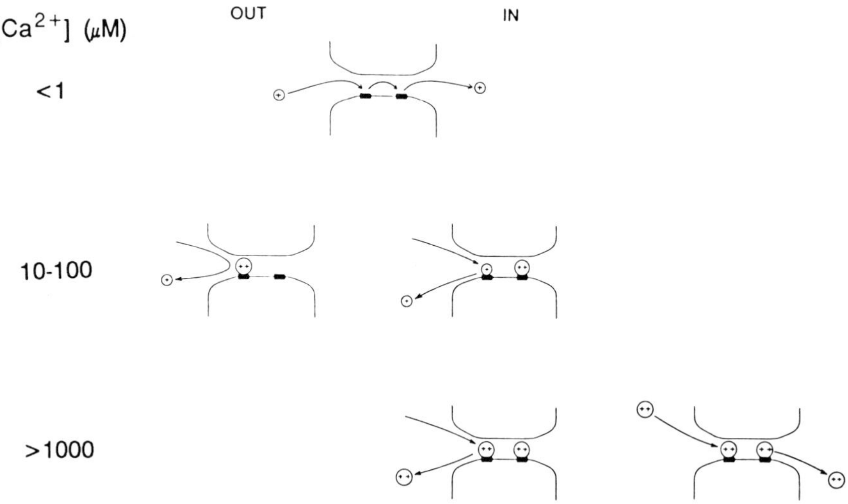

FIG. 2. Cartoon showing occupancy and flux through the two-site model of a Ca^{2+} channel at three different Ca^{2+} concentrations (indicated on the left in micromolar units). Na^{+} passes through the channel at very low Ca^{2+} concentrations, but is blocked above 1 μM Ca^{2+}. Ca^{2+} flux requires double occupancy of the pore, which occurs at millimolar Ca^{2+} concentrations. From Almers *et al.* (1986).

block of monovalent currents at micromolar Ca^{2+}. When doubly occupied, the pore has a low affinity for Ca^{2+}, as measured by saturation of Ca^{2+} flux above 10 m*M*.

An alternative view of the Ca^{2+} channel considers the high-affinity binding site to serve an allosteric function and to be located on the external face of the channel (Kostyuk *et al.*, 1983; Kostyuk and Mironov, 1986). On binding Ca^{2+} to the site, the channel undergoes a conformational change leading to a change in selectivity. When the site is occupied, the channel is Ca^{2+} selective; when the site is unoccupied, the channel is readily permeant to monovalent ions. In addition to the external high-affinity site, the model proposes an intrapore low-affinity site to which ions bind when permeating. This model does not predict the voltage dependence of high-affinity Ca^{2+} binding. Nevertheless, the voltage dependence, although a necessary prediction of the multi-ion pore model, is not sufficient to prove the location of the binding site or to distinguish between the competing models.

III. VOLTAGE-DEPENDENT ACTIVATION

A. The Need for Gating Charge

Activation of a membrane protein by changes in membrane voltage requires the presence of a charge within the protein that moves in response to the changing electric field. By considering the steepness of the relation between Na^+ channel activation and voltage, Hodgkin and Huxley (1952) deduced that there must be six charges moved across the membrane, or a larger number moved partway, in order to open a Na^+ channel. The movement of this charge within the guts of the channel protein must create a small current which Hodgkin and Huxley tried, but failed, to measure with their vacuum tube electronics.

Following the advent of transistor electronics and digital processing techniques, Armstrong and Bezanilla (1974) succeeded at measuring the proposed "gating current" of Na^+ channels. It is a small, outward current which precedes the onset of Na^+ flux through the channel (Fig. 3A). Experiments showed that gating current could be modified by charged agents applied to either side of the membrane, leading Armstrong (1981) to propose that the voltage sensor was a membrane-spanning structure. Armstrong's "ratchet model" (Fig. 3B) envisioned a series of linked charges that would move together one step when there was an appropriate change in the electric field.

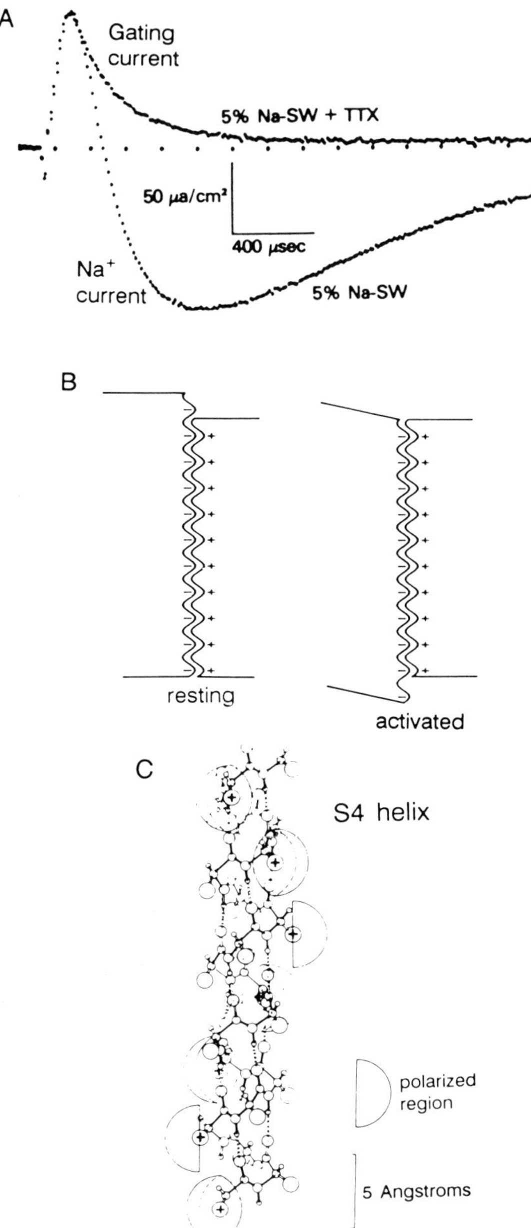

FIG. 3. (A) Gating current in Na^+ channels of squid axons. The upward deflection is gating current and the downward deflection is the small Na^+ current that remains when Na^+ concentration is dropped to 5% of normal. Gating current is recorded in isolation when tetrodotoxin (TTX) blocks the Na^+ current. (B) Armstrong's ratchet model for the voltage sensor. A string of interlinked positive charges interact with a string of negative charges. The

B. The S4 Helix Defines a Family of Voltage-Activated Channels

Voltage-dependent Na^+, K^+, and Ca^{2+} channels must each have mobile gating charge; not surprisingly, they exhibit remarkable homology in their amino acid sequences (Catterall, 1988; Jan and Jan, 1989). The structure of the α_1 subunit of the skeletal muscle Ca^{2+} channel (Tanabe *et al.,* 1987; Ellis *et al.,* 1988) is most like that of the Na^+ channel (Noda *et al.,* 1984, 1986). Like the Na^+ channel, the α_1 subunit is a large protein (170 kDa) consisting of four domains which are homologous with each other. In contrast, the entire sequence of voltage-dependent K^+ channels corresponds to just one of the domains of the Na^+ and Ca^{2+} channels (Papazian *et al.,* 1987). Functional K^+ channels are made of multiple subunits which, together, correspond to the single, multidomain subunit of Na^+ and Ca^{2+} channels.

It may be that the K^+ channel was the evolutionary mother of Na^+ and Ca^{2+} channels. Na^+ channels are found only in multicellular organisms; Ca^{2+} channels are seen in protozoa and higher creatures; but, K^+ channels are seen as far down the evolutionary tree as yeast (Hille, 1984). Perhaps Na^+ and Ca^{2+} channels evolved by linking four copies of a gene for an ancestral channel much like the K^+ channel.

The K^+ channel and each of the domains of the Ca^{2+} and Na^+ channels consist of six or seven membrane-spanning α-helices. The greatest region of homology, by far, exists in a stretch of 26 amino acids called the S4 helix. This stretch consists of 5–7 positively charged amino acids each separated by at least two neutral amino acids. When it was first discovered, it was assumed that this sequence existed outside the membrane because the charge makes it so hydrophilic (Noda *et al.,* 1984). But, the structural similarity to Armstrong's ratchet was soon noticed (Fig. 3C), and the S4 helix was proposed as the membrane-spanning voltage sensor (Noda *et al.,* 1986; Guy and Seetharamulu, 1986; Catterall, 1986). The "sliding helix" model proposes that voltage change induces the S4 helix to slide back or forth through the membrane in explicit analogy to Armstrong's ratchet.

If the S4 helix is the voltage sensor, changing the positively charged amino acids into neutral amino acids should decrease the slope of the curve relating ion channel activation to membrane voltage. Such changes have

change in the electric field causes the chains to move one step relative to each other. The result is the movement of one charge from outside to inside the membrane, though no single charge has moved more than a few angstroms. (C) The S4 helix from domain 4 of the Na^+ channel. The positive charge of the arginine groups wind around the helix and are indicated by the hemispheres. Gating current and ratchet from Armstrong (1981); S4 helix from Benndorf (1989).

been seen in a study of mutated Na^+ channels expressed in oocytes (Stuhmer *et al.*, 1989). The degree of the change in the activation curve correlated with the amount of neutralized charge in the S4 helix. Similar experiments on K^+ channels are less supportive of the simplest interpretation of the sliding helix model. Papazian *et al.* (1991) found that only one of five neutralization mutants of the S4 helix of K^+ channels caused a change in the slope of the activation curve, although four of the five caused a shift in the midpoint of the curve. The shift is not readily interpreted since neutralization of a positively charged residue on a different part of the protein also induced a shift.

Identification of the S4 helix as the universal voltage sensor of voltage-dependent channels is an intellectually satisfying, exceedingly popular idea, but it remains a working hypothesis. In addition to the equivocal data discussed above, an S4 helix has been found in an ion channel which exhibits little or no voltage sensitivity (the cGMP-gated channel of photoreceptors; Kaupp *et al.*, 1989; Jan and Jan, 1990).

C. Expression of Cloned Ca^{2+} Channels

1. Subunits in a Cell Line and Oocytes

The primary subunit of L-type Ca^{2+} channels from skeletal muscle (Tanabe *et al.*, 1987; Ellis *et al.*, 1988) and heart (Mikami *et al.*, 1989) have been cloned. Unfortunately, mutation studies like those done for Na^+ and K^+ channels have not been possible because of the difficulty of expressing functional Ca^{2+} channels in oocytes. The difficulty may involve the need for posttranslational modification or the need to express more than the primary subunit.

Biochemically purified Ca^{2+} channels from skeletal muscle consist of a complex of five polypeptides: α_1 (170 kDa), α_2 (143 kDa), β (54 kDa), δ (27 kDa), and γ (30 kDa) (Catterall, 1988, and references therein). The α_1 subunit is the binding site for the dihydropyridine drugs used to track the protein during purification steps (Sharp *et al.*, 1987); this is the subunit that is homologous to the Na^+ channel. The α_2 and δ subunits are linked by disulfide bonds, resulting in a complex that is similar in size to, and readily confused with, the α_1 subunit. The α_1, α_2, δ and γ subunits each have carbohydrate moieties, indicating that they are exposed to the extracellular surface. Because the β subunit is neither glycosylated nor hydrophobic, it is assumed to reside at the inner face of the membrane.

How much of this complex is necessary to form a functioning channel? Na^+ channels from brain also have many subunits, but the primary subunit is sufficient to form functioning channels in oocytes. Perhaps the difficulty of expressing functioning Ca^{2+} channels after injection of mRNA for the α_1 subunit suggests the need for additional subunits.

Ca^{2+} channels have been successfully transfected and expressed by injection of mRNA for the α_1 subunit into a cell line which was shown not to have the α_2 or δ subunits (Perez-Reyes *et al.*, 1989); thus, the minimum structure is no more than $\alpha_1\beta\gamma$. A low level of Ca^{2+} channel activity is seen in oocytes injected with the mRNA for the cardiac α_1 subunit, and the activity is enhanced if the skeletal muscle α_2 subunit is coinjected (Mikami *et al.*, 1989).

2. Successful Expression in a Mutant Muscle

Muscular dysgenesis is a single recessive defect existing in a carefully inbred colony of mice. A mouse born with the trait asphyxiates at birth because no skeletal muscles can contract. The muscles lack L-type Ca^{2+} current, a defect proposed to be the basis of the failure of excitation–contraction coupling (Beam *et al.*, 1986; Rieger *et al.*, 1987). Use of this mutant muscle as a tissue for Ca^{2+} channel expression is labor-intensive: muscle cells are multinucleated, so many nuclei must be injected to get expression in a single cell.

Successful expression of Ca^{2+} channels in dysgenic muscle was reported in two papers that are landmarks in the field of excitation–contraction coupling (Tanabe *et al.*, 1988, 1990). Both Ca^{2+} current and contraction are restored on expression of the α_1 subunit of the Ca^{2+} channel. Different results are obtained depending on whether cardiac or skeletal channels are expressed. As expected from differences in the native channels, the kinetics of the current from the cardiac clone are faster than that from the skeletal clone. But, the most dramatic effect concerns differences in excitation–contraction coupling. Cells injected with the skeletal clone can contract in the absence of extracellular Ca^{2+}, whereas the cardiac clone results in a contraction which requires extracellular Ca^{2+}. The result indicates that the skeletal muscle α_1 subunit can serve both as a Ca^{2+} channel and as the voltage sensor for release of Ca^{2+} stored in intracellular compartments. This dual role for the skeletal muscle α_1 subunit had been proposed previously, based on physiological studies (Rios and Brum, 1987), and may explain why there are 50 times more dihydropyridine receptors in skeletal muscle than there are functioning Ca^{2+} channels (Schwartz *et al.*, 1985).

IV. INACTIVATION

A. *Slow Inactivation Allows Sustained Responses*

An inactivated channel has entered a state where it is unavailable, or relatively unlikely, to be opened. The physical processes leading to inactivation may vary and there may be multiple inactivated states of a channel. Voltage-dependent Na^+ channels can inactivate either by slight variation of the rest potential of the cell or as a result of the opening of the channel; it is unclear that the same inactivation state is entered from the two different pathways. Voltage-dependent Ca^{2+} channels are more complex still. Different types of Ca^{2+} channels may inactivate due to variation in the resting potential, or in a seemingly timed manner after they have opened, or due to accumulation of Ca^{2+} in the intracellular medium. However, in all cases, Ca^{2+} channels inactivate slowly compared to Na^+ channels. Na^+ channels spontaneously inactivate within milliseconds of being opened, resulting in the brief action potentials of nerve and skeletal muscle (Fig. 4A). The persistence of Ca^{2+} channel activity leads to the sustained action potentials of heart cells (Fig. 4B), which can be hundreds of times longer than the

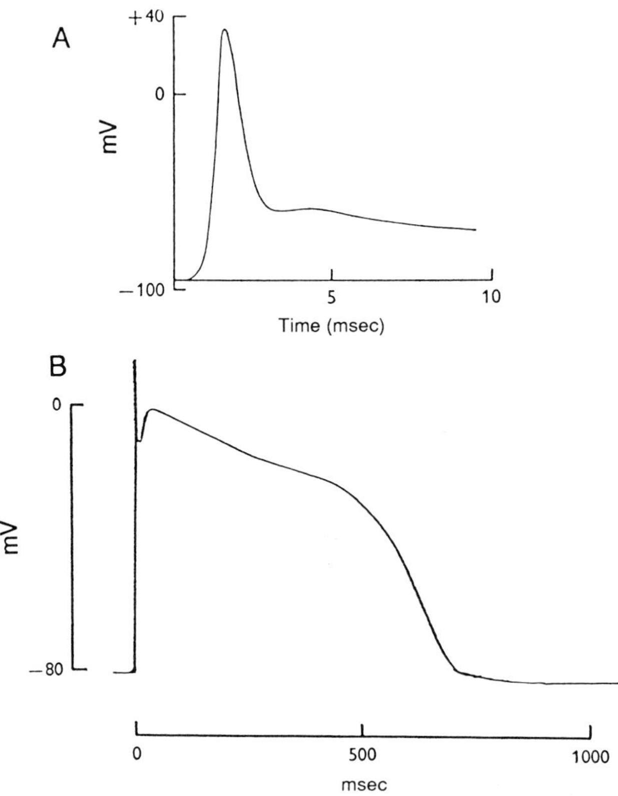

FIG. 4. (A) An action potential from skeletal muscle (Adrian and Peachey, 1973). (B) An action potential from a cardiac Purkinje fiber (Nobel and Tsien, 1969). Note the differences in time scale.

nerve action potential. Inactivation at different rates and by different mechanisms is a distinguishing feature of the various types of Ca^{2+} channels, which are discussed in Section VI. Here, we discuss the mechanisms of Ca^{2+} channel inactivation in general terms.

B. Ca^{2+}-Dependent Inactivation

Many, though not all, types of Ca^{2+} channels inactivate due to Ca^{2+} accumulation within the cell. Thus, the intracellular chemical signal generated by Ca^{2+} channels can feed back to inhibit the channel's activity. This contrasts with Na^{+} channels, which inactivate within a given time after opening regardless of the flux of any ion or the identity of the current carrier. These differences reflect the role of Na^{+} channels to make precisely timed electrical signals, whereas Ca^{2+} channels also create a chemical signal.

Hagiwara and Naka (1964) showed that intracellular injection of Ca^{2+} chelating agents promoted Ca^{2+} channel activity in barnacle muscle, providing the first evidence for the effect of intracellular Ca^{2+} on Ca^{2+} channels. Brehm and Eckert (1978) and Tillotson (1979) showed that the entry of Ca^{2+} during a prior stimulus diminished Ca^{2+} channel activity in a subsequent stimulus. The extent of this inactivation paralleled the amount of Ca^{2+} entry rather than being a function of the conditioning voltage. Moreover, the effect was specific for Ca^{2+} flux and did not occur if Ba^{2+} carried the current through the Ca^{2+} channel (Fig. 5); therefore, the inactivation process is a Ca^{2+}-selective, chemical event. The two other key tests of the theory were (1) injection of Ca^{2+} into the cytoplasm diminished subsequent Ca^{2+} channel activity (Standen, 1981); (2) EGTA, a Ca^{2+} chelator, dimished the effect of Ca^{2+} flux on inactivation (Brehm *et al.*, 1980; Eckert and Tillotson, 1981). The effect of Ca^{2+} on inactivation occurs in a wide variety of cell types, including cardiac cells (Lee *et al.*, 1985). Ca^{2+}-induced inactivation has been seen at the single-channel level and shown to gradually modify the gating kinetics of the channel, consistent with the need for Ca^{2+} accumulation within the cytoplasm rather than an immediate effect of Ca^{2+} as it permeates the pore (Yue *et al.*, 1990a).

The three conceivable locations for the Ca^{2+}-binding site that confers Ca^{2+} sensitivity on the channel are (1) the pore itself; (2) an allosteric site on the channel but remote from the pore; (3) a separate, Ca^{2+}-sensitive molecule that acts on the channel. Chad and Eckert (1986) provided evidence for the third possibility. They showed that perfusion of a cell with calmodulin and the Ca^{2+}-sensitive phosphatase calcineurin speeds the rate of inactivation, and proposed that the essential event is Ca^{2+}-triggered dephosphorylation of the channel.

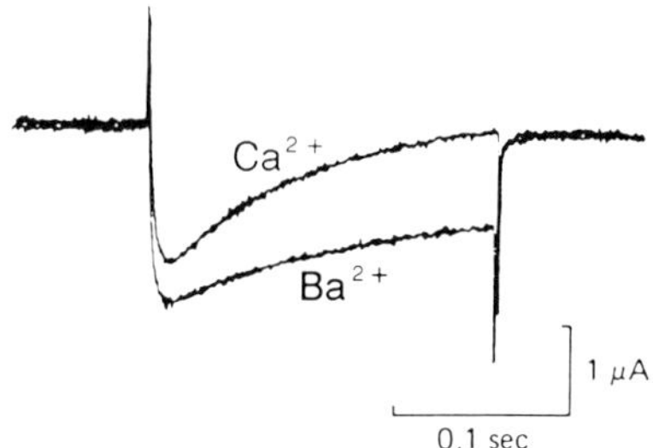

FIG. 5. Ca^{2+}-induced inactivation. Inactivation of Ca^{2+} current is faster than inactivation of Ba^{2+} current through the same channels in an *Aplysia* neuron. From Eckert and Tillotson (1981).

C. Voltage-Dependent Inactivation

As mentioned above, voltage-dependent inactivation of ion channels may occur through two functionally different pathways: (1) channels may become unavailable to open due to slight depolarization of the resting potential without prior opening of the channel; (2) channels may enter an unavailable state as a consequence of opening. Different types of Ca^{2+} channels exhibit these properties to varying degrees. At opposite ends of the spectrum are the L-type and the T-type Ca^{2+} channels that both exist in heart and neurons. The L channel is responsible for the prolonged action potential in Fig. 4B. It inactivates only slowly (and primarily due to Ca^{2+} accumulation) during a prolonged stimulus and does not inactivate on varying the resting potential until one reaches nonphysiologically positive voltages. In contrast, readiness to inactivate is the defining feature of the T channel. During a stimulus, T channels spontaneously inactivate within tens of milliseconds, behaving like a slow Na^+ channel. These differences are discussed fully in Section VI.

V. FACILITATION AND GATING MODES

In chromaffin cells and heart cells, but not in all cells, Ca^{2+} channel activity is increased following appropriate electrical conditioning. Steps to very positive voltages or trains of pulses to more moderate voltages both induce this facilitation. The effect has been analyzed at the single-channel level, showing that the enhanced activity is due to a shift in the mode of Ca^{2+} channel gating. The action of Ca^{2+} channel-enhancing drugs and hormones, some forms of synaptic facilitation, and the increased strength of the heartbeat during rapid heart rates all may result from such shifts in Ca^{2+} channel gating.

A. Chromaffin Cells

Facilitation was first noted by Fenwick *et al.* (1982) in a thorough study of the properties of Ca^{2+} channels in chromaffin cells, the adrenalin-releasing cells of the adrenal gland. Hoshi *et al.* (1984) studied the effect in detail, considering it a model for the facilitation of transmitter release which occurs in some synapses following trains of action potentials. They showed that the facilitation current was recruited by prepulses ranging from + 10 to + 80 mV (Fig. 6) and that trains of brief pulses to + 10 mV were effective.

Artalejo *et al.* (1991a) have shown that the current recruited by facilitating protocols is blocked by dihydropyridine drugs, unlike the Ca^{2+} current before facilitation. The effect has been studied at the single-channel level by Hoshi and Smith (1987) and Artalejo *et al.* (1991b). After a facilitating prepulse, Ca^{2+} channel activity is dominated by long-duration openings of a 27 pS channel that are seen rarely, if at all, without prepulses. Artalejo *et al.* (1991a,b) have shown that there are multiple types of Ca^{2+} channels on chromaffin cells, distinguishable by single-channel conductance and phar-

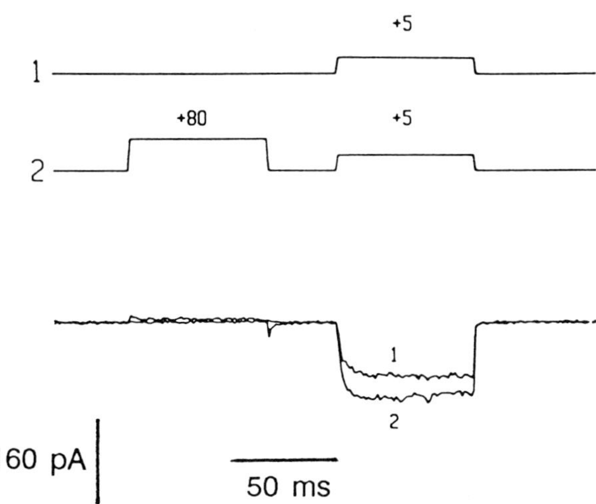

FIG. 6. (A) Facilitation of Ca^{2+} currents in chromaffin cells. When the test pulse is preceded by a large prepulse (trace 2), Ca^{2+} channel activity is enhanced. From Hoshi *et al.* (1984).

macology. The "facilitation channel," having a 27 pS conductance and being sensitive to dihydropyridine drugs, is like the L-type channel of heart and neurons (see Section VI). However, the need for recruiting prepulses to unmask the channel indicates that it rests in a different gating state in chromaffin cells than in the heart.

B. Cardiac Cells

During rapid heart rates, the strength of contraction increases during the first several quick beats, forming the so-called "positive staircase" effect. Lee (1987) has shown that the positive staircase is paralleled by an increase in Ca^{2+} channel activity. The facilitation is a voltage-dependent process, much like in the chromaffin cell. Pietrobon and Hess (1990) have shown that the effect is due to an increase in the open time of L-type channels and have explicitly described the kinetics of the transitions into and out of the short- and long-opening modes. The transitions occur on a time scale of seconds, very long compared to the millisecond rates of normal gating. Thus, the L channel can exist in different gating modes, each with distinguishable kinetics and with clear rates of transition between the modes.

C. Implications for Pharmacology and Modulation

The work of Pietrobon and Hess confirms a proposal made 6 years previously by Hess *et al.* (1984). They suggested that Ca^{2+} channels could exist in different gating modes and drugs could preferentially bind to and promote given modes. Thus, dihydropyridine agonists are proposed to promote the long-opening mode of L-type Ca^{2+} channels, thereby leading to an increase in Ca^{2+} flux.

Two reports indicate that hormone-induced phosphorylation of Ca^{2+} channels can also promote the long-opening mode. In the heart, β-adrenergic stimulation causes a significant increase in long-opening behavior (Yue *et al.*, 1990b); in chromaffin cells, dopamine has the same effect (Artalejo *et al.*, 1990). Both hormones act through cAMP to induce phosphorylation of the channel.

Whether facilitation can occur with other channels besides the L-type and in other cells besides chromaffin and cardiac cells is unknown. Facilitation is clearly not a property seen in all cells (Hoshi *et al.*, 1984). Although facilitation has not been reported in neurons, a strikingly similar phenomenon is seen after depression of neuronal Ca^{2+} channels by a variety of hormones (Lipscombe *et al.*, 1989; Bean, 1989b; Aosaki and Kasai, 1989; Elmslie *et al.*, 1990).

VI. DIFFERENT TYPES OF Ca^{2+} CHANNELS

A. Overview

As discussed in the introduction, Ca^{2+} channels can serve both electrical and chemical functions for a cell. Given the variety of jobs to be performed, it is not surprising that a variety of Ca^{2+} channels have evolved, each with properties specialized for different tasks. In addition to exhibiting distinct kinetic properties, different Ca^{2+} channel types can be localized on different regions of a cell with complex morphology. Finally, Ca^{2+} channels in different tissues display different pharmacological profiles, suggesting the possibility of drugs selective for particular organs.

Multiple types of Ca^{2+} channels can be distinguished on a particular cell through differences in their kinetics, selectivity, and pharmacology. Figure 7 shows Ca^{2+} currents from two different sensory neurons that have

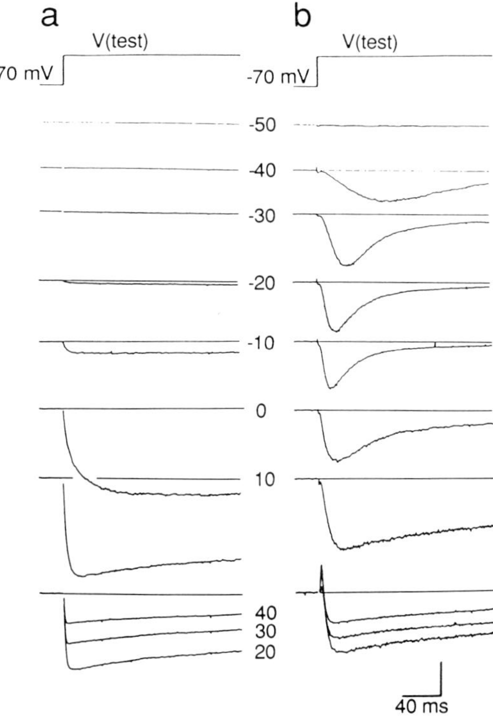

FIG. 7. Low- and high-threshold Ca^{2+} channels on different sensory neurons. Currents are evoked by steps from −70 mV to the indicated voltage. The cell on the left (a) has only high-threshold channels; the cell on the right (b) has both low- and high-threshold channels. From Schroeder *et al.* (1990b).

been stimulated with a series of pulses to the indicated voltages. Ca^{2+} current in Fig. 7a begins to activate at −20 mV and is sustained throughout the 200-msec duration stimulus. In contrast, the cell in Fig. 7b has a Ca^{2+} channel that activates as negative as −40 mV and spontaneously inactivates. Above −10 mV, the sustained current is activated in the same cell. Thus, the cell in Fig. 7b has both low-threshold and high-threshold Ca^{2+} current, whereas the cell in Fig. 7a has only the high-threshold channels. The two types of Ca^{2+} channel differ in their activation range, their tendency to inactivate, and their expression on different sensory neurons. The high-threshold type is the "classic" Ca^{2+} channel, whereas the low-threshold type was discovered recently.

The sections below describe the properties and functions of the low-threshold ("T-type") Ca^{2+} channels and discuss the controversies surrounding multiple types of high-threshold Ca^{2+} channels. Finally, we ask which of these channel types trigger neurosecretion.

B. Low-Threshold (T-Type) Ca^{2+} Channels

The unique properties of low-voltage-activated (T-type) Ca^{2+} channels ensure some physiological roles distinct from those for high-threshold channels. Reports on the existence of low-threshold Ca^{2+} channels in vertebrate neurons and heart cells began appearing in the mid-1980s (Carbone and Lux, 1984; Bean, 1985; Armstrong and Matteson, 1985; Nowycky *et al.*, 1985; Nilius *et al.*, 1985). At the same time, a remarkably similar channel was discovered in the freshwater hypotrich ciliate *Stylonychia mytilus* (Deitmer, 1984). It is of interest that these single-cell organisms and vertebrate nerve cells should possess such similar channels.

Low-threshold Ca^{2+} channels were predicted by Llinas and Yarom (1981) before currents through them were examined directly. They studied the electrical properties of inferior olivary neurons in a slice preparation and observed action potentials consisting of a fast spike and a slow after-depolarizing potential (ADP). The ADP was driven by a calcium flux which exhibited voltage-dependent inactivation between −85 and −65 mV; this was later shown to be characteristic of T-type Ca^{2+} channels.

1. Kinetics and Selectivity

Low-threshold Ca^{2+} channels have been called "T" (Nowycky *et al.*, 1985), "LVA" (Carbone and Lux, 1984), and "SD" (Armstrong and Matteson, 1985). The term LVA (low-voltage-activated) indicates that the channel activates with small depolarizations from rest. The term T, which we will use in this review, refers to the transient nature of the current. Both the low activation threshold and transience are demonstrated in Fig. 7. In addition to inactivating during a maintained stimulus, the availability of T channels is steeply dependent on the resting potential of the cell: T channels are fully available for activation from a holding potential of −95 mV and nearly fully inactivated with a holding potential of −65 mV (Fox *et al.*, 1987a). Thus, neurons often must be hyperpolarized in order to unmask their T channels.

The term SD (slow deactivating) refers to the slow closing kinetics of T channels (Armstrong and Matteson, 1985). Closing kinetics are studied by examining the time course of the current when a cell is instantaneously stepped from a potential where the channels are open (such as −20 mV) to a hyperpolarized potential where the channels will close. Armstrong and Matteson (1985) reported that, at 12°C in GH3 cells, T channels close with a time constant of about 10 msec, whereas high-threshold channels close with a time constant of about 0.3 msec.

Currents through single T channels were first recorded in dorsal root ganglia (DRG) neurons and heart (Carbone and Lux, 1984; Nowycky *et al.*, 1985; Nilius *et al.*, 1985). The single channels were identified as T channels based on activation and inactivation ranges and the transience of their average current. With 110 m*M* Ba^{2+} as the charge carrier, the single-channel conductance in chick DRG is approximately 8 pS.

T and high-threshold Ca^{2+} channels differ in relative permeability to divalent cations. Barium ions pass through T channels about as well as calcium ions (Bean, 1985; Carbone and Lux, 1987a,b). In contrast, single-channel conductance for Ba^{2+} can be twice that for Ca^{2+} in high-threshold channels (Hess *et al.*, 1986).

The basis for Ca^{2+} selectivity in T channels is a high-affinity Ca^{2+}-binding site (Fukushima and Hagiwara, 1985; Carbone and Lux, 1987a,b), much like high-threshold channels (Section II,B). Sodium currents pass through T channels in the absence of calcium and are blocked by micromolar Ca^{2+} in a voltage-dependent manner, suggesting that the site is within the pore.

2. Pharmacology

Various inorganic, multivalent ions (La^{3+}, Cd^{2+}, Co^{2+}, Mg^{2+}, and Ni^{2+}) block both high-threshold and T channels at sufficiently high concentrations, but some show selectivity at lower concentrations. Nickel at a concentrations of 100 μM inhibits T current nearly completely while having a small (~10%) effect on high-threshold Ca^{2+} current (Fox *et al.*, 1987a). Cadmium at 20–50 μM eliminates about 90% of high-threshold currents but only about 50% of T current (Fox *et al.*, 1987a). A more selective blocker for some high-threshold channels is ω-conotoxin (Section VI,C).

Coulter *et al.* (1989) showed that ethosuximide (300–700 μM) and dimethadione (5–9 mM), both anticonvulsants used clinically to treat petit mal epilepsy, selectively inhibit T current in thalamic neurons. The drug concentrations used were in the same range as free serum levels reported at therapeutic anticonvulsant doses in man. Succimide, a compound that is structurally similar to ethosuximide but is inactive in the control of petit mal epilepsy, did not inhibit either the T or high-threshold Ca^{2+} currents.

Other agents that selectively inhibit T channels in some tissues are amiloride and phorbol esters. Amiloride inhibits Na^{+}–H^{+} exchangers in the micromolar concentration range (Aronson, 1985), and is reported to block T channels selectively at low concentrations (30 μM; Tang *et al.*, 1988). An activator of protein kinase C, phorbol ester 12-myristate 13-acetate (PMA), inhibits T channels in rat dorsal root ganglia neurons only when applied at physiological temperatures (Schroeder *et al.*, 1990b).

3. Physiology

Pacemaking tissues, such as certain regions of the heart, fire action potentials in a regular pattern in which the cell repeatedly and spontaneously depolarizes to threshold after each action potential. The T channel has two properties that make it a good candidate to contribute to this spontaneous, subthreshold depolarization: it becomes available to open during the hyperpolarization following an action potential and it is activated by small depolarizations. Hagiwara *et al.* (1988) found that selective block of T channels with low concentrations of Ni^{2+} altered the pace of action potential firing in cells from the sinoatrial node, the tissue responsible for setting the pace of the healthy heart. T channels are also reported to contribute to pacemaker potentials in rat dorsal raphe nucleus neurons (Burlhis and Aghajanian, 1987).

T channels also help create bursts of action potentials in thalamic neurons (Llinas and Jahnsen, 1982; Suzuki and Rogawski, 1989) and in rat DRG neurons (White *et al.*, 1989). Figure 8 shows recordings of both action potentials and Ca^{2+} currents in DRG cells; solutions which block T channels also block the slow depolarization which underlies pairs of action potentials.

Electrical activity of the thalamus appears important in states of sleep and arousal, and T channels are proposed to play a crucial role. When an animal is awake, cells in the thalamus transfer peripheral information to the cortex with high fidelity. When an animal is barbiturized or at times of drowsiness or slow-wave sleep, the thalamic neurons fire in an oscillatory mode which can be detected by electroencephalogram (EEG) or with a

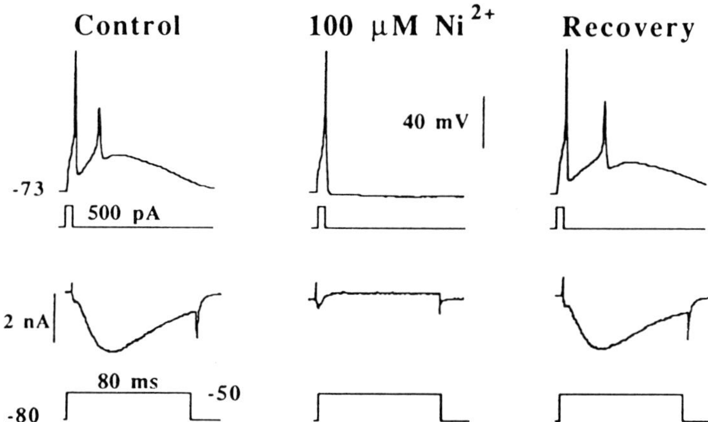

FIG. 8. Both the prolonged depolarization which induces multiple action potentials (upper records) and T-type Ca^{2+} channels (lower traces) are blocked by 100 μM Ni^{2+}. The upper records show action potentials from a sensory neuron evoked by brief pulses of current. The lower records show Ca^{2+} currents evoked by pulses of voltage. From White *et al.* (1989).

microelectrode. It has been proposed that this oscillatory mode makes the brain unresponsive to incoming stimuli (Steriade and Llinas, 1988). This firing pattern, called spindle rhythm, consists of bursts of action potentials at 7–14 Hz repeated at a frequency of 0.1–0.2 Hz (Steriade and Deschennes, 1984). Llinas and Jahnsen (1982) showed that a low-threshold Ca^{2+}-dependent spike, with voltage dependence and pharmacology similar to T channels, underlies the spindle oscillations.

Each of these proposed functions of T channels utilizes the electrical signal produced when Ca^{2+} enters the cell, but does not take advantage of the chemical nature of that signal. The question remains, why evolve a high-affinity Ca^{2+}-binding site in this channel if all you need is influx of positive charge? "Threshold channels," a population of Na^+ channels with activation and inactivation properties like the T channel, have been described in the squid axon (Gilly and Armstrong, 1984). The authors speculated that these channels should influence pacemaking and bursting behavior of autorhythmic cells. Clearly, the electrical roles of T channels could be served by appropriate Na^+ channels, and we might wonder if there are also chemical functions for T channels. There is no report of a Ca^{2+}-induced secretory process inhibited by the pharmacology characteristic of T channels (see Section VI,D). Transmitter release from rat parasympathetic ganglia comes closest: it is inhibited by amiloride but not by low concentrations of Ni^{2+} (Seabrook and Adams, 1989).

C. Controversies about High-Threshold Ca^{2+} Channels

People outside the field often have heard about "N" and "L"-type high-threshold Ca^{2+} channels and consider this terminology the beginning and end of the subject. In fact, there is considerable debate about the appropriate criteria for identifying different high-threshold channels (Table I). The issue is important because the high-threshold channels clearly serve the role of translating electrical signals into chemical signals (Section VI,D), so differences in channel kinetics, localization, and pharmacology may affect the various Ca^{2+}-triggered events. The original criteria and nomenclature for distinguishing different types of high-threshold Ca^{2+} channels were proposed by Tsien and colleagues (Nowycky *et al.*, 1985; Fox *et al.*, 1987a,b): (1) N channels inactivate during a maintained stimulus or with depolarized holding voltages, whereas L channel activity persists; (2) N channels have a single-channel conductance of about 14 pS, whereas L channels have a conductance of 25 pS; (3) L channels are sensitive to dihydropyridine drugs such as nifedipine.

Various aspects of these criteria are debated and the field is in flux. There are two simultaneous trends: one suggests that N and L channels are

TABLE I

Properties That Have Been Proposed to Distinguish T, N, and L Channels[a]

Property	Channel		
	T	N	L
Single channel conductance (isotonic Ba^{2+}) (pS)	8	14	26
Ba^{2+} : Ca^{2+} conductance	$Ba^{2+} \simeq Ca^{2+}$	$Ba^{2+} > Ca^{2+}$	$Ba^{2+} > Ca^{2+}$
Activation range (mV)	−40 to −10	−20 to +10	−20 to + 10
Deactivation rate (msec)	Slow (~10)	Fast (<1)	Fast (<1)
Inactivation range (mV)	−80 to −50	−80 to−20	Positive to −40
Inactivation rate (msec)	Fast (~20)	Fast (~50)	Slow
Selective inorganic blockers (μM)	Ni^{2+} (<100)	Cd^{2+} (<50)	Cd^{2+} (<50)
Selective organic blockers	Amiloride, ethosuximide	ω-Conotoxin	Dihydropyridines (nifedipine, etc.)

[a] Numerical values for rates and concentrations vary between different cells; these values are for rat DRG neurons. Several of the distinctions between N and L channels are debated: (1) single N channels have been recorded with both transient and sustained kinetics (Aosaki and Kasai, 1989); (2) one report suggests that the 14 pS conductance, generally considered diagnostic of N channels, is a subconductance with the main level being close to that of L channels (Plummer *et al.*, 1989). If conductance and inactivation are not distinguishing features between N and L channels, the only difference is the effect of organic blockers.

relatively similar in their kinetics but are still pharmacologically distinct, and the other suggests more than just two types of high-threshold channels.

Aosaki and Kasai (1989) challenged the difference between N and L inactivation rates by showing that 14 pS single channels can exhibit both inactivating and noninactivating kinetics. Plummer *et al.* (1989) claim that N and L channels have similar single-channel conductances and that the 14 pS openings are subconductance levels of the N channel. This high-conductance N channel maintains activity during a prolonged stimulus, so both criteria 1 and 2 become blurred.

The third criterion, dihydropyridine pharamacology, is universally accepted, but even here there are ambiguities. Action of the drugs is voltage dependent (Bean, 1984; Sanguinetti and Kass, 1984): at negative holding potentials, where N channels are available, even the L channel has relatively little drug sensitivity. Another selective pharmacological agent is ω-conotoxin. ω-Conotoxin does not affect Ca^{2+} channels of muscle, but irreversibly blocks neuronal high-threshold Ca^{2+} channels with only small, reversible effects on T channels (McCleskey *et al.*, 1987). The initial report showed that partial blockade by ω-conotoxin did not alter the kinetics of high-threshold current and, therefore, concluded that all components of high-threshold current were equally affected. Subsequent reports, which were accompanied by evidence suggesting that N and L channels have relatively similar kinetics, indicate that the N channel is blocked by ω-conotoxin and the L channel is spared (Aosaki and Kasai, 1989; Plummer *et al.*, 1989).

The upshot of these new studies is that pharmacology is the only clear distinction between N and L channels. N is ω-conotoxin-sensitive, L is dihydropyridine-sensitive. Because Ca^{2+} channels in heart, skeletal, and smooth muscle are insensitive to ω-conotoxin but sensitive to dihydropyridines, high-threshold channels in muscle are L type. However, the picture in neurons is more complex. ω-Conotoxin and dihydropyridines block varying amounts of current in different neurons, and there can be residual current when they are applied together. Evidently, neurons contain a mix of N and L channels as well as pharmacologically unidentified channels (Regan *et al.*, 1991). This heterogeneity of neuronal Ca^{2+} channels is supported by genetic studies which find four distinct families of rat brain cDNAs that have homology to the α_1 subunit of muscle Ca^{2+} channels (Snutch *et al.*, 1990).

Although recent data downplay the kinetic differences between N- and L-type channels, neuronal high-threshold Ca^{2+} current does have both transient and relatively sustained kinetic components. The transient component in sensory neurons is selectively diminished by neuropeptide Y

(Ewald *et al.*, 1988) and by opioids binding at κ receptors (Gross and MacDonald, 1987) or μ receptors (Schroeder *et al.*, 1991). The transient component is also selectively inactivated by bursts of brief pulses which mimic trains of action potentials (Fig. 9) and by prolonged depolarization just above the normal resting potential (Schroeder *et al.*, 1990a). The inactivation depends on the intensity of and interval since a prior burst of electrical activity. In contrast, the slowly inactivating component inactivates only through accumulation of intracellular Ca^{2+}. Thus, there are physiologically important differences. Availability of the transient component is affected by various neuropeptides and depends on the value of the resting potential and the pattern of a prior stimulus. The sustained component is unaffected by any peptides yet studied and is unaware of the electrical history of the cell; it seems to provide a baseline level of voltage-dependent Ca^{2+} entry that is lost only when intracellular Ca^{2+} rises.

D. Ca^{2+} Channel Types Involved in Neurosecretion

Which of the ever-increasing number of Ca^{2+} channel types is present at presynaptic terminals? The issue is important because pharmacological blockade of neurotransmitter release is a potent means of modifying neurological and physiological processes; if Ca^{2+} channel types differ at different synapses, selective pharamacology becomes possible.

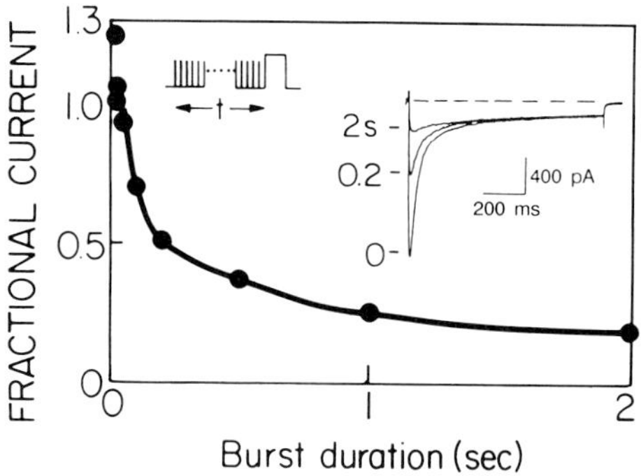

FIG. 9. Bursts of brief pulses selectively inactivate the transient component of high-threshold Ca^{2+} current in rat sensory neurons. The left inset shows the general protocol: a burst of brief pulses is applied to the cell prior to a long pulse which assays the Ca^{2+} current. The right inset shows currents recorded without a prior burst of pulses, following a 0.2-sec burst, or following a 2-sec burst. After the 2-sec burst, only the sustained current remains. Scale: length of line indicates 400 pA or 200 msec. The graph is a plot of the normalized peak current as a function of the burst duration. From Schroeder *et al.* (1990a).

Direct recordings of presynaptic Ca^{2+} currents have been possible in only rare instances. In the squid giant synapse, which is giant enough to poke with multiple microelectrodes, the presynaptic Ca^{2+} currents pass through high-threshold Ca^{2+} channels (Llinas *et al.*, 1981; Augustine *et al.*, 1985). Another preparation allowing direct recording of presynaptic Ca^{2+} currents is the isolated nerve terminal of the mammalian neurohypophysis, which releases oxytocin and vasopressin. Two types of high-threshold channels that differ in single-channel conductance and dihydropyridine pharmacology are found, but there are no low-threshold channels (Lemos and Nowycky, 1989; Lim *et al.*, 1990). Finally, the large presynaptic calyx of the chick ciliary ganglion has also been studied and found to have only high-threshold Ca^{2+} channels (Stanley and Goping, 1991). Thus, all three of the preparations that have been amenable to direct recording use high-threshold Ca^{2+} channels to trigger neurosecretion. But which type of high-threshold channel do they use?

Drugs and toxins which are selective for certain types of Ca^{2+} channels have been tested for their effects on neurotransmission. Kerr and Yoshikami (1984) showed that ω-conotoxin completely and irreversibly inhibited transmitter release at the frog neuromuscular junction. ω-Conotoxin also inhibits release of norepinephrine from cultured rat sympathetic neurons, whereas dihydropyridine blockers do not (Hirning *et al.*, 1988). In contrast, dihydropyridine blockers do inhibit release of substance P from cultured rat sensory neurons (Rane *et al.*, 1987; Perney *et al.*, 1986). Evidently, neurotransmitter release at different synapses can be triggered through different types of Ca^{2+} channels.

It is important to realize that there must be channels other than the N and L types. Lin *et al.* (1990) report a novel spider toxin that blocks Ca^{2+} channels from rat brain that are not blocked by ω-conotoxin or dihydropyridines. Regan *et al.* (1991) report that central neurons are enriched in Ca^{2+} channels that are not sensitive to ω-conotoxin or dihydropyridines, though they are kinetically similar to N and L channels. The possibility that there are pharmacologically distinct Ca^{2+} channels serving the same function on different types of neurons would be exceedingly useful in the clinic and is a major driving force for research in this field.

VII. SUMMARY

Voltage-dependent Ca^{2+} channels can sense, amplify, and terminate electrical signals on cell membranes, while also serving to translate the electrical signal into a cytoplasmic chemical signal. They have an extraordinary selectivity for Ca^{2+} that involves a high-affinity Ca^{2+} binding site.

The ability to sense changes in membrane voltage seems to involve the same amino acid sequence, the S4 helix, as used by other voltage-dependent channels. Inactivation during a maintained stimulus proceeds slowly on most types of Ca^{2+} channels and can be the result of accumulation of intracellular Ca. On some cells, but not all, intense electrical activity facilitates Ca^{2+} channel activity. Much of the variety of Ca^{2+} channel behavior is due to a diversity of Ca^{2+} channel types, the pharmacology of which should be clinically important.

References

Adrian, R. H., and Peachy, L. D. (1973). Reconstruction of the action potential of frog satorius muscle. *J. Physiol. (London)* **235,** 103–131.

Almers, W., and McCleskey, E. W. (1984). Non-selective conductance in calcium channels of frog muscle: Calcium selectivity in a single-file pore. *J. Physiol. (London)* **353,** 585–608.

Almers, W., McCleskey, E. W., and Palade, P. T. (1984). A non-selective cation conductance in frog muscle membrane blocked by micromolar external calcium ions. *J. Physiol. (London)* **353,** 565–583.

Almers, W., McCleskey, E. W., and Palade, P. T. (1986). The mechanism of ion selectivity in calcium channels of skeletal muscle membrane. *Fortschr. der Zool.* Band 33, Luttgau (Hrsg.): Membrane Control-Gustave Fischer Verlag, Stuttgart, New York.

Aosaki, T., and Kasai, H. (1989). Characterization of two kinds of high-voltage-activated Ca-channel currents in chick sensory neurons: Differential sensitivity to dihydropyridines and conotoxin GVIA. *Pfluegers Arch.* **414,** 150–156.

Armstrong, C. M. (1981). Sodium channels and gating currents. *Physiol. Rev.* **61** (3), 644–683.

Armstrong, C. M., and Benzanilla, F. (1974). Charge movement associated with the opening and closing of the activation gates of the Na channels. *J. Gen. Physiol.* **63,** 533–552.

Armstrong, C. M., and Matteson, D. R. (1985). Two distinct populations of calcium channels in a clonal line of pituitary cell. *Science* **227,** 65–67.

Armstrong, D. L. (1989). Calcium channel regulation by calcineurin, a Ca^{2+}-activated phosphatase in mammalian brain. *Trends Neurosci.* **12** (3), 117–122.

Aronson, P. S. (1985). Kinetic properties of the plasma membrane Na^+-H^+ exchanger. *Annu. Rev. Physiol.* **47,** 545–560.

Artalejo, C. R., Ariano, M. A., Perlman, R. L., and Fox, A. P. (1990). Activation of facilitation calcium channels in chromaffin cells by D_1 dopamine receptors through a cAMP/protein kinase A-dependent mechanism. *Nature (London)* **348,** 239–242.

Artalejo, C. R., Mogul, D. J., Perlman, R. L., and Fox, A. P. (1991b). Bovine chromaffin cells exhibit three types of calcium channels: Facilitation, induced by large pre-depolarizations or repetitive activity, is due to the increased opening probability of a 27 ps channel. *J. Physiol. (London)*, in press.

Artalejo, C. R., Dahmer, M. K., Perlman, R. L., and Fox, A. P. (1991a). Two types of Ca^{2+} currents are found in bovine chromaffin cells: Facilitation is due to the recruitment of one type. *J. Physiol. (London)* **432,** 681–707.

Augustine, G. J., Charlton, M. P., and Smith, S. J. (1985). Calcium entry and transmitter release at voltage-clamped nerve terminals of squid. *J. Physiol. (London)* **369,** 163–181.

Beam, K. G., Knudson, C. M., and Powell, J. A. (1986). A lethal mutation in mice eliminates the slow calcium current in skeletal muscle cells. *Nature (London)* **320,** 168–170.

Bean, B. P. (1984). Nitrendipine block of cardiac calcium channels: High affinity binding to the inactivated state. *Proc. Natl. Acad. Sci. U.S.A.* **81,** 6388–6392.
Bean, B. P. (1985). Two kinds of calcium channels in canine atrial cells. *J. Gen. Physiol.* **86,** 1–30.
Bean, B. P. (1989a). Classes of calcium channels in vertebrate cells. *Annu. Rev. Physiol.* **51,** 367–385.
Bean, B. P. (1989b). Neurotransmitter inhibition of neuronal calcium currents by changes in channel voltage dependence. *Nature* (*London*) **340,** 153–156.
Bean, B. P. (1990). Gating for the physiologist. *Nature* (*London*) **348,** 192–193.
Benndorf, K. (1989). A reinterpretation of NA channel gating and permeation in terms of a phase transition between a transmembrane S4 α-helix and a channel-helix. *Eur. Biophys. J.* **17,** 257–271.
Brehm, P., and Eckert, R. (1978). Calcium entry leads to inactivation of calcium channels in Paramecium. *Science* **202,** 1203–1206.
Brehm, P., Eckert, R., and Tillotson, D. (1980). Calcium-mediated inactivation of calcium current in Paramecium. *J. Physiol.* (*London*) **306,** 193–203.
Burlhis, T. M., and Aghajanian, G. K. (1987). Pacemaker potentials of serotonergic dorsal raphe neurons: Contribution of a low-threshold Ca^{2+} conductance. *Synapse* **1,** 582–588.
Carbone, E., and Lux, H. D. (1984). A low voltage-activated, fully inactivating Ca channel in vertebrate sensory neurones. *Nature* (*London*) **310,** 501–503.
Carbone, E., and Lux, H. D. (1987a). Kinetics and selectivity of a low-voltage-activated calcium current in chick and rat sensory neurones. *J. Physiol.* (*London*) **386,** 547–570.
Carbone, E., and Lux, H. D. (1987b). Single low-voltage-activated calcium channels in chick and rat sensory neurones. *J. Physiol.* (*London*) **386,** 571–601.
Catterall, W. A. (1986). Voltage-dependent gating of sodium channels: Correlating structure and function. *Trends Neurosci.* **9,** 7–10.
Catterall, W. A. (1988). Structure and function of voltage-sensitive ion channels. *Science* **242,** 50–61.
Chad, J. E., and Eckert R. (1986). An enzymatic mechanism for calcium current inactivation in dialysed Helix neurones. *J. Physiol.* (*London*) **378,** 31–51.
Coulter, D. A., Huguenard, J. R., and Prince, D. A. (1989). Specific petit mal anticonvulsants reduce calcium currents in thalamic neurons. *Neurosci. Lett.* **98,** 74–78.
Dascal, N., Snutch, T. P., Lubbert, H., Davidson, N., and Lester, H. A. (1986). Expression and modulation of voltage-gated calcium channels after RNA injection in Xenopus oocytes. *Science* **231,** 1147–1150.
Deitmer, J. W. (1984). Evidence for two voltage-dependent calcium currents in the membrane of the ciliate stylonychia. *J. Physiol.* (*London*) **355,** 137–159.
Dupont, J.-L., Bossu, J.-L., and Feltz, A. (1986). Effect of internal calcium concentration on calcium currents in rat sensory neurones. *Pfluegers Arch.* **406,** 433–435.
Eckert, R., and Tillotson, D. L. (1981). Calcium-mediated inactivation of the calcium conductance in caesium-loaded giant neurons of Aplysia californica. *J. Physiol.* (*London*) **314,** 265–280.
Ellis, S. B., Williams, M. E., Ways, N. R., Brenner, R., Sharp, A. H., Leung, A. T., Campbell, K. P., McKenna, E., Koch, W. J., Hui, A., Schwartz, A., and Harpold, M. M. (1988). Sequence and expression of mRNAs encoding the α_1 and α_2 subunits of a DHP-sensitive calcium channel. *Science* **241,** 1661–1664.
Elmslie, K. S., Zhou, W., and Jones, S. W. (1990). LHRH and GTP-S modify calcium current activation in bullfrog sympathetic neurones. *Neuron* **5,** 75–80.
Ewald, D. A., Sternweis, P. C., and Miller, R. J. (1988). Guanine nucleotide-binding protein

G_0-inducing coupling of neuropeptide Y receptors to Ca^{2+} channels in sensory neurons. *Proc. Natl. Acad. Sci. U.S.A.* **85,** 3633–3637.

Fatt, P., and Ginsborg, B. L. (1958). The ionic requirements for the production of action potentials in crustacean muscle fibres. *J. Physiol. (London)* **142,** 516–543.

Fenwick, E. M., Marty, A., and Neher, E. (1982). Sodium and calcium channels in bovine chromaffin cells. *J. Physiol. (London)* **331,** 599–635.

Fox, A. P., Nowycky, M. C., and Tsien, R. W. (1987a). Kinetic and pharmacological properties distinguishing three types of calcium currents in chick sensory neurones. *J. Physiol. (London)* **394,** 149–172.

Fox, A. P., Nowycky, M. C., and Tsien, R. W. (1987b). Single-channel recordings of three types of calcium channels in chick sensory neurones. *J. Physiol. (London)* **394,** 173–200.

Fukushima, Y., and Hagiwara, S. (1985). Currents carried by monovalent cations through calcium channels in mouse neoplastic B lumphocytes. *J. Physiol. (London)* **358,** 255–284.

Gilly, W. F. and Armstrong, C. M. (1984). Threshold channels—a novel type of sodium channel in squid giant axon. *Nature (London)* **309,** 448–450.

Gross, R. A., and MacDonald, R. L. (1987). Dynorphin A selectively reduces a large transient (N-type) calcium current of mouse dorsal root ganglion neurons in cell cultures. *Proc. Natl. Acad. Sci. U.S.A.* **84,** 5469–5473.

Guy, H. R., and Seetharamulu, P. (1986). Molecular model of the action potential sodium channel. *Proc. Natl. Acad. Sci. U.S.A.* **83,** 508–512.

Hagiwara, S. (1983). "Membrane Potential-Dependent Ion Channels in Cell Membrane. Phylogenetic and Developmental Approaches." Raven Press, New York.

Hagiwara, S., and Byerly, L. (1981). Calcium channel. *Annu. Rev. Neurosci.* **4,** 69–125.

Hagiwara, S., and Naka, K. I. (1964). The initiation of spike potential in barnacle muscle fibers under low intracellular Ca^{++}. *J. Gen. Physiol.* **48,** 141–161.

Hagiwara, S., and Takahashi, K. (1967). Surface density of calcium ion and calcium spikes in the barnacle muscle fiber membrane. *J. Gen. Physiol.* **50,** 583–601.

Hagiwara, S., Fukuda, J., and Eaton, D. C. (1974). Membrane currents carried by Ca, Sr, and Ba in barnacle muscle fiber during voltage clamp. *J. Gen. Physiol.* **63,** 564–578.

Hagiwara, N., Irisawa, H., and Kameyama, M. (1988). Contribution of two types of calcium currents to the pacemaker potentials of rabbit sino-atrial node cells. *J. Physiol. (London)* **395,** 233–253.

Hess, P. (1990). Calcium channels in vertebrate cells. *Annu. Rev. Neurosci.* **13,** 337–56.

Hess, P., and Tsien, R. W. (1984). Mechanism of ion permeation throught calcium channels. *Nature (London)* **309,** 453–456.

Hess, P., Lansman, J. B., and Tsien, R. W. (1984). Different modes of Ca channel gating behaviour favoured by dihydropyridine Ca agonists and antagonists. *Nature (London)* **311,** 538–544.

Hess, P., Lansman, J. B., and Tsien, R. W. (1986). Calcium channel selectivity for divalent and monovalent cations. *J. Gen. Physiol.* **88,** 293–319.

Hille, B. (1984) "Ionic Channels of Excitable Membranes." Sinauer Associates, Sunderland, Massachusetts.

Hirning, L. D., Fox, A. P., McCleskey, E. W., Olivera, B. M., Thayer, S. A., Miller, R. J., and Tsien, R. W. (1988). Dominant role of N-type Ca^{2+} channels in evoked release of norepinephrine from sympathetic neurons. *Science* **239,** 57–61.

Hodgkin, A. L., and Huxley, A. F. (1952). A quantitative description of membrane current and its application to conduction and excitation in nerve. *J. Physiol. (London)* **117,** 500–544.

Hoshi, T., and Smith, S. J. (1987). Large depolarization induces long openings of voltage-dependent calcium channels in adrenal chromaffin cells. *J. Neurosci.* **7** (2), 571–580.

Hoshi, T., Rothlein, J., and Smith, S. J. (1984). Facilitation of Ca^{2+}-channel currents in bovine adrenal chromaffin cells. *Proc. Natl. Acad. Sci. U.S.A.* **81,** 5871–5875.

Jan, L. Y., and Jan Y. N. (1989). Voltage-sensitive ion channels. *Cell (Cambridge, Mass.)* **56,** 13–25.

Jan, L. Y., and Jan, Y. N. (1990). A superfamily of ion channels. *Nature (London)* **345,** 672.

Kasai, H., and Aosaki, T. (1989). Modulation of Ca-channel current by an adenosine analog mediated by a GTP-binding protein in chick sensory neurons. *Pfluegers Arch* **414,** 145–149.

Kass, R. S., and Sanguinetti, M. C. (1984). Inactivation of calcium channel current in the calf cardiac Purkinje fiber. *J. Gen. Physiol.* **84,** 705–726.

Kaupp, U. B., Niidome, T., Tanabe, T., Terada, S., Bonigk, W., Stuhmer, W., Cook, N. J., Kangawa, K., Matsuo, H., Hirose, T., Miyata, T., and Numa, S. (1989). Primary structure and functional expression from complementary DNA of the rod photoreceptor cyclic GMP-gated channel. *Nature (London)* **342,** 762–766.

Kerr, L. M., and Yoshikami, D. (1984). A venom peptide with a novel presynaptic blocking action. *Nature (London)* **308**(5956), 282–284.

Kostyuk, P. G., and Krishtal, O. A. (1977). Effects of calcium and calcium-chelating agents on the inward and outward current in the membrane of mollusc neurones. *J. Physiol. (London)* **270,** 569–580.

Kostyuk, P. G., and Mironov, S. L. (1986). Some predictions concerning the calcium channel model with different conformational states. *Gen. Physiol. Biophys.* **6,** 649–659.

Kostyuk, P. G., Mironov, S. L., and Shuba, Y. M. (1983). Two ion-selecting filters in the calcium channel of the somatic membrane of mollusc neurons. *J. Membr. Biol.* **76,** 83–93.

Lansman, J. B., Hess, P., and Tsien, R. W. (1986). Blockade of current through single calcium channels by Cd^{2+}, Mg^{2+}, and Ca^{1+}. *J. Gen. Physiol.* **88,** 321–347.

Lee, K. S. (1987). Potentiation of the calcium-channel currents of internally perfused mammalian heart cells by repetitive depolarization. *Proc. Natl. Acad. Sci. U.S.A.* **84,** 3941–3945.

Lee, K. S., and Tsien, R. W. (1982). Reversal of current through calcium channels in dialysed single heart cells. *Nature (London)* **297,** 498–501.

Lee, K. S., Marban, E., and Tsien, R. W. (1985). Inactivation of calcium channels in mammalian heart cells: Joint dependence on membrane potential and intracellular calcium. *J. Physiol. (London)* **364,** 395–411.

Lemos, J. R., and Nowycky, M. C. (1989). Two types of calcium channels coexist in peptide-releasing vertebrate nerve terminals. *Neuron* **2,** 1419–1426.

Lim, N. F., Nowycky, M. C., and Bookman, R. J. (1990). Direct measurement of exocytosis and calcium currents in single vertebrate nerve terminals. *Nature (London)* **344,** 449–451.

Lin, J.-W., Rudy, B., and Llinas, R. (1990). Funnel-web spider venom and a toxin fraction block calcium current expressed from rat brain mRNA in Xenopus oocytes. *Proc. Natl. Acad. Sci. U.S.A.* **87,** 4538–4542.

Lipscombe, D., Kongsamut, S., and Tsien, R. W. (1989). α-Adrenergic inhibition of sympathetic neurotransmitter release mediated by modulation of N-type calcium-channel gating. *Nature (London)* **340,** 639–642.

Llinas, R., and Jahnsen, H. (1982). Electrophysiology of mammalian thalamic neurones in vitro. *Nature (London)* **297,** 406–408.

Llinas, R., and Yarom, Y. (1981). Properties and distribution of ionic conductances generating electroresponsiveness of mammalian inferior olivary neurones in vitro. *J. Physiol. (London)* **315,** 569–584.

Llinas, R. Steinberg, I. Z., and Walton, K. (1981). Relationship between presynaptic calcium current and postsynaptic calcium current and postsynaptic potential in squid giant synapse. *Biophys. J.* **33,** 323–352.

McCleskey, E. W., and Almers, W. (1985). The Ca channel in skeletal muscle is a large pore. *Proc. Natl. Acad. Sci. U.S.A.* **82,** 7149–7153.

McCleskey, E. W., Fox, A. P., Feldman, D. H., Cruz, L. J., Olivera, B. M., Tsien, R. W., and Yoshikami, D. (1987). Conotoxin: Direct and persistent blockade of specific types of calcium channels in neurons but not muscle. *Proc. Natl. Acad. Sci. U.S.A.* **84,** 4327–4331.

Mikami, A., Imoto, K., Tanabe, T., Niidome, T., Mori, Y., Takeshima, H., Narumiya, S., and Numa, S. (1989). Primary structure and functional expression of the cardiac dihydropyridine-sensitive calcium channel. *Nature (London)* **340,** 230–233.

Nilius, B., Hess, P., Lansman, J. B., and Tsien, R. W. (1985). A novel type of cardiac calcium channel in ventricular cells. *Nature (London)* **316,** 443–446.

Noble, D., and Tsien, R. W. (1969). Reconstruction of the repolarization process in cardiac Purkinje fibres based on voltage clamp measurements of the membrane current. *J. Physiol.* **200,** 233–254.

Noda, M., Shimizu, S., Tanabe, T., Takai, T., Kayano, T., Ikeda, T., Takahashi, H., Nakayama, H., Kanaoka, Y., Minamino, N., Kangawa, K., Matsuo, H., Raftery, M. A., Hirose, T., Inayama, S., Hayashida, H., Miyata, T., and Numa, S. (1984). Primary structure of Electrophorus electricus sodium channel deduced from cDNA sequence. *Nature (London)* **312,** 121–127.

Noda, M., Ikeda, T., Kayano, T., Suzuki, H., Takeshima, H., Kurasaki, M., Takahashi, H., and Numa, S. (1986). Existence of distinct sodium channel messenger RNAs in rat brain. *Nature (London)* **320,** 188–192.

Nowycky, M. C., Fox, A. P., and Tsien, R. W. (1985). Three types of neuronal calcium channel with different calcium agonist sensitivity. *Nature (London)* **316,** 440–443.

Papazian, D. M., Schwarz, T. L., Tempel, B. L., Jan, Y. N., and Jan, L. Y. (1987). Cloning of genomic and complementary DNA from Shaker, a putative potassium channel gene from Drosophila. *Science* **237,** 749–753.

Papazian, D. M., Timpe, L. C., Jan, Y. N., and Jan, L. Y. (1991). Alteration of voltage-dependence of Shaker potassium channel by mutations in the S4 sequence. *Nature (London)* **349,** 305–310.

Perez-Reyes, E., Kim, H. S., Lacerda, A. E., Horne, W., Wei, X., Rampe, D., Campbell, K. P., Brown, A. M., and Birnbaumer, L. (1989). Induction of calcium currents by the expression of the α_1-subunit of the dihydropyridine receptor from skeletal muscle. *Nature (London)* **340,** 233–236.

Perney, T. M., Hirning, L. D., Leeman, S. E., and Miller, R. J. (1986). Multiple calcium channels mediate neurotransmitter release from peripheral neurons. *Proc. Natl. Acad. Sci. U.S.A.* **83,** 6656–6659.

Pietrobon, D., and Hess, P. (1990). Novel mechanism of voltage-dependnet gating in L-type calcium channels. *Nature (London)* **346,** 651–655.

Plummer, M. R., Logethetis, D. E., and Hess, P. (1989). Elementary properties and pharmacological sensitivities of calcium channels in mammalian peripheral neurons. *Neuron* **2,** 1453–1463.

Rane, S. G., Holz, G. G. III., and Dunlap, K. (1987). Dihydropyridine inhibition of neuronal calcium current and substance P release. *Pfleugers Arch.* **409,** 361–366.

Regan, L. J., Sah, D. W. Y., and Bean, B. P. (1991). Calcium channels in rat central and peripheral neurons: High-threshold current resistant to dihydropyridine blockers and ω-conotoxin. *Neuron* **6,** 269–280.

Rieger, F., Bournaud, R., Shimahara, T., Garcia, L., Pincon-Raymond, M., Romey, G., and

Lazdunski, M. (1987). Restoration of dysgenic muscle contraction and calcium channel function by co-culture with normal spinal cord neurons. *Nature* (*London*) **330,** 563–566.

Rios, E., and Brum, G. (1987). Involvement of dihydropyridine receptors in excitation–contraction coupling in skeletal muscle. *Nature* (*London*) **325,** 717–720.

Salkoff, L., and Wyman, R. (1981). Genetic modification of potassium channels in Drosophila Shaker mutants. *Nature* (*London*) **293,** 228–230.

Sanguinetti, M. C., and Kass, R. S. (1984). Voltage-dependent block of calcium channel current in the calf cardiac Purkinje fibre by dihydropyridine calcium channel antagonists. *Circ. Res.* **55,** 336–348.

Satin, L. S., and Cook, D. L. (1989). Calcium current inactivation in insulin-secreting cells is mediated by calcium influx and membrane depolarization *Pfleugers Arch.* **414,** 1–10.

Schroeder, J. E., Fischbach, P. S., Mamo, M., and McCleskey, E. W. (1990a). Two components of high-threshold Ca^{++} current inactivate by different mechanisms. *Neuron* **5,** 445–452.

Schroeder, J. E., Fischbach, P. S., and McCleskey, E. W. (1990b). T-type calcium channels: Heterogeneous expression in rat sensory neurons and selective modulation by phorbol esters. *J. Neurosci.* **10**(3), 947–951.

Schroeder, J. E., Fischbach, P. S., Zheng, D., and McCleskey, E. W. (1991). Activation of μ opioid receptors inhibits transient high- and low-threshold Ca^{2+} currents, but spares a sustained current. *Neuron* **6,** 13–20.

Schwartz, L. M., McCleskey, E. W., and Almers, W. (1985). Dihydropyridine receptors in muscle are voltage-dependent but most are not functional calcium channels. *Nature* (*London*) **314,** 747–751.

Schwarz, T. L., Tempel, B. L., Papazian, D. M., Jan, Y. N., and Jan, L. Y. (1988). Multiple potassium-channel components are produced by alternative splicing at the Shaker locus in Drosophila. *Nature* (*London*) **331,** 137–142.

Seabrook, G. R., and Adams, D. J. (1989). Inhibition of neurally-evoked transmitter release by calcium channel antagonists in rat parasympathetic ganglia. *Br. J. Pharmacol.* **97,** 1125–1136.

Sharp, A. H., Imagawa, T., Leung, A. T., and Campbell, K. P. (1987). Identification and characterization of the dihydropyridine-binding subunit of the skeletal muscle dihydropyridine receptor. *J. Biol. Chem.* **262,** 12309–12315.

Snutch, T. P., Leonard, J. P., Gilbert, M. M., Lester, H. A., and Davidson, N. (1990). Rat brain expresses a heterogeneous family of calcium channels. *Proc. Natl. Acad. Sci. U.S.A.* **87,** 3391–3395.

Standen, N. B. (1981). Ca channel inactivation by intracellular CA injection into Helix neurones. *Nature* (*London*) **293,** 158–159.

Stanley, E. F., and Goping, G. (1991). Characterization of a calcium current in a vertebrate cholinergic presynaptic nerve terminal. *J. Neurosci.,* **11,** 985–993.

Steriade, M., and Deschennes, M. (1984). The thalamus as a neuronal oscillator. *Brain Res. Rev.* **8,** 1–63.

Steriade, M., and Llinas, R. R. (1988). The functional states of the thalamus and the associated neuronal interplay. *Physiol. Rev.* **68**(3), 649–742.

Stuhmer, W., Conti, F., Suzuki, H., Wang, X., Noda, M., Yahagi, N., Kubo, H., and Numa, S. (1989). Structural parts involved in activation and inactivation of the sodium channel. *Nature* (*London*) **339,** 597–603.

Suzuki, S., and Rogawski, M. A. (1989). T-type calcium channels mediate the transition between tonic and phasic firing in thalamic neurons. *Proc. Natl. Acad. Sci. U.S.A.* **86,** 7228–7232.

Tanabe, T., Takeshima, H., Mikami, A., Flockerzi, V., Takahashi, H., Kangawa, K.,

Kojima, M., Matsuo, H., Hirose, T., and Numa, S. (1987). Primary structure of the receptor for calcium channel blockers from skeletal muscle. *Nature (London)* **328,** 313–318.

Tanabe, T., Beam, K. G., Powell, J. A., and Numa, S. (1988). Restoration of excitation–contraction coupling and slow calcium current in dysgenic muscle by dihydropyridine receptor complementary DNA. *Nature (London)* **336,** 134–139.

Tanabe, T., Mikami, A., Numa, S., and Beam, K. G. (1990). Cardiac-type excitation–contraction coupling in dysgenic skeletal muscle injected with cardiac dihydropyridine receptor cDNA. *Nature (London)* **344,** 451–453.

Tang, C.-M., Presser, F., and Morad, M. (1988). Amiloride selectively blocks the low threshold (T) calcium channel. *Science* **240,** 213–215

Tempel, B. L., Papazian, D. M., Schwarz, T. L., Jan, Y. N., and Jan, L. Y. (1987). Sequence of a probable potassium channel component incoded at Shaker Locus of Drosophilia. *Science* **237,** 770–775.

Tillotson, D. (1979). Inactivation of Ca conductance dependent on entry of Ca ions in molluscan neurons. *Proc. Natl. Acad. Sci. U.S.A.* **76**(3), 1497–1500.

Timpe, L. C., Schwarz, T. L., Tempel, B. L., Papazian, D. M., Jan, Y. N., and Jan, L. Y. (1988). Expression of functional potassium channels from Shaker cDNA in Xenopus oocytes. *Nature (London)* **331,** 143–145.

Tsien, R. W., Hess, P., McCleskey, E. W., and Rosenberg, R. L. (1987). Calcium channels: Mechanisms of selectivity, permeation, and block. *Annu. Rev. Biophys. Chem.* **16,** 265–290.

White, G., Lovinger, D. M., and Weight, F. F. (1989). Transient low-threshold Ca^{++} current triggers burst firing through an afterdepolarizing potential in an adult mammalian neuron. *Proc. Natl. Acad. Sci. U.S.A.* **86,** 6802–6806.

Yue, D. T., Backx, P. H., and Imredy, J. P. (1990a). Calcium-sensitive inactivation in the gating of single calcium channels. *Science* **250,** 1735–1738.

Yue, D. T., Herzig, S., Marban, E. (1990b). β-adrenergic stimulation of calcium channels occurs by potentiation of high-activity gating modes. *Prac. Natl. Acad. Sci.* **87,** 753–757.

CHAPTER 10

Potassium Channels in Developing Excitable Cells

Maria Isabel Behrens* **and Ramon Latorre***,†

*Centro de Estudios Cientificos de Santiago, Santiago, Chile

†Departamento de Biologia Facultad de Ciencias, Universidad de Chile, Santiago, Chile

I. INTRODUCTION

An important event in the overall process of cytodifferentiation of many cells is the expression of membrane channels responsible for active ion currents. Ion channels are present in excitable cells at very early stages of development and much effort has been expended to understand the process of development of excitability. It has been shown that the ionic dependence of the responses to electrical stimuli at early stages of development is often different from that seen in mature cells. By studying the developmental timetable of the appearance of different ion channels in membranes, it is assumed that a useful step toward the discovery of the general rules of development will be achieved.

There are several technical problems in the study of ion channels in developing systems (Spitzer, 1979, 1985). At the embryonic stages of interest, the cells are small and anatomically undifferentiated; thus, they are difficult to locate and unequivocally identify. Furthermore, they are fragile and rapidly damaged by impalement with microelectrodes. The ability to culture dissociated cells in a fully defined medium has stimulated new approaches to the study of cell development under controlled environmental conditions. This approach is justified because cells in culture express membrane properties comparable to those observed *in situ* (Spitzer, 1979; Fischbach and Nelson, 1984; Gruol and Franklin, 1987).

Excitable cells from many different species and tissues have been studied *in vivo* and *in vitro* from nonmammalian and mammalian preparations. A general conclusion that has been obtained from studies in developing neurons is that the location, time course, and sequence of appearance of the voltage-dependent currents are specific for a given population of cells. In general, the expression of outward currents precedes that of inward currents (Warner, 1973; Spitzer, 1979; Bader *et al.,* 1983, 1985; Ahmed, 1988a,b). A similar conclusion is obtained in developing flight muscles in *Drosophila* (Salkoff and Wyman, 1981; Wei and Salkoff, 1986).

In this review, we discuss current experiments on K^+ channels in developing neurons and muscle cells from vertebrates and invertebrates. The regulation of a specific type of muscle K^+ channel with development and innervation is also reviewed. A general conclusion that emerges from the study of different cells in development is that K^+ channels seem to be the first channels to be expressed. Furthermore, K^+ channels are expressed in these cells before the acquisition of excitability. This suggests that their role may be related to metabolic functions of the cell.

A. Diversity of K^+ Channels in Cells

In their original description of the ionic origin of the action potential, Hodgkin and Huxley (1952) included a potassium current as giving rise to the repolarizing phase of the action potential. Since then, several tens of different K^+ channels playing an immense diversity of physiological roles ranging from volume regulation to heart beat (see, e.g., Latorre and Miller, 1983; Hille, 1984; Rudy, 1988; Jan and Jan, 1990) have been characterized with the help of the voltage and patch–clamp techniques. The dissection of the different K^+ channels has been carried out using pharmacological and, more recently, genetic tools. These channels have been classified according to their type of kinetics of activation (and inactivation), channel conductance, and sensitivity to different toxins and blockers.

The primary structure of several K^+ channels has been elucidated and, in this regard, the pioneering experiments were done with the *Drosophila* mutant *Shaker*. The *Shaker* mutation, so-called because of the leg-shaking phenotype induced under ether anesthesia, promotes lacks of motor control, synaptic deficiency, as well as several other problems related with the normal functioning of excitable cells. All these problems are a consequence of a defect in the "A" currents that, since they are repolarizing currents, promotes in the mutant changes in the threshold, frequency, and duration of the action potentials of nerve and muscle as well as producing postsynaptic potentials that are longer and larger than in the wild-type strain of fly (Jan *et al.,* 1977). The "A" channel is codified by the *Shaker* locus (Jan *et al.,* 1977; Salkoff and Wyman, 1981). The genomic DNA that forms part of the mutation site was cloned by Papazian *et al.* (1987). The first reported sequence of amino acids codified by a Shaker clone (Tempel *et al.,* 1987) allowed a direct comparison with other peptides able to form ion channels. In contrast to other voltage-dependent channels (e.g., Na^+ and Ca^{2+} channels), the peptide forming the "A" channel has a molecular weight and amino acid sequence resembling one of the four homologous domains of Na^+ or Ca^{2+} channels (Fig. 1). Jan and Jan (1990) have discussed the means to generate the great K^+ channel diversity. Both multiplicity of genes (Buttler *et al.,* 1989; Baker and Salkoff, 1990; Wei *et al.,* 1990) and alternative splicing (Iverson *et al.,* 1988; Timpe *et al.,* 1988) can generate different K^+ channel-forming polypeptides. Because these polypeptides are smaller than those forming other voltage-dependent channels, it is likely that K^+ channels are formed by putting together several (4?) subunits. If the subunits are different, K^+ channel diversity can arise because heteromultimeric forms may have electrical properties different from those of homomultimeric channels. Isacoff *et al.* (1990) have tested this hypothesis by coinjecting messenger RNA obtained from different *Shaker* variants into *Xenopus* oocytes. Oocytes coinjected with two different *Shaker* mRNAs expressed macroscopic "A" currents kinetically different to both parent "A" currents. Similar results were obtained at the single-channel level, lending strong support to the hypothesis that heteromultimeric "A" channels can be formed by association of different poly-

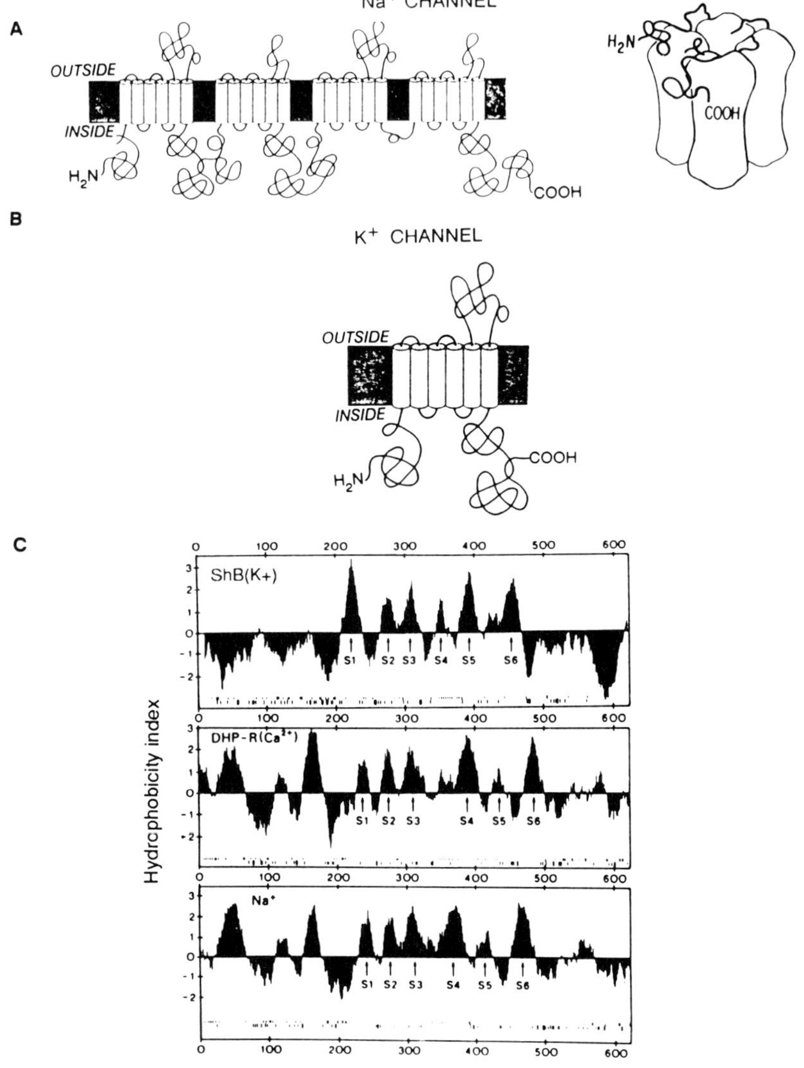

FIG. 1. Proposed transmembrane topology for voltage-dependent channels. (A) Na^+ channel α subunit. The polypeptide contains four homologous pseudosubunits (domains). The channel topology is based on hydrophobicity analysis (C) and secondary structure prediction. The folding in the membrane to form the ion channel is shown at the right (Miller, 1989). (B) K^+ channel. K^+ channels are probably tetrameric assemblies of subunits. (C) Aligned hydrophobicity profiles of subunit sequences of voltage-dependent ion channels. The plots are aligned according to the transmembrane segment S2 of the Shaker B protein [ShB(K^+)]. The repeat *I* of the dihydropyridine receptor [DPH-R(Ca^{2+})] and the repeat *I* of rat brain Na^+ channel are shown for comparison. (Modified from Maelicke, 1988.)

peptides (see also Ruppersberg *et al.,* 1990). Different channel glycosylation or phosphorylation levels may also contribute to K^+ channel diversity.

B. K^+ Channels and Cell Development

Several voltage-dependent outward K^+ currents have been recorded from both invertebrate and vertebrate excitable developing cells (see, e.g., Wei and Salkoff, 1986; Ahmed, 1988b; Wilson and Chiu, 1990; for recent reviews on K^+ channels that the reader should consult Hille, 1984; Rudy, 1988; McManus, 1991). These include a transient outward current, I_A (Connors and Stevens, 1971a,b; Neher, 1971; Salkoff and Wyman, 1981), the delayed rectifier K^+ current, I_K (Hodgkin and Huxley, 1952), a transient Ca^{2+}-activated K^+ current, $I_{K(Ca)}$ (Adams *et al.,* 1982b; MacDermott and Weight, 1982), a slowly activated $I_{K(Ca)}$ current (Meech and Standen, 1975; Barret *et al.,* 1980), a noninactivating "M" outward current (Adams *et al.,* 1982a; Brown *et al.,* 1982), and an inward rectifier current (Adrian, 1969; Reuter, 1984). I_A currents are blocked by millimolar amounts of 4-aminopyridine (4-AP) added to the external medium and 4-AP sensitivity is often used as a criterion to identify these type of currents. However, extreme caution should be taken in this regard since there are other types of K^+ currents that are even more sensitive to 4-AP than "A" currents. Furthermore, given the great diversity of "A" currents, a dose–response curve should be done in each case inasmuch as sensitivity varies between the different "A" channels. A blockade induced by 4-AP has been observed at concentrations as low as 100 μM in CA3 hippocampal neurons (Segal and Barker, 1984) and as high as 10 mM in *Drosophila* muscle (Rudy, 1988). The delayed rectifier type of channel is blocked by internal and, in some cases, by externally applied tetraethylammonium (TEA) and 4-AP. $I_{K(Ca)}$ currents mediated by the high-conductance K(Ca) channel are blocked in general by internal and external TEA, the external TEA site having a much larger affinity for this organic cation. On the other hand, the small-conductance K(Ca) channel is insensitive to this quaternary ammonium ion (Blatz and Magleby, 1987; Latorre *et al.,* 1989). The main caveat of these organic blockers is that they are not very specific for a given K^+ channel and great efforts have been made to find specific toxins able to inhibit them specifically (see, e.g., Moczydlowski *et al.,* 1988; Garcia *et al.,* 1991). The scorpion toxins charibdotoxin and iberiotoxin block the high-conductance K(Ca) channel, the bee venom apamin specifically blocks the small-conductance K(Ca) channel, and the scorpion toxin noxiustoxin blocks the delayed rectifier rather specifically (Garcia *et al.*, 1991). To our knowledge, these toxins have not been used in developmental studies. From all the classes of K^+ channels discussed above, the two most frequent K^+ channels observed in developing cells are the I_A channel and the I_K channel.

II. K^+ CHANNELS IN DEVELOPING NEURONS

Early studies in developing neurons have been mainly performed in nonmammalian preparations. The work on intact *Xenopus* embryos, primarily in embryonic Rohon-Beard and dorsal root ganglion neurons (Spitzer, 1979, 1985), has led to three general conclusions: (1) outward currents precede inward currents; (2) long-lasting Ca^{2+}-dependent action potentials tend to differentiate with maturation into more rapid Na^+-dependent action potentials (Spitzer, 1979); and (3) there is an intermediate period of cell maturation in which the action potential shows a dependence on Ca^{2+} and Na^+ ions.

Most of the recent work on K^+ channels in developing neurons has been performed in cultured cells. Of great value in recent years has been the possibility to remove and dissociate portions of the embryonic nervous system into single cells, with the aim of studying the differentiation process in isolated neurons. In a series of studies, Spitzer (1979, 1985) and Warner (1981) have shown that cells isolated from neural plate-stage amphibian embryos will differentiate *in vitro* into neurite-bearing cells that generate action potentials when stimulated, and that can form junctional synapses onto muscle target cells. Differentiating neurons isolated from the neural plate undergo developmental changes *in vitro*. With development, changes in their sensitivity to neurotransmitters and in the ionic dependence of their action potentials appear. These changes are similar to those of their counterparts in intact embryos (Spitzer and Lamborghini, 1976; Bixby and Spitzer, 1984). These studies suggest that some ectodermal cells previously identified during neural plate induction will differentiate relatively autonomously, and that during their development these cells will express certain properties common to spinal neurons *in situ* (Barish, 1986). Dissociated cell cultures provide the opportunity to characterize and isolate specific ion currents using the voltage-clamp or patch–clamp techniques. The examination of individual currents and the evaluation of their contribution to the whole cell current may help understanding of the mechanism behind developmental changes in neurons.

A. Amphibian Neurons

Barish (1986) studied K^+ currents in embryonic amphibian (*Ambystoma*) spinal neurons during differentiation in culture. He observed the presence of Na^+, Ca^{2+}, and voltage-gated K^+ currents in these neurons at the earliest stages of development at which the cells could be morphologically identified (2 to 3 days in culture). As differentiation proceeds *in vitro,* the pattern of total membrane current becomes increasingly dominated by outward currents, and the duration of the action potential changed from 100 msec (2–3 days after being placed in culture) to 1 msec (about 10 days after neurite outgrowth). During this period, the action potential also changed from a Ca^{2+}-dependent spike to a Na^+-dependent action potential. The examination of K^+ currents in isolation from inward currents shows an increase in magnitude and in the rates of activation and deactivation. These results suggest that the increase in the K^+ current amplitude and activation rate would be sufficient to change the duration and ionic sensitivity of the action potential. An important conclusion of this work is that different K^+ conductance systems are present in the cell membrane very early in development. However, no evidence of a transient "A" current has been found in embryonic amphibian spinal neurons or in adult cat motoneurons (Schwindt and Crill, 1982). Maturation appears to imply a modification in channel kinetics and density, although a change in channel conductance cannot be discarded at present. Regarding the kinetic change of I_K, Barish (1986) pointed out that neuron maturation also implied a shift to the left along the voltage axis of the activation curve, and suggested a possible change in surface potential near the gating machinery of the delayed rectifier. Complex changes in squid axon I_K amplitude and kinetics have been found by the group of Bezanilla (Augustine and Bezanilla, 1990; Perozo and Bezanilla, 1990) upon channel phosphorylation. However, most of their results can be explained in terms of a simple model that considers a change in the fixed charge density in the neighborhood of the gating system (Perozo and Bezanilla, 1990). It is tempting to suggest that Barish's (1986) findings may be due to changes in channel phosphorylation with cell maturation. Development may also imply expression of kinetically different delayed rectifiers using mechanisms such as those discussed in Section I,A.

As discussed above, *Xenopus* neurons *in vivo* as well as *in vitro,* initially fire Ca^{2+}-dependent action potentials and, with maturation, these slow action potentials are transformed into fast Na^+ spikes. Blair and Dionne (1985) have proposed an interesting mechanism for the action potential shortening in *Xenopus* spinal neurons that involves a change in Ca^{2+} sensitivity of a large-conductance Ca^{2+}-activated K^+ channel with maturation (maxi K^+ channel; Latorre and Miller, 1983; Latorre *et al.*, 1989; McManus, 1991). Using the patch–clamp technique, they recorded a 155 pS K^+-selective channel that was insensitive to Ca^{2+} in early developmental stages (7–11 hr). At 7–11 hr Blair and Dionne (1985) reported that the 155 pS channel was active even in the absence of intracellular Ca^{2+} (Fig. 2B). However, at 1–2 days, channel activity was sensitive to the internal Ca^{2+} concentration and channel opening was not detected at $[Ca^{2+}] < 10^{-10}$ *M* (Fig. 2A). The fraction of time that the channel remains in the open state was voltage dependent at early and late stages of maturation. The different Ca^{2+} sensitivity of this channel can be accounted for by the mechanisms of alternative splicing, a multigene family, and/or post-translational modifications as discussed earlier in this review. However,

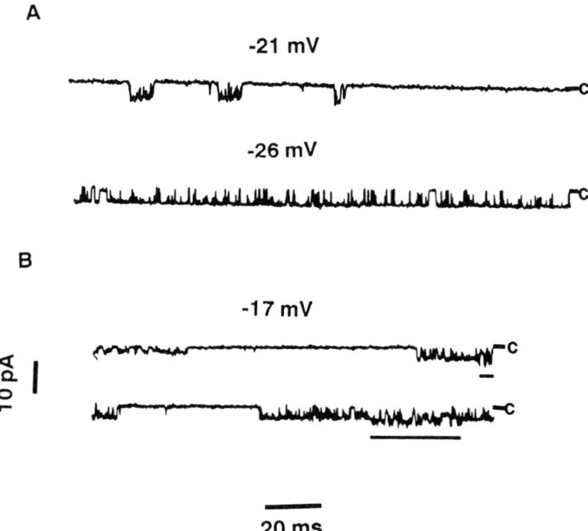

FIG. 2. Development of Ca^{2+} sensitivity by K^+ channels in spinal neurons from *Xenopus* neural plates. (A) Calcium-activated K^+ channel activity in an inside-out patch from a neuron after 23.5 hr in culture. The top record was taken in a solution containing 1 μM internal Ca^{2+} at the indicated membrane voltage. The second record was obtained in the presence of 10 μM internal Ca^{2+}. Notice that the probability of the channel being open for the record obtained at −26 mV (10 μM Ca^{2+}) is much larger than the one obtained at −21 mV (1 μM Ca^{2+}). Increasing Ca^{2+} concentration increased burst duration and decreased interburst intervals. Average channel conductance was 156 pS. (B) In younger neurons (8.5 hr in culture), channel activity is present in the absence of internal Ca^{2+}. In the particular inside-out patch shown in B, a small-conductance channel is also present. Simultaneous activity of the 155 pS channel and the small channel is marked by the horizontal bar below the records. Increasing the $[Ca^{2+}]$ to a final concentration of 100 μM did not change channel activity (not shown). (Modified from Blair and Dionne, 1985.)

we point out here that we do not know at present whether or not Ca^{2+}-activated K^+ channels share structural similarities with other K^+ channels, or if they belong to a different family. On the other hand, Ewald *et al.* (1985) have shown that Ca^{2+} sensitivity of K(Ca) channels is increased upon application of the catalytic subunit of protein kinase together with Mg^{2+} and ATP to the inner surface of the membrane.

B. Mammalian Neurons

Recently, a number of studies have been reported in mammalian neuron cells. Rat cortical neurons grown in a serum-free culture have been investigated using the gigaohm-seal whole-cell voltage-clamp and single-channel recording techniques (Ahmed, 1988a). According to a previous study (Ahmed and Fellows, 1988), the majority of the cells dissociated from neocortex of fetal rats (E18) are in G1 phase of the cell cycle and thus, provide a synchronized population of cells suitable for developmental studies. The earliest appearing membrane current in the soma of these pyramidal-shaped neurons grown in serum-free culture medium is a voltage-dependent K^+ current. This current is composed of a rapidly rising transient outward current with similar pharmacological characteristics to I_A and another current that is similar to the delayed rectifier. The magnitudes of both currents increased with time in culture, but the ratio between them (0.3) remained unchanged at all membrane voltages. This is in contrast to the development of K^+ currents in neural crest cells (Bader *et al.,* 1985), where only the delayed rectifier current increases with time, although both I_A and the delayed rectifier channels are present in neurons cultured for 24 hr. Ahmed (1988a) also reported the presence of 32 pS and 120 pS K^+ channels in neocortical neurons. The larger conductance channel was selective for K^+ and sensitive to internal Ca^{2+} and voltage, and probably belongs to the class of maxi K^+ channels. The small-conductance channel was voltage dependent but, due to its apparent low density, was less well characterized. Both channels were found in the neurons between 24 hr and 10 days in culture. It is not clear what is the contribution of the maxi K^+ channel to the total macroscopic current in the neocortical neurons used in Ahmed's (1988b) studies because they do not exhibit any voltage-dependent Ca^{2+} conductance (Ahmed, 1988b). However, embryonic rat cortical cells in *serum* culture express at least two different Ca^{2+}-selective currents (Zona *et al.,* 1986).

Dorsal root ganglion neurons dissected from 12-day-old (E12) mouse embryos were analyzed 3 hr after their isolation with patch–clamp recording techniques (Valmier *et al.,* 1989). A homogeneous subpopulation of neurons was identified by morphological criteria and by coexpression of vimentin and neurofilament triplet protein. Two types of K^+ currents were already present at this stage of postnatal development: one that can be compared with the transient I_A current and another, similar to I_K, found in other mammalian preparations. An inward Na^+ current was also present, displaying properties similar to those reported in mature mammalian neurons. An interesting facet of this work is that the preparation allows for the isolation of a homogeneous subset of neurons and that currents can be measured within a few hours of isolation, avoiding the risk of changes in ionic currents that occur with maturation.

Avian limb motoneurons recorded at the earliest time at which they can be identified show both types of K^+ currents, the delayed-rectifier-type K^+ current (I_K) and a more rapidly activating, transient K^+ current (I_A), together with Na^+ and Ca^{2+} currents. In these cells, the I_A current shows a great increase, with maturation paralleling the decrease in action potential duration, whereas the I_K current shows only a moderate increase with development (McCobb *et al.,* 1990). These results are similar to those described in rat sympathetic neurons, where a relatively late appearance of I_A currents that affect action potentials has been identified (Nerbonne *et al.,* 1986).

The sequence of appearance of K^+ channels has been studied in cultured rat cerebellar Purkinje neurons with the aim of correlating it with the observed changes in spontaneous activity of these cells (Yool *et al.,* 1988). The advantage of cerebellar Purkinje neurons in culture is that they provide a suitable model for investigating mechanisms of central nervous system neuronal development (Gruol and Franklin, 1987).

Purkinje neurons *in vivo* originate on about day 15 of gestation (E15) (Altman and Bayer, 1985), but delay differentiation until after birth (E21). Therefore, the establishment of cerebellar cultures of rat embryos at day 1 before birth allows the study of these neurons during their entire developmental process. These cells are electrically unexcitable when immature, but acquire excitable membrane properties according to a programmed developmental sequence. Using patch–clamp techniques, Yool *et al.* (1988) characterized the predominant classes of active K^+ channels from days 5 to 29 *in vitro*. They observed four classes of K^+ channels that were identified by their conductances (27, 44, 70, and 100 pS) and other properties, including voltage dependence, sensitivity to TEA, mean open time, and time of appearance during development. The 27 and 44 pS K^+ channels were present beginning at day 5 *in vitro*. The 70 pS K^+ channel appeared at day 7 and the 100 pS at day 9 *in vitro*. Maturation involved differential changes in the frequency of occurrence of the different K^+ channels in the membrane patches. The activity of the 27 pS type of K^+ channel decreased with maturation of the Purkinje neuron. On the other hand, the frequency of occurrence during development increased both for the 70 and 100 pS K^+ channels, with no changes in the 44 pS type. However, the 44 pS channel type undergoes a 10-fold increase in mean open time during development. The change in sensitivity to TEA of the spontaneous activity observed with maturation parallels the presence of the larger conductance K^+ channels (70 and 100 pS).

Yool *et al.* (1988) suggested that developmental regulation of the activity of K^+-selective channels contributes to the ionic mechanism underlying maturation of spontaneous activity patterns. Immature neurons showed a slow pattern of single spontaneous spikes having a low sensitivity to TEA (Fig. 3A). By day 13, the pattern of action potential firing became more complex, concomitant with an increase in maximal firing rate and an increased sensitivity to TEA (Fig. 3B). Also, with maturation, spontaneous activity became dependent on the large-conductance K^+ chan-

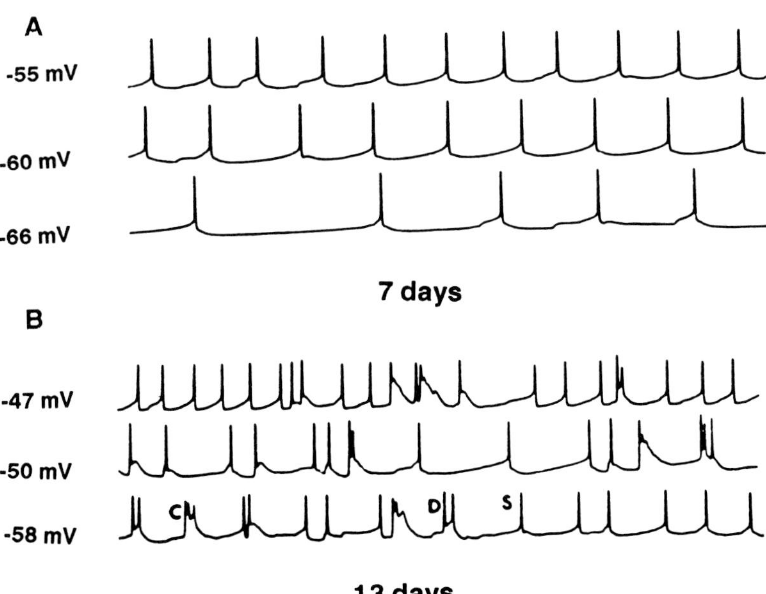

FIG. 3. Spontaneous activity changes with development in Purkinje neurons. (A) Spontaneous electrical activity in immature (7 days) neurons. Notice that membrane hyperpolarization decreases the frequency of spontaneous firing of action potentials. Immature neurons fired simple spikes. Neuron electrical activity becomes more complex with development (B) At 13 days, simple spikes (S) are mixed with double (D) and complex (C) action potentials. Voltages at the left of the records are holding potentials obtained by current injection. (Modified from Yool *et al.*, 1988.)

nels. The 100 pS channel appears to belong to the maxi K channel class since it is blocked by submillimolar TEA concentrations and activated by internal Ca^{2+}. Therefore, it is probable that this channel regulates Ca^{2+} conductances in the neuron. Another important point in this regard is that Ca^{2+} currents play an increasingly more notorious role in generating spontaneous activity as cell maturation progresses. This observation may explain the relatively greater importance of the 100 pS channel in the mature activity pattern. It would be interesting to characterize these channels pharmacologically with specific toxins such as charybdotoxin (Miller *et al.*, 1985) or iberiotoxin and to grow the cells in the presence of these toxins to see how the development of electrical activity changes under these conditions.

Rat inferior olivary neurons at different stages of postnatal development have been studied by current-clamp measurements of slices *in vitro* (Pettigrew *et al.*, 1988). At the earliest time of recording, i.e., postnatal day 2–3, all voltage-dependent conductances seen in older animals were already present in these cells. Pettigrew *et al.* (1988) observed Na^+-dependent action potentials, low-threshold and high-threshold Ca^{2+} spikes, Ca^{2+}-dependent afterhyperpolarizations, instantaneous and time-dependent inward rectifications at hyperpolarized levels of membrane potential, and inactivation with time of an I_A-like K^+ conductance at depolarizing levels of membrane potential. It has not been possible to demonstrate a sequential development of ionic conductances in these cells, probably because brain stem neurons mature earlier than other cells such as cortical neurons. Pettigrew *et al.* (1988) suggested that the balance between K^+ and Ca^{2+} channels changes throughout maturation and is largely in favor of K^+ channels at very immature stages. They reached this conclusion based on results showing that blockade of K^+ channels with 4-AP or Cs^+ unmasks Ca^{2+} currents during the first postnatal week.

C. Mammalian Myelinated Nerve Fibers

At the Ranvier nodes of myelinated fibers, Na^+ currents show a much faster time course for the process of inactivation and a larger leakage current as compared with the squid giant axon. Moreover, early calculations that take into account these facts (Frankenhauser and Huxley, 1964) suggested that membrane repolarization after an action potential can take place in the absence of K^+ currents. Although in the frog node of Ranvier K^+ currents are present, these are very small or absent in mammalian nodes, indicating the absence or a very low density of K^+ channel in this region of the nerve fiber, as predicted by Frankenhauser and Huxley (1964) computations (Horackova *et al.*, 1968; Brismar, 1979; Chiu *et al.*, 1979; Kocsis and Waxman, 1980). However, experimentally promoted demyelinization induced the appearance of delayed outward currents which are blocked by TEA, Cs^+, and 4-AP (Chiu and Ritchie, 1980, 1981). It is of interest therefore, to look for K^+ channels in developing mammalian myelinated nerves because nerve myelination is only complete in mature fibers.

Ritchie (1982) found that both regenerating and developing myelinated fibers in rabbit sciatic nerve show a pronounced sensitivity to 4-AP (but see Waxman and Foster, 1980). 4-Aminopyridine increased the duration of the compound action potential and induced a prolonged refractoriness following an impulse. In completely developed nerves, this contribution of the K^+ currents to the action potential is virtually absent (see above). A simple explanation for these results is that the K^+ currents seen by Ritchie (1982) in regenerating and developing nerve fibers originate from channels located in the paranodal area (the area between the node and the internode). However, it is also possible that, in immature fibers, K^+ channels are also present in the node. More recently, mammalian myelinated axons of rat sciatic nerve have been studied *in situ* over the course of development using *in vitro* sucrose gap and intraaxonal recording techniques (Eng *et al.*, 1988). In agreement with previous studies, sensitivities to the K^+ channel blockers 4-AP and TEA develop with different time courses during sciatic nerve maturation (Waxman and Foster, 1980; Foster *et al.*, 1982; Ritchie, 1982; Kocsis *et al.*, 1983). The effect of 4-AP is much more pronounced in immature nerves and decreases during maturation. This result is in agreement with the observation that the fibers become sensitive again to 4-AP following demyelination of mature myelinated axons (Bostock *et al.*, 1981; Chiu and Ritchie, 1981; Ritchie and Chiu, 1981; Targ and Kocsis, 1984; Bostock and Grafe, 1985). The TEA-sensitive afterhyperpolarization following repetitive firing remains accessible to this blocker after myelination and is present in both young and mature nerves. In the light of these results, Eng *et al.* (1988) proposed that 4-AP sensitive channels may have an internodal location and thus become masked during maturation as they are covered by myelin. TEA-sensitive channels, on the other hand, may be located mainly in the nodal region, although they may have some representation in the internodal axon membrane.

D. K^+ Channels, Schwann Cells, and Development

Although Schwann cells have traditionally been considered a sort of passive support that insulates the nerve fiber with myelin, more recently it has become clear that mammalian Schwann cells in culture contain voltage-dependent ion channels (Chiu *et al.*, 1984; Chiu and Wilson, 1989). Development of the myelin sheath appears to imply changes in the ionic currents present in Schwann cells. In particular, Na^+ currents are present in Schwann cells of small nonmyelinated axons, but cannot be detected in large myelinated axons from sciatic nerve of rabbit (Chiu, 1987). Wilson and Chiu (1990) studied the change in K^+ channel expression that occurs in Schwann cells in developing sciatic nerves of newborn rats using the patch–clamp technique. During the first postnatal week, they found that Schwann cells contained two different types of K^+ channels. One is an inward rectifier (K_{IR}) channel with a conductance of 35 pS in the −150 to −90 mV voltage range relative to the resting potential (Fig. 4A). K_{IR} chan-

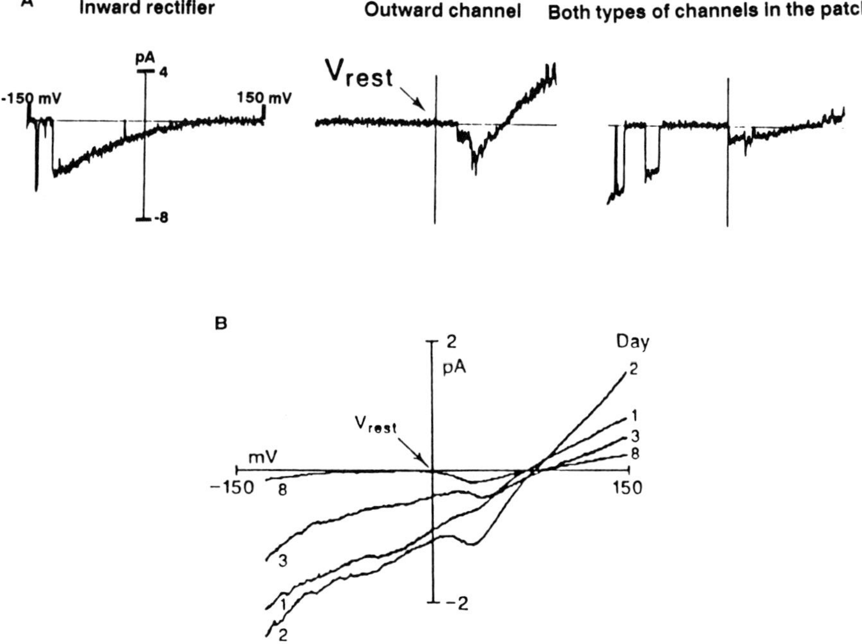

FIG. 4. K^+ channel activity in soma of newborn Schwann cells. (A) Current–voltage (I–V) relationship for a cell-attached patch containing an inward rectifier channel. Single-channel current was collected during a linear applied voltage ramp from −150 to +150 mV relative to the resting potential V_{rest} (left panel). An I–V relationship for the outward K^+ channel is shown in the central panel. The right panel shows an I–V curve taken in a cell-attached patch containing simultaneously inward and outward K^+ channels. The pipette contained a 160 m*M* KCl solution. The bath contained an NaCl-rich solution (Locke). (B) Average of all ensemble currents obtained from different cell-attached recordings at each indicated day. The minimum number of cells examined was 29. Solutions as in A. (Modified from Wilson and Chiu, 1990.)

nels carry preferentially inward current and they are blocked by Cs^+ and (surprisingly, see Rudy, 1988) by 4-AP. The second channel found in the early stages of Schwann cell development is a voltage-dependent K^+ (K_o channel) with a 13 pS conductance (Fig. 4A). Depolarizing voltages activate the K_o channel. In a series of elegant experiments, Wilson and Chiu (1990) showed that the current mediated by the K_{IR} channel almost vanishes from the Schwann cell soma by day 8, being maximal at day 2, whereas I_{Ko} currents are more resilient to decrease with developmental cell changes (Fig. 4B). By day 8, 94 and 82% of the average ensemble I_{IR} and I_{Ko} currents have disappeared, respectively. Chiu and Wilson (1990) have argued that since neither the I_{IR} channel conductance nor the probability of opening changed with development, the number of channels must be decreasing with development [recalling that $I = NP_o g(V - V_K)$, where I is the macroscopic current, N the total number of channels, P_o the probability of opening, V the membrane potential, and V_K the equilibrium potential for K^+]. Furthermore, direct measurements indicate that the number of channels per patch decreased 14-fold from day 2 to day 8. Thus, it appears that the K_{IR} channels, which are open at the resting potential, are needed in early myelinogenesis to prevent K^+ accumulation that occurs after prolonged nerve activity (see, e.g., Frankenhauser and Hodgkin, 1956; Connors *et al.*, 1982). Inward rectifier channels would not play the same role in fully myelinated axons since activity is restricted to the Ranvier nodes. On the other hand, the K_o channel disappearance with maturation parallels the decrease in proliferation. We note here that when neonatal Schwann cells in culture are stimulated to divide, an outward K^+ current is enhanced (Wilson and Chiu, 1988).

III. SMOOTH MUSCLE CELLS AND K(Ca) CHANNELS

In smooth muscle cells from human aorta, most of the voltage-dependent K^+ current is transported by maxi K(Ca) channels. Maturation of these K^+ currents in fetal and adult human aorta smooth muscle cells in culture has been studied using the single-channel and the whole-cell recording variants of the patch–clamp technique (Bregestovski *et al.*, 1988). Both fetal and adult K(Ca) macroscopic currents showed similar kinetics, Ca^{2+} sensitivity, and TEA blockade. However, K(Ca) currents are 4.3-fold larger in adult when compared with those measured in fetal smooth muscle cells. In order to have a better insight on the mechanism that originates this difference, Bregestovski *et al.* (1988) measured the K(Ca) single-channel properties in both types of smooth muscle cells. In symmetrical 140 mM K^+, channel conductance is about 240 pS in fetal and adult

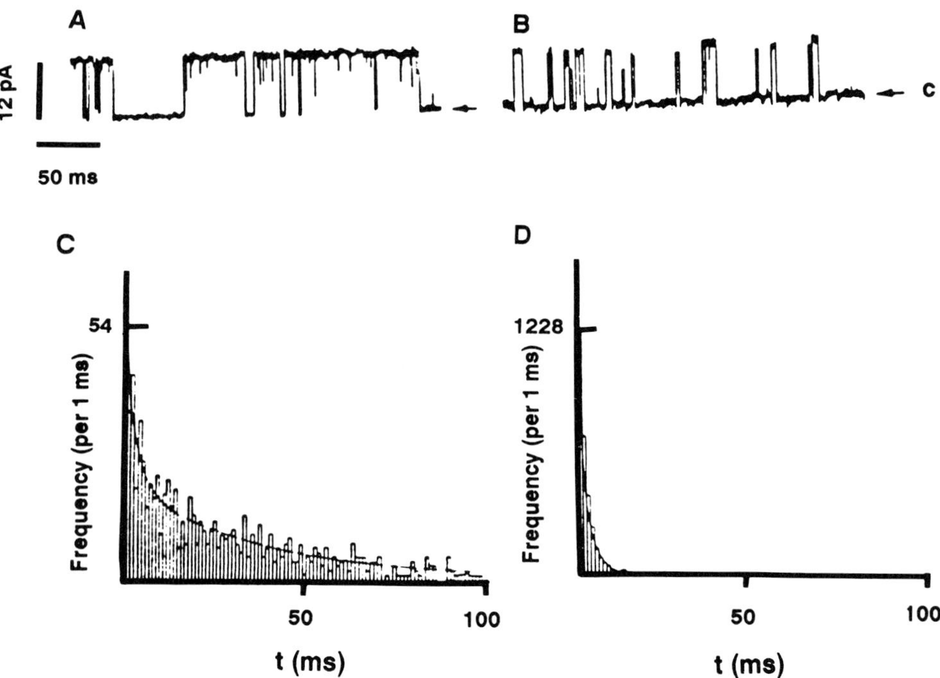

FIG. 5. Development of a Ca^{2+}-activated K^+ channel in smooth muscle cells from human aorta. (A) Channel current record obtained in an inside-out patch from adult smooth muscle cell. The record is characterized by the long burst periods. The arrow at the right side of the records indicates the closed state. An analysis of the channel open dwell times is given in the form of a dwell time histogram in C. Mean open times of 2.65 and 36.7 msec were obtained from the fit with two exponential components to the data. (B) Channel current record obtained in an inside-out patch from an embryonic smooth muscle cell. Burst duration is shortened and the kinetic analysis of the open times indicates a single component with a mean open time of 2.3 msec (D). Voltage was 50 mV and the internal $[Ca^{2+}]$ was 0.1 μM. (Modified from Bregestovski *et al.*, 1988.)

muscle (Fig. 5A and B). This conductance agrees well with that found for maxi K^+ channels in other preparations (Latorre *et al.*, 1989; McManus, 1991). Moreover, fetal and adult muscle channels have similar ion selectivities and are equally sensitive to blockers such as TEA and quinine. The distribution of open dwell times showed, however, that a long time constant kinetic component is lost in the K(Ca) fetal channels (cf. Fig. 5C and D). For example, at a membrane potential of 50 mV, the open dwell time distribution for the fetal K(Ca) channel can be fitted with a single exponential with a time constant of 2.3 msec, whereas the open dwell time distribution for the adult channel needed two exponential components with time constants of 2.7 and 36.7 msec. The area occupied by the slow exponential component present in adult smooth muscle cells is an appreciable fraction of the total area described by the open dwell time distribution. Therefore, the disappearance of the slow component in the fetal K(Ca) channel results in a decrease in the open probability and hence in the total macroscopic current mediated by this channel when compared with the adult channel. Both fetal and adult K(Ca) channels are found to be sensitive to Ca^{2+}, a finding that implies that the mechanism for K(Ca) current maturation is different from that found by Blair and Dionne (1985) in *Xenopus* spinal neurons (see above). As proposed by Bregestovski *et al.* (1988), the nature of the differences between fetal and adult smooth muscle K(Ca) channel kinetics may reside in a change in genetic expression with development. An example of this type of mechanism is the changes in acetylcholine receptor channel activity induced by expression of a new subunit (Mishina *et al.*, 1986). Changes in membrane lipid composition can also induce modifications of channel activity. The concentration of cholesterol in the smooth muscle membrane has been shown to promote reversible alterations in K(Ca) channel gating (Bolotina *et al.*, 1989).

IV. K^+ CHANNELS IN DEVELOPING *Drosophila*

Drosophila flight muscles display a combination of advantages for the study of electrical properties of developing cells in a more comprehensive way than in other insect neuromuscular systems (Salkoff, 1983; Salkoff, 1985). These advantages include (1) large isopotential muscle cells which may be voltage-clamped using the two-microelectrode technique, and (2) the ability to isolate the different conductance systems selectively by the use of genetic tools and the developmental properties of the system. In this regard, *Drosophila* represents an organism that is suitable for the use of both genetics and neurophysiological techniques (Benzer, 1973; Salkoff and Wyman, 1981; Salkoff and Tanouye, 1986; Rao *et al.*, 1990).

At the midpoint of pupal development in *Drosophila,* there appears a single ion channel type that carries a fast transient K^+ current (Salkoff and Wyman, 1981). This current corresponds to I_A. Maturation is followed by the development of a slower, voltage-dependent K^+ current, of the delayed rectifier type. The two outward currents are different not only because of the different time of appearance and kinetics, but also because it has been demonstrated that the I_A current is absent from the *Drosophila* mutant *Shaker* (see Section I and Solc *et al.*, 1987). At the end of pupal development, a large Ca^{2+} current, I_{Ca} current, appears abruptly (Salkoff and Wyman, 1981, 1983).

Later studies in these muscle cells have demonstrated the presence of a large Ca^{2+}-sensitive K^+ current (I_C; Wei and Salkoff, 1986). In the presence of high external Ca^{2+} concentrations, an outward current with a rate of activation similar to the voltage-dependent I_K that is blocked by TEA can be unmasked at very early stages of development (72-hr pupae). The I_C current is different from I_A and from I_K inasmuch as it is present in the mutant *Shaker* and is not present in the absence of Ca^{2+} in the external medium. Furthermore, Wei and Salkoff (1986) demonstrated that the I_C current appears before I_A in flight muscles. The presence of an early developing Ca^2-activated K^+ current also indicates the existence of Ca^{2+} channels at this stage of maturation. However, Ca^{2+} currents are not evident in normally developing muscles until the adult eclosion stage (Salkoff and Wyman, 1981; Salkoff, 1985), which is much later than the period in which the I_C currents are observed. Wei and Salkoff (1986) investigated the possibility of the presence of less prominent Ca^{2+} currents early in muscle cell development. When EGTA or ATP is injected in order to lower the intracellular Ca^{2+} concentration, a large inward Ca^{2+} current is unmasked in cells that had no trace of inward current before the EGTA or ATP injections. These results clearly demonstrated the presence of Ca^{2+} channels in flight muscles more than 2 days before the appearance of outward K^+ currents. However, the fact that this inward Ca^{2+} current is activated only after a decrease in intracellular Ca^{2+} by the injection of EGTA or ATP indicates that Ca^{2+} currents are normally inactive at early stages of muscle development. The significance of the presence of Ca^{2+} inward currents at early stages of development is unknown, but is probably due to the fact that the intracellular concentration of Ca^{2+} in early cells is high enough to inhibit Ca^{2+} channel activity. As pointed out by Wei and Salkoff (1986), the role of I_C at this developmental stage of *Drosophila* muscle is also unclear and they proposed that this channel may be preventing unwanted Ca^{2+} influx.

A new K^+ conductance appears in the first day of the adult *Drosophila*. This is a transiently Ca^{2+}-activated K^+ current (I_{Acd}), with a time course similar to that of I_A, and is capable of reducing membrane excitability (MacDermott and Weight, 1982; Salkoff and Tanouye, 1986). A mutation of *Drosophila*, slowpoke (*slo*), specifically abolishes I_{Acd} without affecting I_A (Elkins *et al.*, 1986). The mutant *slo* was used to demonstrate directly that I_{Acd} is the current primarily responsible for the action potential repolarization. This current is blocked by TEA and charybdotoxin, but is unaffected by quinine and apamin (Elkins *et al.*, 1986). More recently, Komatzu *et al.* (1990) found that there are no differences between normal and heterozygous (*slo*/+) I_{Acd} channels, suggesting that the *slo* mutation can be a defect in a modulation mechanism specific for appropriate functioning of the I_{Acd} channel-forming protein.

V. CELL INTERACTION AND ION CHANNEL DEVELOPMENT

Most of the studies on ion channels in development discussed above are oriented to the sequence of appearance of the different K^+ channels and their evolution during development of individual cells. Also, most experiments are performed on isolated cells in culture. This gives the advantage of studying the cells in defined medium under controlled environmental conditions. However, under natural conditions, cells during development are exposed to contacts with cells of different kinds and, most probably, the expression and properties of ion channels depend on such interactions.

A. Ca^{2+}-Activated K^+ Channels of Small Conductance

An interesting example of the effect of interaction of cells from different tissues on the expression of K^+ channels is observed with the small-conductance Ca^{2+}-activated K^+ channels (SK channels) of skeletal muscle. SK channels are specifically blocked by apamin, a toxin from bee venom (Haberman, 1972; Romey and Lazdunski, 1984; Moczydlowski *et al.*, 1988). At concentrations of 1–10 n*M*, apamin specifically inhibits SK channels (Burgess *et al.*, 1981; Romey and Lazdunski, 1984; Pennefather *et al.*, 1985). Another toxin, Lq VIII, which is isolated from venom of the scorpion *Leirus quinquestriatus*, has been isolated recently and competes with apamin for binding sites in liver (Castle and Strong, 1986).

Single-channel studies in skeletal muscle cells in culture show that SK channels have a small conductance (6–18 pS), compared to the maxi channel (200 pS). Their opening probability has a weak voltage dependence, they possess a higher Ca^{2+} sensitivity (half-maximal activation is obtained at 200–500 n*M* Ca^{2+}) than the maxi channel, and they are not affected by 5 m*M* external TEA (Blatz and Magleby, 1986). Similar single-channel properties have been demonstrated in guinea pig hepatocytes (Capiod and Ogden, 1989), except that the range of Ca^{2+} concentration that activates the channel is narrower (0.3–1 μM) and the open probability versus $[Ca^{2+}]$ curve is steeper than that observed in rat myotubes (Blatz and Magleby, 1986) or in sympathetic ganglion neurons (Gurney *et al.*, 1987).

^{125}I-Labeled apamin-binding experiments have shown that apamin-binding sites (and presumably SK channels) are present in cells from many different tissues, such as liver (Cook and Haylett, 1985), myotubes (Hugues *et al.*, 1982d), smooth muscle of the gut (Hugues *et al.*, 1982b), undifferentiated pheochromocytoma cells (Schmid-Antomarchi *et al.*, 1986), brain synaptosomes (Hugues *et al.*, 1982a), neuroblastoma cells (Hugues *et al.*, 1982c), and embryonic neurons (Seagar *et al.*, 1984). From the effect of the injection of apamin on animals, it is clear that apamin transverses the blood–brain barrier and has actions in the central nervous system (Habermann and Cheng-Raude, 1975; Haberman, 1972). Autoradiographic and equilibrium binding experiments show that apamin-binding sites occur in many regions of the central nervous system, particularly in the limbic forebrain (Mourre *et al.,* 1986).

SK channels are involved in neurotransmitter- and hormone-induced secretions (Banks *et al.,* 1979; Capiod and Ogden, 1989), and also in the hyperpolarization that follows action potentials in many excitable cells, such as sympathetic ganglion neurons (Pennefather *et al.,* 1985), spinal cord motoneurons (Zhang and Krnjevic, 1987), dopamine-containing neurons of the substantia nigra (Shepard and Bunney, 1988), supraoptic hypothalamic neurons (Buorque and Brown, 1987), sensorimotor cortical neurons (Schwindt *et al.,* 1988), and muscle spindle neurons (Kruse and Poppele, 1989).

B. Down-Regulation of SK Channels by Innervation

Myoblast cells in culture show a slow hyperpolarization that follows action potentials. Voltage-clamp studies in these cells have determined that this afterhyperpolarization is promoted by a $I_{K(Ca)}$ current that is insensitive to TEA and blocked by apamin (Romey and Lazdunski, 1984). On the other hand, in adult skeletal muscle, ^{125}I-labeled apamin binding is negligible, indicating a very low density or the absence of apamin-sensitive SK channels. However, apamin receptors are detectable after denervation of the muscles (Schmidt-Antomarchi *et al.,* 1985). Concomitant with the appearance of apamin receptors, denervated muscles show an afterhyperpolarization that is inhibited by apamin. Furthermore, apamin receptors detected with ^{125}I-labeled apamin binding are present in rat fetal muscle. As maturation proceeds, the receptors decrease in number and disappear within the first week of postnatal life (Schmid-Antomarchi *et al.,* 1985). The application of inhibitors of protein synthesis concomitant with denervation prevents the expression of apamin receptors (Schmid-Antomarchi *et al.,* 1985), suggesting that this process requires protein synthesis.

The role of innervation on SK channels in muscle is further demonstrated by the experiments on myotubes in culture. In these cells, the apamin-sensitive afterhyperpolarization disappears when myotubes are cocultured with spinal cord neurons under the conditions of *in vitro* innervation (Schmidt-Antomarchi *et al.,* 1985; Suarez-Isla *et al.,* 1986). Furthermore, the apamin-sensitive afterhyperpolarization can be inhibited when the culture medium of spinal neuron cells *in vitro* is added to myotubes in culture. A low-molecular-mass fraction (<4 kDa) isolated from the culture medium of spinal neurons is also effective in suppressing the afterhyperpolarization. This effect seems to be specific for spinal cord neurons, because the addition of the incubation medium of retinal neurons in culture to myocytes had no effect on the apamin-sensitive afterhyperpolarization (Suarez-Isla *et al.*, 1986).

The results discussed above suggest that the expression of apamin receptors in skeletal muscle membranes is regulated by nerve–muscle interactions. It is possible that the nerve secretes a neurotrophic factor which could somehow interact with the membrane of skeletal muscles and inhibit the expression of SK channels. In accordance with this possibility, the blockade of axonal flow by the application of colchicine around rat sciatic nerves induces the appearance of ^{125}I-labeled apamin-binding sites in the membranes of skeletal muscles of the hind leg. ^{125}I-Labeled apamin binding is detected at day 4 after the application of colchicine to the nerve, is maximum around day 8–15, and then decreases, being negligible at day 35. No binding is detected in control muscles, in which a buffer solution was applied to the nerve (M. I. Behrens and C. Vergara, unpublished observations). This suggests that the factor that may regulate the expression of SK channels in muscle travels along the nerve by axonal flow.

VI. RECAPITULATION AND CODA

A general conclusion of the results described above is that K^+ channels are present in the membrane of cells from very early stages of development. Previously, this had been observed mainly in nonmammalian tissues (Spitzer, 1979). Recent results, in mammalian neurons and *Drosophila* flight muscles, confirm this statement. In *Drosophila* flight muscles, the sequence of appearance of ion channels during maturation has been studied with great precision. The first functional channel to appear seems to be a Ca^{2+}-activated K^+ channel. The existence of *Drosophila* mutants lacking a given type of ion channel has allowed the cloning of a gene called *Shaker* that encodes a family of K^+ channel components that correspond to the A current (Papazian *et al.*, 1987; Rao *et al.*, 1990; Schwarz *et al.*, 1990). The availability of mutants in other cell types will certainly be of extraordinary value for the study of ion channels in development (see, e.g., Komatzu *et al.*, 1990).

In developing neurons, K^+ channels are visualized in the membrane at the earliest time at which these cells can be recorded. In the soma of rat neocortex neurons in culture, K^+ channels are the first channels to be expressed. They are expressed before any inward current is measured;

Na^+ currents are observed in these neurons several hours later. In other neurons, such as spinal neurons in culture, dorsal root ganglion cells in culture, and rat inferior olivary neuron slices, they are observed together with Na^+ or Ca^{2+} inward currents. The simultaneous presence of outward and inward currents in these cells is most probably due to the difficulty in obtaining very early recordings during development. It may be that K^+ channels are the first channels to be expressed in these cells too. The presence of very early inward Ca^{2+} currents in neurons in development, like those demonstrated in *Drosophila* flight muscles, cannot be discarded at the moment.

The presence of K^+ channels early in neuronal development, before the appearance of any inward current in the cell body, poses the interesting question of the functional role of these channels. They are probably not related to repolarization of the membrane or modulation of excitability. Furthermore, K^+ channels are not only restricted to excitable cells in development. They seem to play an important role in the development of nonexcitable cells too. Studies with the gigaohm-seal patch–clamp technique have revealed the presence of voltage-gated K^+ channels, similar to the delayed rectifier, in human T lymphocytes and natural killer cells. Furthermore, K^+ channels seem to play a role in the induction of proliferation of T and cytotoxic lymphocytes (DeCoursey *et al.,* 1984; Sharma, 1988). A recent study in murine thymocytes shows that these cells express three types of voltage-gated K^+ channels, which vary with the cell developmental state and can be used as cell-surface markers to identify precursors of the cytotoxic suppressor T cell lineage (Lewis and Cahalan, 1988).

Calcium-activated K^+ channels have the virtue of being activated by the intracellular Ca^{2+} concentration and some of them also by membrane depolarization (Latorre *et al.,* 1989). In this sense, they have been proposed as the natural link between the electrical activity of the plasma membrane and the cell metabolism (Meech, 1978). For example, maxi K^+ channels have been described in many excitable and nonexcitable cells (Behrens *et al.,* 1988; Latorre *et al.,* 1989). In nonexcitable cells, the function of maxi K channels is related to the regulation of secretion of K^+ and other ions and water (Petersen and Findlay, 1987; Gitter *et al.,* 1987). Another function that has been proposed is the protection of the cell against adverse external conditions by restoring the membrane potential to its normal value (Christensen and Zeuthen, 1987). The channels described in *Xenopus* neurons, Purkinje neurons, neocortical neurons, and smooth muscle most probably belong to the maxi K^+ class. *Drosophila* flight muscles also express K(Ca)s early in development. The early presence of K(Ca) channels in neurons and *Drosophila* flight muscles suggests that these currents may be related to metabolic functions of the cells at this stage of development.

Acknowledgments

This work was financed in part by the Fondo Nacional de Investigacion, grants 451-88 and 296-89, NIH GM-35981, and by the Tinker Foundation. RL is a recipient of a John D. Guggenheim fellowship. RL wish to thank the Dreyfus Bank for generous support from a private foundation that they made available to him.

References

Adams, P. R., Brown, D. A., and Constanti, A. (1982a). M-currents and other potassium currents in bullfrog sympathetic neurons. *J. Physiol. (London)* **330,** 537–572.

Adams, P. R., Constanti, A., Brown, D. A., and Clark, R. B. (1982b). Intracellular Ca^{2+} activates a fast voltage sensitive K^+ current in vertebrate sympathetic neurones. *Nature (London)* **296,** 746–749.

Adrian, R. H. (1969). Rectification in muscle membrane. *Prog. Biophys. Mol. Biol.* **19,** 340–369.

Ahmed, Z. (1988a). Expression of membranes currents in rat neocortical neurones in serum free culture. I. Inward currents. *Dev. Brain Res.* **40,** 285–295.

Ahmed, Z. (1988b). Expression of membrane currents in rat neocortical neurons in serum free culture. II. Outward currents. *Dev. Brain Res.* **40,** 297–305.

Ahmed, Z., and Fellows, R. E. (1988). Determination of the birth date and proliferative state of dissociated cells from fetal rat brain. *Dev. Brain Res.* **37,** 77–87.

Altman, J., and Bayer, S. A. (1985). Embryonic development of rat cerebellum. III. Regional differences in the time of origin, migration, and settling of Purkinje cells. *J. Comp. Neurol.* **232,** 42–65.

Augustine, C. K., and Bezanilla, F. (1990). Phosphorylation modulates potassium conductance and gating current of perfused giant axons of squid. *J. Gen. Physiol.* **95,** 245–271.

Bader, C. R., Berhans, D., Duprin, E., and Kato, A. C. (1983). Development of electrical membrane properties in cultured avian neural crest. *Nature (London)* **305,** 808–810.

Bader, C. R., Berhans, D., and Duprin, E. (1985). Voltage dependent potassium currents in developing neurones from quail mesencephalic neural crest. *J. Physiol. (London)* **366,** 129–151.

Baker, K., and Salkoff, L. (1990). The Drosophila *Shaker* gene codes for a distinctive K^+ current in a subset of neurons. *Neuron* **2,** 129–140.

Banks, B. E. C., Brown, C., Burgess, G. M., Burnstock, G., Claret, M., Cocks, T. M., and Jenkinson, T. H. (1979). Apamin blocks certain neurotransmitter-induced increases in potassium permeability. *Nature (London)* **282,** 415–417.

Barret, E. F., Barret, J. N., and Crill, W. E. (1980). Voltage sensitive outward currents in cat motoneurones. *J. Physiol. (London)* **304,** 251–276.

Barish, M. E. (1986). Differentiation of voltage-gated potassium current and modulation of excitability in cultured amphibian spinal neurones. *J. Physiol. (London)* **375,** 229–250.

Behrens, M. I., Vergara, C., and Latorre, R. (1988). Calcium-activated potassium channels of large unitary conductance. *Braz. J. Med. Biol. Res.* **21,** 1101–1117.

Benzer, S. (1973). Genetic dissection of behavior. *Sci. Am.* **29,** 34–37.

Bixby, J. L., and Spitzer, N. C. (1984). The appearance and development of neurotransmitter sensitive in *Xenopus* embryo on spinal neurons *in vitro*. *J. Physiol.* (*London*) **353,** 143–155.

Blair, L. A., and Dionne, V. E. (1985). Developmental acquisition of Ca^{2+}-sensitivity by K^+ channels in spinal neurons. *Nature* (*London*) **315,** 329–331.

Blatz, A. L., and Magleby, K. L. (1986). Single apamin-blocked Ca^{2+}-activated K^+ channels of small conductance in cultured rat skeletal muscle. *Nature* (*London*) **323,** 718–720.

Blatz, A. L., and Magleby, K. L. (1987). Calcium-activated potassium channels. *Trends Neurosci.* **10,** 463–467.

Bolotina, V., Omelyanenko, V., Heyes, B., Ryan, U., and Bregestovski, P. (1989). Variations of membrane cholesterol alter the kinetics of Ca^{2+}-dependent K^+ channels and membrane fluidity in vascular smooth muscle cells. *Pfluegers Arch.* **415,** 262–268.

Bostock, H., and Grafe, P. (1985). Activity-dependent excitability changes in normal and demyelinated rat spinal root axons. *J. Physiol.* (*London*) **365,** 239–257.

Bostock, H., Searl, T. A., and Sherratt, R. M. (1981). The effect of 4-amino pyridine and tetraethylammonium ions on normal and demyelinated mammalian nerve fibers. *J. Physiol.* (*London*) **313,** 301–315.

Bregestovski, P. D., Pritseva, O. Y., Serebryakov, V., Stinnakre, J., Turmin, A., and Zamoyski, V. (1988). Comparison of Ca^{2+}-dependent K^+ channels in the membrane of smooth muscle cells isolated from adult and foetal aorta. *Pfluegers Arch.* **413,** 8–13.

Brismar, T. (1979). Potential clamp experiments on myelinated nerve fibres from alloxan diabetic rats. *Acta Physiol. Scand.* **105,** 384–386.

Brown, D. H., Adams, P. R., and Costanti, A. (1982). Voltage-sensitive K^+ currents in sympathetic neurones and their modulation by neurotransmitter. *J. Auton. Nerv. Syst.* **6,** 23–35.

Buorque, C. W., and Brown, D. A. (1987). Apamin and *d*-tubocurarine block the after-hyperpolarization of rat supraoptic neurosecretory neurons. *Neurosci. Lett.* **82,** 185–190.

Burgess, G. M., Claret, M., and Jenkinson, D. H. (1981). Effects of quinine and apamin on the calcium dependent potassium permeability of mammalian hepatocytes and red cell. *J. Physiol.* (*London*) **317,** 67–90.

Buttler, A., Wei, A., Baker, K., and Salkoff, L. (1989). A family of putative potassium channels genes in *Drosophila*. *Science* **243,** 943–947.

Capiod, T., and Ogden, D. C. (1989). The properties of calcium-activated potassium ion channels in guinea-pig isolated hepatocytes. *J. Physiol.* (*London*) **409,** 285–295.

Castle, N. A., and Strong, P. N. (1986). Identification of two toxins from scorpion (*Leiurus quinquestriatus*) venom which block distinct classes of calcium-activated potassium channel. *FEBS Lett.* **209,** 117–121.

Chiu, S. Y. (1987). Sodium currents in axon-associated Schwann cells from adult rabbits. *J. Physiol.* (*London*) **386,** 181–203.

Chiu, S. Y., and Ritchie, J. M. (1980). Potassium channels in nodal and internodal axonal membrane of mammalian myelinated fiber. *Nature* (*London*) **284,** 170–171.

Chiu, S. Y., and Ritchie, J. M. (1981). Evidence for the presence of potassium channels in the paranodal region of acutely demyelinated mammalian nerve fibers. *J. Physiol.* (*London*) **313,** 415–437.

Chiu, S. Y., and Wilson, G. F. (1989). The role of potassium channels in Schwann cell proliferation in Wallerian degeneration of explant rabbit sciatic nerves. *J. Physiol.* (*London*) **408,** 199–222.

Chiu, S. Y., Ritchie, J. M., Rogart, R. B., and Stagg, D. (1979). A quantitative description of membrane currents in rabbit myelinated nerve. *J. Physiol.* (*London*) **292,** 149–166.

Chiu, S. Y., Shrager, P., and Ritchie, J. M. (1984). Neuronal-type Na^+ and K^+ channels in rabbit cultured Schwann cells. *Nature* (*London*) **311,** 156–157.

Christensen, O., and Zeuthen, T. (1987). Maxi K^+ channels in leaky epithelia are regulated by intracellular Ca^{2+}, pH, and membrane potential. *Pfluegers Arch.* **408,** 249–259.

Connors, J. A., and Stevens, C. F. (1971a). Voltage clamp studies of a transient outward membrane currents in gastropod neural somata. *J. Physiol. (London)* **213,** 21–30.

Connors, J. A., and Stevens, C. F. (1971b). Prediction of repetitive firing behavior from voltage clamp data on an isolated neurone soma. *J. Physiol. (London)* **213,** 31–53.

Connors, B. W., Ransom, B. R., Kunis, D. M., and Gutnick, M. J. (1982). Activity-dependent K accumulation in the developing rat optic nerve. *Science* **216,** 1341–1343.

Cook, N. S., and Haylett, D. G. (1985). Effects of apamin, quinine and neuromuscular blockers on K(Ca) channels in guinea-pig hepatocytes. *J. Physiol. (London)* **358,** 373–394.

DeCoursey, T. E., Chandy, K. G., Gupta, S., and Cahalan, M. D. (1984). Voltage-gated K channels in human T lymphocytes: A role in mitogenesis. *Nature (London)* **307,** 465–468.

Elkins, T., Ganetzky, B., and Wu, C.-F. (1986). A *Drosophila* mutation that eliminates a calcium-dependent potassium current. *Proc. Natl. Acad. Sci. U.S.A.* **83,** 8415–8419.

Eng, D. L., Gordenl, T. R., Kocsis, J. D., and Waxman, S. B. (1988). Development of 4-AP and TEA sensitivities in mammalian myelinated nerve fibers. *J. Neurophysiol.* **60,** 2168–2178.

Ewald, D., Williams, A., and Levitan, I. B. (1985). Modulation of single Ca^{++}-dependent K^+ channel activity by protein phosphorylation. *Nature (London)* **315,** 503–506.

Fischbach, G. D., and Nelson, P. G. (1984). Cell culture in neurobiology. *In* "Handbook of Physiology. Sect. I: The Nervous System" (E. R. Kandel, ed.), Vol. 1, pp. 719–774. Am. Physiol. Soc., Washington, D.C.

Foster, R. E., Connors, B., and Waxman, S. G. (1982). Rat optic nerve electrophysiological, pharmacological and anatomic studies during developmental. *Dev. Brain Res.* **3,** 371–386.

Frankenhaeuser, B., and Hodgkin, A. L. (1956). The after effects of impulses in the giant nerve fibres of *Loligo*. *J. Physiol. (London)* **131,** 341–376.

Frankenhaeuser, B., and Huxley, A. F. (1964). The action potential in the myelinated nerve fibre of *Xenopus laevis* as computed on the basis of voltage clamp data. *J. Physiol. (London)* **171,** 302–315.

Garcia, M. L., Galvez, A., Garcia-Calvo, M., King, V. F., Vazquez, J., and Kaczorowski, G. J. (1991). Use of toxins to study potassium channels. *J. Bioenerg. Biomembr.* (in press).

Gitter, A. H., Beyenbach, K. W., Chadwick, C. W., Gross, P., Minuth, W. W., and Fromter, E. (1987). High conductance potassium channels in apical membranes of principal cells cultured from rabbit renal cortical collecting duct anlagen. *Pfluegers Arch.* **408,** 282–290.

Gruol, D. L., and Franklin, C. L. (1987). Morphological and physiological differentiation of Purkinje neurons in cultures of rat cerebellum. *J. Neurosci.* **7,** 1271–1293.

Gurney, A. M., Tsien, R. W., and Lester, H. A. (1987). Activation of a potassium current by rapid photochemically generated step increases of intracellular calcium in rat symphathetic neurons. *Proc. Natl. Acad. Sci. U.S.A.* **84,** 3496–3500.

Haberman, E. (1972). Bee and wasp venoms. *Science* **177,** 314–322.

Haberman, E., and Cheng-Raude, D. (1975). Central neurotoxicity of apamin, crotamin, phospholipase A, and alpha-amanitin. *Toxicon* **13,** 465–473.

Hille, B. (1984). "Ionic Channels of Excitable Membranes." Sinauer, Sunderland, Massachusetts.

Hodgkin, A. L., and Huxley, A. F. (1952). A quantitative description of membrane current and its application to conduction and excitation in nerve. *J. Physiol. (London)* **117,** 500–544.

Horackova, M., Nonner, W., and Stampfli, R. (1968). Action potentials and voltage clamp currents of single rat Ranvier nodes. *Proc. Int. Union Physiol. Sci.* **7,** 198.

Hugues, M., Duval, O., Kitabgi, P., Lazdunski, M., and Vicent, J. P. (1982a). Preparation of a pure monoiodo derivative of the bee venom neurotoxin apamin and its binding properties to rat brain synaptosomes. *J. Biol. Chem.* **257,** 2762–276.

Hugues, M., Duval, D., Schmid, H., Kitabgi, P., Lazdunski, M., and Vincent, J. P. (1982b). Specific binding and pharmacological interactions of apamin, the neurotoxin from bee venom, with guinea-pig colon. *Life Sci.* **31,** 437–443.

Huges, M., Romey, G., Duval, O., Vincent, J. P., and Lazdunski, M. (1982c). Apamin as a selective blocker of Ca^{2+} dependent K^+ channel in neuroblastoma cell: Voltage clamp and biochemical characterization of the toxin receptor. *Proc. Natl. Acad. Sci. U.S.A.* **79,** 1308–1312.

Hugues, M., Schmid, H., Romey, G., Duval, D., Frelin, C., and Lazdunski, M. (1982d). The Ca-dependent slow K conductance in cultured muscle cells: characterization with apamin. *EMBO J.* **9,** 1039–1042.

Isacoff, E. Y., Jan, Y. N., and Jan, L. Y. (1990). Evidence for the formation of heteromultimeric potassium channels in *Xenopus* oocytes. *Nature (London)* **345,** 530–534.

Iverson, L. E., Tanouye, M. A., Lester, H. A., Davison, N., and Rudy, B. (1988). A-Type potassium channels expressed from *Shaker* locus cDNA. *Proc. Natl. Acad. Sci. U.S.A.* **85,** 5723–5727.

Jan, L. Y., and Jan, Y. N. (1990). How might the diversity of potassium channels be generated. *Trends Neurosci.* **13,** 415–419.

Jan, Y. N., Jan, L. Y., and Dennis, M. J. (1977). Two mutation of synaptic transmission in *Drosophila. Proc. R. Soc. London, Ser. B* **198,** 87–108.

Kocsis, J. D., and Waxman, S. G. (1980). Absence of potassium conductance in central myelinated axons. *Nature (London)* **287,** 348–349.

Kocsis, J. D., Ruiz, J. A., and Waxman, S. G. (1983). Maturation of mammalian myelinated fibers: Changes in action potential characteristics following 4-aminopyridine application. *J. Neurophysiol.* **50,** 449–463.

Komatzu, A., Singh, S., Rathe, P., and Wu, C.-F. (1990). Mutational and gene dosage analysis of calcium activated potassium channels in Drosophila: Correlation of micro- and macroscopic currents. *Neuron* **4,** 313–321.

Kruse, M. N., and Poppele, R. E. (1989). Electrical tuning in muscle spindles receptors. *Brain Res.* **496,** 303–306.

Latorre, R., and Miller, C. (1983). Conduction and selectivity in K^+ channels. *J. Membr. Biol.* **71,** 11–30.

Latorre, R., Oberhauser, A., Labarca, P., and Alvarez, O. (1989). Varieties of calcium-activated potassium channels. *Annu. Rev. Physiol.* **51,** 385–399.

Lewis, R. S., and Cahalan, M. D. (1988). Specific expression of potassium channels in developing murine T lymphocytes. *Science* **239,** 771–775.

McCobb, D. P., Best, P. M., and Beam, K. G. (1990). The differentiation of excitability in embryonic chick limb motoneurons. *J. Neurosci.* **10,** 2974–2984.

MacDermott, A. B., and Weight, F. F. (1982). Action potential repolarization may involve a transient Ca^{2+} sensitive outward current in a vertebrate neurone. *Nature (London)* **300,** 185–188.

Maelicke, A. (1988). Structural similarities between ion channel proteins. *Trends Neurosci.* **11,** 199–202.

McManus, O. (1991). Calcium-activated potassium channels: Regulation by calcium. *J. Bioenerg. Biomembr.* (in press).

Meech, R. W. (1978). Calcium-dependent potassium activation in nervous tissues. *Annu. Rev. Biophys. Bioeng.* **7,** 1–18.

Meech, R. W., and Standen, N. B. (1975). Potassium activation in Helix aspersa neurones under voltage clamp: A component mediated by calcium influx. *J. Physiol.* (*London*) **249,** 211–249.

Miller, C. (1989). Genetic manipulation of ion channels: A new approach to structure and mechanism. *Neuron* **2,** 1195–1205.

Miller, C., Moczydlowski, E., Latorre, R., and Phillips, M. (1985). Charybdotoxin, a protein inhibitor of single Ca^{2+}-activated K^+ channels from skeletal muscle. *Nature* (*London*) **313,** 316–318.

Mishina, M., Takai, T., Imoto, K., Noda, M., Takahashi, T., Numa, S., Methfessel, C., and Sakmann, B. (1986). Molecular distinction between fetal and adult forms of muscle acetylcholine receptor. *Nature* (*London*) **321,** 406–411.

Moczydlowski, E., Lucchesi, K., and Ravindran, A. (1988). An emerging pharmacology of peptide toxins targeted against potassium channels. *J. Membr. Biol.* **105,** 95–111.

Mourre, C., Hugues, M., and Lazdunski, M. (1986). Quantitative autoradiographic mapping in rat brain of the receptor of apamine, a polypeptide toxin specific for one class of Ca^{2+} dependent K^+ channels. *Brain Res.* **382,** 239–249.

Neher, E. (1971). Two fast transient current components during voltage clamp on snail neurones. *J. Gen. Physiol.* **58,** 36–53.

Nerbonne, J. M., Gurney, A. M., and Rayburn, H. B. (1986). Development of the fast transient outward K^+ current in embryonic sympathetic neurons. *Brain Res.* **378,** 197–202.

Papazian, D., Schwarz, T., Tempel, B., Jan, Y. N., and Jan, L. Y. (1987). Cloning of genomic and complementary DNA fron Shaker, a putative potassium channel. *Science* **237,** 749–753.

Pennefeather, P., Lancaster, B., Adams, P. R., and Nicole, R. A. (1985). Two distinct Ca-dependent K currents in bullfrog sympathetic ganglion cells. *Proc. Natl. Acad. Sci. U.S.A.* **82,** 3040–3044.

Perozo, E., and Bezanilla, F. (1991). Phophorylation of K^+ channels in the squid giant axon: A mechanistic analysis. *J. Bioenerg. Biomembr.* (in press).

Petersen, O. H., and Findlay, I. (1987). Electrophysiology of the pancreas. *Physiol. Rev.* **67,** 1054–1116.

Pettigrew, A. G., Crepel, F., and Krupa, M. (1988). Development of ionic conductances in neurons of the inferior olive in the rat: An *in vitro* study. *Proc. R. Soc. London, Ser. B* **234,** 199–218.

Rao, Y., Jan, L. Y., and Jan, Y. N. (1990). Similarity of the product of the *Drosophila* neurogenic gene *big brain* to transmembrane channel proteins. *Nature* (*London*) **345,** 163–167.

Reuter, H. (1984). Ion channels in cardiac membranes. *Annu. Rev. Physiol.* **46,** 473–484.

Ritchie, J. M. (1982). Sodium and potassium channels in regenerated and developing mammalian myelinated nerves. *Proc. R. Soc. London, Ser. B* **215,** 273–287.

Ritchie, J. M., and Chui, S. Y. (1981). Distribution of sodium and potassium channels in mammalian myelinated nerve. *In* "Demyelinating Decease: Basic and Clinical Electrophysiology" (S. G. Waxman and J. M. Ritchie, eds.), pp. 329–342. Raven, New York.

Romey, G., and Lazdunski, M. (1984). The coexistence in rat muscle cells of two distinct classes of Ca^{2+}-dependent K channels with different pharmacological properties and different physiological functions. *Biochem. Biophys. Res. Commun.* **118,** 669–674.

Rudy, B. (1988). Diversity and ubiquity of K channels. *Neuroscience* **25,** 729–749.

Ruppersberg, P., Schroter, H. K., Sakmann, B., Stocker, M., Sewing, S., and Pongs, O. (1990). Heteromultimeric channels formed by rat brain potassium-channel proteins. *Nature* (*London*) **345,** 535–537.

Salkoff, L. B. (1983). Genetic and voltage-clamp analysis of a *Drosophila* potassium channel. *Cold Spring Harbor Symp. Quant. Biol.* **48,** 221–231.

Salkoff, L. B. (1985). Development of ion channels in the flight muscles of *Drosophila. J. Physiol. (Paris)* **80,** 275–282.

Salkoff, L. B., and Tanouye, M. A. (1986). Genetics of ion channels. *Physiol. Rev.* **66,** 301–329.

Salkoff, L. B., and Wyman, R. J. (1981). Outward currents in developing *Drosophila* flight muscle. *Science* **212,** 461–463.

Salkoff, L. B., and Wyman, R. J. (1983). Ion currents in *Drosophila* flight muscles. *J. Physiol. (London)* **337,** 687–709.

Schmid-Antomarchi, H., Renaud, J.-F., Romey, G., Hugues, M., Schmid, A., and Lazdunski, M. (1985). The all-or-none response of innervation in expression of apamin receptor and apamin-sensitive Ca^{2+}-activated K^+ channel in mammalian skeletal muscle. *Proc. Natl. Acad. Sci. U.S.A.* **82,** 2188–2191.

Schmid-Antomarchi, H., Hugues, M., and Lazdunski, M. (1986). Properties of the apamin-sensitive Ca^{2+}-activated K channel in PC12 pheochromocytoma cells which hyper produce the apamin receptor. *J. Biol. Chem.* **261,** 8633–8637.

Schwarz, T. L., Papazian, D. M., Canetto, R. C., Jan, Y. N., and Jan, L. Y. (1990). Immunological characterization of K^+ channel components from the *Shaker* locus and differential distribution of splicing variants in *Drosophila. Neuron* **21,** 119–127.

Schwindt, P. C., and Crill, W. E. (1982). Factors influencing motoneuron rhythmic firing: Results from a voltage clamp study. *J. Neurophysiol.* **48,** 875–890.

Schwindt, P. C., Spain, W. J., Foehring, C. E., Stafstrom, C. E., Chubb, M. C., and Crill, W. E. (1988). Multiple potassium conductances and their functions in neurons from cat sensorimotor cortex *in vitro. J. Neurophysiol.* **59,** 424–449.

Seagar, M. J., Granier, C., and Courand, F. (1984). Interactions of the neurotoxin apamin with a Ca^{2+}-activated K^+ channel in primary neuronal cultures. *J. Biol. Chem.* **259,** 1491–1495.

Segal, M., and Barker, J. L. (1984). Rat hippocampal neurons in culture: Potassium conductances. *J. Neurophysiol.* **51,** 1409–1433.

Sharma, B. (1988). Inhibition of the generation of cytotoxic lymphocytes by potassium ion channel blockers. *Immunology* **65,** 101–105.

Shepard, P. D., and Bunney, B. S. (1988). Effect of apamin on the discharge properties of putative dopamin-containing neurons *in vitro. Brain Res.* **463,** 380–384.

Solc, C. K., Zagota, F. N., and Aldrich, R. W. (1987). Single-channel and genetic analysis reveal two distinct A-type potassium channels in *Drosophila. Science* **236,** 1094–1098.

Spitzer, N. C. (1979). Ion channels in development. *Annu. Rev. Neurosci.* **2,** 363–397.

Spitzer, N. C. (1985). The control of development of neuronal excitability. *In* "Molecular Basis of Neural Development" (G. M. Edelman, W. E. Gall, and W. M. Cowan, eds.), pp. 67–88. Neurosci. Res. Found., New York.

Spitzer, N. C., and Lamborghini, J. E. (1976). The development of the action potential mechanism of amphibian neurons isolated in culture. *Proc. Natl. Acad. Sci. U.S.A.* **73,** 1641–1645.

Suarez-Isla, B. A., Cosgrove, J. W., Thompson, J. M., and Rapoport, S. I. (1986). A soluble factor (<4000 Da) from chick spinal cord blocks slow hyperpolarizing afterpotentials in cultured rat muscle cells. *Dev. Brain Res.* **30,** 274–277.

Targ, E. G., and Kocsis, J. D. (1984). 4-Aminopyridine leads to restoration of conduction in demyelinated rat sciatic nerve. *Brain Res.* **328,** 358–361.

Tempel, B. L., Papazian, D. M., Schwarz, T. L., Jan, Y. N., and Jan, L. Y. (1987). Sequence of a probable potassium channel component encoded at the *Shaker* locus of *Drosophila. Science* **237,** 770–775.

Timpe, L. C., Jan, Y. N., and Jan, L. Y. (1988). Four cDNA clones from *Shaker* locus of *Drosophila* induce kinetically distinct A-type potassium currents in *Xenopus* oocytes. *Neuron* **1,** 659–667.

Valmier, J., Simonneau, M., and Boisseau, S. (1989). Expression of voltage-dependent sodium and transient K currents in an identified subpopulation of dorsal root ganglion cells acutely isolated from 12-days-old mouse embryos. *Pfluegers Arch.* **414,** 360–368.

Warner, A. E. (1973). The electrical properties of the ectoderm in the amphibian embryo during induction and early development of the nervous system. *J. Physiol.* (*London*) **235,** 267–286.

Warner, A. E. (1981). The early development of the nervous system. *In* "Development of the Nervous System" (P. R. Garrod and J. Feldman, eds.), pp. 109–127. Cambridge Univ. Press, London.

Waxman, S. G., and Foster, R. E. (1980). Ionic channel distribution and heterogeneity of the axon membrane in myelinated fibers. *Brain Res. Dev.* **2,** 205–234.

Wei, A., and Salkoff, L. (1986). Occult *Drosophila* calcium channel and twinning of calcium and voltage activated potassium channels. *Science* **233,** 780–782.

Wei, A., Covarruvias, M., Butler, A., Baker, K., Pak, M., and Salkoff, L. (1990). K^+ current diversity is produced by an extended gene family conserved in *Drosophila* and mouse. *Science* **248,** 599–603.

Wilson, G. F., and Chiu, S. Y. (1988). Effects of mitogens on Schwann cell ion channels. *Soc. Neurosci. Abstr.* **14,** 64.4.

Wilson, G. F., and Chiu, S. Y. (1990). Potassium channel regulation in Schwann cells during early developmental myelinogenesis. *J. Neurosci.* **10,** 1615–1625.

Yool, A. J., Dionne, V. E., and Groul, D. L. (1988). Developmental changes in K^+-selective channel activity during differentiation of the Purkinje neuron in culture. *J. Neurosci.* **8,** 1971–1980.

Zhang, L., and Krnjevic, K. (1987). Apamin depresses selectively the afterhyperpolarization of cat spinal motoneurons. *Neurosci. Lett.* **74,** 58–62.

Zona, C., Dichter, M., Pironne, G., and Ayala, G. F. (1986). Calcium currents in rat cortical neurons in cell culture. *Soc. Neurosci. Abstr.* **12,** 1345.

CHAPTER 11

Potassium Channels in Development, Activation, and Disease in T Lymphocytes

M. D. Cahalan, K. G. Chandy, and S. Grissmer
Department of Physiology and Biophysics, University of California at Irvine, Irvine, California 92717

I. INTRODUCTION

The development of the patch-clamp technique (Hamill *et al.*, 1981) has made it possible to characterize ion channels in small cells, including those of the immune system. In the nervous system, voltage-gated Na^+ and K^+ channels effect the upstroke and repolarizing phases, respectively, of the action potential. Cells of the immune system do not normally generate action potentials, although voltage-gated Na^+ channels have been observed infrequently in T cells, usually in low numbers per cell. Moreover, voltage-gated Ca^{2+} channels have not been detected in most lymphoid cells. Thus, action potentials are not involved in generating Ca^{2+} signals during the early stages of T cell activation. Indeed, depolarization, which induces Ca^{2+} signals in electrically excitable nerve and muscle cells, inhibits Ca^{2+} signaling and the subsequent cascade of cell-activation events in lymphocytes. Several distinct types of voltage-gated and second-messenger-operated K^+ and Cl^- channels do, however, exist in T and B lymphocytes, including channel types also expressed in the nervous system, as well as novel channels not described in other cell types. It appears that in these electrically inexcitable cells, ion channels mediate cellular functions involving intracellular biochemical signaling, rather than through rapid electrical signaling. The presence of K^+ channels, in particular, is apparently required for several basic functions in T lymphocytes, such as activation, the secretion of lymphokines, mitogenesis, and the regulation of cell volume.

There is considerable diversity of ion channels within cells of the hematopoietic lineage. In T lymphocytes, at least eight distinct types of K^+, Cl^-, and Ca^{2+} channels have been characterized (reviewed in Lewis and Cahalan, 1990). Investigators in our laboratory and several others have also recorded from bone marrow cells, thymocytes, B lymphocytes, monocytes, macrophages, neutrophils, red blood cells, eosinophils, and even platelets, as well as numerous related cell lines including a human T-leukemia line (Jurkat), a mouse T-lymphoma line (S49), a variety of antibody-secreting hybridomas, and a mast-cell line derived from a rat basophilic leukemia (RBL). In addition to the channel types found in T lymphocytes, these studies have revealed (in selected cell types): transient or T-type voltage-gated Ca^{2+} channels (hybridomas), inward-rectifying K^+ channels (macrophages, RBL cells), large-conductance Ca^{2+}-activated K^+ channels (macrophages), Ca^{2+}-activated Cl^- current (mast cells), and G-protein-activated K^+ channels (RBL cells). Clearly, cells of the blood can express a diversity of channel types rivaling that of the nervous system, although ligand-activated channels have not yet been found in blood cells.

In this review, we focus primarily on K^+ channels in T lymphocytes, one subtype of which was the first channel to be described in lymphocytes. We summarize the properties of known voltage- and Ca^{2+}-activated K^+ channels in lymphocytes and describe what is presently understood of their molecular characteristics, functions, and expression during normal development, mitogen activation, and in autoimmune disorders. In addition, we describe the coordinated activity of Ca^{2+} and K^+ channels that underlies mitogen-stimulated Ca^{2+} signaling.

II. PROPERTIES OF K^+ CHANNELS IN LYMPHOCYTES

Over the past 5 years, it has become clear that there are several K^+ channel subtypes in lymphocytes. In addition to at least three distinct types of voltage-gated K^+ channels, there are at least two types of Ca^{2+}-activated K^+ channels.

A. Voltage-Gated K^+ Channels

1. Human T Cells

Voltage-gated K^+ channels have been extensively investigated in lymphocytes. Independently, Matteson and Deutsch and our laboratory initially characterized the voltage dependence and ion selectivity of a K^+ channel in T cells separated from human peripheral blood (DeCoursey *et al.*, 1984; Matteson and Deutsch, 1984), while Hagiwara and colleagues described the same channel type in a murine cytotoxic T-cell line (Fukushima *et al.*, 1984a). A family of voltage-clamp currents from a human T cell is illustrated in Fig. 1A. The channels open during membrane depolarization, with half-maximal activation occurring at about -40 mV. If the depolarization is maintained, the channel subsequently inactivates with a time constant of 100–200 msec, intermediate between inactivation of the classical transient outward "A current" and "delayed rectifier" current. Recovery from inactivation is quite slow, requiring 30–60 sec at -80 mV, resulting in "use-dependent" inactivation during repetitive voltage-clamp pulses.

Kinetic modeling permitted a calculation of K^+ efflux through voltage-gated K^+ channels in human T cells, where they number approximately 400 per cell. The calculated flux corresponded well with radioactive tracer flux measurements and thus predicted that these channels constitute the major pathway for passive K^+ efflux and thereby support the resting membrane potential (Cahalan *et al.*, 1985). Studies employing voltage-

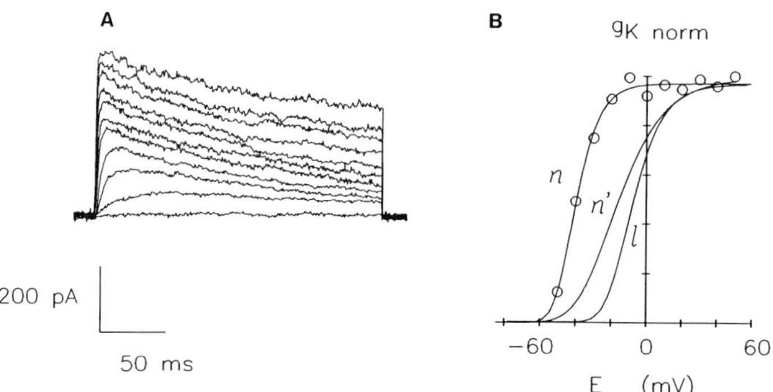

FIG. 1. Voltage-gated K^+ currents in a human $CD4^+CD8^-$ peripheral blood T lymphocyte. (A) A family of type *n* K^+ currents elicited by voltage steps from −50 to +50 mV in 10-mV increments, applied from a holding potential of −80 mV. (B) Normalized conductance–voltage relation for the currents shown in A. For each current record, the conductance was calculated by dividing the peak current by the difference between the membrane potential and the K^+ equilibrium potential. The line through the data points (circles) represents a least squares fit of a Boltzmann function raised to the fourth power (cf. to Cahalan *et al.*, 1985). The potential, $V_{1/2}$, at which the conductance is half-maximal is −39.8 mV, a value specific for type *n* K^+ channels. Two other lines represent Boltzmann functions specific for type *n′* and type *l* K^+ channels and are included in the graph for comparison (see Table I).

sensitive indicator dyes and pharmacological blockers have substantiated this functional role, by finding that K^+ channel blockers cause membrane depolarization (Wilson and Chused, 1985; Grinstein and Smith, 1990). At rest, only a few K^+ channels are required on the average to support a negative membrane potential, usually −50 to −70 mV (Deutsch *et al.*, 1979; Wilson and Chused, 1985). The term "resting potential" may be a misnomer, since in these small cells, with input resistances $>10^{10}$ Ω, fluctuations in membrane potential due to the opening and closing of single K^+ channels have been observed in single cells with current-clamp whole-cell recording. Because the channels are opened by depolarization, they serve to "protect" the cell from depolarization.

2. Channel Subtypes: *n*, *n′*, and *l*

As in human T cells, patch-clamp recordings from thymocytes and mature T cells isolated from mice have revealed voltage-gated K^+ channels as the predominant type. The most commonly seen K^+ channel closely matches the human T cell K^+ channels described above; we have termed this type *n* for its abundance in *normal* human T cells. Type *n* channels have also been found in other cells of the immune system, including helper and cytotoxic clonal T cell lines, macrophages, and B cells (reviewed in DeCoursey *et al.*, 1985; Lewis and Cahalan, 1990). Two

TABLE I

Comparison of Type *n*, *l*, and *n'* K^+ Currents from T Cells[a]

Type	Gating			Single channel conductance	Blockers (K_d)				
	V_n (mV)	τ_h at 40 mV (msec)	τ_{tail} at −60 mV (msec)	γ (pS)	TEA (m*M*)	4-AP (m*M*)	Quinine (μ*M*)	Verapamil (μ*M*)	CTX (n*M*)
Type *n* (human)	−36	178	44[b]	9, 16[c]	8	0.19	14	6	<1.0
Type *n* (mice)	−36	107	49	12, 18[c]	10–13	<0.2	4–40	4–40	<1.0
Type *l* (mice)	−9, 0[d]	340	2.6	21, 27[c]	0.08	<0.2	4–40	4–40	no effect
Type *n'* (mice)	−10	>300	~30–60	18	~100	n.d.[e]	n.d.	n.d.	<<5.0

[a] Data taken from DeCoursey *et al.* (1987a), Cahalan *et al.* (1985), Lewis and Cahalan (1988a), and Cahalan and Lewis (1988).

[b] τ_{tail} for human type *n* channels was measured from tail current data obtained from five human peripheral blood T cells.

[c] In human T cells two distinct single channel events have been described with conductances of 9 and 16 pS, which may represent two types of channels or one channel with two conductance states (Cahalan *et al.*, 1985). Mouse type *n* and type *l* channels each exhibit only one single channel conductance state, which has been variously reported to be 12 pS (DeCoursey *et al.*, 1987*a*) and 18 pS (Lewis and Canalan, 1988*a*) for type *n*, and 21 pS (DeCoursey *et al.*, 1987*a*) and 27 pS (Lewis and Cahalan, 1988*a*) for type *l*.

[d] V_n for type *l* channels has been reported to be −9 and 0 mV by DeCoursey *et al.* (1987) and Cahalan and Lewis (1988).

[e] Not determined.

additional channel types, termed *l* (for *large* conductance) and *n′* (for similar to *n*), have been found, with a developmental pattern of expression that is normally restricted to resting cells with the cytotoxic/suppressor phenotype (Lewis and Cahalan, 1988a,b). In addition, type *l* channels are numerous in lymphocytes from several murine models for autoimmunity and, in fact, were first described in T cells from the *lpr* mouse (Chandy *et al.*, 1986; DeCoursey *et al.*, 1987a). Thus, as described in greater detail below, the three types of K^+ channels are selectively expressed in a characteristic pattern during differentiation and mitogenic activation, and in autoimmune disorders. These three K^+ channel subtypes can be distinguished based on their different kinetics of opening (activation) and closing (deactivation or inactivation), their voltage dependence, their ion-permeation properties, and pharmacological sensitivities.

3. Activation and Deactivation

Whereas type *n* K^+ channels can be half-activated at potentials of approximately −40 mV, type *l* K^+ channels must be depolarized by an additional 30 mV in order to open to the same degree (DeCoursey *et al.*, 1987a; Lewis and Cahalan, 1988a); i.e., *l*-channel activation is displaced to the right along the voltage axis (Fig. 1B). Channel closing (deactivation) is also dramatically different in type *l* channels (Fig. 2). Following a conditioning depolarizing prepulse to open the channels, type *n* and *n′* K^+ channels close slowly, with a time constant of ~30 msec at −60 mV. Type *l* K^+ channels close much more rapidly, with a time constant of ~2 msec at −60 mV. Quantitatively, the difference in channel-closing rates is even larger than the difference in the voltage dependence for opening the channels; described as a voltage shift, type *l* channel closing kinetics are displaced by 70–80 mV in the depolarized direction along the voltage axis, relative to the closing of *n* channels. Neutralizing negative surface charges near Na^+ channels of nerve, induced by raising pH or divalent cation concentration, can account for the displacement of activation parameters to the right along the voltage axes (Hille *et al.*, 1975). However, this model would not be able to account for a shift of 70–80 mV. Thus, opening and closing kinetics indicate major differences in the response of the type *n* and type *l* K^+ channel proteins to membrane potential; a simple change in the "environment" surrounding the channel protein's "voltage sensor," due, for example, to altered surface charge, cannot account for the difference between the activation of *n* and *l* channels.

4. Inactivation

The degree of use-dependent inactivation during repetitive (1 Hz) depolarizing pulses also provides a convenient way to distinguish between

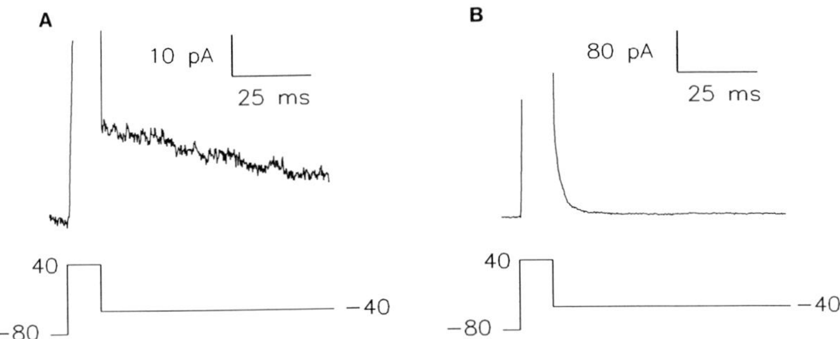

FIG. 2. K^+ tail currents: type *n* in a $CD4^+CD8^-$ (A) and type *l* in a $CD4^-CD8^-$ (B) T lymphocyte from *gld* mice. Tail currents were elicited by voltage steps to −40 mV after a 15-msec depolarizing prepulse to 40 mV to open the channels.

diverse types of K^+ channels (Fig. 3). For the type *n* channels, recovery during the interpulse interval is incomplete, resulting in a substantial reduction of the K^+ current. In contrast, *l*- and *n'*-type channels do not exhibit use dependence; thus, no reduction of the K^+ current is seen during repetitive pulses delivered at 1 Hz in cells expressing these channels.

In summary, three types of voltage-gated K^+ channels can be distinguished in lymphocytes based on their "gating" properties of activation, deactivation, and inactivation. Their properties are summarized in Table I. Single-channel recording, pharmacological probes, and molecular biological studies of channel expression have reinforced this characterization, as described below.

5. Single-Channel Conductance and Ion Permeation

Each type of voltage-gated K^+ channel in T cells has its own characteristic single-channel conductance. This conductance is a measure of the number of ions passing through one open channel in a given time. In mammalian Ringer solution, the single-channel conductances for the different channel types are 18, 27, and 17 pS for *n, l,* and *n'* K^+ channels, respectively (Lewis and Cahalan, 1988a). These values are higher when the concentration of extracellular K^+ ($[K^+]_o$) is elevated.

In addition to raising the conductance, external K^+ results in altered channel gating of type *n* channels by slowing the deactivation of K^+ channels (Cahalan *et al.,* 1985). This behavior, which has been described previously for K^+ channels in axons, is especially dramatic in type *n* channels. A second permeant ion, rubidium, is even more effective than K^+ in reducing the channel-closing rate. One possible explanation is that occupancy of an open channel by a permeant ion inhibits channel closure.

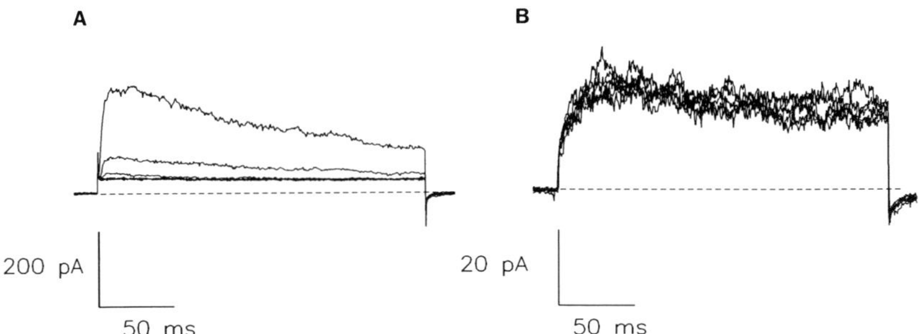

FIG. 3. Use-dependent inactivation in a $CD4^+CD8^-$ (A) and in a $CD4^-CD8^+$ (B) T lymphocyte from *gld* mice. Repetitive pulses of 200 msec duration were applied once every second to 40 mV from a holding potential of −80 mV. The corresponding type *n* K^+ currents (A) decrease dramatically in amplitude from largest to smallest, whereas type *n′* (B) and type *l* (not shown) K^+ currents do not decrease in amplitude.

6. Pharmacology

Channel blockers are important tools for investigating ion channels. Careful analysis of the channel-blocking mechanism can often provide information about such biophysical properties as gating or permeation. The agents, in some instances, allow for different channel subtypes to be readily distinguished. High-affinity binding of pharmacological agents may provide biochemical handles for channel purification. In addition, binding sites and channel topology can be revealed through functional analysis of site-directed mutations. Finally, these agents may help in assessing the functional roles of ion channels in lymphocytes.

A wide variety of agents can block type *n* channels, including the classical K^+ channel blockers tetraethylammonium (TEA) and 4-aminopyridine (4-AP); agents formerly considered blockers of Ca^{2+}-activated K^+ channels, such as quinine, quinidine, cetiedil, and a scorpion toxin, charybdotoxin (CTX); and Ca^{2+}-channel antagonists such as verapamil, dihydropyridines, diltiazem, and polyvalent cations. Many of the blockers exhibit complexities in their effects that can be described in terms of altered channel inactivation. For example, verapamil appears to enhance K^+ channel inactivation (Chandy *et al.*, 1984; DeCoursey *et al.*, 1985).

Experimental protocols employing the channel blockers TEA and CTX provide a ready means to distinguish the three K^+ channel subtypes pharmacologically (DeCoursey *et al.*, 1987a; Lewis and Cahalan, 1988b). Both TEA and CTX act by occluding the channel opening from the outside. The rank order for potency of channel block by TEA is (approximate K_i values listed in m*M*, from most sensitive to least): *l* (0.1) > *n* (10) > *n′*

(100). Figure 4 illustrates TEA block of type *n* and *l* channels. Type *n* and *n'* channels are blocked by CTX in the nanomolar range, whereas type *l* appears to be entirely insensitive to CTX. Thus, substantial channel block by 1 m*M* TEA along with insensitivity to CTX is diagnostic for the *l*-type channels. The presence of *n* or *n'* channel subtypes is indicated by CTX sensitivity, while sensitivity to 10 m*M* TEA distinguishes *n* from *n'*.

Block of type *n* K^+ channels by external TEA results in an apparent reduction in the single-channel conductance, consistent with the idea that blocking and unblocking by TEA is fast compared with channel opening (DeCoursey *et al.*, 1987a). However, individual blocking events for the block of type *l* K^+ channels by TEA can be resolved. At the single-channel level, the events can be visualized as a "flickery" block when TEA hops in and out of its binding site on the *l* channel.

In addition to blocking, TEA slows the time course of the type *n* K^+ current inactivation, resulting in cross-over of macroscopic current records in the presence and absence of TEA (Grissmer and Cahalan, 1989a). The integral of the macroscopic K^+ current during a prolonged depolarizing pulse is unaltered in solutions with different TEA concentrations. Our results can be explained with a simple kinetic scheme in which open channels blocked by TEA cannot inactivate. Thus, the conventional view that TEA blocks K^+ conductance without affecting gating does not apply to inactivation.

Charybdotoxin was initially described as a blocker of Ca^{2+}-activated "maxi-" K^+ channels from muscle (Miller *et al.*, 1985), but it also blocks type *n* voltage-gated K^+ channels in human T lymphoid cells (Sands *et al.*, 1989). After binding to its external site at or near the channel mouth, CTX typically occludes the channel for about 10 sec before unbinding. Of the three types of K^+ channels present in murine thymocytes, types *n* and *n'* are blocked by CTX at nanomolar concentrations. The third variety, type *l*, is unaffected by CTX. Thus, CTX is a high-affinity blocker of selected lymphocyte K^+ channels, as well as a useful tool for biochemical purification of channel components. The pharmacological studies indicate that block by CTX should not be considered diagnostic for Ca^{2+}-activated K^+ channels, as it applies to a subset of voltage-gated K^+ channels as well. The existence of high-affinity sites for CTX binding may help to classify structurally related K^+ channels. Lymphocyte K^+ channels, *Shaker* (from *Drosophila*), and mammalian K^+ channels expressed in oocytes, and the Ca^{2+}-activated maxi-K^+ channel are likely to share a common extracellular "receptor" for CTX (Christie *et al.*, 1989; MacKinnon and Miller, 1989; Sands *et al.*, 1989). The molecular nature of this receptor is currently being mapped by site-directed mutagenesis experiments (MacKinnon and Miller, 1989).

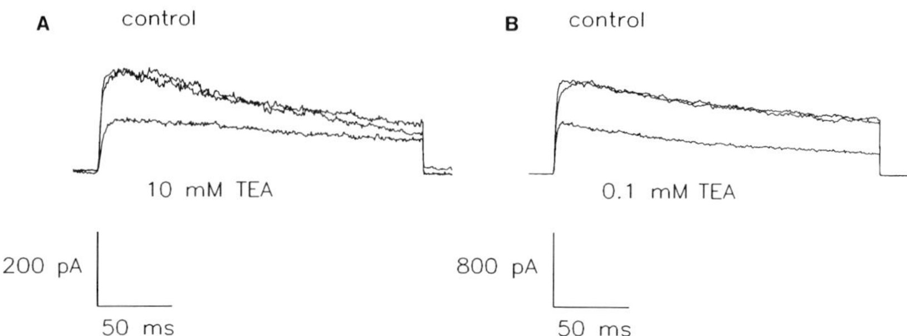

FIG. 4. Effect of tetraethylammonium (TEA) on (A) type *n* and (B) type *l* K^+ currents in T cells from *gld* mice. K^+ currents were elicited by depolarizing pulses of 200 msec duration to +40 mV from a holding potential of −80 mV. Currents are shown before, during, and after application of TEA. (A) Current in a $CD4^-CD8^-$ T cell blocked about half by 0.1 m*M* TEA. (B) Current in a $CD4^+CD8^-$ T cell blocked about half by 10 m*M* TEA.

7. Modulation of K^+ Channels

In a variety of cell types, ion channels are important targets for modulation by receptor-triggered signal transduction pathways. In a number of cases, ion channels have been shown to be directly modulated by GTP-binding proteins, nucleotides, intracellular ions, or enzymes; each of these elements of intracellular signal transduction appears to become activated during mitogen stimulation of lymphocytes. Mitogen stimulation initiates multiple signals within lymphocytes, including phospholipase C activation and phosphorylation by one or more tyrosine kinases (Schwartz, 1990). The search for ion channel involvement in these pathways has begun, with initial work focusing on type *n* channels.

Mediated primarily by influx through a conductive pathway, the rise in Ca^{2+} following mitogen stimulation initiates a variety of events within the cell, including changes in the activity of K^+ channels. Cytosolic Ca^{2+} inhibits the type *n* voltage-gated K^+ channels by promoting the process of inactivation (Bregestovski *et al.*, 1986; Choquet *et al.*, 1987; Grissmer and Cahalan, 1989c). Our studies show that raising either external or internal $[Ca^{2+}]$ (or adding Ba^{2+} in the bath solution) accelerates the rate of K^+ channel inactivation in a similar way. If $[Ca^{2+}]_i$ is raised experimentally using an EGTA-buffered solution, inactivation becomes faster, to the same limiting extent as with external Ca^{2+}, but at micromolar levels instead of millimolar levels. The common effect of raising either external or internal $[Ca^{2+}]$ suggests that the binding site for divalent ions is within the open channel's transport pathway. External K^+ can compete with divalent ions for the site. If the channel closes while a divalent ion is bound, the ion can become trapped inside the channel. Ba^{2+} can remain trapped inside a closed channel for more than 15 min, whereas Ca^{2+} appears to cross the membrane to the inside during the interval between test pulses. These experiments suggest that voltage-gated K^+ channels become inactivated by a rise in $[Ca^{2+}]_i$ to the micromolar range that can occur at the peak of the $[Ca^{2+}]_i$ signal (Grissmer and Cahalan, 1989b). Before $[Ca^{2+}]_i$ reaches such high levels, however, a Ca^{2+}-activated K^+ conductance begins to supplant the voltage-gated K^+ conductance (see below). As a result, the cell will not depolarize during the Ca^{2+} influx, even though type *n* channels are inactivated. The interplay between K^+ channels and the rise in $[Ca^{2+}]_i$ will be discussed further (see Section V,B).

Voltage-gated K^+ channels are also sensitive to changes in pH and temperature. The normal pH inside lymphocytes is 7.1. K^+ current magnitudes are increased at alkaline pH and inhibited at acid pH, with no significant changes in kinetics (Deutsch and Lee, 1989). The biological significance of this pH dependence is uncertain. When temperature is elevated, the magnitude and rates of activation, deactivation, and inactivation are all increased (Lee and Deutsch, 1990; Pahapill and Schlichter, 1990). The recovery from inactivation is particularly temperature dependent, resulting in a lower degree of use dependence during repetitive stimulation. These studies reinforce the conclusion that the voltage-gated K^+ channel is the predominant channel type in resting cells.

Cross-talk between signal transduction pathways may be an important mechanism of modulating cellular responses. For example, elevated levels of cAMP, through activation of β-adrenergic receptors, inhibit mitogen-stimulated T-lymphocyte activation (Coffey and Hadden, 1985). One possibility, in analogy with known inhibitory effects of K^+ channel blockers (see below), is that cAMP might decrease the K^+ conductance. This channel modulation hypothesis for growth inhibition by cAMP is consistent with tantalizing, but as yet inconclusive, data. Forskolin, an activator of adenylyl cyclase, blocks and accelerates inactivation in type *n* K^+ channels. However, this appears to be a direct effect rather than being mediated through adenylyl cyclase, as indicated by the rapidity of the effect and the identical action of an inactive forskolin derivative (Krause *et al.*, 1988). Choquet and colleagues have found that type *n* channels in immortalized B lymphoma cell lines inactivate more rapidly following elevation of [cAMP] (Choquet *et al.*, 1987), but these data have not been confirmed in T cells (Lee and Deutsch, 1990). It is known that the rate and extent of use-dependent inactivation increase during dialysis with pipette solutions containing F^- as the major anion; these solutions may contain contaminating amounts of aluminum and may activate G proteins during dialysis (Cahalan *et al.*, 1985), but these effects have not been seen with receptor stimulation. More experiments are needed to test the hypothesis that inhibitory effects of cAMP in T cells are due to inactivation of type *n* K^+ channels. Modulation of K^+ channels by serotonin and by β-endorphin has also been reported, but the intracellular signaling mechanisms await elucidation (Choquet and Korn, 1988; Hough *et al.*, 1990).

B. Ca^{2+}-Activated K^+ Channels

Before the patch-clamp technique became available, measurements of membrane potential and cytosolic Ca^{2+} had suggested the presence of Ca^{2+}-activated K^+ channels in lymphocytes. Quinine, which blocks Ca^{2+}-activated K^+ channels in other cells, induces membrane depolarization in lymphocytes (Lew and Ferreira, 1978). Considering the nonselective actions of quinine on other channels, including voltage-gated K^+ channels, this result must be considered as ambiguous. Mitogenic activation of human T cells results in increased $[Ca^{2+}]_i$ accompanied by membrane hyperpolarization (Tsien *et al.*, 1982; Felber and Brand, 1983; Wilson and Chused, 1985; Tatham *et al.*, 1986; MacDougall *et al.*, 1988; Grinstein and Dixon, 1989). Moreover, pharmacological studies indicate that 4-AP, a blocker of the voltage-gated channels, could not prevent the hyperpolarization induced by Ca^{2+} ionophores, indicating that some other K^+ channel type can become activated by the rise in $[Ca^{2+}]_i$ (Wilson and Chused, 1985).

Patch-clamp studies have only recently confirmed the presence of Ca^{2+}-activated K^+ channels in lymphocytes and related cell lines (Mahaut-Smith and Schlichter, 1989a,b; Grissmer and Cahalan, 1989b). Three types of experiment have been done: cell-attached recordings of single-channel activity in response to agents which result in elevated Ca^{2+}; whole-cell dialysis using EGTA-buffered pipette solutions with defined Ca^{2+}; and simultaneous Ca^{2+} and current measurement during exposure to a Ca^{2+} ionophore or mitogenic stimulus using the whole-cell or perforated-patch method. As is the case with the family of voltage-gated K^+ channels, the available evidence, summarized below, indicates that there are at least two types of Ca^{2+}-activated K^+ channels that can be expressed in lymphocytes.

1. Ca^{2+} Sensitivity

In human B cells and rat thymocytes, raising $[Ca^{2+}]_i$, either by application of ionomycin or by replacing external Ca^{2+} following incubation in Ca^{2+}-free medium, was found to activate two types of channels with single-channel conductances of ~8 and ~25 pS (Mahaut-Smith and Schlichter, 1989b). The open-channel probability was virtually independent of membrane potential, and the threshold [Ca^{2+}] for activation of the larger Ca^{2+}-activated K^+ channels was estimated to be ~200–300 n*M*. The Ca^{2+} dependence was explored further in experiments using whole-cell recordings from Jurkat T cells dialyzed with different Ca^{2+} concentrations. The resulting Ca^{2+}-activation curve indicates that even small rises above the resting $[Ca^{2+}]_i$ will activate K^+ channels. In addition, the dose–response curve for Ca^{2+} to activate the channel is much steeper than expected for a single Ca^{2+} ion activating the channel, suggesting that there are multiple independent binding sites for Ca^{2+} which must be occupied in order to activate the channel (S. Grissmer *et al.*, unpublished observations).

2. Pharmacology

Pharmacological evidence indicates that at least two distinct types of Ca^{2+}-activated K^+ channels are present in different lymphoid cells, but their distribution according to cell type is not yet completely determined. Apamin, a component of bee venom, is known to block Ca^{2+}-activated K^+ channels in neuroblastoma cells (Lazdunski, 1983), GH3 cells (Ritchie, 1987), and cultured rat muscle cells (Hugues *et al.*, 1982; Blatz and Magleby, 1986). In Jurkat cells, 1 n*M* apamin completely and irreversibly blocked the Ca^{2+}-activated K^+ channels without affecting the voltage-gated K^+ channels (Grissmer and Cahalan, 1989b). CTX is a potent blocker of voltage-gated type *n* K^+ channels in human T lymphocytes (Sands *et al.*, 1989) and of Ca^{2+}-activated K^+ channels in skeletal muscle (Miller *et al.*, 1985). However, the predominant (apamin-sensitive) Ca^{2+}-activated K^+ channels in Jurkat T cells are not affected by CTX. 4-AP selectively blocks the voltage-gated K^+ channels, without affecting the Ca^{2+}-activated K^+ channel. TEA affects both of these types of K^+ channels, however, with different potency. The voltage-gated K^+ channels are about half-blocked by 10 m*M* TEA, whereas the Ca^{2+}-activated K^+ channels are about 5 times more sensitive to block by TEA (K_d ~2 m*M*) (Grissmer *et al.*, 1992).

The second type of Ca^{2+}-activated K^+ channel, which has been only partially characterized, *is* sensitive to block by CTX. Using a voltage-sensitive dye, Grinstein and Smith (1990) have reported that CTX blocks not only the resting K^+ conductance, but also a K^+ conductance activated by a Ca^{2+} ionophore. Mahaut-Smith and Schlichter (1989a) have reported that the Ca^{2+}-activated K^+ current ($I_{K(Ca)}$) in human B lymphocytes and rat thymocytes was blocked by CTX. In their cell-attached experiments, channel activity could be induced by 1 μM ionomycin with 10 n*M* CTX in the patch-pipette. However, no activity could be induced with 50 n*M* CTX. In Jurkat cells, after complete block of the apamin-sensitive Ca^{2+}-activated K^+ channels, elevated Ca^{2+} induces a small CTX-sensitive conductance, suggesting that Jurkat cells have mostly apamin-sensitive channels, but also a few CTX-sensitive channels. Some cell types, however, can express much larger CTX-sensitive Ca^{2+}-activated K^+ currents (D. Choquet and M. D. Cahalan, unpublished observations). Much work remains to be done in order to clarify the relative expression of the two types of Ca^{2+}-activated K^+ channels.

In summary, the available evidence indicates that at least two types of Ca^{2+}-activated K^+ channels with different sensitivities to apamin and CTX are present in lymphoid cells. Now that specific blockers of Ca^{2+}-activated K^+ channels have been developed, it should be possible to assess the roles of these channels in cellular functions. Given the importance of intracellular Ca^{2+} in mediating fundamental processes such as signal transduction, it can be anticipated that future studies on Ca^{2+}-activated K^+ channels will yield valuable insights into the regulation of cellular behavior.

III. MOLECULAR BASIS OF K^+ CHANNEL DIVERSITY

What are the molecular properties of the different types of K^+ channels in lymphocytes, and how is their expression regulated? Based on similarities in sensitivity to CTX and channel kinetics, types *n* and *n'* channels are likely to be closely related. Recently, molecular cloning has begun to reveal the sequences of voltage-gated K^+ channels, thus providing considerable insight into the structure of the genes encoding K^+ channels, and, in turn, making it possible to deduce information about the protein structures of the channels as well. This section reviews progress in this area and describes our initial efforts to define the molecular basis for K^+ channel diversity in lymphocytes. Thus far, no clones for Ca^{2+}-activated channels have been isolated, and information about their molecular properties is very scanty. The discovery of apamin and CTX as high-affinity blockers for these channels should facilitate the isolation of these channel proteins.

A. K^+ Channel Genes in Drosophila

The molecular mechanisms generating diversity of voltage-gated K^+ channels were initially studied in *Drosophila* by taking advantage of chromosomal mapping of a behavioral defect, the *Shaker* mutation (Kamb *et al.*, 1987; Papazian *et al.*, 1987; Pongs *et al.*, 1988; Schwartz *et al.*, 1988). In the fruit fly, expression of at least four distinct genes, *Shaker, Shaw, Shab,* and *Shal,* gives rise to diverse K^+ channel proteins (Papazian *et al.*,

1987; Kamb *et al.*, 1987; Baumann *et al.*, 1987; Pongs *et al.*, 1988; Schwartz *et al.*, 1988; Butler *et al.*, 1989; Wei *et al.*, 1990). Alternative splicing of a single primary transcript derived from the *Shaker* locus generates at least four K^+ channel proteins that exhibit distinct physiological properties (Papazian *et al.*, 1987; Kamb *et al.*, 1988; Pongs *et al.*, 1988; Schwartz *et al.*, 1988). The *Shab* and *Shal* genes also undergo alternative splicing of primary RNA transcripts, each giving rise to two distinct K^+ channel proteins (Butler *et al.*, 1989; Wei *et al.*, 1990). Expression of these genes in *Xenopus* oocytes results in voltage-gated K^+ channels which can be distinguished on the basis of biophysical and pharmacological criteria (Timpe *et al.*, 1988a,b; Wei *et al.*, 1990). The four *Drosophila* K^+ channel genes share sequence relatedness, especially in a region upstream to S1, suggesting that they resulted from duplication of a primordial channel gene; however, they are located on separate chromosomes (Wei *et al.*, 1990), suggesting a duplicating event distinct from that which generated the major histocompatibility or hemoglobin gene loci.

B. K^+ Channel Structure

Based on hydropathy plots, and in analogy with voltage-gated Na^+ and Ca^{2+} channels, it has been suggested that a functional K^+ channel consists of four polypeptide chains, each with six membrane helices, designated S1 through S6 (Noda *et al.*, 1986; Pongs *et al.*, 1988). Tempel *et al.* (1988) suggest the presence of an additional membrane-spanning helix located between S5 and S6. In both models, the S4 transmembrane segment is thought to act as a voltage sensor, since in this region every third residue is either an arginine or a lysine, as in the voltage sensor of Na^+ and Ca^{2+} channels. Future analyses should determine the correct topology of these proteins and further our understanding of the mechanism of action of the channel proteins. Because a single mRNA species is sufficient to give functional K^+ current after injection in *Xenopus* oocytes, it is believed that K^+ channels can assemble as homomultimers. However, recent evidence indicates that heteromultimers of different K^+ channel subunits can give rise to additional diversity of K^+ channels. Channels with properties distinct from the homomultimers resulted from coinjection of two mRNA species (Christie *et al.*, 1989; Isacoff *et al.*, 1990; Ruppersberg *et al.*, 1990).

Site-directed mutagenesis experiments are beginning to define functionally significant regions of the protein. Work is currently focusing on voltage-dependent activation, inactivation, and toxin binding. For example, the inactivation rate is slowed in *Shaker*-B mutants that have deletions in the NH_2-terminus beginning at the sixth amino acid. Treatment of these channels in inside-out patches with the NH_2-terminal peptide restores inactivation (Toshi *et al.*, 1988). Similarly, mutational approaches have been used to localize the CTX-binding site to the S5–S6 interdomain in the *Shaker*-B channel, close to glutamate 422 (MacKinnon and Miller, 1989).

C. K^+ Channel Genes in Vertebrates

Each of the four *Drosophila* genes for K^+ channels, *Shaker, Shaw, Shab,* and *Shal,* has one or more mammalian homologs, making it possible to group these vertebrate K^+ channel genes into subfamilies, (Tempel *et al.*, 1988; Baumann *et al.*, 1988; Christie *et al.*, 1989; McKinnon, 1989; Stühmer *et al.*, 1989; Kamb *et al.*, 1989; Frech *et al.*, 1989; Yokoyama *et al.*, 1989; Chandy *et al.*, 1990a,b; Wei *et al.*, 1990; Beckh and Pongs, 1990; Betsholtz *et al.*, 1990; Swanson *et al.*, 1990; Grupe *et al.*, 1990; Ghanshani *et al.*, 1990; Ribera, 1990). However, the names for these genes are often confusing, and a simplified nomenclature is suggested (Table II), to make it easier for the reader to follow the subsequent description of the genes encoding T cell K^+ channels. The genes are grouped into four families based on sequence relatedness with each other and to the *Drosophila* genes. Each family is given a number; as additional families are discovered, they could be numbered based on their date of discovery. In the table, a name is proposed for each gene along with its current synonyms. For example, the proposed name for MBK1, is $K_v1.1$; i.e., a K^+ channel gene (K) which is voltage-dependent ($_v$) and is the first identified member of the *Shaker*-related family 1 (1.1). Blanks in the table indicate that gene homologs have not yet been identified in a particular species. Families could include genes that share at least 65% sequence relatedness. The species and tissue origin are excluded from the name since these genes are expressed in diverse tissues and in many different species. The proposed nomenclature could be used for other types of K^+ channels as well. For example, ligand-gated K^+ channels could be similarly named for the particular ligand (e.g., Ca^{2+}, ATP being substituted for V).

Several vertebrate K^+ channel genes have been recently isolated and characterized. In contrast to *Drosophila* K^+ channels, many of these genes have coding regions that are encoded by single uninterrupted exons (Chandy *et al.*, 1990a; Takumi *et al.*, 1988; Murai *et al.*, 1989; Swanson *et al.*, 1990; Ribera, 1990; Grupe *et al.*, 1990), but have introns in the noncoding regions. We have identified three other genes in mice, $K_v1.7$ (Chandy *et al.*, 1990b), and $K_v3.3$ and $K_V3.4$ (Ghanshani *et al.*, 1990), that each have at least one intron in the coding region. In the $K_V3.3$ and $K_v3.4$ genes, the region from S1 to the 3′ end of the protein is encoded in one exon, and a

TABLE II

Simplified Nomenclature for a Family of Vertebrate Voltage-Dependent K^+ Channel Genes

Proposed	Xenopus	Mouse	Rat	Human
Shaker-related subfamily 1				
$K_v1.1$	—	MBK1[1,a]	RCK1[2]	HK1[3,4]
		MK1[5]	RBK1[6,7]	
$K_v1.2$	XSha2[8]	MK2[5]	RBK2[6]	HK4[3,4]
			RCK5[9]	
			NGK1[10]	
$K_v1.3$	—	MK3[5,11]	RCK3[9]	HPCN3[12]
			RGK5[13]	
			KV3[14]	
$K_v1.4$	—		RCK4[9]	HK2[3,4]
			RHK1[15]	HPCN2[12]
				HK1[12a]
$K_v1.5$	—		KV1[14]	HPCN1[12]
				HK2[12a]
$K_v1.6$	—	—	KV2[14]	HBK2[16]
			RCK2[9]	
K()1.7	—	MK6[17]	RK6[17]	HaK6[17]
		MK4[18]		
Shab-related subfamily 2				
$K_v2.1$	—	Mshab[19]	DRK1[20]	—
Shaw-related subfamily 3				
$K_v3.1$	—	NGK2[10]	K_v4[21,b]	—
		Mshaw22[22]		
$K_v3.2$	—	MShaw12[22]	RKShIIIA[23]	—
			Rshaw12[22]	
K()3.3	—	$K_v3.3$[18,24]	—	—
		Mshaw19[22]		
$K_v3.4$	—	$K_v3.4$[18,24]	Raw3[25]	—
Shal-related family 4				
$K_v4.1$	—	Mshal1[26]	—	—
$K_v4.2$			RK5[27]	

[a] Superscript numbers refer to the following references: (1) Tempel *et al.* (1988); (2) Baumann *et al.* (1988); (3) Kamb *et al* (1989); (4) Mathew *et al.* (1989); (5) Chandy *et al.* (1990a); (6) McKinnon (1989); (7) Christie *et al.* (1989); (8) Ribera (1990); (9) Stühmer *et al.* (1989); (10) Yokoyama *et al.* (1989); (11) Grissmer *et al.* (1990); (12) Philipson *et al.* (1991); (12a) Tamkun *et al.* (1991); (13) Douglass *et al.* (1990); (14) Swanson *et al.* (1990); (15) Tseng-Crank *et al.* (1990); (16) Grupe *et al.* (1990); (17) Betsholtz *et al.* (1990); (18) Chandy *et al.* (1990b); (19) Pak *et al.* (1991a); (20) Frech *et al.* (1989); (21) Luneau *et al.* (in press); (22) Pak , M., and Salkoff, L. A., unpublished data; (23) McCormack *et al.* (1990); (24) Ghanshani *et al.* (1990); (25) Schroter *et al.* (1991); (26) Pak *et al.* (1991b); (27) Roberds and Tamkun (1991).

[b] K_v4 is an alternatively spliced version of NGK2.

separate exon encodes the NH_2-terminus; if additional 5′ exons exist, then multiple forms of the protein may be generated by alternative splicing. $K_V1.7$ has an exon encoding the region spanning the S2 to the 3′ end; studies are ongoing to find the remainder of the gene (K. G. Chandy *et al.*, unpublished observations). Recently, S. Buhrow (personal communication) has demonstrated that $K_V3.1$ is alternatively spliced at the 3′ end. Thus, vertebrates appear to generate diverse forms of K^+ channels by encoding genes at multiple distinct genomic loci and by alternative splicing.

D. Genes Encoding Lymphocyte Channels

Expression of the $K_V1.3$ gene in *Xenopus* oocytes results in channels with properties that are indistinguishable from those of type *n* K^+ channels (Grissmer *et al.*, 1990). Furthermore, using the polymerase chain reaction technique, we have demonstrated that mRNA for $K_V1.3$ can be detected in the EL4 T cell line, which expresses about 200 type *n* K^+ channels per cell (Cahalan *et al.*, 1987). Based on these data, we conclude that MVK1.3 encodes the type *n* K^+ channel in T cells. Using MVK1.2- and MVK1.3-specific noncoding probes, we have localized these genes to human chromosomes 12 and 13, respectively (Grissmer *et al.*, 1990). Recently, Douglass *et al.* (1990) have suggested that the rat genomic equivalent of $K_V1.3$ (RGK5) encodes the type *n* K^+ channel and hybridizes with a 9.5-kb transcript in the mouse thymus.

Expression of the *Shaw*-related K^+ channel cDNA, $K_V3.1$ (NGK2), in *Xenopus* oocytes produces K^+ channels with biophysical properties that resemble the type *l* K^+ channel in T cells. These channels have a single-channel conductance of about 26 pS, open at −10 to 0 mV, and appear to be half-blocked by 0.1 m*M* TEA (Yokoyama *et al.*, 1989). Future studies will determine whether this gene does indeed encode the type *l* K^+ channel.

IV. VOLTAGE-GATED K^+ CHANNELS IN DEVELOPMENT

T Cell development takes place in the thymus, where T cell precursors divide rapidly, giving rise to functionally defined classes of T cells. Using specific monoclonal antibodies which recognize two different surface glycoproteins (CD4 and CD8), four main T cell subsets can be identified: $CD4^-CD8^-$ (DN or double negative), $CD4^+CD8^+$ (double positive), $CD4^+CD8^-$ (helper T cells), and $CD4^-CD8^+$ (killer T cells). Precursor cells in the thymus express neither CD4 or CD8 markers, nor the T cell receptor (TCR). These $CD4^-CD8^-TCR^-$ cells express ~200 voltage-gated type *n* K^+ channels per cell (Schlichter *et al.*, 1986; Lewis and Cahalan, 1988a,b). Precursor thymocytes acquire CD4, CD8, and TCR during development, giving rise to the most abundant population in the thymus, the double-positive ($CD4^+CD8^+TCR^+$) cortical thymocytes. These thymocytes, which express large numbers of type *n* K^+ channels (~200/cell), later undergo a selection process in which self-reactive cells are eliminated. Further maturation leads to the loss of either CD4 or CD8 expression; this results in the formation of helper ($CD4^+CD8^-$) and killer ($CD4^-CD8^+$) T cells. Interestingly, the pattern of K^+ channel expression is linked to the functional class of lymphocyte. Helper cells exhibit low numbers of type *n* K^+ channels (~20/cell), whereas killer cells display moderate numbers of type *l* or *n′* K^+ channels (~100/cell). During maturation, some cells acquire TCR without concomitant expression of CD4 or CD8; the functional role of these DN ($CD4^-CD8^-TCR^+$) T cells is not clear. Mature DN T cells exhibiting TCRs have small numbers of voltage-gated K^+ channels of either type *n, l,* or *n′*. However, when resting T or B cells from mouse spleen are stimulated by appropriate polyclonal mitogens, the number of type *n* K^+ channels per cell increases dramatically, perhaps due to new synthesis and functional assembly of channels (McKinnon and Ceredig, 1986; DeCoursey *et al.*, 1987b; Sutro *et al.*, 1988, 1989; Amigorena *et al.*, 1990). Thus, the number of type *n* channels per cell parallels cell proliferation, both during development in the thymus and during activation of mature cells. An interpretation of this correlation and the functional role of K^+ channels during development and activation are discussed below.

In summary, K^+ channel expression changes with respect to both type and number of channels during T cell development. Immature, rapidly dividing thymocytes ($CD4^-CD8^-TCR^-$, $CD4^+CD8^+$) and mature, activated T cells display large numbers of type *n* K^+ channels. On the other hand, mature, unactivated T cells express smaller numbers of voltage-gated K^+ channels; the type of channel expressed is correlated with the cell surface phenotype. Figure 5 provides a visual scheme for K^+ channel expression during T cell development and illustrates the aberrations in channel expression which are associated with autoimmune diseases (see below). Studies with channel blockers, summarized below, reinforce the idea that the presence of type *n* channels is functionally significant for cell proliferation.

V. K^+ CHANNELS AND T CELL FUNCTION

Early events in the activation of lymphocytes, platelets, mast cells, and several other cell types include increased Ca^{2+} and K^+ fluxes, changes in membrane potential, and second messenger generation. The hydrolysis of phosphatidylinositol 4,5-bisphosphate (PIP_2) gives rise to inositol phos-

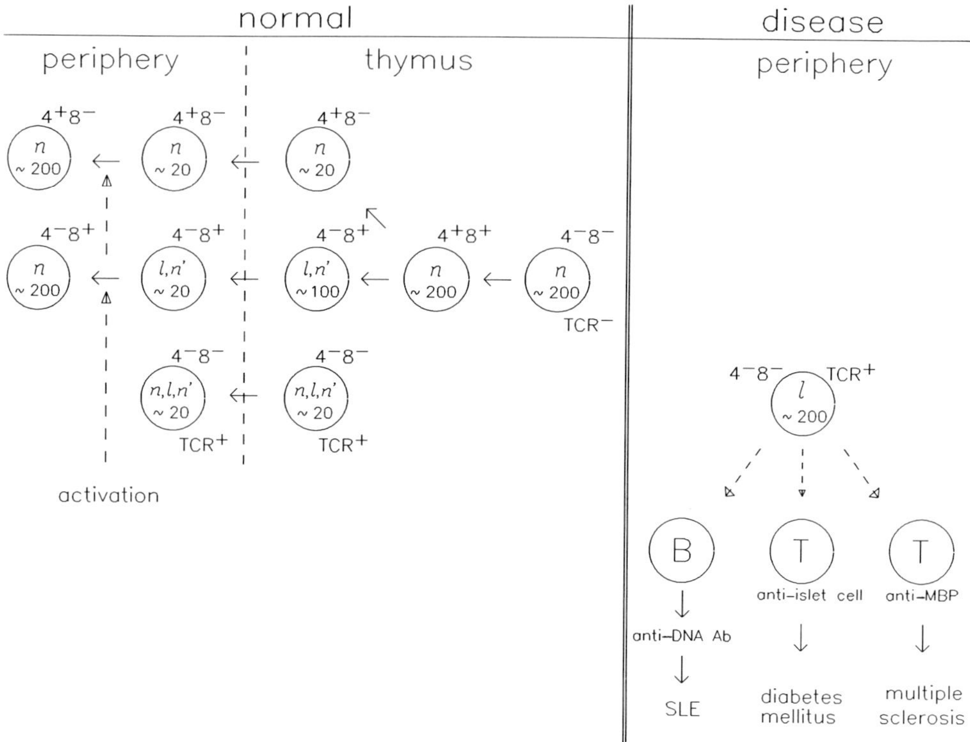

FIG. 5. Voltage-gated K^+ channel expression in development, activation, and disease. TCR, T cell receptor; Ab, antibodies; SLE, systemic lupus erythematosus; MBP, myelin basic protein.

phates and diacylglycerol, and $[Ca^{2+}]_i$ rises. Commitment to the activation pathway following antigen binding is incompletely understood but is widely considered to involve both a $[Ca^{2+}]_i$ rise and activation of protein kinase C (Crabtree, 1989). Additional cellular events include the appearance of mRNA for cellular protooncogenes *c-fos* and *c-myc* within 1–30 min of stimulation, phosphorylation of several substrates, cell enlargement during a phase of increased protein synthesis, secretion of interleukins, and, ultimately, cell division (Schwartz, 1990). Events elicited *in vitro* by polyclonal mitogens, such as lectins or cross-linking antibodies to components of the T cell receptor, are likely to occur *in vivo* following recognition and binding by the T cell receptor of antigen complexed with a major histocompatability (MHC) protein at the surface of an antigen-presenting cell. The ensuing cell proliferation represents the major mechanism for directing an immune response specifically against an antigenic challenge.

A. T Cell Activation

Agents that block type *n* K^+ channels inhibit selected events of lymphocyte activation. We have used seven compounds (TEA, 4-AP, quinine, diltiazem, verapamil, cetiedil, trifluoperazine, and chlorpromazine) in pharmacological studies of mitogen-stimulated cell proliferation, protein synthesis, and interleukin-2 secretion (DeCoursey *et al.,* 1984, 1985; Chandy *et al.,* 1984, 1986). The compounds, although differing substantially in their chemical structure, block type *n* K^+ channels in T cells and, in a parallel potency sequence, inhibit the proliferation of T cells stimulated by mitogenic lectins, by mitogenic antibodies, or by allogeneic cells. K^+ channel blockers inhibit thymidine uptake measured at 60 hr only if they are added during the first 20–30 hr of mitogen stimulation. This implies that the K^+ channel is required for some early event(s) in T cell activation. K^+ channel blockers also inhibit protein synthesis, secretion of interleukin-2, induction of *c-fos* and *c-myc,* and transport of metabolites, but not expression of interleukin-2 receptors (Chandy *et al.,* 1984; DeCoursey *et al.,* 1985; Schell *et al.,* 1987; Hough *et al.,* 1990). Many of these same blockers also inhibit mitogenic activation of B lymphocytes (Sutro *et al.,* 1988, 1989; Amigorena *et al.,* 1990), cytotoxic killing of target cells (Schlichter *et al.,* 1986), and cell volume regulation (reviewed in Lewis and Cahalan, 1990).

The use of pharmacological agents is one approach in assessing the functional role of ion channels. Specific blockers of K^+ channels have been shown to affect fundamental and characteristic cellular behaviors. CTX, possibly the most specific blocker of type *n* channels in lymphocytes, also inhibits mitogenesis and interleukin-2 secretion when tested in a serum-free culture system (Price *et al.,* 1989). The possible involvement of Ca^{2+}-activated K^+ channels sensitive to CTX must still be evaluated. Parallels in the expression of type *n* channels and the state of cell proliferation during T cell development and activation provide further evidence in support of the hypothesis that these channels are required for some stage during cell-cycle progression. A caveat to bear in mind is that even agents thought to be specific for a particular channel type may later be demonstrated to affect other channel types, as has been the case for quinine, CTX, and the Ca^{2+} antagonists. The weight of evidence implicates a requirement for functional type *n* channels in T cell activation, but Ca^{2+}-activated K^+ channels with overlapping pharmacological properties, including sensitivity to CTX, may also contribute. A possible mechanism by which K^+ channels participate in signal transduction is presented below.

B. Calcium Signaling

In a variety of cell types, Ca^{2+} can serve as a current carrier across membranes through a variety of distinct voltage-dependent and ligand-gated channel types, and thereby initiate intracellular events by activating enzymes, cytoskeletal elements, or ion channels. In lymphocytes, cytosolic Ca^{2+} rises during mitogen stimulation. Part of the initial rise in $[Ca^{2+}]_i$ is generated by a mechanism involving 1,4,5-inositol-trisphosphate (IP_3) and release of Ca^{2+} from internal stores, while the remainder comes from the outside (Weiss and Imboden, 1987; Berridge *et al.*, 1988; Meyer and Stryer, 1988; Lewis and Cahalan, 1989). The $[Ca^{2+}]_i$ rise in single cells can be seen to be oscillatory. In this section, we review the efforts to investigate Ca^{2+} signaling in individual T lymphocytes with patch-clamp and indicator-dye techniques. Among the possible contenders for mediating mitogen-stimulated Ca^{2+} influx we consider, in turn, voltage-gated Ca^{2+} channels, voltage-gated K^+ channels, IP_3-activated Ca^{2+}-permeable cation channels, and mitogen-regulated Ca^{2+}-selective channels.

1. Are Voltage-Gated Ca^{2+} Channels Found in T Cells?

The presence of voltage-gated Ca^{2+} channels has been suggested in lymphocytes, because several known blockers of Ca^{2+} channels can inhibit T cell activation (Birx *et al.*, 1984), albeit at levels higher than those known to inhibit bona fide voltage-gated Ca^{2+} channels from cardiac cells. We found, however, that the same agents inhibit voltage-gated type *n* K^+ channels. The list now includes verapamil, nifedipine and related dihydropyridines, diltiazem, and polyvalent cations such as Cd^{2+}, Co^{2+}, or La^{3+}. The inhibition of K^+ channels by these agents is direct and cannot be explained by their inhibition of an undetected inward Ca^{2+} current. Type *l* channels are also blocked by several of the "Ca^{2+} antagonists," but appear to be not as sensitive to block by the polyvalent cations. In summary, the pharmacological evidence in favor of the presence of voltage-gated Ca^{2+} channels is not sustained, due to nonspecific effects of Ca^{2+} antagonists.

Several groups using patch-clamp methods have independently reached the conclusion that voltage-gated Ca^{2+} channels are not present in lymphoid or related cells. Our efforts to examine this question have included whole-cell, perforated-patch, and cell-attached recordings, employing protocols which successfully reveal current through Ca^{2+} channels in other cell types. In lymphocytes, whole-cell capacitance is low and input resistance is high (1–10 pF and $>10^{10}\ \Omega$, respectively), making the limits of detection highly sensitive—often to the level of one channel per cell (far beyond the normal resolution of cell-surface molecules using fluorescently labeled monoclonal antibodies in a FACS analyzer). Only one group has reported the presence of voltage-gated Ca^{2+} channels in a subline of Jurkat cells (Dupuis *et al.*, 1989), but their results have not yet been confirmed.

The absence of voltage-gated Ca^{2+} channels in Jurkat T cells is also indicated by measurements of fura-2 Ca^{2+}-indicator fluorescence. In cells expressing voltage-gated Ca^{2+} channels, depolarization by raising K^+ normally would result in a rise in $[Ca^{2+}]_i$, detectable as a fura-2 signal. However, in Jurkat cells, depolarization by K^+ does not raise $[Ca^{2+}]_i$; rather, it either does little to the already low $[Ca^{2+}]_i$ in resting cells, or it reduces $[Ca^{2+}]_i$ dramatically in mitogen-stimulated cells undergoing Ca^{2+} oscillations (Lewis and Cahalan, 1989). This result corresponds to the inhibitory effects of elevated K^+ on biochemical measures of T cell activation (Gelfand *et al.*, 1984). Further experiments need to be done to test for the presence of voltage-gated Ca^{2+} channels, but the weight of evidence suggests that they are not present in most T or B lymphocyte-related cells. One definite exception is antibody-secreting hybridoma cells, which do express transient (T-type) voltage-gated Ca^{2+} channels (Fukushima *et al.*, 1984b; Bosma and Sidell, 1988), an attribute that is likely to be derived from the parental myeloma cells, which also express these channels (Fukushima and Hagiwara, 1983).

2. K^+ Channels and Ca^{2+} Influx

Since conventional voltage-gated Ca^{2+} channels appear not to be present in T cells, alternative pathways of Ca^{2+} entry must exist. In whole-cell patch-clamp experiments, addition of a mitogenic lectin, phytohemagglutinin (PHA) or concanavalin A (Con A), was found acutely to increase the rate of K^+ channel opening, sometimes resulting in a net increase in channel activity (DeCoursey *et al.*, 1984, 1985). We therefore investigated the possibility that, as has been demonstrated for other K^+ channels (Inoue, 1981), K^+ channels might be sparingly permeable to divalent ions and thereby provide the influx pathway during mitogen stimulation. Although Ca^{2+} does appear to be sparingly permeant in the type *n* channel, from studies of Ca^{2+} and Ba^{2+} block summarized above, it is unlikely to contribute to Ca^{2+} signaling directly, because estimated fluxes through the K^+ channel are much lower than those that occur during Ca^{2+} signaling (Grissmer and Cahalan, 1989b). Moreover and more directly, in fura-2 measurements of single-cell $[Ca^{2+}]_i$, depolarization to open the type *n* channels does not result in increased $[Ca^{2+}]_i$ (Lewis and Cahalan,

1988c). This voltage-clamp result does not mean, however, that in intact cells the K^+ channels have no influence on the Ca^{2+} signal. We have found that K^+ channel blockers, if applied at levels that block nearly all of the channels, do inhibit mitogen-stimulated oscillations in $[Ca^{2+}]_i$. This is not consistent with a previous report that TEA had little or no effect on the Ca^{2+} signal in T cells (Gelfand *et al.,* 1986). The discrepancy can be resolved by the difference in the fraction of K^+ channels blocked, or by differences in cell type and stimulus. Our results suggest that K^+ channels can indirectly regulate the Ca^{2+} signal in Jurkat T cells by affecting influx through a mitogen-regulated Ca^{2+} channel. A functionally significant Ca^{2+} influx pathway has recently been identified in Jurkat T cells, by combining patch-clamp measurements with fura-2 ratiometric measurement of $[Ca^{2+}]_i$, as described below.

3. Mitogen-Regulated Ca^{2+} Current

The small cell volume of lymphocytes means that micromolar levels of intracellular Ca^{2+} can, in principle, arise from a Ca^{2+}-selective inward current of less than 1 pA amplitude. The dynamics of Ca^{2+} signaling in single Jurkat T cells have been visualized using a video-imaging system to monitor fura-2 fluorescence. PHA, a mitogenic lectin, induces repetitive, periodic $[Ca^{2+}]_i$ oscillations peaking in the micromolar range with a period of about 100 sec (Lewis and Cahalan, 1989). Ca^{2+} oscillations last for hours following mitogen stimulation, and appear to depend primarily on influx of extracellular Ca^{2+}, as they require extracellular Ca^{2+} and are eliminated by Ca^{2+}-channel blockers and membrane depolarization.

Phospholipase C activity can generate several metabolites of phosphotidylinositol following mitogen stimulation. Particular attention has been paid to IP_3 as a potential mediator of influx, in addition to its established role of releasing Ca^{2+} by binding to receptors on intracellular membrane compartments (reviewed in Gardner, 1989). Using mainly single-channel measurements from inside-out patches excised from Jurkat cells, Gardner and colleagues found that IP_3 can activate a Ca^{2+}-permeable cation channel (Kuno and Gardner, 1987). This channel is difficult to study because it is small and flickery, and because it occurs infrequently from patch to patch.

Because of these difficulties, our laboratory has taken an alternative approach to identify the Ca^{2+} influx mechanism. The correlation of particular components of membrane current with the $[Ca^{2+}]_i$ rise in an individual cell has permitted the identification of a small (0.5–5 pA per cell) membrane current underlying Ca^{2+} signaling evoked by PHA. The current is highly selective for Ca^{2+} over Na^+, K^+, and Cl^-, and is blocked by Ni^{2+} at concentrations that block the $[Ca^{2+}]_i$ rise in PHA-stimulated cells. From the current–voltage relation, depolarization diminishes the current, while hyperpolarization enhances it. The high degree of Ca^{2+} selectivity and absence of conductance fluctuations, which indicate extremely low single-channel conductance, distinguish this current from IP_3-activated single channels described by Gardner's group, although both appear to be inhibited by a rise in $[Ca^{2+}]_i$. These results suggest that mitogen-evoked $[Ca^{2+}]_i$ oscillations arise from the periodic opening and closing of Ca^{2+} channels in the plasma membrane. However, the intracellular mechanisms that activate this current remain an open question.

4. Channel Activity during Ca^{2+} Signaling

Studies combining patch-clamp currents with measurement of $[Ca^{2+}]_i$ suggest a possible mechanism by which blockade of K^+ channels may indirectly inhibit T cell activation by interfering with the mitogen-stimulated Ca^{2+} signal. We hypothesize that coordinated activity of Ca^{2+} channels, voltage-gated K^+ channels, Ca^{2+}-activated K^+ channels, and release of Ca^{2+} from intracellular stores generates the mitogen-stimulated Ca^{2+} signal. Initially, $[Ca^{2+}]_i$ is low (100 n*M*) and the resting potential is determined by voltage-gated K^+ channels. Following mitogen or antigen binding, signals initiated by the T cell receptor complex activate second-messenger pathways which result in the release of Ca^{2+} from intracellular stores, along with the activation of inward Ca^{2+} current. The resultant rise in $[Ca^{2+}]$ is initially supported by the negative membrane potential developed by type *n* K^+ channels. As soon as the $[Ca^{2+}]_i$ rises, the opening of Ca^{2+}-activated K^+ channels may further boost the inward Ca^{2+} current by hyperpolarizing the cell. At still higher $[Ca^{2+}]_i$, both the Ca^{2+} current itself and the voltage-gated K^+ channels appear to be inhibited. We have demonstrated that membrane depolarization, resulting from elevated $[K^+]_o$, inhibits the Ca^{2+} signal, presumably by reducing the electrical driving force favoring Ca^{2+} influx (Lewis and Cahalan, 1989). Possibly via the same mechanism, blocking K^+ channels with CTX and apamin suppresses Ca^{2+} oscillations in a manner similar to the effect of increasing $[K^+]_o$ (Grissmer *et al.*, 1992). Inhibition of mitogen-stimulated cell proliferation by K^+ channel blockers or by depolarization may both be due to consequent inhibition of Ca^{2+} signaling.

VI. VOLTAGE-GATED K^+ CHANNELS IN AUTOIMMUNE DISEASES

Autoimmune diseases, including type-1 diabetes mellitus, rheumatoid arthritis, myasthenia gravis, multiple sclerosis, and systemic lupus erythematosus (SLE), affect several hundred million people worldwide.

These disorders are caused by self-reactive lymphocytes which damage tissues within the body. Murine models have provided important insights into the mechanisms that underlie autoimmunity (Theofilopoulos and Dixon, 1985). We have recently found that aberrant double-negative (DN) ($CD4^-CD8^-TCR^+$) T cells from mice with juvenile diabetes, SLE, type II collagen arthritis, or a multiple sclerosis-like disease, experimental allergic encephalomyelitis (EAE), exhibit abnormally large numbers of type *l* K^+ channels (Chandy *et al.*, 1990c; Grissmer *et al.*, 1990). The possible significance of this finding is discussed below in Section VI,G. Helper and killer T cells from diseased mice retain their normal pattern of ion channel expression.

A. T Cell Subsets in Systemic Lupus Erythematosus

Systemic lupus erythematosus is characterized by polyclonal B cell activation which leads to autoantibody production (e.g., against double-stranded DNA). The autoantibodies form immune complexes that generate an inflammatory response in various tissues, including the kidney. The immunological abnormalities leading to B cell hyperactivity have not been elucidated, although T cells appear to play a role in this process (Theofilopoulos and Dixon, 1985; Shultz and Sidman, 1987). Treatment with anti-CD4 antibodies retards the development of autoimmune diseases in mouse models, suggesting that $CD4^+CD8^-$ helper T cells play a significant role in the pathogenesis of SLE (Wofsy and Seaman, 1985; Santoro *et al.*, 1988), type-1 diabetes mellitus (Miller *et al.*, 1988), EAE (Waldor *et al.*, 1985), type II collagen arthritis (Ranges *et al.*, 1985), and myasthenia gravis (Christadoss and Dauphinee, 1986). Treatment with anti-CD4 also retards the expansion of DN T cells in *lpr* mice and simultaneously slows the onset and lessens the severity of SLE (Santoro *et al.*, 1988). This suggests that DN T cell proliferation is dependent on CD4 cells, and may be a link in the chain of events that lead to the development of SLE. Several other lines of evidence (to be discussed in Section VI,G) implicate DN T cells with helper activity in the development of autoimmunity. In the ensuing paragraphs, we demonstrate that the DN T cells associated with autoimmunity exhibit large numbers of type *l* K^+ channels, a K^+ channel phenotype that appears unique to these cells.

B. Type l Channels in Murine Models for Systemic Lupus Erythematosus

Several inbred strains of mice spontaneously develop a disease resembling human SLE (Theofilopoulos and Dixon, 1985; Shultz and Sidman,

1987). Unlike normal DN T cells which display about 10–20 K^+ channels of types *n, n'*, or *l*, DN T cells from all the mouse models for SLE express exclusively type *l* K^+ channels in large (~180/cell) numbers. The increase in type *l* channel expression parallels the onset of disease (Chandy *et al.*, 1986, 1990c; Grissmer *et al.*, 1988) and appears to correlate with the ability of these cells to induce autoantibody production by B cells (Datta *et al.*, 1987; Sainis and Datta, 1988). Helper ($CD4^+CD8^-$) and cytotoxic ($CD4^-CD8^+$) T cells from mice with SLE displayed a small number of K^+ channels, as do their phenotypic counterparts from normal mice.

C. Nonobese Diabetic Mice

Nonobese diabetic (NOD) mice spontaneously develop a disease resembling juvenile diabetes, apparently due to T cell-mediated destruction of pancreatic islet cells (Miller *et al.*, 1988). Interestingly, the majority of DN T cells from NOD mice exhibit large numbers of type *l* K^+ channels, averaging ~200 channels/cell (Chandy *et al.*, 1990c).

D. Chronic Experimental Allergic Encephalomyelitis

Inoculation of SJL/PLJ mice with myelin basic protein induces an immune response, leading to demyelination. Some mice progress into a chronic relapsing form of encephalomyelitis which resembles multiple sclerosis in humans (Waldor *et al.*, 1985). A large fraction of DN T cells from mice with chronic relapsing encephalomyelitis display numerous type *l* K^+ channels (Chandy *et al.*, 1990c).

E. Type II Collagen Arthritis

Some DBA-J/LacJ mice immunized with type II collagen and complete Freund's adjuvant develop a disease resembling rheumatoid arthritis (Stuart *et al.*, 1984, 1988; Trentham, 1988). About 40% of DN T cells from diseased mice displayed large numbers of type *l* K^+ channels, whereas DN T cells from unimmunized mice had a normal pattern of channel expression (Grissmer *et al.*, 1990b). Other T cell subsets retained their normal pattern of channel expression.

F. Nonspecific Immune Response

DN T cells from normal mice immunized with either heat-killed *Escherichia coli* or complete Freund's adjuvant express small numbers of K^+ channels of all three types, like their phenotypic counterparts in normal unimmunized mice (S. Grissmer *et al.*, unpublished observations). Thus, excessive expression of type *l* K^+ channels appears to be correlated with the presence of chronic autoimmune diseases, rather than with a normal immune response.

G. Double-Negative T Cells, Heat-Shock Proteins, Type *l* Channels, and the Autoimmune Process

Several lines of evidence suggest that DN T cells with helper activity may induce autoantibody production by B cells in both mice and humans with SLE. Symptomatic SLE mice and patients with SLE have increased numbers of DN T cells in their blood; these DN T cells induce cationic anti-DNA autoantibody production by B cells *in vitro* (Datta *et al.*, 1987; Sainis and Datta, 1988; Shivakumar *et al.*, 1989; Adams *et al.*, 1990). Lymphoproliferation of DN T cells in mice with the *lpr* or *gld* mutations appears to accelerate the onset of SLE (Theofilopoulos and Dixon, 1985). These proliferating DN T cells produce interleukin-6, tumor necrosis factors α and β, and interferon γ (Murray and Martens, 1989; Murray *et al.*, 1990; Umland *et al.*, 1989; Klinman, 1990), cytokines that may contribute to the autoimmune process. In fact, interferon γ has been recently shown to activate a large proportion of B cells from diseased *lpr* mice (but not from normal controls), resulting in increased autoantibody formation (Klinman, 1990).

DN T cells may also play a role in the induction of other types of autoimmune and chronic inflammatory disorders that are associated with the development of an immune response against heat-shock proteins (van Eden *et al.*, 1988; Res *et al.*, 1988; Modlin *et al.*, 1989; Holoshitz *et al.*, 1989). Heat-shock proteins are well conserved in evolution. Mycobacteria express a 65-kDa heat-shock protein (Hsp65) that shares about 50% similarity with a similar molecule in humans (Jindal *et al.*, 1989). The Hsp65 molecule has many different epitopes (Lamb *et al.*, 1987; Gaston *et al.*, 1990) that may be similar to antigenic epitopes in joints (van Eden *et al.*, 1988; Res *et al.*, 1988; Holoshitz *et al.*, 1989; Gaston *et al.*, 1990), the skin (Modlin *et al.*, 1989), and in the pancreas (Elias *et al.*, 1990). Immune responses mounted against Hsp65 could cross-react with self-antigens, resulting in tissue damage. Many lines of evidence support this hypothesis. Exposure of normal mouse spleen cells to heat-shock (42°C for 30 min) results in the proliferation of autoreactive DN T cells that recognize heat-shock proteins on the stressed cells (Rajasekar *et al.*, 1990). Immune reactivity to Hsp65 is associated with the development of adjuvant arthritis in rats and rheumatoid arthritis in humans (van Eden *et al.*, 1988; Res *et al.*, 1988; Gaston *et al.*, 1990). In fact, mycobacterial-responsive DN T cells have been recently isolated from the joints of patients with rheumatoid arthritis (Holoshitz *et al.*, 1989) and from the diseased skin of patients infected with *Mycobacterium leprae* (Modlin *et al.*, 1989); these cells appear to induce monocyte aggregation (Modlin *et al.*, 1989). $CD4^+CD8^-$ T cell clones that recognize an Hsp65-like molecule on pancreatic β cells cause type-1 diabetes mellitus within a few days after injection into young NOD/lt mice (Elias *et al.*, 1990); uninjected NOD/lt mice develop diabetes at about 5 months of age. These diabetes-inducing $CD4^+CD8^-$ T cells may be derived from DN T cells, since mitogens can induce differentiation of normal DN T cells into $CD4^+CD8^-$ autoreactive T cells (Morisset *et al.*, 1988; Reimann *et al.*, 1988). Thus, disparate types of environmental stimuli might activate heat-shock or stress-protein-responsive DN T cells that may cross-react with self-antigens and cause tissue damage. Some of these DN T cells may differentiate into autoreactive $CD4^+CD8^-$ T cells, further contributing to the immune assault on self tissues. In addition, activated DN T cells may secrete lymphokines (Murray and Martens, 1989; Murray *et al.*, 1990; Umland *et al.*, 1989; Klinman, 1990; Gijbels *et al.*, 1990) which may recruit other immune cells into the vicinity and stimulate both B and T cells; some of these activated B or T cells may react with self-antigens and enhance tissue damage.

Altered K^+ channel expression in DN T cells provides a link between SLE, juvenile diabetes, multiple sclerosis, and type II collagen arthritis, and focuses attention on the role of DN T cells in autoimmunity. The augmented expression of type *l* K^+ channels in the autoimmune-associated DN T cells may reflect the activated status of the cells and their ability to secrete specific types of lymphokines. Further study of channel expression, normal and abnormal, will hopefully clarify the role of these channels in the pathogenesis of autoimmune disorders.

VII. CONCLUSIONS

A diverse repertoire of ion channels, including K^+, Cl^-, and Ca^{2+} channels, has been identified in T and B lymphocytes through patch-clamp techniques. At least three types of voltage-gated K^+ channels and two types of Ca^{2+}-activated K^+ channels contribute to channel diversity within the family of K^+ channels. The expression of these channels appears to be regulated during development, differentiation, and activation. It has become clear that ion channels mediate a variety of cellular behaviors. K^+-channel blockade inhibits lymphocyte functions, including mitogen-stimulated secretion of lymphokines, cell proliferation, and volume regulation in response to hypotonic osmotic challenge. Video-imaging techniques have enabled the dynamics of Ca^{2+} signaling to be visualized in individual T lymphocytes, with oscillations depending on the coordinated activity of mitogen-regulated Ca^{2+} channels, voltage-gated K^+ channels, and Ca^{2+}-activated K^+ channels. Drugs that modulate K^+ channels are useful for the treatment of several diseases (Cook, 1988; Robertson and Steinberg, 1990). Therefore, T cell channel abnormalities occurring during autoimmune disorders may provide a focus for diagnosis and therapeutic intervention in human patients. Molecular biological approaches have begun to reveal the structure of lymphocyte K^+ channels and have much to contribute toward understanding cellular mechanisms for altered channel expression during development and activation, and in disease.

Acknowledgments

We thank Ruth Davis for helpful suggestions on the manuscript and Dr. Erwin Neher for his hospitality at the Max Planck Institut für Biophysikalische Chemie, Göttingen, Germany. Work in the authors' laboratories has been supported by the National Institutes of Health, the Office of Naval Research, and Pfizer, Inc. We are also grateful to the Alexander von Humboldt Foundation for support (MDC). Dr. George Gutman and Mr. Calvin Williams helped in formulating the nomenclature for potassium channel genes.

References

Adams, S., Zordan, T., Sainis, K., and Datta, S. (1990). T cell receptor V-beta genes expressed by IgG anti-DNA autoantibody-inducing T cells in lupus nephritis: Forbidden receptors and double-negative T cells. *Eur. J. Immunol.* **20,** 1435–1443.

Amigorena, S., Choquet, D., Tcillaud, J.-L., Korn, H., and Fridman, W. H. (1990). Ion channel blockers inhibit B cell activation at a precise stage of the G1 phase of the cell cycle. *J. Immunol.* **144,** 2038–2045.

Baumann, A., Krah-Jentzens, I., Müller, R., Müller-Hotkamp, F., Seidel, R., Kecskemetry, N., Casal, J., Ferrus, A., and Pongs, O. (1987). Molecular organization of the maternal effect region of the *Shaker* complex of *Drosophila:* Characterization of an I_A channel transcript with homology to the sodium channel. *EMBO J.* **6,** 3419–3429.

Baumann, A., Grupe, A., Ackermann, A., and Pongs, O. (1988). Structure of the voltage dependent potassium channel is highly conserved from *Drosophila* to vertebrate central nervous systems. *EMBO J.* **7,** 2457–2463.

Beckh, S., and Pongs, O. (1990). Members of the RCK potassium channel family are differentially expressed in the rat nervous system. *EMBO J.* **9,** 777–782.

Berridge, M. J., Cobbold, P. H., and Cuthbertson, K. S. R. (1988). Spatial and temporal aspects of cell signalling. *Philos. Trans. R. Soc. London, Ser. B* **320,** 325–343.

Betsholtz, C., Baumann, A., Kenna, S., Ashcroft, F. M., Ashcroft, S. J. H., Berggren, P.-O., Grupe, A., Pongs, O., Rorsmann, P., Sandblom, J., and Welsh, M. (1990). Expression of voltage-gated K^+ channels in insulin-producing cells. Analysis by polymerase chain reaction. *FEBS Lett.* **263,** 121–126.

Birx, D. L., Berger, M., and Fleisher, T. A. (1984). The interference of T cell activation by calcium channel blocking agents. *J. Immunol.* **133,** 2904–2909.

Blatz, A. L., and Magleby, K. L. (1986). Single apamin-blocked Ca^{2+}-activated K^+ channels of small conductance in cultured rat skeletal muscle. *Nature (London)* **323,** 718–720.

Bosma, M., and Sidell, N. (1988). Retinoic acid inhibits Ca^{2+} currents and cell proliferation in a B-lymphocyte cell line. *Cell. Immunol.* **115,** 299–309.

Bregestovski, P., Redkozubov, A., and Alexeev, A. (1986). Elevation of intracellular calcium reduces voltage-dependent potassium conductance in human T cells. *Nature (London)* **319,** 776–778.

Butler, A., Wei, A., Baker, K., and Salkoff, L. (1989). A family of K^+ channel genes in *Drosophila. Science* **243,** 943–947.

Cahalan, M. D., and Lewis, R. S. (1988). Role of potassium and chloride channels in volume regulation by T lymphocytes. *In* "Cell Physiology of Blood" (R. B. Gunn and J. C. Parker, eds.), pp. 281–301. Rockefeller Univ. Press, New York.

Cahalan, M. D., Chandy, K. G., DeCoursey, T. E., and Gupta, S. (1985). A voltage-gated potassium channel in human T lymphocytes. *J. Physiol. (London)* **358,** 197–237.

Cahalan, M. D., Chandy, K. G., DeCoursey, T. E., Gupta, S., Lewis, R. S., and Sutro, J. (1987). Ion channels in T lymphocytes. *Adv. Exp. Med. Biol.* **213,** 85–102.

Chandy, K. G., DeCoursey, T. E., Cahalan, M. D., McLaughlin, C., and Gupta, S. (1984). Voltage-gated K^+ channels are required for T lymphocyte activation. *J. Exp. Med.* **160,** 369–385.

Chandy, K. G., DeCoursey, T. E., Fischbach, M., Talal, N., Cahalan, M. D., and Gupta, S. (1986). Altered K^+ channel expression in abnormal T lymphocytes from mice with the *lpr* gene mutation. *Science* **233,** 1197–1200.

Chandy, K. G., Williams, C. B., Spencer, R. H., Aguilar, B. A., Ghanshani, S., Tempel, B. L., and Gutman, G. A. (1990a). A family of three mouse potassium channel genes with intronless coding regions. *Science* **247,** 973–979.

Chandy, K. G., Williams, C. B., Spencer, R. H., Aguilar, B. A., Ghanshani, S., Chandy, G., Tempel, B. L., and Gutman, G. A. (1990b). Multiple genes contribute to the K^+ channel diversity in the mouse. *Biophys. J.* **57,** 11a. (Abstr.)

Chandy, K. G., Cahalan, M. D., and Grissmer, S. (1990c). Autoimmune diseases linked to abnormal K^+ channel expression in double-negative $CD4^-CD8^-$ T cells. *Eur. J. Immunol.* **20,** 747–751.

Choquet, D., and Korn, H. (1988). Dual effects of serotonin on a voltage-gated conductance in lymphocytes. *Proc. Natl. Acad. Sci. U.S.A.* **85,** 4557–4661.

Choquet, D., Sarthou, P., Primi, D., Cazenave, P.-A., and Korn, H. (1987). Cyclic AMP-modulated potassium channels in murine B cells and their precursors. *Science* **235,** 1211–1214.

Christadoss, P., and Dauphinee, M. J. (1986). Immunotherapy for myasthenia gravis: A murine model. *J. Immunol.* **136,** 2437–2440.

Christie, M. J., Adelman, J. P., Douglass, J., and North, R. A. (1989). Expression of a cloned rat brain potassium channel in *Xenopus* oocytes. *Science* **244,** 221–224.

Coffey, R. G., and Hadden, J. W. (1985). Neurotransmitters, hormones, and cyclid nucleotides in lymphocyte regulation. *Fed. Proc.* **44,** 112–117.

Cook, N. S. (1988). The pharmacology of potassium channels and their therapeutic potential. *Trends Pharmacol.* **9,** 21–28.

Crabtree, G. R. (1989). Contingent genetic regulatory events in T lymphocyte activation. *Science* **243,** 355–361.

Datta, S. K., Patel, H., and Berry, D. (1987). Induction of a cationic shift in IgG anti-DNA autoantibodies: Role of T helper cells with classical and novel phenotypes in three murine models of lupus nephritis. *J. Exp. Med.* **165,** 1252–1268.

DeCoursey, T. E., Chandy, K. G., Gupta, S., and Cahalan, M. D. (1984). Voltage-gated K^+ channels in human T lymphocytes: A role in mitogenesis? *Nature (London)* **307,** 465–468.

DeCoursey, T. E., Chandy, K. G., Gupta, S., and Cahalan, M. D. (1985). Voltage-dependent ion channels in T-lymphocytes. *J. Neuroimmunol.* **10,** 71–95.

DeCoursey, T. E., Chandy, K. G., Gupta, S., and Cahalan, M. D. (1987a). Two types of potassium channels in murine T lymphocytes. *J. Gen. Physiol.* **89,** 379–404.

DeCoursey, T. E., Chandy, K. G., Gupta, S., and Cahalan, M. D. (1987b). Mitogen induction of ion channels in murine T lymphocytes. *J. Gen. Physiol.* **89,** 405–420.

Deutsch, C., and Lee, S. C. (1989). Modulation of K^+ currents in human lymphocytes by pH. *J. Physiol. (London)* **413,** 399–413.

Deutsch, C. J., Holian, A., Holian, S. K., Daniele, R. P., and Wilson, D. F. (1979). Transmembrane electrical and pH gradients across human erythrocytes and human peripheral lymphocytes. *J. Cell. Physiol.* **99,** 79–94.

Douglass, J., Osborne, P. B., Cai, Y.-C., Wilkinson, M., Christie, M. J., and Adelman, J. P. (1990). Characterization of RGK5, a genomic clone encoding a lymphocyte channel. *J. Immunol.* **144,** 4841–4850.

Dupuis, G., Heroux, J., and Payet, M. D. (1989). Characterization of Ca^{2+} and K^+ currents in the human Jurkat T cell line: Effects of phytohaemagglutinin. *J. Physiol. (London)* **412,** 135–154.

Elias, D., Markovits, D., Reshef, T., van der Zee, R., and Cohen, I. R. (1990). Induction and therapy of autoimmune diabetes in the non-obese diabetic (NOD/Lt) mouse by a 65 kDa heat shock protein. *Proc. Natl. Acad. Sci. U.S.A.* **87,** 1576–1580.

Felber, S. B., and Brand, M. D. (1983). Early plasma-membrane-potential changes during stimulation of lymphocytes by concanavalin A. *Biochem. J.* **210,** 885–891.

Frech, G., VanDongen, M. J., Schuster, G., Brown, A. M., and Joho, R. H. (1989). A novel potassium channel with delayed rectifier properties isolated from rat brain expression cloning. *Nature (London)* **340,** 642–646.

Fukushima, Y., and Hagiwara, S. (1983). Voltage-gated Ca^{2+} channel in mouse myeloma cells. *Proc. Natl. Acad. Sci. U.S.A.* **80,** 2240–2242.

Fukushima, Y., Hagiwara, S., and Henkart, M. (1984a). Potassium current in clonal cytotoxic T lymphocytes from the mouse. *J. Physiol. (London)* **351,** 645–656.

Fukushima, Y., Hagiwara, S., and Saxton, R. E. (1984b). Variation of calcium current during the cell growth cycle in mouse hybridoma lines secreting immunoglobulins. *J. Physiol. (London)* **355,** 313–321.

Gardner, P. (1989). Calcium and T lymphocyte activation. *Cell* **59,** 15–20.

Gaston, J. S. H., Life, P. F., Jenner, P. J., Colston, M. J., and Bacon, P. A. (1990). Recognition of a mycobacteria-specific epitope in the 65 kD heat-shock protein by synovial fluid derived T cell clones. *J. Exp. Med.* **171,** 831–841.

Gelfand, E. W., Cheung, R. K., and Grinstein, S. (1984). Role of membrane potential in the regulation of lectin-induced calcium uptake. *J. Cell. Physiol.* **121,** 533–599.

Gelfand, E. W., Cheung, R. K., and Grinstein, S. (1986). Mitogen-induced changes in Ca^{2+} permeability are not mediated by voltage-gated K^+ channels. *J. Biol. Chem.* **261,** 11520–11523.

Ghanshani, S., Chandy, G., Kerlin, J., Dethlefs, B., Strong, M., Gutman, G. A., and Chandy, K. G. (1990). Mouse genomic clones encoding *Shaker*- and *Shaw*-related voltage-gated K^+ channel genes. *Tenth Int. Biophys. Congr.,* p. 378.

Gijbels, K., Van Damme, J., Proost, P., Put, W., Carton, H., and Billiau, A. (1990). Interleukin 6 production in the central nervous system during experimental autoimmune encephalomyelitis. *Eur. J. Immunol.* **20,** 233–235.

Grinstein, S., and Dixon, S. J. (1989). Ion transport, membrane potential and cytoplasmic pH in lymphocytes: Changes during activation. *Physiol. Rev.* **69,** 417–482.
Grinstein, S., and Smith, J. D. (1990). Calcium-independent cell volume regulation in human lymphocytes. *J. Gen. Physiol.* **95,** 97–120.
Grissmer, S., and Cahalan, M. D. (1989a). TEA prevents inactivation while blocking open K^+ channels in human T lymphocytes. *Biophys. J.* **55,** 203–206.
Grissmer, S., and Cahalan, M. D. (1989b). Ionomycin activates a potassium-selective conductance in human T lymphocytes. *Biophys. J.* **55,** 245a. (Abstr.)
Grissmer, S., and Cahalan, M. D. (1989c). Divalent ion trapping inside potassium channels of human T lymphocytes. *J. Gen. Physiol.* **93,** 609–630.
Grissmer, S., Cahalan, M. D., and Chandy, K. G. (1988). Abundant expression of type *l* K^+ channels: A marker for lymphoproliferative diseases? *J. Immunol.* **141,** 1137–1142.
Grissmer, S., Dethlefs, B., Wasmuth, J. J., Goldin, A. L., Gutman, G. A., Cahalan, M. D., and Chandy, K. G. (1990a). Expression and chromosomal localization of a lymphocyte K^+ channel gene. *Proc. Natl. Acad. Sci. U.S.A.* **87,** 9411–9415.
Grissmer, S., Hanson, D., Natale, P., Cahalan, M. D., and Chandy, K. G. (1990b). Double negative ($CD4^-CD8^-$) T cells from mice with type II collagen arthritis display aberrant potassium channel expression. *J. Immunol.* **145,** 2105–2109.
Grissmer, S., Lewis, R. S., and Cahalan, M. D. (1992). In preparation.
Grupe, A., Schroter, K. H., Ruppersberg, J. P., Stocker, M., Drewes, T., Beckh, S., and Pongs, O. (1990). Cloning and expression of a human voltage-gated potassium channel. A novel member of the RCK potassium channel family. *EMBO J.* **9,** 1749–1756.
Hamill, O. P., Marty, A., Neher, E., Sakmann, B., and Sigworth, F. J. (1981). Improved patch-clamp techniques for high-resolution current recording from cells and cell-free membrane patches. *Pfluegers Arch.* **391,** 85–100.
Hille, B., Woodhull, A. M., and Shapiro, B. J. (1975). Negative surface charge near sodium channels of nerve: Divalent ions, monovalent ions, and pH. *Philos. Trans. R. Soc. London, Ser. B* **270,** 301–318.
Holoshitz, J., Konig, F., Coligan, J. E., De Bruyn, J., and Strober, S. (1989). Isolation of $CD4^-CD8^-$ mycobacteria-reactive T lymphocyte clones from rheumatoid arthritis synovial fluid. *Nature (London)* **339,** 226–229.
Hough, C. J., Halperin, J. I., Mazorow, D. L., Yeandle, S. I., and Millar, D. B. (1990). Beta-endorphin modulates T-cell intracellular calcium flux and c-*myc* expression via a potassium channel. *J. Neuroimmunol.* **27,** 163–171.
Hugues, M., Schmid, H., Romey, G., Duval, D., Frelin, C., and Lazdunski, M. (1982). The Ca^{2+}-dependent slow K^+ conductance in cultured rat muscle cells: Characterization with apamin. *EMBO J.* **1,** 1039–1042.
Inoue, I. (1981). Activation-inactivation of potassium channels and development of the potassium channel spike in internally perfused squid giant axons. *J. Gen. Physiol.* **78,** 43–61.
Isacoff, E. Y., Jan, Y. N., and Jan, L. Y. (1990). Evidence for the formation of heteromultimeric potassium channels in *Xenopus* oocytes. *Nature (London)* **345,** 530–534.
Jindal, S., Dudani, A. K., Harley, C. B., Singh, B., and Gupta, R. S. (1989). Primary structure of a human mitochondrial protein homologous to the bacterial and plant chaperonins and to the 65-kilodalton mycobacterial antigen. *Mol. Cell. Biol.* **9,** 2279–2283.
Kamb, A., Iverson, L., and Tanouye, M. A. (1987). Molecular characterization of *Shaker,* a *Drosophila* gene that encodes a potassium channel. *Cell* **50,** 405–413.
Kamb, A., Tseng-Clark, J., and Tanouye, M. A. (1988). Multiple products of the *Drosophila Shaker* gene may contribute to potassium channel diversity. *Neuron* **1,** 421–430.

Kamb, A., Weir, M., Rudy, B., Varmus, H., and Kenyon, C. (1989). Identification of genes from pattern formation, tyrosine kinase, and potassium channel families by DNA amplification. *Proc. Natl. Acad. Sci. U.S.A.* **86,** 4372–4376.

Klinman, D. (1990). IgG1 and IgG2a production by autoimmune B cells treated *in vitro* with IL4 and IFN_{gamma}. *J. Immunol.* **144,** 2529–2534.

Krause, D., Lee, S. C., and Deutsch, C. (1988). Forskolin effects on the voltage-gated K^+ conductance of human T cells. *Pfluegers Arch.* **412,** 133–140.

Kuno, M., and Gardner, P. (1987). Ion channels activated by inositol 1,4,5-trisphosphate in plasma membrane of human T-lymphocytes. *Nature (London)* **328,** 301–304.

Lamb, J. R., Ivanyi, J., Rees, A. D. M., Rothbard, J. B., Holland, K., Young, R. A., and Young, D. B. (1987). Mapping of T cell epitopes using recombinant antigens and synthetic peptides. *EMBO J.* **6,** 1245–1249.

Lazdunski, M. (1983). Apamin, a neurotoxin specific for one class of Ca^{2+}-dependent K^+ channels. *Cell Calcium* **4,** 421–428.

Lee, S. C., and Deutsch, C. (1990). Temperature dependence of K^+ channel properties in human T lymphocytes. *Biophys. J.* **57,** 49–62.

Lew, V. L., and Ferreira, H. G. (1978). Calcium transport and the properties of a calcium-activated potassium channel in red cell membranes. *Curr. Top. Membr. Transp.* **10,** 217–277.

Lewis, R. S., and Cahalan, M. D. (1988a). Subset-specific expression of potassium channels in developing murine T lymphocytes. *Science* **239,** 771–775.

Lewis, R. S., and Cahalan, M. D. (1988b). The plasticity of ion channels: Parallels between the nervous and immune system. *TINS* **11,** 214–218.

Lewis, R. S., and Cahalan, M. D. (1988c). Voltage-dependent calcium signalling in single T lymphocytes. *Soc. Neurosci. Abstr.* **14,** 298.

Lewis, R. S., and Cahalan, M. D. (1989). Mitogen-induced oscillations of cytosolic Ca^{2+} and transmembrane Ca^{2+} current in human leukemic T cells. *Cell Regul.* **1,** 99–112.

Lewis, R. S., and Cahalan, M. D. (1990). Ion channels and signal transduction in lymphocytes. *Annu. Rev. Physiol.* **52,** 415–430.

Luneau, C. J., Williams, J. B., Marshall, J., Levitan, E. S., Oliva, C., Smith, J. S., Antanavage, J., Folander, K., Stein, R. B., and Swanson, R. (1991). Alternative splicing contributes to K^+ channel diversity in the mammalian nervous system. *Proc. Natl. Acad. Sci. U.S.A.* **88,** 3932–3936.

MacDougall, S. L., Grinstein, S., and Gelfand, E. W. (1988). Activation of Ca^{2+}-dependent K^+ channels in human B lymphocytes by anti-immunoglobulin. *J. Clin. Invest.* **81,** 449–454.

MacKinnon, R., and Miller, C. (1989). Mutant potassium channels with altered binding of charybdotoxin, a pore-blocking peptide inhibitor. *Science* **245,** 1382–1385.

Mahaut-Smith, M. P., and Schlichter, L. C. (1989a). Ca^{2+}-activated K^+ channels in lymphocytes. *Pfluegers Arch.* **414,** s164–s165. (Abstr.)

Mahaut-Smith, M. P., and Schlichter, L. C. (1989b). Ca^{2+}-activated K^+ channels in human B lymphocytes and rat thymocytes. *J. Physiol. (London)* **415,** 69–83.

Mathew, M., Ramaswami, M., Gautam, M., Kamb, C. A., Rudy, B., and Tanouye, M. (1989). Cloning and characterization of human potassium channel genes. *Soc. Neurosci. Abstr.* **15,** 540.

Matteson, D. R., and Deutsch, C. (1984). K^+ channels in T lymphocytes: A patch clamp study using monoclonal antibody adhesion. *Nature (London)* **307,** 468–471.

McCormack, T., Vega-Saenz de Miera, E. C., and Rudy, B. (1990). Molecular cloning of a member of a third class of *Shaker* family K^+ channel genes in mammals. *Proc. Natl. Acad Sci. U.S.A.* **87,** 5227–5231.

McKinnon, D. (1989). Isolation of a cDNA clone coding for a putative second potassium channel indicates the existence of a gene family. *J. Biol. Chem.* **264,** 8230–8236.

McKinnon, D., and Ceredig, R. (1986). Changes in the expression of potassium channels during mouse T cell development. *J. Exp. Med.* **164,** 1846–1861.

Meyer, T., and Stryer, L. (1988). Molecular model for receptor-stimulated calcium spiking. *Proc. Natl. Acad. Sci. U.S.A.* **85,** 5051–5055.

Miller, B., Appel, M. C., O'Neill, J. J., and Wicker, L. S. (1988). Both the Lyt-2^+ and L3T4^+ T cell subsets are required for the transfer of diabetes in nonobese diabetic mice. *J. Immunol.* **140,** 52–58.

Miller, C., Moczydlowski, E., Latorre, R., and Phillips, M. (1985). Charybdotoxin, a protein inhibitor of single Ca^{2+}-activated K^+ channels from mammalian skeletal muscle. *Nature* (*London*) **313,** 316–318.

Modlin, R. L., Pirmez, C., Hoffman, F. M., Torigian, V., Uyemura, K., Rea, T. H., Bloom, B. R., and Brenner, M. B. (1989). Lymphocytes bearing antigen-specific gamma/delta T-cell receptors accumulate in human infectious disease lesions. *Nature* (*London*) **339,** 544–548.

Morisset, J., Trannoy, E., De Talance, A., Spinella, S., Debre, P., Godet, P., and Seman, M. (1988). Genetics and strain distribution of concanavalin A-reactive Ly-2^-, L3T4^- peripheral precursors of autoreactive cells. *Eur. J. Immunol.* **18,** 387–394.

Murai, T., Kakizuka, A., Takumi, T., Ohkubo, H., and Nakanishi, S. (1989). Molecular cloning and sequence analysis of human genomic DNA encoding a novel membrane protein which exhibits a slowly activating potassium channel activity. *Biochem. Biophys. Res. Commun.* **161,** 176–181.

Murray, L. J., and Martens, C. (1989). The abnormal lymphocytes in *lpr* mice transcribe interferon-gamma and tumor necrosis factor genes spontaneously *in vivo*. *Eur. J. Immunol.* **19,** 563–565.

Murray, L. J., Lee, R., and Martens, C. (1990). *In vivo* cytokine gene expression in T cell subsets of the autoimmune MRL/Mp-*lpr*/*lpr* mouse. *Eur. J. Immunol.* **20,** 163–170.

Noda, M., Ikeda, T., Suzuki, H., Takeshima, H., Takahashi, H., Kuno, M., and Numa, S. (1986). Expression of functional sodium channel from cloned cDNAs. *Nature* (*London*) **322,** 826–828.

Pahapill, P. A., and Schlichter, L. C. (1990). Modulation of potassium channels in human T lymphocytes: effects of temperature. *J. Physiol.* (*London*) **422,** 103–126.

Pak, M. D., Covarrubias, M., Ratcliffe, A., and Salkoff, L. (1991a). A mouse brain homolog of the *Drosphila Shab* K^+ channel with conserved delayed-rectifier properties. *J. Neurosci.* **11,** 869–880.

Pak, M. D., Baker, K., Covarrubias, M., Butler, A., Ratcliffe, A., and Salkoff, L. (1991b). *mShal,* a family of A-type K^+ channel cloned from mammalian brain. *Proc. Natl. Acad. Sci. U.S.A.* **88,** 4386–4390.

Papazian, D. M., Schwarz, T. L., Tempel, B. L., Jan, Y. N., and Jan, Y. L. (1987). Cloning of genomic and complementary DNA from *Shaker,* a putative potassium channel gene from *Drosophila. Science* **237,** 749–753.

Philipson, L. H., Hice, R. E., Schaefer, K., LaMendola, J., Bell, G. I., Nelson, D. J., and Steiner, D. F. (1991). Sequence and functional expression in Xenopus oocytes of a human insulinoma and islet potassium channel. *Proc. Natl. Acad. Sci. U.S.A.* **88,** 53–57.

Pongs, O., Kecksmethy, N., Mueller, R., Krak-Jentzens, I., Baumann, A., Kiltz, H. H., Canal, I., Llamazares, S., and Ferrus, A. (1988). *Shaker* encodes a family of putative potassium chanel proteins in the nervous system of *Drosophila*. *EMBO J.* **7,** 1087–1096.

Pragnell, M., Snay, K. J., Trimmer, J. S., McClusky, N. J., Naftolin, F., Kaczmarek, L. K., and Boyle, M. B. (1990). Estrogen induction of a small, putative K^+ channel mRNA in rat uterus. *Neuron* **4,** 807–812.

Price, M., Lee, S. C., and Deutsch, C. (1989). Charybdotoxin inhibits proliferation and interleukin 2 production in human peripheral blood lymphocytes. *Proc. Natl. Acad. Sci. U.S.A.* **86,** 10171–10175.

Rajasekar, R., Sim, G. K., and Augustin, A. (1990). Self heat shock and gamma-delta T-cell reactivity. *Proc. Natl. Acad. Sci. U.S.A.* **87,** 1767–1771.

Ranges, G. E., Sriram, S., and Copper, S. M. (1985). Prevention of type II collagen arthritis by *in vivo* treatment with anti-L3T4. *J. Exp. Med.* **162,** 1105–1110.

Reimann, J., Bellan, A., and Conradt, P. (1988). Development of autoreactive L3T4$^+$ T cells from double-negative (L3T4$^-$/Ly-2$^-$) Thy-1$^+$ spleen cells of normal mice. *Eur. J. Immunol.* **18,** 989–999.

Res, P. C. M., Schaar, C. J., Breedeveld, F. C., van Eden, W., van Embden, J. D. A., Cohen, I. R., and de Vries, R. R. P. (1988). Synovial fluid T cell reactivity against 65 kD heat shock protein of mycobacteria in early chronic arthritis. *Lancet* **ii,** 478–480.

Ribera, A. B. (1990). A potassium channel gene is expressed at neural induction. *Neuron* **5,** 691–701.

Ritchie, A. K. (1987). Two distinct calcium-activated potassium currents in a rat anterior pituitary cell line. *J. Physiol.* (*London*) **385,** 591–609.

Roberds, S. L., and Tamkum, M. M. (1991). Cloning and tissue-specific expression of five voltage-gated potassium channel cDNAs expressed in rat heart. *Proc. Natl. Acad. Sci. U.S.A.* **88,** 1798–1802.

Robertson, D. W., and Steinberg, M. I. (1990). Potassium channel modulators: Scientific applications and therapeutic promise. *J. Med. Chem.* **33,** 1529–1541.

Ruppersberg, J. P., Schröter, K. H., Sakmann, B., Stocker, M., Sewing, S., and Pongs, O. (1990). Heteromultimeric channels formed by rat brain potassium-channel proteins. *Nature* (*London*) **345,** 535–537.

Sainis, K., and Datta, S. K. (1988). CD4$^+$ T cell lines with selective patterns of autoreactivity as well as CD4$^-$/CD8$^-$ T helper cell lines augment the production of idiotypes shared by pathogenic anti-DNA autoantibodies in NZB × SWR model of murine lupus nephritis. *J. Immunol.* **140,** 2215–2224.

Sands, S. B., Lewis, R. S., and Cahalan, M. D. (1989). Charybdotoxin blocks voltage-gated K^+ channels in human and murine T lymphocytes. *J. Gen. Physiol.* **93,** 1061–1074.

Santoro, T. J., Portanova, J. P., and Kotzin, B. L. (1988). The contribution of L3T4$^+$ T cells to lymphoproliferation and autoantibody production in MRL-*lpr*/*lpr* mice. *J. Exp. Med.* **167,** 1713–1718.

Schell, S. R., Nelson, D. J., Fozzard, H. A., and Fitch, F. W. (1987). The inhibitory effects of K^+ channel-blocking agents on T lymphocyte proliferation and lymphokine production are "nonspecific". *J. Immunol.* **139,** 3224–3230.

Schlichter, L., Sidell, N., and Hagiwara, S. (1986). K^+ channels are expressed early in human T-cell development. *Proc. Natl. Acad. Sci. U.S.A.* **83,** 5625–5629.

Schröter, K.-H., Ruppersberg, J. P., Wunder, F., Rettig, J., Stocker, M., and Pongs, O. (1991). Cloning and functional expression of a TEA-sensitive A-type potassium channel from rat brain. *FEBS Lett.* **278,** 211–216.

Schwartz, R. H. (1990). A cell culture model for T lymphocyte clonal anergy. *Science* **248,** 1349–1356.

Schwartz, T. L., Tempel, B. L., Papazian, D. M., Jan, Y. N., and Jan, L. Y. (1988). Multiple potassium-channel components are produced by alternative splicing at the *Shaker* locus in *Drosophila*. *Nature* (*London*) **331,** 137–142.

Shivakumar, S., Tsokos, G., and Datta, S. K. (1989). T cell receptor alpha/beta expressing double-negative ($CD4^-CD8^-$) and $CD4^+$ T helper cells in humans augment the production of pathogenic anti-DNA autoantibodies associated with lupus nephritis. *J. Immunol.* **143,** 101–112.

Shultz, L. D., and Sidman, C. L. (1987). Genetically determined murine models of immunodeficiency. *Annu. Rev. Immunol.* **5,** 367–403.

Stuart, J. M., Townes, A. S., and Kang, A. H. (1984). Collagen autoimmune arthritis. *Annu. Rev. Immunol.* **2,** 199–218.

Stuart, J. M., Watson, W. C., and Kang, A. H. (1988). Collagen autoimmunity and arthritis. *Fed. Proc.* **2,** 2950–2956.

Stühmer, W. J., Ruppersberg, J. P., Schröter, K. H., Sakman, B., Stocker, M., Giese, K. P., Perschke, A., Baumann, A., and Pongs, O. (1989). Molecular basis of functional diversity of voltage-gated potassium channels in mammalian brain. *EMBO J.* **8,** 3235–3244.

Sutro, J. B., Vayuvegula, B. S., Gupta, S., and Cahalan, M. D. (1988). Up-regulation of voltage-sensitive K^+ channels in mitogen-stimulated B lymphocytes. *Biophys. J.* **53,** 460a. (Abstr.)

Sutro, J. B., Vayuvegula, B. S., Gupta, S., and Cahalan, M. D. (1989). Voltage-sensitive ion channels in human B lymphocytes. *Adv. Exp. Med. Biol.* **254,** 113–122.

Swanson, R., Marshall, J., Smith, J. S., Williams, J. B., Boyle, M. B., Folander, K., Luneau, C. J., Antanavage, J., Oliva, C., Buhrow, S. A., Bennet, C., Stein, R. B., and Kaczmarek, L. K. (1990). Cloning and expression of cDNA and genomic clones encoding three delayed rectifier potassium channels in rat brain. *Neuron* **4,** 929–939.

Takumi, T., Ohkubo, H., and Nakanishi, S. (1988). Cloning of a membrane protein that induces a slow voltage-gated potassium current. *Science* **242,** 1042–1045.

Tamkun, M. M., Knoth, K. M., Walbridge, J. A., Kroemer, H., Roden, D. M., and Glover, D. M. (1991). Molecular cloning and characterization of two voltage-gated K^+ channel cDNAs from human ventricle. *FASEB J.* **5,** 331–337.

Tatham, P. E. R., O'Flynn, K., and Linch, D. C. (1986). The relationship between mitogen-induced membrane potential change and intracellular free calcium in human T lymphocytes. *Biochim. Biophys. Acta* **856,** 202–211.

Tempel, B. L., Jan, Y. N., and Jan, L. (1988). Cloning of a probable potassium channel gene from mouse brain. *Nature* (*London*) **332,** 837–839.

Theofilopoulos, A. N., and Dixon, F. (1985). Murine models of systemic lupus erythematosus. *Adv. Immunol.* **37,** 269–391.

Timpe, L. C., Schwartz, T. L., Tempel, B. L., Papazian, D. M., Jan, Y. N., and Jan, L. Y. (1988a). Expression of functional K^+ channels from *Shaker* cDNA. *Nature* (*London*) **331,** 143–145.

Timpe, L. C., Jan, Y. N., and Jan, L. Y. (1988b). Four cDNA clones from the *Shaker* locus of *Drosophila* induce kinetically distinct A-type potassium currents in *Xenopus* oocytes. *Neuron* **1,** 659–667.

Toshi, T., Zagotta, W. N., and Aldrich, R. W. (1988). Mutations in the amino terminal variable domain alter inactivation of *Shaker* B potassium channels in *Xenopus* oocytes. *Soc. Neurosci. Abstr.* **15,** 338.

Trentham, D. E. (1988). Collagen arthritis in rats, arthritogenic lymphokines and other aspects. *Int. Rev. Immunol.* **4,** 25–33.

Tseng-Crank, J. C., Tseng, G. N., Schwartz, A., and Tanouye, M. A. (1990). Molecular cloning and functional expression of a potassium channel cDNA isolated from a rat cardiac library. *FEBS Lett.* **268,** 63–68.

Tsien, R. Y., Pozzan, T., and Rink, T. J. (1982). T-Cell mitogens cause early changes in cytoplasmic free Ca^{2+} and membrane potential in lymphocytes. *Nature (London)* **295,** 68–71.

Umland, S., Lee, R., Howard, M., and Martens, C. (1989). Expression of lymphokine genes in splenic lymphocytes of autoimmune mice. *Mol. Immunol.* **26,** 649–656.

van Eden, W., Thole, J. E. R., van der Zee, R., Noordzij, A., Embden, J. D. A., Hensen, E. J., and Cohen, I. R. (1988). Cloning of the mycobacterial epitope recognized by T lymphocytes in adjuvant arthritis. *Nature (London)* **331,** 171–173.

Waldor, M. K., Sriram, S., Hardy, R., Herzenberg, L. A., Lanier, L., Lim, M., and Steinman, L. (1985). Reversal of experimental allergic encephalomyelitis with monoclonal antibody to a T-cell subset marker. *Science* **227,** 415–418.

Wei, A., Covarrubias, M., Butler, A., Baker, K., Pak, M., and Salkoff, L. (1990). K^+ current diversity is produced by an extended gene family conserved in *Drosophila* and mouse. *Science* **248,** 599–603.

Weiss, A., and Imboden, J. B. (1987). Cell surface molecules and early events involved in human T lymphocyte activation. *Adv. Immunol.* **41,** 1–38.

Wilson, H. A., and Chused, T. M. (1985). Lymphocyte membrane potential and Ca^{2+}-sensitive potassium channels described by oxonol dye fluorescence measurements. *J. Cell. Physiol.* **125,** 72–81.

Wofsy, D., and Seaman, W. E. (1985). Successful treatment of autoimmunity in NZB/NZW F_1 mice with monoclonal antibody to L3T4. *J. Exp. Med.* **161,** 378–391.

Yokoyama, S., Imoto, K., Kawamura, T., Higashida, H., Iwabe, N., Miyata, T., and Numa, S. (1989). Potassium channels from NG108-15 neuroblastoma-glioma hybrid cells: Primary structure and functional expression from cDNAs. *FEBS Lett.* **259,** 37–43.

CHAPTER 12

Development of Epithelial Na^+ Channels and Regulation by Guanine Nucleotide Regulatory (G) Proteins and Phospholipids

Horacio F. Cantiello and Dennis A. Ausiello
Renal Unit, Massachusetts General Hospital, Harvard Medical School, Boston, Massachusetts 02114

I. INTRODUCTION

In 1945, a student at the University of Copenhagen first demonstrated that antidiuretic hormone, also known as vasopressin, increases NaCl reabsorption through the isolated amphibian skin preparation (Jorgensen *et al.*, 1946; Ussing, 1949). Ussing's techniques to study transepithelial ionic movements have since been used to develop the paradigms of ion transport physiology. However, more than 40 years later, the molecular mechanisms involved in vasopressin modulation of epithelial Na^+ transport are still unknown. Nevertheless, the past 20 years have seen an increase in our understanding of intracellular regulatory processes and the coupling of hormone receptors to effector systems. A new family of signal transducers, called guanine nucleotide-binding (G) proteins, has been discovered and shown to play a central role in the regulation of ion transport. This chapter does not intend to address all aspects relevant to epithelial ion channel physiology since excellent comprehensive reviews are already available (Garty and Benos, 1988; Van Driessche and Zeiske, 1985); it instead summarizes some of the more recent advances on the expression of epithelial Na^+ channels and how their activity is functionally controlled by regulatory mechanisms, both dependent and independent of hormone–receptor and second-messenger signaling events taking place at the basolateral pole of an epithelial cell. In particular, we focus on recent findings demonstrating the presence of apically located regulatory mechanisms that center on G protein control of epithelial Na^+ channel activity. Fundamental to the understanding of the regulation of Na^+ channels is the nature of the methods used for assessing channel activity. Briefly described below are some of the major techniques used to generate data on ion channel regulation. In addition, amiloride and its analogs have provided fundamental information on Na^+ channel activity pertinent to most Na^+ channel studies. A brief description of amiloride action is also given.

A. Electrophysiological Techniques

1. Short Circuit Current

The classical *in vitro* preparation of Ussing (the Ussing chamber) was the first *in vitro* system that allowed a systematic study of ion transport through epithelia (Ussing and Zerahn, 1951; Koefoed-Johnsen and Ussing, 1958). By voltage-clamping the spontaneous transepithelial potential to zero when the epithelium is bathed by identical solutions, a net ionic current could be measured across the epithelium as the short-circuit current, I_{sc}. Because the I_{sc} was found to be equivalent to the unidirectional radioisotopic Na^+ influx in selected epithelia, Ussing's technique allowed for the electrophysiological study of the transepithelial movement of Na^+ in tight epithelia (Ussing, 1949).

In the early 1960s, Diamond made another important discovery. He demonstrated that epithelia, such as the gallbladder, which do not display a sizeable transepithelial membrane potential, also have high rates of Na^+

and Cl^- movement (Barry *et al.,* 1971; Diamond, 1978; Frömter and Diamond, 1972). Ussing's technique allowed Frömter and Diamond to classify epithelial membranes based on their transepithelial electrical resistance (R_m) (Frömter, 1972; Frömter and Diamond, 1972; Diamond, 1978). This electrical property relates to the expression of tight junctional structures which seal epithelial cells together and delimit the apical and basolateral domains. The terms ''leaky'' and ''tight'' epithelia were thus coined, establishing that epithelial tissues such as the renal proximal convoluted tubule, with an R_m of less than 100 Ω cm^{-2}, were ''leaky'' epithelium and the rabbit urinary bladder, with an R_m of more than 10^4 Ω cm^{-2}, is a ''tight'' epithelium (Diamond, 1978).

A further understanding of apical Na^+ movement through such epithelial barriers resulted from the use of selective inhibitors of this process. In this regard, the use of amiloride, a benzylimidazoyl-guanidinium diuretic exhibiting a potent and selective inhibitory effect on epithelial Na^+ channels, has been of seminal importance (Baer *et al.,* 1967; Benos, 1982). Nevertheless, the fact that only macroscopic transepithelial currents are followed under I_{sc} conditions created the problem of assessing the kinetic steps and actual molecular mechanisms inherent to the transepithelial movement of Na^+. In the late 1970s, Lindemann and Van Driessche applied the technique of current fluctuations (or noise) analysis to deduce molecular information about the Na^+ entry mechanism (Lindemann and Van Driessche, 1977; Van Driessche and Borghgraef, 1975).

2. Noise Analysis

An extension of the short circuit technique was developed to study the actual fluctuations in the macroscopic I_{sc} from voltage-clamped epithelia (Van Driessche and Borghgraef, 1975). If the I_{sc} is observed in the frequency domain, the data obtained may disclose, after certain assumptions are made, information on the single-molecular events involved in the epithelial ion translocation (Lindemann, 1984; Van Driessche and Zeiske, 1985).

Noise analysis strictly depends on the kinetic models to which the channel blocker-induced current fluctuations are fitted in the frequency domain. In order to gather information from epithelia, the assumption has been made that a channel only exists in two conformations, open or closed. A third substate can be induced by the addition of an ion-channel blocker, which, in the case of Na^+ channels, may be achieved by the addition of amiloride or triamterene (Christensen and Bindslev, 1982; Helman *et al.,* 1983; Lindemann and Van Driessche, 1977). Because a channel spontaneously fluctuates between open and closed states, the macroscopic current density, I, under voltage-clamping conditions is

$$I = NiP_o \quad (1)$$

where N is the total number of channels, i is the unitary channel current, and the open probability P_o is the inverse of the mean open time. The reciprocal of P_o represents the closed probability P_c and its inverse the mean closed time, respectively. By analyzing the fluctuations of the macroscopic current by fast Fourier transformation, the variance of the current fluctuations or power spectrum, S is displayed in the frequency (f) domain (double log-plot) from which a "Lorentzian,"

$$S(f) = S_o/[1 + (f/f_c)^2] \quad (2)$$

can be obtained, where S_o is the plateau at low frequencies and f_c, the so-called corner frequency, is obtained at $S_o/2$. This value is equivalent to the sum of the open to close and reciprocal rate constants. In order to convey useful information, the $S(f)$ has to be fitted to a best nonlinear approximation of the above theoretical model. Thus, a relationship might be obtained between (1) and (2) such that

$$S_o = 2Ni^2P_oP_c/\pi f_c \quad (3)$$

The power spectrum from which S_o is "peeled-off" is usually masked by other sources of noise such as $1/f$ or "shot" noise and white noise, which is independent of the frequency. It is thus to be kept in mind that all parameters derived from noise analysis will relay and be indirectly obtained from the kinetic model from which the data are extracted (Van Driessche and Zeiske, 1985).

By applying this approach, Lindemann and Van Driessche made the first direct demonstration that the apical Na^+ movement across an epithelial apical barrier is indeed electrodiffusional and mediated through pores or channels (Lindemann and Van Driessche, 1977, 1978). Noise analysis has allowed the characterization of several other ion-selective channels including K^+ (Gögelein and Van Driessche, 1981; Lindemann and Van Driessche, 1977, 1978), and Cl^- (Zeiske *et al.*, 1983) channels from various epithelia including leaky epithelia. Although providing new insights, this indirect approach needed confirmation from other techniques that directly access single molecular events.

3. Patch-Clamp Technique

A major breakthrough in ion channel physiology was the development by Neher and co-workers in the early 1980s of the patch-clamp technique for the measurement of the kinetic properties of single ion channel conductance (Hamill *et al.*, 1981). A micropipette deposited on the surface of a cell membrane is kept in close contact with the cell membrane by suction. Further suction induces the interface between the glass and the cell membrane to undergo a "fusion event" in which the electrical resistance becomes extremely high, $>10^9\ \Omega$, thus isolating a very small area or patch of cell membrane which is surrounded by the rim of the pipette.

Because this methodology allows a direct acquisition of single-channel data, the actual possibility of obtaining such information relies on the dynamic signal/noise ratio of the data acquisition system, which includes a current/voltage (i/v) converter, several amplifying steps, and a high-frequency or "boost" circuit to optimize the bandwidth response (Rae and Levis, 1984; Sigworth, 1984). Several sources of noise have to be taken into account, such as that from the amplifier and pipette seal (Rae and Levis, 1984). Provided that a suitable signal/noise ratio is met, the technique is the only available experimental approach that allows one to follow single molecular transitions directly. In contrast to noise analysis, where most information arises from modeling the macroscopic current fluctuations and assumes a single-channel species, the characteristic properties of single molecular ionic channel currents, i, such as conductance, γ (in picosiemens, pS, where 1 Siemen = 1/ohm), the open and closed probabilities P_o and P_c, and several other properties such as gating and selectivity, can be directly assessed by the patch-clamp technique.

B. Amiloride and Analogs

Amiloride (*N*-amidino-3,5-diamino-6-chloropyrazinecarboxamide) (Baer *et al.*, 1967) is a potent and specific inhibitor of Na^+ transport mechanisms in a wide variety of cellular and epithelial transport systems (Benos, 1982), including Na^+ channels (Benos *et al.*, 1979; Benos and Watthey, 1981; Cuthbert and Shum, 1974), Na^+/H^+ exchanger (Cantiello *et al.*, 1986; Grinstein and Furuya, 1984), Na^+/Ca^{2+} antiporter (Schellenberg *et al.*, 1983; Siegl *et al.*, 1984), Na^+-coupled cotransport mechanisms (Cook *et al.*, 1986, 1987), and the Na^+/K^+-ATPase (Soltoff and Mandel, 1983). Amiloride is a pyrazinoyl-guanidium with NH_2 substituents at the C_3 and the C_5 positions and a Cl^- substitution at the C_6 position of the pyrazine ring. The inhibitory effect of amiloride on apical Na^+ channel from such tight epithelia as the abdominal frog skin and the toad urinary bladder is a rapid and reversible phenomenon. The pK_a of amiloride is 8.7, rendering the drug positively charged under physiological conditions (Benos *et al.*, 1980). This feature has important physiological consequences because only the protonated form of amiloride is inhibitory toward all the aforementioned transport systems. This may explain the strong effect of pH on the activity of this diuretic in Na^+ channel-mediated I_{sc} of tight epithelia such as the frog skin (Li *et al.*, 1987; Li and Lindemann, 1982). Both substitutions at the pyrazine ring as well as the acyl-guanidium moieties modify the inhibitory affinity of this compound on Na^+ channel activity. Substitutions which shorten the guanidinium side chain severely reduce the molecule's ability to inhibit the Na^+ channel activity (Benos *et al.*, 1980). Furthermore, alterations on the 5-amino group eliminate amiloride's effectiveness to inhibit Na^+ channels. Substitutions of H^+ or I^- for Cl^- but not Br^- also severely reduce the inhibitory effect of the molecule. Common derivatives with high affinity as inhibitors of electrogenic Na^+ transport in renal epithelia are (in order of potency) phenamil > benzamil > amiloride ≫ ethylisopropylamiloride (Benos, 1982; Frelin *et al.*, 1987). Since haloger substitutions of position 6 modify the k_{off} without affecting the k_{on} for inhibition, it has been proposed that the binding of amiloride to the channel is mediated by an interaction of a charged side chain to the protein. It is still not clear whether there exits a direct competition between Na^+ and amiloride for its binding site. Competition studies between amiloride and external Na^+ have led to results in which pure competition (Aronson *et al.*, 1983), pure noncompetition (Moran, 1987), or mixed inhibition (Ives *et al.*, 1983) have been reported. Furthermore, amiloride at low concentrations tends to increase the I_{sc} (Li *et al.*, 1987; Li and DeSousa, 1979). This is most clearly evident on nonselective cationic channels of the larval bullfrog skin (Hillyard *et al.*, 1982a,b), although in this case the possibility of entirely different channel species cannot yet be ruled out. This is also consistent with previously reported results by us and others where both high and low affinities for amiloride have been reported for Na^+ channels from LLC-PK_1 epithelial cells (Cantiello *et al.*, 1987; Asher *et al.*, 1988). Since Na^+ channels have been reported with different affinities for amiloride, it is as yet unknown whether nonselective channels may change selectivity in combination with their affinity for Na^+. The existence of Na^+ channel subtypes with different pharmacological affinities for compounds such as tetrodotoxin and sea anemone toxins has been established for brain voltage-dependent Na^+ channels. Interestingly, these compounds share an acyl-guanidium moiety similar to that present in the structure of amiloride (Shimizu, 1986). Furthermore, compounds such as benzimidazole-2-guanidine and *p*-chloromercuriphenyl sulfonate are known to abolish the self-inhibitory state of epithelial Na^+ channels which is responsible for the saturation effect of external Na^+ (Dick and Lindemann, 1975; Garcia-Romeu, 1974; Zeiske and Lindemann, 1974). These compounds increase I_{sc}; thus, compounds which share amiloride's guanidinium moiety may screen a binding site which alters the electric field profile surrounding the channel's conduction site (Li *et al.*, 1987; Zeiske and Lindemann, 1975). It is important to realize, however, that high concentrations of amiloride produce a variety of effects, not only on various ion transport mechanisms, but also on several different enzymes and cell components, including inhibition of protein kinases (Besterman *et al.*, 1985), alteration of G protein function, and binding to DNA (Besterman *et al.*, 1987; Stiernberg *et al.*, 1983) and high-affinity imidazoline-binding proteins, all with potential deleterious effects on cell function.

II. DEVELOPMENTAL EXPRESSION OF EPITHELIAL Na^+ CHANNELS

A. Conductive and Ion Selectivity Properties

The "two-electrode" model originally postulated by Koefoed-Joensen and Ussing establishes that the apical membrane of tight epithelial cells is almost exclusively selective to Na^+, in contrast to the basolateral membrane, which is exclusively selective for K^+ (Koefoed-Johnsen and Ussing, 1958). This highly selective behavior of the apical domain of tight epithelia, $P_{Na}/P_K > 1000$, was first demonstrated by Palmer *et al.* (1980; Palmer, 1982). The transapical membrane potential in conjunction with the electrochemical Na^+ concentration gradient offers a favorable electrochemical driving force for Na^+ movement into the cell, regardless of the selectivity of the apical ion channels (Yoshitomi and Frömter, 1985). In contrast, the electrochemical gradient for K^+ is nearly zero. Thus, selectivity, although important, may not be essential in the process of Na^+ channel-mediated reabsorption under physiological conditions. In contrast, under conditions where the external solution contains very little Na^+, only highly Na^+-selective channels would "behave" properly. This may be the case for the amphibian abdominal skin, where the environmental medium contains extremely low Na^+ concentrations.

An early observation was that the transepithelial Na^+ movement displayed saturation kinetics as a function of the external apical Na^+ concentration (Cereijido *et al.*, 1964; Garty and Lindemann, 1984). These saturation kinetics may be indicative of a regulatory site at the Na^+ channel "mouth," probably involving Ca^{2+} or an external guanidinium-binding site (Li and Lindemann, 1982; Li *et al.*, 1987; Palmer and Frindt, 1987; Zeiske and Lindemann, 1974). Although noise analysis data have helped simplify the model for understanding the underlying molecular mechanisms inherent to Na^+ channel movement, no saturation of single Na^+ channel conductance was seen (Lindemann, 1984; Van Driessche and Zeiske, 1985). In order for noise analysis models to apply, the assumption has to be made that a single population of amiloride-sensitive channels is present at the apical membrane of epithelia such that a Lorentzian-shaped spectrum is obtained. Thus, this analysis is influenced by the kinetic model under which the spectral analysis is done. Further, in such experiments, the basolateral membrane was depolarized with a high serosal K^+ concentration. It has, instead, been demonstrated in patch-clamp experiments in A6 cells, toad urinary bladder, and rat cortical collecting duct cells that the assumption of a single population of apical Na^+ channels largely underestimates the actual functional status of channels in the tight epithelial apical membrane (Hamilton and Eaton, 1985).

Na^+ channels come in different "flavors" and it is probable that there is a family of channels present at the apical domain of epithelial cells. The apical membrane of A6 cells, for example, displays at least two different kinds of Na^+ channels with different conductive properties, including voltage sensitivity and ionic selectivity (Hamilton and Eaton, 1986b). The fact that both ion channels are blocked by amiloride with the same affinity may be an indication of an understated ion channel conductance as deduced from noise analysis. Na^+ channels display conductances within the range of 3–40 pS with a selectivity from less than 1 to selectivities greater than 1000 (Loo *et al.*, 1983; Palmer, 1982; Sariban-Sohraby *et al.*, 1984). Again, in these cases, data acquisition relied on the effect of low concentrations of amiloride ($<10^{-6}$ *M*) on Na^+ channel activity. One inconsistency is the flickering channel noise induced by amiloride, interpreted as a single population of channels in which the kinetic behavior cannot be reconciled with saturation kinctics of the channel as a function of the external Na^+ concentration (Läuger, 1984). It is possible that one or more Na^+ channels coexist in tight epithelial membranes. This has been recently demonstrated by patch-clamping of the toad urinary bladder in which a variety of ion channels were observed, ranging from 5 to 59 pS (Frings *et al.*, 1988). At least two different Na^+ channels were found, both with a conductance of 5 pS but only one sensitive to amiloride. Interestingly, it was not this channel species but larger ones (>17–58 pS) that were induced by either vasopressin or cAMP.

In order to explain some of the above results, alternative hypotheses need to be sought which rely on a dynamic expression of various channel species depending on the developmental status of epithelial cells. A working hypothesis we forward relies on an inverse relationship existing between the conductivity properties of an ion channel species and its selectivity for Na^+ over K^+. An original approach to this idea was forwarded by Hamilton and Eaton to explain conductivity–selectivity differences by channel "aging" (Hamilton and Eaton, 1986a). In studies where apical channels were induced in tight model epithelia upon hormonal activation, the "new" population of channels displayed different characteristics to the ones already present (Lewis *et al.,* 1986; Loo *et al.,* 1983). Thus, this hypothesis is also based on an as yet unexplained mutual exclusion of the various ion channel species in which, as the cell ages, the particular average apical channel population also changes. In order to support this contention and, as indicated in Fig. 1, where data have been gathered from various sources, including amphibian skin (Palmer, 1982) and the mammalian brain endothelium (Vigne *et al.,* 1989), a clear inverse correlation between ionic conductance and cation selectivity can be observed. These data, accumulated over the past 20 years, imply that the more Na^+ selective an epithelial ion channel is, the lower its conductance becomes. As seen in Fig. 1, only nonselective cation amiloride-sensitive channels are observed, with conductances between 15 and 40 pS. These results include kallikrein-degraded Na^+ channels from rabbit urinary bladder which have lost their Na^+ selectivity (Zweifach and Lewis, 1988). In this preparation, it is postulated that Na^+ channels which have been expelled from the apical membrane of the urinary bladder into the urine lose their selectivity and amiloride-binding properties, although they gain a higher conductive status. This is consistent with other data also displayed in Fig. 1 from the recently observed Na^+ channel of brain endothelial cells (Vigne *et al.,* 1989). Although not a typical epithelium, endothelial layers originate in the same mesenchymal sheet of the embryo and thus may contain the same morphological information to express epithelial-like characteristics. This channel is slightly selective for Na^+ ($P_{Na}/P_K \approx 1.5:1$), has high amiloride sensitivity, and a high conductance ($\gamma = 23$ pS). Channels with similar characteristics have been recently found to mediate apical Na^+ transport in the inner-medullary collecting duct of the mammalian kidney (Light *et al.,* 1988) and may reflect developmental changes similar to those observed in the ontogenesis of the adult frog skin.

B. Developmental Expression

The adult abdominal amphibian skin expresses highly selective Na^+ channels (Palmer *et al.,* 1980; Ussing and Zerahn, 1951). In contrast, the I_{sc} of the epidermis of larval stages X–XIX of the bull frog *Rana catesbeiana* is mainly mediated by K^+-selective and Ba^{2+}-sensitive channels with no selectivity for either Na^+ or Li^+ ions (Hillyard *et al.,* 1982a,b). Ba^{2+} also blocks the apical and basolateral K^+ channels of adult skins, although the contribution of these channels to the I_{sc} is minimal (De Wolf and Van Driessche, 1986; Van Driessche and Hillyard, 1985). Interestingly, amiloride increases K^+-mediated I_{sc}, a phenomenon similar to that

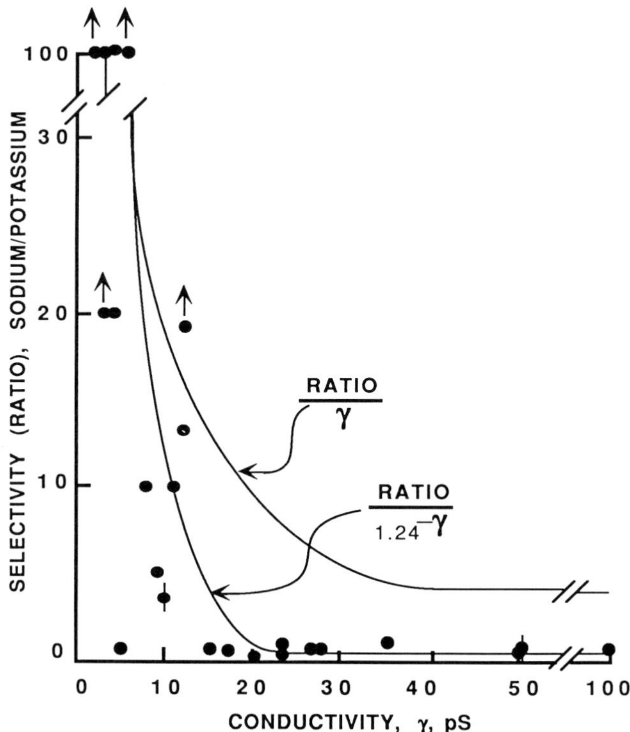

FIG. 1. Sodium channel conductance versus selectivity ratio in various transporting epithelia. Data from different epithelial preparations were obtained from a variety of studies and indicated as ion channel conductance versus selectivity ratio; the following are the range of tissue ratios utilized. Amphibian lens epithelium (The only epithelium from which no information is available as to whether channels are amiloride sensitive), 25–30 : 1, 50 : 1.2–1.8, 100 : 1, 12 : 13 (Rae and Levis, 1984); adult frog skin, 5.5 : 100, 0.8–3.0, 3.3–10 : >100 (Lindemann and Van Driessche, 1977, 1978; Van Driessche and Zeiske, 1985 and other refs. therein); rat cortical collecting duct, 8 : 10 (Palmer and Frindt, 1986); rabbit proximal tubule (pars recta) 12 : >19 (Gögelein and Greger, 1986); subconfluent A6 cells, 9 : 5 (Cantiello *et al.*, 1989); confluent A6 cells, 10 : 3–4, 2.8 : 20 (Hamilton and Eaton, 1985, 1986b); mammalian brain endothelial cells, 23 : 1.5 (Vigne *et al.*, 1989); rat inner medullary collecting duct cells, 28 : 1 (Light *et al.*, 1988); reconstituted channels from A6 cells, 35 ± 8 : 2 (Sariban-Sohraby *et al.*, 1984); confluent LLC-PK_1 cells, 15–23 : 1 (Moran and Moran, 1984; Cantiello *et al.*, 1987; also Cantiello, unpublished data); toad urinary bladder cells, 4.8 : >100, 5 : 1, 17 : 1 (Frings *et al.*, 1988); toad urinary bladder, 3 : >100 (Palmer, 1982); partially degraded mammalian urinary bladder channels, 20 : 0.5 (Zweifach and Lewis, 1988); hen coprodeum 4 : >20 (Christensen and Bindsler, 1982).

induced by addition of a benzimidazole-2-guanidine at concentrations which normally render Na^+ channel activity low in the adult skin (Hillyard *et al.,* 1982a). Furthermore, only high concentrations of amiloride, $>100\ \mu M$, are effective in inhibiting Na^+ channel activity. As tadpoles reach stages XIX–XXIV, the appearance of Na^+-selective channels is increasingly evident. It is not known whether the normal population of apical channels changes selectivity as an ontogenical trait or whether the entire population of epithelial cells is replaced by molting.

The original assumption that the apical Na^+ permeability of most tight epithelia is a constant parameter for the species and/or the epithelial cell considered is challenged by dynamic changes which occur during development and adaptation, both of which may introduce functional modifications in the selectivity of epithelial ion channels. A low-Na^+ diet, for example, induces the hen coprodaeum to express amiloride-sensitive Na^+ channels not normally present in control animals fed with a normal diet (Bindslev, 1979; Christensen and Binslev, 1982). This may be equivalent to what occurs in the mammalian distal colon, where changes in selectivity of ion channels can be induced by aldosterone treatment (Clauss *et al.,* 1984; Frizzell and Schultz, 1978; Will *et al.,* 1980). Most colonic epithelia, such as rabbit, human, and toad colon, express a large, and amiloride-sensitive, transepithelial potential of the order of 30 mV, consistent with the presence of amiloride-sensitive Na^+ channels. In contrast, the normal rat colon only generates potentials in the order of 5 mV (Will *et al.,* 1980). This transepithelial potential is not amiloride-sensitive, thus implying the lack of epithelial Na^+ channels. Various maneuvers, including a Na^+-deficient diet, diets which involve high K^+, dehydration, furosemide, and mineralocorticoid treatment all induce diuresis as well as modify the expression of amiloride-sensitive Na^+-selective channels in apical colonic epithelia (Clauss *et al.,* 1984; Crabbé, 1964; Cuthbert *et al.,* 1974; Frizzell and Schultz, 1978; Nagle and Crabbé, 1980; Palmer *et al.,* 1982). Chronically aldosterone-treated rats can increase the colonic I_{sc} up to sixfold (Fimognari *et al.,* 1967). This effect of aldosterone has been observed not only on the amiloride-sensitive but also on amiloride-insensitive pathways in the rat distal colon (Will *et al.,* 1980). In contrast, the rabbit distal colon of Na^+-depleted animals shows a reversal of the amiloride-sensitive to amiloride-insensitive mechanisms that is assumed to be a consequence of prolonged aldosterone stimulation. Recent findings indicate that the effect of mineralocorticoids such as aldosterone may be mediated by expression of a pool of otherwise quiescent Na^+ channels that are already present in apical epithelial membranes. Kleyman and co-workers observed that the binding of amiloride analogs to the apical membrane of A6 cells was not modified by addition of aldosterone, although the transepithelial electrical parameters had increased fourfold, thus indicating that aldosterone mediates a response which, in turn, activates otherwise quiescent Na^+ channels (Kleyman *et al.*, 1989).

Developmental studies pertaining to the electrophysiology of polarized epithelial cells are still scanty. Some preliminary information indicates that, depending on their location, epithelial cells from the Lieberkühn crypt of the intestinal tract change their functional activity from secretory to absorptive (Welsh *et al.*, 1982). Since the cells lining the epithelial surface do not change, this change in function may be indicative of an aging process, a case which is comparable to those changes observed in the ion channel selectivity during the larval stages of the amphibian skin. In this regard, studies from epithelial cells in culture have provided information which may indicate a dynamic correlation between the cell biological status of an epithelial cell and the functional activity of the various Na^+ transport mechanisms. For example, the sodium phosphate cotransporter of epithelial cells in culture is only expressed before cells reach confluency and demonstrate an apical cell domain (Mohrmann *et al.*, 1986). In contrast, the apical Na^+–glucose cotransporter and the Na^+/H^+ antiporter have been observed only as postconfluent events (Cantiello *et al.*, 1986; Mohrmann *et al.*, 1987b), thus indicating that the expression of the various Na^+ transport mechanisms may be related to the functional events which take place during the various stages of cell development. Contrary to the dogma in epithelial physiology, we have observed Na^+ channel activity in exponentially growing cells in the mammalian LLC-PK_1 cell line (Cantiello and Ausiello, 1986; Cantiello *et al.*, 1987) and, more recently, by immunolocalization of Na^+ channel antibodies and the use of the patch-clamp technique in amphibian A6 cells. These data strongly suggest that, before cells become a transporting epithelium, they are able to express ion channel activity that may change as the cell ages.

III. REGULATION OF Na^+ CHANNEL ACTIVITY

The ability of epithelia to respond to environmental signals relies on their capacity to induce intracellular signals in the form of changes in the concentration of secondary messengers, including Ca^{2+}, pH, and cyclic nucleotides such as cAMP and cGMP. Such signals, which are triggered by hormone–receptor interactions at the basolateral membranes of epithelial cells, enable a resultant physiological response at the apical membrane, such as an increase in transepithelial Na^+ transport. The actual cascade of events is assumed to develop from the targeting of such second-messenger systems to biochemical pathways which involve transducers such as protein kinases, resulting in the phosphorylation of one or more apical membrane proteins, including the ion channels themselves. Recent evidence from studies of purified epithelial Na^+ channels suggest, for example, that one of the proteins in the copurified complex (of at least six proteins) is phosphorylated by cAMP-dependent protein kinase (Sariban-Sohraby *et al.*, 1988). Thus, changes in the cAMP concentration induced by vasopressin activation of adenylyl cyclase in tight epithelia may increase Na^+ transport through such molecular mechanisms. This more classical picture of second-messenger-mediated ion transport regulation which, in epithelia, links the apical with the basolateral membrane, has recently been extended by the discovery of novel regulatory mechanisms which involve the coupling of G proteins to ion channel activity in pathways which may be independent of the cytosolic components of the cell.

In 1971, Rodbell and colleagues made the first demonstration that GTP was essential for promoting the stimulation of adenylyl cyclase, which led to the identification of G proteins as important regulators of hormone action (Rodbell *et al.*, 1971a,b). Since then, an increasingly large family of ubiquitous signal transduction G proteins have been implicated in various physiological events, including the regulatory control of hormone action (Manning and Gilman, 1983; Bokoch and Gilman, 1984; Gilman, 1984) and intracellular signaling (Bourne *et al.*, 1990), and as both direct and indirect regulators of ion channels (Birnbaumer and Brown, 1987; Brown and Birnbaumer, 1990) and other ion transport proteins such as the Na^+/H^+ exchanger (Satoh *et al.*, 1985), the Na^+/Ca^{2+} antiporter (Brechler *et al.*, 1990), and the Na^+/K^+-ATPase (Bertorello and Aperia, 1989).

G proteins exhibit a conserved amino acid sequence identity that has been preserved across the phyla from bacteria to mammalian cells. Following the demonstration that G proteins modulate adenylyl cyclase activity, it was demonstrated that two distinct G proteins are directly linked to the adenylyl cyclase hormone–receptor complex which are implicated in the stimulation (G_s) and inhibition (G_i) of adenylyl cyclase (Harwood *et al.*, 1973). Subsequently, the demonstration of G_o (o = other) and other "classical" G proteins has made this group only a subset of an evergrowing superfamily of G proteins (Gilman, 1987).

Heterotrimeric G proteins are membrane-bound, not integral membrane proteins, composed of α, β, and γ subunits. The $\beta\gamma$ complex is probably functionally interchangeable in the various systems thus far tested, implying that the α subunit, carrying the GTP binding and GTPase activity, is the subunit which confers G protein activity and selectivity. G proteins cycle between their inactive form, the GDP-bound α subunit coupled to the $\beta\gamma$ complex, to an active form that results from GDP displacement and Mg^{2+}-dependent GTP binding (Birnbaumer, 1990; Gilman, 1987). This latter reaction is stimulated by a hormone–receptor complex binding to a G protein. A conformational change in the G protein occurs in which the GTP-bound α subunit is spatially displaced from the $\beta\gamma$ complex and can interact with effector molecules. The cycle is ended by GTP hydrolysis to GDP and the α subunit then reassociates with $\beta\gamma$.

G proteins are selectively sensitive to nicotinamide-adenine dinucleotide (NAD)-dependent ADP-ribosylating agents which transfer an ADP-ribose from NAD to the α subunits of the G protein (Alouf *et al.*, 1984). Bacterial toxins, such as cholera toxin and the heat-labile enterotoxin of *Escherichia coli,* stimulate ADP-ribosylation of α_s at an internal arginine residue. The exotoxin from *Bordetella pertussis,* pertussis toxin, stimulates ADP-ribosylation of α_i on a cysteine residue near the carboxy terminus. ADP-ribosylation by toxins results in inhibition of GTPase activity (and thus persistent activation) of α_s, or uncoupling of hormone–receptor complexes and inhibition of the formation of the active state of α_i (Gilman, 1984).

A. G Protein Regulation of Ion Transport

G protein regulation of ion transport has opened a new chapter in the way intracellular signals are understood to control ionic movements across cell membranes. G protein control of ion transport was originally observed in the regulation of the Na^+/H^+ exchanger (Satoh *et al.*, 1985). This activity is through "classical" coupling of G proteins to enzymes, such as phospholipases, involved in phospholipid signal transduction.

Light-activation of rhodopsin and the production of retinol are also linked to a phosphodiesterase (Kuhn, 1986). This interaction is mediated by another G protein, transducin (G_t) that, by activating a phosphodiesterase, cleaves cGMP and controls the excitability of mammalian rods and cones by modulating a G protein-sensitive, voltage-independent cation channel (Watkins *et al.*, 1987). A potential role for G_t activation of phospholipase A_2 in the regulation of outer rod segments has also been implied, but in this case the $\beta\gamma$ and not the α subunit is the putative regulator.

G protein control of ion channels may reflect a widespread intracellular regulatory mechanism which seems to be proximal to the channels themselves and may be the molecular link to receptor-mediated regulation of channel activity (Birnbaumer and Brown, 1987; Brown and Birnbaumer, 1990). Several heterotrimeric G proteins are involved in such mechanisms. G_i proteins activate K^+ channels in cardiocytes (Codina *et al.*, 1987; Yatani *et al.*, 1987a, 1988) and neuronal cells (Andrade *et al.*, 1986). One of the G_i proteins, $\alpha_{i\text{-}3}$, is also involved in the activation of Na^+ channels in

A6 epithelial cells (Cantiello *et al.,* 1989), nonselective cation channels in the renal inner medulla (Light *et al.,* 1989a), and Cl^- channels in the cortical collecting duct (Light *et al.,* 1990a). Another G_i-like protein, G_o, also a pertussis toxin substrate, activates Ca^{2+} channels (Hescheler *et al.,* 1987; Tsunoo *et al.,* 1986) and K^+ channels in neuronal cells (Andrade *et al.,* 1986). G_s, instead, inhibits Na^+ channels in the heart (Schubert *et al.,* 1989) but stimulates Ca^{2+} channel activity in striated and cardiac muscles (Rosenthal *et al.,* 1988; Yatani and Brown, 1989; Yatani *et al.,* 1987a). Although in most cases this coupling event may be the reflection of a receptor–channel linking mechanism, a "direct" G protein regulation of channels has also been proposed (Brown and Birnbaumer, 1988). This type of regulation is discussed in detail below.

In receptor-mediated channel activation, the reaction is initiated by hormone binding to its endogenous receptor. The most thoroughly studied G-protein-gated channel to date is the muscarinic, receptor-sensitive K^+ channel that regulates heart rate by hyperpolarizing atrial muscle cells (Brown and Birnbaumer, 1990; Soejima and Noma, 1984). The activation of this inwardly rectifying K^+ channel can be differentiated by its acetylcholine stimulation from several other K^+ channels also present in these cells. The first link between G proteins and the activation of the muscarinic K^+ channel was demonstrated by the essential role of GTP in its activation, which was also pertussis toxin sensitive (Pfaffinger *et al.,* 1985). Channel activity was stimulated with the nonhydrolyzable analogs of GTP, Gpp(NH)p or GTPγS, and was independent of muscarinic receptor occupancy (Breitwieser and Szabo, 1985, 1988). These analogs also can overcome the inhibitory effect of pertussis toxin (Endoh *et al.,* 1985; Pfaffinger *et al.,* 1985; Sorota *et al.,* 1985). In excised, inside-out patches of atrial myocytes, it was observed that activated G_i (but not G_s) reinstated channel activity after pertussis toxin treatment (Codina *et al.,* 1987; Yatani *et al.,* 1988). The three subtypes of the α_i subunit of the G_i protein have been observed to activate muscarinic K^+ channels with the same affinity (Yatani *et al.,* 1988), thus indicating that the α subunit may carry the regulatory activity, but that α_is are promiscuous with regard to K^+ channel regulation. It has recently been demonstrated that purified $\beta\gamma$ subunits are inhibitors of K^+ channels in myocytes, further demonstrating the role of α_i as the activating subunit (Okabe *et al.,* 1990). More recently, a different role for the $\beta\gamma$ subunit has been established in the modulation of this channel, which may involve activation of phospholipase A_2, independent of the α_i activation (Kim *et al.,* 1989; Neer and Clapham, 1988).

Brown and Birnbaumer have put forward the hypothesis that the gating of ion channels can be directly achieved by G proteins (Brown and Birnbaumer, 1988; Brown *et al.,* 1989). This contention awaits experimental proof and need not be the only role of G proteins in the control of ion channel activity. Although "classical" G protein regulation of an ion channel may involve the coupling of receptors to G proteins and, in turn, the G protein directly coupled to channels, it is now known that alternative, indirect pathways exist for G protein regulation of ion channels (Kim *et al.*, 1989; Neer and Clapham, 1988). For example, G_i regulation of apical epithelial Na^+ channels may not be linked to an apical receptor mechanism and may offer the first demonstration of G protein–ion channel coupling which is independent of receptor occupancy, but possibly responsive to intracellular signal transduction mechanisms such as protein phosphorylation (Sariban-Sohraby *et al.*, 1988; H. F. Cantiello *et al.*, unpublished observations). In addition, as discussed below, G protein regulation of epithelial Na^+ channels is not direct, but rather mediated through phospholipid metabolites (Cantiello *et al.*, 1990).

Several ion-selective channels are controlled by G proteins, in particular, Ca^{2+} channels from sarcolemmal preparations, neurons, and neurosecretory cells (Dolphin, 1990; Hescheler *et al.*, 1987; Holz *et al.*, 1986; Trautwein and Hescheler, 1990; Tsunoo *et al.*, 1986; Yatani *et al.*, 1987a). In GH_3 rat anterior pituitary cells, for example, somatostatin inhibits secretion, reduces intracellular Ca^{2+}, and induces a cAMP-independent hyperpolarization, effects which are similar to acetylcholine binding to the muscarinic receptor present in the heart (Tsunoo *et al.*, 1986). In this preparation, G_i but neither G_s nor the purified $\beta\gamma$ subunits is effective in reactivating channel activity (Yatani *et al.*, 1987b). G protein regulation of Ca^{2+} channels by guanine nucleotide modification of the dihydropyridine binding to receptors is known to activate Ca^{2+} channels in cardiac and skeletal muscles (Hescheler *et al.*, 1987; Trautwein and Hescheler, 1990). In reconstituted Ca^{2+} channels from T tubule and the cardiac sarcolemma, for example, the β-adrenergic-receptor coupling to the channels was regulated by G_s or the purified α_s whose effect was independent of adenylyl cyclase and cAMP-dependent protein kinase activity, which also plays an additional role in the control of Ca^{2+} channel activity (Trautwein and Hescheler, 1990). This is in contrast to the effect on neurons and neurosecretory cells, where G proteins seem to exert an inhibitory effect on ion channels (Dolphin, 1990). γ-Aminobutyric acid (GABA) and α-adrenergic receptor activation, for example, reduce Ca^{2+} currents in ganglion cells whose effect is blocked by GDPβS. Somatostatin-mediated inhibition of Ca^{2+} currents in these cells is also prevented by pertussis toxin, thus implying a role of a G_i in the receptor-coupling to the channel (Andrade *et al.*, 1986). Reversal of the pertussis toxin effect by addition of GTPγS induced a transient inhibition of Ca^{2+} currents. Pertussis toxin-sensitive Ca^{2+} channel regulation by opioid receptors has also been demonstrated in neuroblastoma–glioma cells which is restored by injection of G_i or G_o (Hescheler *et al.*, 1987). The G_o was 10 times more effective than the G_i, an effect that is consistent with the abundance of G_o in the brain and related tissues. G_t, in contrast, was completely devoid of activity. One interesting finding was the complete lack of effect of the $\beta\gamma$ complex (without α subunit) to modify the α_o reactivation of the Ca^{2+} currents (Hescheler *et al.*, 1987). In this particular case, the effect can be mimicked by phorbol esters or oleylacyl-glycerol, indicating that Ca^{2+}-dependent protein kinase may be involved in the regulation of these channels and also implying that the G protein effect may be indirectly mediated through protein kinases. Thus, the role of G proteins on Ca^{2+} channels has been found to be stimulatory in muscle and inhibitory in neurons and neurosecretory cells.

The molecular events that actually link G protein conformational changes with ion channel activity have not been established. In studies performed with cultured myocytes from rats of different ages containing muscarinic-sensitive K^+ channels, it was found that the pertussis toxin-mediated ADP-ribosylation of membrane preparations revealed clear differences among different age groups (Moscona-Amir *et al.*, 1988). Young cultures expressed a 40-kDa substrate that was recognized by $\alpha_{i\text{-}3}$ antibodies as well as two other novel pertussis toxin substrates of 42 and 28 kDa. Aged cultures only expressed the 40-kDa protein, implying that the interaction between G proteins and the muscarinic receptors is strongly dependent on age. More interestingly, the aging phenomenon that modulates the expression of the pertussis toxin substrates for ADP-ribosylation and the coupling of G proteins to the muscarinic receptors was found to be controlled by the amount of phosphatidylcholine present in the preparation. It had been previously observed that the lipidic composition of myocyte membranes was dramatically changed with age (Yechiel and Barenholz, 1985). In experiments in which phosphatidylcholine-containing liposomes were added to cardiac receptor preparations of aged cells, the elevation of the phosphatidylcholine/sphingomyelin ratio, which reduces the cholesterol level to the values observed in young tissues (Yechiel *et al.*, 1985), modified the ADP-ribosylation pattern in such a way that the aged treated cultures became similar to the young tissues, expressing the three substrates of 42, 40, and 28 kDa. Thus, phosphatidylcholine treatment of tissues modified both the expression of pertussis toxin substrates and the binding affinity for carbamylcholine to the muscarinic receptor. Thus, there appears to be an interaction among structural membrane phospholipids, G proteins, and phospholipases. It is possible that a signal transduction pathway involving G protein activation of phospholipases produces one or more phospholipid byproducts which in turn regulate ion channels. Evidence for this pathway is presented below.

B. Role of Phospholipids in the G Protein Control of Ion Channels

G proteins have been implicated in the activation of several phosphodiesterase reactions, including activation of cGMP-activated phosphodiesterase in rods and cones (Kuhn, 1986), phosphatidylinositol (Cockcroft, 1987) and Ca^{2+}-dependent phospholipase C (Bradford and Rubin, 1985; Paris and Poyssegur, 1987), and phospholipase A_2 (Axelrod *et al.*, 1988; Fain *et al.*, 1988; Jelesma and Axelrod, 1987). The activation of phospholipase A_2 in cardiac muscle results in the activation of G protein-sensitive K^+ channels (Kim *et al.*, 1989). The addition of arachidonic acid or 5-lipoxygenase products such as the leukotrienes (LT) B_4 and C_4 mimicked the effect of phospholipase activation. The fact that antibodies against phospholipase A_2 or the addition of nordihydroguaiaretic acid (NDGA), a 5-lipoxygenase inhibitor, prevented the effect of arachidonic acid on channel activity implies that both phospholipase and free fatty acids are involved in K^+ channel regulation. Further support for this pathway was obtained by the use of 5-hydroperoxy-5,8,10,14-eicosatetraenoic acid (5-HPETE) and 12-hydroxy-5,8,10,14-eicosatetraenoic acid (12-HETE), by-products of the 5-lipoxygenase pathway. Both were effective in activating K^+ channels. Evidence supporting a more direct role of fatty acids on cardiac K^+ channel activation has also been demonstrated by Kim and Clapham (1989), who found that two different inwardly rectifying K^+ channels from cardiocytes were directly activated by arachidonic or linolenic acids. In this case, arachidonic acid did not work from outside of the cells, as was shown for the receptor-regulated channel described above, but readily induced channel activity once exposed to the cytosolic side of the cell. This effect was linked to a direct effect of arachidonic acid, since neither 5-HETE, 5-HPETE, nor several derivatives of the 5-lipoxygenase pathway were effective in inducing channel activity.

The above data suggest that muscarinic K^+ channels can be activated directly by α_i subunits of G_i, or indirectly by $\beta\gamma$ activation of phospholipase A_2. However, as will be discussed for the epithelial Na^+ channel, it is possible that α_i mediates its effects through phospholipases as well (Cantiello *et al.*, 1990). Scherer and Breitwieser (1990) have recently found that arachidonic acid modifies the affinity of activation of the acetylcholine-mediated K^+ channel activity, an effect which is also present when activation is achieved by GTPγS turning on the G α_i subunit, independent of muscarinic receptor occupancy. This inhibitory effect of arachidonic acid was blocked by NDGA and reestablished by LTC_4. Thus, a regulatory role of the lipoxygenase pathways in the coupling of G_i to the K^+ channel must also be considered.

C. Regulation of Epithelial Na^+ Channels by G Proteins

Leaky epithelial cells such as the proximal tubule contain Na^+ channels which may be involved in the reabsorption of Na^+ (Gögelein and Greger, 1986). The LLC-PK_1 cell provides a model for this pathway. We have observed that the apical plasma membrane of LLC-PK_1 cells contains a high-affinity amiloride-sensitive Na^+ channel (Cantiello *et al.*, 1987) whose activity is regulated by a guanylyl cyclase coupled to a receptor for atrial natriuretic peptides (ANP) (Cantiello and Ausiello, 1986; Mohrmann *et al.*, 1987a). This was the first demonstration of a direct effect of this hormone on a Na^+-transporting epithelium and provided an explanation for a tubular effect of this peptide, later confirmed by Zeidel *et al.* (1987) and Light *et al.* (1989b) in rat inner medullary collecting duct cells. In an attempt to define the mechanisms involved in this pathway, we found that activation of protein kinase C inhibited the amiloride-sensitive, Na^+ channel-mediated uptake in LLC-PK_1 cells (Mohrmann *et al.*, 1987a). Furthermore, down-regulation of protein kinase C prevented the effect of ANP, thus indicating that the ability of ANP and/or GMP to inhibit Na^+ uptake requires the presence of an active protein kinase C. In addition, pertussis toxin was also found to inhibit the amiloride-sensitive Na^+ transport to the same extent as ANP, cGMP, or protein kinase C, demonstrating that a G_i protein was involved in the modulation of Na^+ channel activity in LLC-PK_1 cells. This hypothesis was further supported by the finding that pertussis toxin was able to ADP-ribosylate a 41-kDa membrane protein, consistent with G_i (Mohrmann *et al.*, 1987a).

These observations have been extended to tight epithelial preparations, where we have recently found that pertussis toxin completely abolishes the amiloride-sensitive Na^+ uptake in A6 toad kidney cells, a finding that supports the contention that the control of the Na^+ uptake through amiloride-sensitive channels in tight epithelia can be entirely accounted for by a G protein-regulated mechanism (Cantiello *et al.*, 1989). Further support for this hypothesis was obtained from studies with the purified Na^+ channel complex of 700 kDa from A6 cells (six major bands observed on reducing gels with molecular masses ranging from 35 to 300 kDa), which was found to contain a 41-kDa substrate for pertussis toxin ADP-ribosylation, consistent with the $\alpha_{i\text{-}3}$ subunit of $G_{i\text{-}3}$ (D. A. Ausiello *et al.*, unpublished observations). Western blot analysis of this pertussis toxin substrate confirmed that it was $\alpha_{i\text{-}3}$. Both $\alpha_{i\text{-}3}$ and the Na^+ channel were localized in A6 cells by confocal imaging with specific antibodies which were found in distinct but adjacent domains of the apical cell surface. These data provided a structural correlate to recent observations from our laboratory which directly addressed the regulatory role of $G\alpha_{i\text{-}3}$ on Na^+ channel activity.

By utilizing the patch-clamp technique, we were able to obtain single-channel currents from the apical surface of A6 epithelial cells grown to partial confluence in which spontaneous Na^+ channel activity was present in approximately 60% of successful patches. Na^+ channels had no detectable rectifying properties, a selectivity for $Na^+ : K^+$ of 5 : 1, and a conductance of 9 pS in symmetrical Na^+ solutions (Cantiello *et al.*, 1989). This channel was the most abundant species which displayed amiloride sensitivity [apparent $K_i < 10^{-7}$ *M*, in close agreement with previously reported data by Hamilton and Eaton (1985)]. Interestingly, these authors have also reported Na^+ channels in postconfluent cells grown on permeable supports that display a lower conductance, 3 pS, which are also amiloride sensitive. In contrast, more recent patch-clamp studies indicate that the average Na^+ channel conductance of confluent A6 cells is 5 pS, which, upon reconstitution into liposomes, increases to 10 pS (Sariban-Sohraby and Fisher, 1990). The possibility exists that, as the cells age, changes in the conductive properties of a population of Na^+ channels also change (see earlier discussion). Although the putative mechanisms for such changes are as yet unknown, these findings are consistent with our previous hypothesis of selectivity changes as a function of the age of cells, a phenomenon that could be present in cells in culture.

In order to assess the possible implication of a functional G protein in the regulation of the 9 pS Na^+ channel, the effect of GTP, its analogs, and pertussis toxin was investigated. The percent open time of spontaneous channels increased up to 40% following addition of GTP. Furthermore, Na^+ channel activity was also observed on addition of GTP to patches which had not displayed spontaneous channel activity. In contrast, GDPβS inhibited Na^+ channel activity from spontaneously active channels by approximately 80% without affecting the single-channel conductance. Addition of preactivated pertussis toxin completely inhibited spontaneous Na^+ channel activity in approximately 3 min without affecting the single-channel conductance. Removal of pertussis toxin did not reverse its effect, in contrast to the addition of GTPγS (but not GTP), which readily restored Na^+ channel activity in patches previously exposed to pertussis toxin.

To demonstrate that the pertussis toxin effect was indeed mediated through a G_i-like protein, picomolar concentrations of purified and GTPγS-activated human $\alpha_{i\text{-}3}$ subunit were added which selectively reversed the effect of pertussis toxin on Na^+ channel activity in a dose-dependent manner (Cantiello *et al.*, 1989). This effect was selective for $\alpha_{i\text{-}3}$ because similar concentrations of $\alpha_{i\text{-}2}$ did not reverse the effect of pertussis toxin. These data are in agreement with our recent observations indicating the spatial localization of $\alpha_{i\text{-}3}$ at the subapical side of epithelial cells, in contrast to the $\alpha_{i\text{-}2}$ subunit, which is only present at the basolateral pole, where it participates in adenylyl cyclase regulation (Ercolani *et al.*, 1990). Our findings of a G_i protein-regulated Na^+ channel in A6 cells may be indicative of a more general regulatory pathway in tight epithelial cells since Garty *et al.* (1989) recently found that GTPγS is able to increase up to threefold amiloride-sensitive $^{22}Na^+$ transport into apical vesicles obtained from the toad urinary bladder.

Recently, it has been found that various apical epithelial ion channels are also under G_i protein regulation. The apical Na^+ transport from rat inner medullary collecting duct cells can be entirely accounted for by a high-affinity, amiloride-sensitive Na^+ pathway which is cGMP-dependent and sensitive to activation of guanylyl cyclase coupled to atrial natriuretic peptide receptors (Blumenfeld *et al.*, 1988; Zeidel *et al.*, 1987). Light *et al.* (1988) were successful in applying the patch-clamp technique to these cells. Their results indicate that the apical membrane of rat inner medullary collecting duct cells contains an amiloride-sensitive, nonselective, 28 pS cationic channel which is cGMP-inhibitable and also regulated by an apically located G protein (Light *et al.*, 1989a,b). Their studies also indicate that this nonselective ion channel is under tonic stimulation of a G_i protein; thus, the channel can be blocked by pertussis toxin, whose effect is reversed by picomolar concentrations of $\alpha_{i\text{-}3}$. Although no receptor has been yet identified to couple to this G protein, the channel is inhibited by activation of a guanylyl cyclase coupled to the atrial natriuretic peptide receptor at the basolateral membrane of these polar epithelial cells that results in an increase in intracellular cGMP, similar to the results observed in LLC-PK_1 cells (Light *et al.*, 1990b). Interestingly, cGMP exerts a dual effect on channel activity. cGMP activates a cGMP-dependent protein kinase and this activation selectively blocks the channel activity, although not the effect of GTPγS stimulation. Because this response is ATP dependent, the cGMP–protein kinase effect is thought to be mediated through the phosphorylation of one or more regulatory proteins. On the other hand, after pertussis toxin treatment, cGMP also blocks the spontaneous as well as the GTPγS or $\alpha_{i\text{-}3}$-activated channel in a process that the authors postulate to be an allosteric site on the channel.

Another epithelial channel seems to be regulated by the same apical G_i system, a voltage-dependent, fast-inactivating 300 pS Cl^- channel from rabbit cortical collecting duct which displays several substates and can be inhibited by pertussis toxin and reactivated by either the addition of GTPγS or the activated $\alpha_{i\text{-}3}$ subunit of $G_{i\text{-}3}$ (Light *et al.*, 1990a). This channel has also been observed in several types of epithelial cell including

Maden–Darby canine kidney (MDCK) (Kolb *et al.*, 1985) and A6 (Nelson *et al.*, 1984). Although the G_i regulation of this channel shares similarities with the Na^+ channel from A6 cells and the cation channel described above, Cl^- channels in this cell model seem to be activated by Ca^{2+}-dependent mechanisms. This process of activation raises the possibility of an as yet unidentified mechanism of coupling to other events such as protein kinase C activation. Thus, a general mechanism for apical coupling between a G_i system and ion channels in epithelia could be considered as an important regulatory pathway in the control of the apical ionic permeability.

D. Role of Phospholipids in the G Protein Control of Na^+ Channels

Limited information is available concerning the regulation by fatty acids of ion transport in epithelia. The regulation by aldosterone of epithelial ion transport has been shown to require the transcription of proteins that may be involved in the regulation of the expression of ion channels of apical membrane of tight epithelia (Lifschitz *et al.*, 1973). Yet, the idea has been forwarded that aldosterone action in epithelia may involve membrane lipid metabolic changes (Goodman *et al.*, 1971; Hestrin-Lerner and Hokin, 1964). Crabbé (1964) first demonstrated that aldosterone induced an increase in the incorporation of ^{32}P into phosphatidic acid, phosphatidylinositol, and phosphatidylcholine in toad urinary bladder. In contrast, incubation with oxytocin did not result in this incorporation. In similar preparations, Goodman *et al.* (1975) demonstrated that, as early as 30 min after aldosterone incubation, toad urinary bladder increased the phospholipid metabolism in response to the hormone. In particular, oleic acid metabolism was enhanced. Aldosterone also induced phospholipid and reacylation cycles (Lien *et al.*, 1975). More delayed effects were observed after 4 hr of aldosterone treatment in which oxidation of all fatty acids preincubated with the preparation was enhanced. For example, the elongation and saturation of oleic acid as well as the recycling of acetyl-coA derived from the oleic acid metabolism and the metabolism of other fatty acids were observed. These studies were the first to indicate that alterations in the lipid metabolism may be of significance in the role of aldosterone in epithelial cell physiology. Furthermore, evidence from this particular study clearly implied a role for phospholipase A_2 in the reduction of the latent period between hormone action and the onset of a physiological response (Yorio and Bentley, 1978). These results were supported by the effect of acetyl-coA-carboxylase inhibitors, which inhibit the aldosterone-stimulated Na^+ transport as well as the hormonally induced lipid synthesis

(Lien *et al.*, 1975). In comparable studies where amiloride was used to block Na^+ transport, it was observed that the metabolic changes in the lipid composition took place prior to the changes in the Na^+ transport (Lien *et al.*, 1975).

Some 13 years ago, Yorio and Bentley (1978) postulated that the role of aldosterone may be related to an increase in phospholipids that increases the apical membrane permeability of these epithelial cells to Na^+. A role for aldosterone in the regulation of phospholipase A_2 activity was then sought. In the lipidic domain of cellular plasma membranes, phospholipase A_2 hydrolyzes phosphatidylcholine into lysophosphatidylcholine which eventually induces the production of L-α-glycerol-phosphatidylcholine in tissues such as the toad bladder. This pathway may be responsible for the availability of free fatty acids and the activation of the lipoxygenase and cyclooxygenase pathways. Yorio and Bentley showed that mepacrine decreased phospholipase A_2 activity by 35% and completely inhibited the aldosterone-stimulated Na^+ transport in the toad urinary bladder (although interestingly, not the AVP effect on Na^+ transport).

As discussed above, the regulation of apical epithelial ion channels is under control of a colocalized G_i transducing system which seems to complement the hormone–receptor-coupled events at the basolateral membrane. Due to the important role of G protein modulation of phospholipase-mediated reactions, most consistently ascribed to the α_i subunits rather than the $\beta\gamma$, a set of studies was undertaken in our laboratory to determine whether the $G\alpha_{i\text{-}3}$-mediated regulation of A6 Na^+ channels was a component of an apically located, phospholipid signal transduction pathway that was also involved in the modulation of ion channel activity. In the apical membrane of A6 epithelial cells, we had demonstrated that a G_i protein tonically maintains Na^+ channels open. This apically located signal transduction pathway may be sensitive to, or independent of, classical second messengers known to be responsible for modulation of Na^+ channel activity in epithelia.

Again, the patch-clamping of excised inside-out apical membranes of A6 partially confluent cells was used (Cantiello *et al.*, 1990). It was observed that mepacrine, a nonspecific phospholipase inhibitor, decreased GTPγS-stimulated activity of the 9 pS Na^+ channel by 88%. However, the addition of arachidonic acid in the presence of mepacrine induced Na^+ channel activity. Two different enzymatic pathways have been implicated in the release of arachidonic acid (Axelrod *et al.*, 1988; Nakamura and Ui, 1985; Samuelsson, 1983; Samuelsson and Funk, 1989): direct mobilization of arachidonic acid by phospholipase A_2 from the catalysis of phospholipids such as phosphatidylcholine and phosphatidylethanolamine, and the activation of polyphosphoinositide-specific phospholipase C that leads to the formation of arachidonic acid-containing diacylglycerol. Cleavage of diacylglycerol is then mediated by diacylglyceride lipases, with the subsequent mobilization of arachidonic acid. Both pathways are regulated by G proteins (Axelrod *et al.,* 1988; Nakamura and Ui, 1985). The GTPγS-induced increase in epithelial Na^+ channel activity was unaffected by the phospholipase C inhibitor neomycin, in contrast to the effect of mepacrine. This observation is consistent with an effect of GTPγS on the activation of phospholipase A_2 to generate arachidonic acid. In patches exposed to preactivated pertussis toxin to eliminate spontaneous Na^+ channel activity, the phospholipase agonist melittin also increased the number of open channels and increased the percent open time, whose effect was inhibited by mepacrine. To assess whether the melittin-activation of Na^+ channels is mediated through phospholipase A_2, experiments were conducted in Ca^{2+}-free solution containing 5 m*M* EGTA. Under these conditions, melittin did not activate Na^+ channels, although addition of 5 m*M* Ca^{2+} to the bathing solution immediately increased channel activity, further supporting an effect of melittin on a Ca^{2+}-sensitive phospholipase A_2.

To demonstrate the specific requirement for arachidonic acid, the effect of fatty acids with various degrees of polyunsaturation was tested. Addition of eicosapentaenoic, docosahexaenoic, or eicosatetraynoic acids to mepacrine-treated patches had no effect on Na^+ channel activity. It thus seemed unlikely that the arachidonic acid response is the result of a direct effect on the Na^+ channel shared by mono-, di-, or polyunsaturated fatty acids as has been shown for K^+ channels in myocytes (Kim and Clapham, 1989; Kurachi *et al.,* 1989; Ordway *et al.,* 1989). This suggested that a specific metabolite of arachidonic acid might be necessary for its effect on the Na^+ channel, possibly lipoxygenase products, which are known to regulate K^+ channels in nonpolar cells (Kim and Clapham, 1989; Kurachi *et al.,* 1989; Ordway *et al.,* 1989). GTPγS or arachidonic acid-induced channel activity was inhibited by the addition of the 5-lipoxygenase inhibitor nordihydroguaiaretic acid (NDGA). In the presence of NDGA, Na^+ channel activity was restored to stimulated levels with the addition of 100 n*M* leukotriene D_4, which increased the percent open time and the average channel number. Washout of the leukotriene reversed its effect. Melittin-stimulated channel activity was also blocked by NDGA. Since NDGA should decrease the availability of hydroperoxides and the production of leukotrienes by the 5-lipoxygenase, addition of 5(*S*)-HPETE, the parental substrate for the synthesis of leukotrienes, should reverse the effect of NDGA. In the presence of 5(*S*)-HPETE the total number of active channels also increased.

Direct support for a role of G proteins at a step proximal to the phospholipase and lipoxygenase pathways in the activation of apical Na^+ channels was then obtained by the addition of purified, activated human $\alpha_{i\text{-}3}$ to the cytosolic side of excised patches. As discussed above, the addition of activated $\alpha_{i\text{-}3}$ increases the percent open time of Na^+ channels in pertussis toxin-treated patches. In contrast, in the presence of pertussis toxin plus NDGA, the $\alpha_{i\text{-}3}$ subunit was no longer effective in activating Na^+ channels, but Na^+ channel activity was reinstated upon addition of LTD_4. These data support the conclusion that apical Na^+ channels can be activated by picomolar concentrations of the $\alpha_{i\text{-}3}$ subunit of the G_i protein that stimulates phospholipase activity. The resultant arachidonic acid undergoes a conversion to lipoxygenase products which either directly, or through additional pathways that have yet to be defined, activate amiloride-sensitive Na^+ channels. The identity of the specific endogenous enzymes and phospholipid products in A6 cell membranes responsible for Na^+ channel regulation remain to be defined, although the nature of this response bears strong similarities to that reported for mineralocorticoid-activation of epithelial Na^+ channels.

Aldosterone induces an increase in Na^+ channel activity in A6 cells that is not mediated by incorporation of new channels but by activation of an otherwise quiescent pool of already present Na^+ channels in the apical membrane (Kleyman *et al.*, 1989). Both basal and aldosterone-stimulated epithelial Na^+ channel activity is inhibited by mepacrine in another tight epithelium, the toad urinary bladder, (Yorio and Bentley, 1978), implying phospholipase A_2 activity in the regulation of this channel (Goodman *et al.*, 1975). Thus, our data provide an explanation for the potential molecular mechanism for the activation of Na^+ channels by phospholipid products that might also be a component of aldosterone action (see Fig. 2). Further support for this hypothesis comes from the study of the gene for $\alpha_{i\text{-}3}$. The 5′ regulatory segment of the $\alpha_{i\text{-}3}$ gene has steroid-responsive elements consistent with the ability of mineralocorticoids to induce its expression (E. J. Holtzman *et al.*, unpublished observations). Thus, $\alpha_{i\text{-}3}$ could be an aldosterone-induced protein. The above model is consistent with the short-term effect of aldosterone, which involves the activation rather than the synthesis of otherwise quiescent apical ion channels.

E. Cytoskeleton–Membrane Interactions

The possibility exists that intracellular in addition to extracellular, receptor-mediated, ion channel-regulatory signals can exist at the plasma membrane of epithelia. In this regard, the cortical cytoskeleton, in particular actin filament organization, may play a key role.

The cytoskeletal matrix of eukaryotic cells is a prominent intracellular network consisting of actin filaments, intermediate filaments, and microtu-

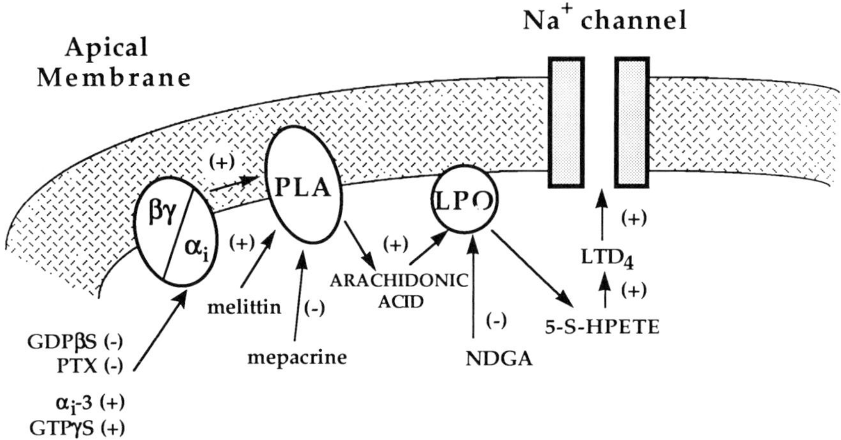

FIG. 2. Schematic representation of metabolic pathways linking apical G_i proteins to the epithelial Na^+ channel. This pathway is based on experimental evidence as per Cantiello *et al.*, 1989, 1990. G_i is represented by the membrane bound heterotrimer $\beta\gamma + \alpha_i$ whose endogenous activation by binding of GTP stimulates the production of arachidonic acid via activation of phospholipase A_2 (PLA). This metabolic pathway also involves the 5-lipoxygenase (LPO), hydroxyperoxidation of arachidonic acid into 5-S-HPETE, and the final product(s) leukotriene D_4 (LTD_4).

bules (Small, 1989; Stossel, 1982). Fundamental aspects of cellular function require the integrity and coordinated dynamic changes of these different cytoskeletal elements, including cellular shape and movement, cell–cell interactions, cell–substrate adhesion, migration, and hormone action (Stossel, 1982; Hall, 1984; Small, 1989). Of particular importance is the actin filament network, a structure that is localized in the cortical cytoplasm, an area of the cytosol facing the inner side of the plasma membrane. Actin filaments, for example, have been implicated in the activation of human neutrophils, the stimulation of gastric parietal cells and the onset of secretion (Mercier *et al.*, 1989), the vectorial delivery of apical membrane proteins in epithelial cells (Achler *et al.*, 1989), and the hydro-osmotic response to vasopressin in tight epithelia (Ding *et al.*, 1991; Taylor *et al.*, 1973). Nevertheless, little is known about the molecular mechanisms which link the cytoskeletal matrix with the cellular membrane.

A current view of the interaction between actin filament networks and cellular membranes involves the linkage of actin filaments to transmembrane proteins, mediated by actin-binding proteins such as spectrin, fodrin, and ankyrin (Geiger, 1985, 1989; Hartwig *et al.*, 1989; Louvard, 1989; Stossel, 1982). Recently, studies have shown that these proteins are also linked to ion transport proteins. Drenckhahn *et al.* (1985) observed the colocalization of ankyrin and spectrin to the anion transport protein band 3 in renal intercalated cells. Ankyrin links fodrin, another actin-binding protein, to the α subunit of the Na^+/K^+-ATPase in the MDCK cells (Nelson and Veshnock, 1987). More recently, Edelstein *et al.* (1988) observed that a 33-kDa cytoskeletal protein tightly binds to the voltage-sensitive rat brain Na^+ channel.

A possible interaction between cytoskeletal proteins and the plasma membrane may be the modification of such transmembrane protein function by membrane lipids (Burn, 1988). Several proteins actively involved in actin filament organization, including the actin-severing/capping proteins gelsolin and profilin, and proteins which anchor actin to the plasma membrane, such as α-actinin, vinculin, and spectrin, bind lipids in a noncovalent manner (Burn, 1988; Burn *et al.*, 1985; Geiger, 1985, 1989). This may be true for actin as well (Saint Onge and Gicquaud, 1990; Schwartz and Luna, 1988). The actin-binding protein calpactin and the actin-severing protein gelsolin not only bind lipids but their biological activity is strongly dependent on Ca^{2+} and phospholipid metabolism (Kawamoto and Hidaka, 1984; Yin *et al.*, 1980). Covalent modification by binding of long-chain fatty acids, such as myristic and palmitic acids, has been shown to occur with the anchoring proteins vinculin, ankyrin, and p36-actin (Burn, 1988; Burn *et al.*, 1985; Geiger, 1985, 1989; Lassing and Lindberg, 1985). Furthermore, myosin I, a single-headed form of myosin present in nonmuscle cells that also interacts with actin, was found to bind phospholipids *in vitro* (Adams and Pollard, 1989). The phospholipids involved in this reaction include phosphatidylserine and phosphatidylinositol-4,5-bisphosphate but not phosphatidylcholine. Since this binding affinity is strongly dependent on the ionic strength of the medium, it is thought that myosin I binds to negatively charged surfaces provided by anionic phospholipids. These data suggest that the cytoskeletal coupling to specific lipids may serve as a means for the transduction of information through lipidic environments.

The dynamics of the cortical and actin-filament-based cytoskeleton may involve several critical steps which are oriented toward the maintenance of an equilibrium in the actin organization. Changes in actin polymerization, such as seen upon the G protein-mediated *N*-formyl peptide receptor activation of human neutrophils, are known to follow a biphasic response in which there is a rapid rise in F actin that peaks a few seconds after stimulation and comprises an activation response which raises the basal level of F-actin from 20 to 75% of the total amount of intracellular actin (White *et al.*, 1982; Yassin *et al.*, 1985). This is followed by a secondary slow response which peaks and decays to the basal level around 3 min after activation. This biphasic response may be linked to Ca^{2+}-related events. Interestingly, less than 100 *N*-formyl peptide receptors are required to initiate such response in neutrophils. Considering that actin represents about 10% of the cell protein in neutrophils, 50% of which is stimulated in a few seconds, then at least 5×10^6 monomeric actin molecules are assembled into F actin in a period of 10 sec by the action of very few receptors. This suggests an amplification of the G-protein–receptor linkage to monomeric actins of $1:10^5$ (Omann *et al.*, 1987).

More recently, both fMet-Leu-Phe and phorbol ester activation of neutrophils have been also linked to a Ca^{2+}-stimulated activation of a Cl^- conductance (Myers *et al.*, 1990). Although the demonstration of a Cl^- channel coupled to *N*-formyl receptor activation in neutrophils and actin polymerization is lacking, the information presented in previous sections of this chapter supports the presence of a link between ion channels, G proteins, and possibly the cytoskeleton in receptor-mediated coupled events. Grinstein *et al.* have found that pertussis toxin-insensitive G protein-mediated events are also involved in actin polymerization, and may be linked to proliferation (Grinstein *et al.*, 1985; Lu and Grinstein, 1990). This raises the possibility of a pertussis toxin-insensitive G protein which is different from receptor-coupled G proteins, and which would be located downstream of the protein kinase C pathway. This may be consistent with mechanisms which mimic toxin-mediated ADP-ribosylation of actin but which are independent of pertussis toxin. For example, the toxins from *Clostridium botulinum* or *C. perfringens* are known to ADP-ribosylate actin and cap and prevent actin polymerization *in vitro* (Aktories and Wegner, 1989; Aktories *et al.*, 1988). This may be a novel mechanism for actin modulation and establishes a possible role for endogenous ADP-ribosyltransferases in the control of cell function including control of the G proteins (Tanuma *et al.*, 1988). The pertussis toxin-insensitive G protein system in neutrophils could be related to mechanisms other than cAMP-induced responses. Elevation of cAMP in intact neutrophils, for example, attenuates the fMet-Leu-Phe-induced actin organization, and cAMP added directly to permeabilized neutrophils results in actin disassembly. Thus, this system may represent an integration of several regulatory pathways, all of which involve G proteins. The receptor-mediated response is coupled not only to activation of phospholipases such as A_2 and C but also to the production of phospholipids and intracellular Ca^{2+} that may directly or indirectly regulate actin polymerization (Cockcroft and Gomperts, 1985; Gomperts *et al.*, 1986; Molski *et al.*, 1984; Volpi *et al.*, 1985; Yassin *et al.*, 1985). In this sense, we have three important events: activation of ion channels and other ion transport mechanisms, coupling of G proteins to receptors with activation of phospholipid metabolism, and actin polymerization. This may be the reflection of an integrated mechanism through which signal transduction, phospholipids, and actin organization modulate membrane events including ion channel activity.

F. Cytoskeletal Regulation of Ion Channels

Actin filaments, as well as actin-binding proteins are enriched in the cortical cytoplasm that faces the inner side of the apical membranes of epithelial cells (Louvard, 1989). This actin filament organization interfaces the apical membrane with the subjacent cytoplasm and provides a support for the spatial distribution of integral membrane proteins. Another novel role for the cortical cytoskeleton may be the regulation of membrane transport proteins such as ion channels. Sachs *et al.* were the first to observe an interaction between cortical cytoskeleton and the so-called stretch-activated channel activity in muscle (Guharay and Sachs, 1984; Sachs, 1987).

One of the most useful tools in experimentally approaching the functional role of the cytoskeleton in membrane-related events, and in particular ion channel regulation, has been the use of a family of fungal toxins, the cytochalasins, which are isoindole-substituted macrocyclic rings known to interact with cell membranes and, in particular, actin filaments (Cribbs *et al.*, 1982; Forscher and Smith, 1988; Schliwa, 1982). Cytochalasin D can increase or decrease actin polymerization depending on ionic strength and the Mg^{2+} concentration (Goddette and Frieden, 1986).

In order to assess the possible role of the actin filament network on Na^+ channel activity, a series of experiments was conducted in our laboratory in which cytochalasin D was used in either cell-attached or excised apical membrane patches of A6 cells. Acute addition of cytochalasin D induced, almost immediately, the opening of Na^+ channels, thus indicating the presence of an actin filament network in excised patches of A6 epithelial cells (Cantiello *et al.*, 1991). This was also shown in cell-attached patches. Filamin, known to cross-link actin filaments, was added to patches exposed to cytochalasin D. Filamin completely reversed the cytochalasin D effect, returning channel activity to control levels. The effects of cytochalasin D and filamin on channel activity raise the possibility that the cross-linking status of the actin network may be involved in the regulation of channels. The effect of cytochalasin D was mimicked by the addition of exogenous actin in the presence of ATP, thus indicating that the presence of free actin filaments may indeed modulate ion channel activity (Prat *et al.*, 1991). The cytoskeletal regulation of Na^+ channels is also supported by the immuno-colocalization of apical Na^+ channels with actin filaments in A6 cells as well as the presence of ankyrin and actin in the purified Na^+ channel complex (Cantiello *et al.*, 1991; Smith *et al.*, 1990). Since the cytochalasin D activation of Na^+ channels is inhibited by the phospholipase inhibitor mepacrine (H. F. Cantiello *et al.*, unpublished observations), it seems likely that the cytoskeletal control of ion channel activity may involve the coupling of actin filamental structures to the metabolic pathways observed in the apical membrane of A6 epithelial cells, as described above.

The above data support the contention that apical epithelial Na^+ channels not only colocalize with a G_i protein, the phospholipase A_2, and the lipoxygenase pathways but also with cortical actin networks, thus providing functional evidence for a potentially novel regulatory mechanism that may combine the above molecular steps in an either sequential or simultaneous interactive sequence. This evidence on Na^+ channel regulatory function is consistent with recent evidence indicating the colocalization of such proteins as the $\alpha_{i\text{-}3}$ subunit of the G_i complex (D. J. Benos *et al.*, unpublished observations; Smith *et al.*, 1990), the actin-binding protein ankyrin, and actin itself with the purified Na^+ channel complex. Together, the data support the presence of a novel regulatory pathway of apical Na^+ channels in which the various enzymatic reactions may display a concerted complex of regulatory steps which link the membrane proteins coupled to the Na^+ channel on one hand and the actin-based cortical network on the other, which would be the basis for the intracellular control of channel activity.

IV. CONCLUSIONS AND PERSPECTIVE

The past 30 years have been extremely productive in the development of techniques suitable to the understanding of epithelial ion channel physiology, and particularly fruitful in the advancement of our knowledge of epithelial Na^+ channel regulation. Hormone-sensitive epithelia have been studied to the extent that our current understanding of vasopressin action is thought to involve intracellular second messengers such as cyclic AMP and Ca^{2+}, which is consistent with classical receptor-targetting events thoroughly studied in the various nonpolar tissues.

In contrast, recent information has prompted the studies of newly characterized regulatory proteins such as G proteins, which are a superfamily of novel information–transduction mediators that interact with such effectors as ion channels in a membrane-delimited fashion. Such events may be complimentary to second-messenger-related events and may represent the ultimate target for both intra- and extracellular signals to control integral membrane proteins.

In this regard, G proteins seem to play a fundamental role in the control of ion channels in general and, in particular, epithelial Na^+ channel activity. This mechanism(s) may involve the control of phosphodiesterase activity such as that of phospholipases and the onset of lipid metabolic pathways which control such activity. The regulatory pathways described above may also be the target of hormonal action such as that of aldosterone. It is appealing to postulate that, due to the membrane-delimited domain of interactions of such pathways, novel regulatory mechanisms complimentary to conventional second messengers may also serve as an anchoring step for intracellular control mechanisms, which include critical changes in apical membrane lipid composition, lipid charge modifications, and lipid phase transitions, as well as the linkage of cytoskeletal networks to the apical plasma membrane. Actin networks, which colocalize with Na^+ channels, have thus far been thought to play an important structural role in the spatial distribution of transmembrane proteins within the plasma membrane. More recent information raises the interesting possibility that these dynamic structures also control epithelial ion channel function. With the advent of modern technologies, such as patch-clamping of epithelial cells, whereby the cytosolic domain of the apical epithelial membrane is made readily available to experimental exploration, new findings are bound to increase which will bridge the gap between different regulatory systems at the various cellular levels and the apical plasma membrane. Particularly appealing is the possibility that the actin-based cortical network may be a regulatory pathway of apical G protein control mechanisms. This potential role of the cytoskeleton may also be extended to the binding and regulation of the structural membrane lipidic composition that is also under the regulatory control of G proteins. Furthermore, such interactive networks may also be the basis of selectivity–conductivity differences observed in epithelial ion channel activity. The next few years await new findings to elucidate these intriguing postulations.

References

Achler, C., Filmer, D., Merte, C., and Drenckhahn, D. (1989). Role of microtubules in polarized delivery of apical membrane proteins to the brush border of the intestinal epithelium. *J. Cell Biol.* **109,** 179–189.

Adams, R. J., and Pollard, T. D. (1989). Binding of myosin I to membrane lipids. *Nature (London)* **340,** 565–568.

Aktories, K., and Wegner, A. (1989). ADP-ribosylation of actin by clostridial toxins. *J. Cell Biol.* **109,** 1385–1387.

Aktories, K., Just, I., and Rosenthal, W. (1988). Different types of ADP-ribose protein bonds formed by botulinum C_2 toxin, botulinum ADP-ribosyltransferase C_3 and pertussis toxin. *Biochem. Biophys. Res. Commun.* **156,** 361–367.

Alouf, J. E., Fehrenbach, F. J., Freer, J. H., and Jeljaszewicz, J. (1984). "Bacterial Protein Toxins." Academic Press, New York.

Andrade, R., Malenka, R. C., and Nicoll, R. A. (1986). A G protein couples serotonin and GABA receptors to the same channels in hippocampus. *Science* **234,** 1261–1265.

Aronson, P. S., Suhm, M. A., and Nee, J. (1983). Interaction of external H^+ with the Na^+/H^+ exchanger in renal microvillus membrane vesicles. *J. Biol. Chem.* **258,** 6767–6771.

Asher, C., Moran, A., Rossier, B. C., and Garty, H. (1988). Sodium channels in membrane vesicles from cultured toad bladder cells. *Am. J. Physiol.* **254,** C512–C518.

Axelrod, J., Burch, R. M., and Jelesma, C. (1988). Receptor-mediated activation of phospholipase A_2 via GTP-binding proteins: Arachidonic acid and its metabolites as second messengers. *Trends Neurosci.* **11,** 117–123.

Baer, J. E., Jones, C. B., Spitzer, S. A., and Russo, H. F. (1967). The potassium-sparing and natriuretic activity of *N*-amidino-3,5-diamino-6-chloropyrazinecarboxamide hydrochloride dihydrate (amiloride hydrochloride). *J. Pharmacol. Exp. Ther.* **157,** 472–485.

Barry, P. H., Diamond, J. M., and Wright, E. M. (1971). The mechanism of cation permeation in rabbit gallbladder. Dilution potentials and biionic potentials. *J. Membr. Biol.* **4,** 358–394.

Benos, D. J. (1982). Amiloride: A molecular probe for sodium transport in tissues and cells. *Am. J. Physiol.* **242,** C131–C145.

Benos, D. J., and Watthey, J. W. M. (1981). Inferences on the nature of the apical sodium entry site in frog skin epithelium. *J. Pharmacol. Exp. Ther.* **219,** 481–488.

Benos, D. J., Mandel, L. J., and Balaban, R. S. (1979). On the mechanism of the amiloride–sodium entry site interaction in anuran skin epithelia. *J. Gen. Physiol.* **73,** 307–326.

Benos, D. J., Mandel, L. J., and Simon, S. A. (1980). Effects of chemical group specific reagents on sodium entry and the amiloride binding site in frog skin: Evidence for separate sites. *J. Membr. Biol.* **56,** 149–158.

Bertorello, A., and Aperia, A. (1989). Regulation of Na^+-K^+-ATPase activity in kidney proximal tubules: Involvement of GTP binding proteins. *Am. J. Physiol.* **256,** F57–F62.

Besterman, J. M., Stratford May, W., Jr., LeVine, H., III, Cragoe, E. J., Jr., and Cuatrecasas, P. (1985). Amiloride inhibits phorbol ester-stimulated Na^+/H^+ exchange and protein kinase C. An amiloride analog selectively inhibits Na^+/H^+ exchange. *J. Biol. Chem.* **260,** 1155–1159.

Besterman, J. M., Elwell, L. P., Blanchard, S. G., and Cory, M. (1987). Amiloride intercalates into DNA and inhibits DNA topoisomerase II. *J. Biol. Chem.* **262,** 13352–13358.

Bindslev, N. (1979). Sodium transport in the hen lower intestine. Induction of sodium sites in the brush border by a low-sodium diet. *J. Physiol. (London)* **288,** 449–466.

Birnbaumer, L. (1990). Transduction of receptor signal into modulation of effector activity by G proteins: The first 20 years or so. *FASEB J.* **4,** 3178–3188.

Birnbaumer, L., and Brown, A. M. (1987). G protein opening of K^+ channels. *Nature (London)* **327,** 21–22.

Blumenfeld, J. B., Cantiello, H. F., and Lechene, C. (1988). Ion transport properties of rat inner medullary collecting duct cells (IMCD) in primary culture. *J. Gen. Physiol.* **92,** 28a. (Abstr.)

Bokoch, G. M., and Gilman, A. G. (1984). Inhibition of receptor-mediated release of arachidonic acid by pertussis toxin. *Cell* **39,** 301–308.

Bourne, H. R., Sanders, D. A., and McCormick, F. (1990). The GTPase superfamily: A conserved switch for diverse cell functions. *Nature (London)* **348,** 125–132.

Bradford, P., and Rubin, R. (1985). Pertussis toxin inhibits chemotactic factor-induced phospholipase C stimulation and lysosomal enzyme secretion in rabbit neutrophils. *FEBS Lett.* **183,** 317–320.

Brechler, V., Pavoine, C., Lotersztajn, S., Garbarz, E., and Pecker, F. (1990). Activation of Na^+/Ca^{2+} exchange by adenosine in ewe heart sarcolemma is mediated by a pertussis toxin-sensitive G protein. *J. Biol. Chem.* **265,** 16851–16855.

Breitwieser, G. E., and Szabo, G. (1985). Uncoupling of cardiac muscarinic and β-adrenergic receptors from ion channels by a guanine nucleotide analogue. *Nature (London)* **317,** 538–540.

Breitwieser, G. E., and Szabo, G. (1988). Mechanism of muscarinic receptor-induced K^+ channel activation as revealed by hydrolysis-resistant GTP analogues. *J. Gen. Physiol.* **91,** 469–493.

Brown, A. M., and Birnbaumer, L. (1988). Direct G protein gating of ion channels. *Am. J. Physiol.* **254,** H401–H410.

Brown, A. M., and Birnbaumer, L. (1990). Ionic channels and their regulation by G protein subunits. *Annu. Rev. Physiol.* **52,** 197–213.

Brown, A. M., Yatani, A., Codina, J., and Birnbaumer, L. (1989). G protein-gated channels: A third major category of ionic channels. *Am. J. Hypertens.* **2,** 124–127.

Burn, P. (1988). Amphitropic proteins: A new class of membrane proteins. *Trends Biol. Sci.* **13,** 79–83.

Burn, P., Rotman, A., Meyer, R. K., and Burger, M. M. (1985). Diacylglycerol in large α-actinin/actin complexes and in the cytoskeleton of activated platelets. *Nature (London)* **314,** 469–474.

Cantiello, H. F., and Ausiello, D. A. (1986). Atrial natriuretic factor and cGMP inhibit amiloride-sensitive Na^+ transport system in the cultured renal epithelial cell line, LLC-PK_1. *Biochem. Biophys. Res. Commun.* **134,** 852–860.

Cantiello, H. F., Scott, J. A., and Rabito, C. A. (1986). Polarized distribution of the Na^+/H^+ exchange system in a renal cell line (LLC-PK_1). *J. Biol. Chem.* **261,** 3252–3258.

Cantiello, H. F., Scott, J. A., and Rabito, C. A. (1987). Characteristics of a sodium-conductive pathway in an epithelial cell line (LLC-PK_1). *Am. J. Physiol.* **252,** F590–F597.

Cantiello, H. F., Patenaude, C. R., and Ausiello, D. (1989). G protein subunit, $\alpha_{i\text{-}3}$, activates a pertussis toxin-sensitive Na^+ channel from the epithelial cell line, A6. *J. Biol. Chem.* **264,** 20867–20870.

Cantiello, H. F., Patenaude, C. R., Codina, J., Birnbaumer, L., and Ausiello, D. A. (1990). G-alpha-i-3 regulates epithelial Na^+ channels by activation of phospholipase A_2 and lipoxygenase pathways. *J. Biol. Chem.* **265,** 21624–21628.

Cantiello, H. F., Stow, J. L., and Ausiello, D. A. (1991). Cortical actin filaments co-localize with and regulate apical epithelial Na^+ channels in A6 cells. *FASEB J.* (Abstr.) **5,** A690.

Cereijido, M., Herrera, F. C., Flanigan, W. J., and Curran, P. F. (1964). The influence of Na^+ concentration on Na^+ transport across frog skin. *J. Gen. Physiol.* **47,** 879–893.

Christensen, O., and Bindslev, N. (1982). Fluctuation analysis of short-circuit current in a warm-blooded sodium-retaining epithelium: Site current, density, and interaction with triamterene. *J. Membr. Biol.* **65,** 19–30.

Clauss, W., Aranson, S., Munck, B., and Skadhauge, E. (1984). Aldosterone-induced sodium transport in lower intestine: Effects of varying NaCl intake. *Pfluegers Arch.* **401,** 354–360.

Cockcroft, S. (1987). Polyphosphoinositide phosphodiesterase: Regulation by a novel guanine nucleotide binding protein, G_p. *Trends Biochem. Sci.* **12,** 75–78.

Cockcroft, S., and Gomperts, B. D. (1985). Role of guanine nucleotide binding protein in the activation of polyphosphoinositide phosphodiesterase. *Nature (London)* **314,** 534–536.
Codina, J., Yatani, A., Grenet, D., Brown, A. M., and Birnbaumer, L. (1987). The α subunit of the GTP binding protein G_K opens atrial potassium channels. *Science* **236,** 442–445.
Cook, J. S., Shaffer, C., and Cragoe, E. J., Jr. (1986). Inhibition of Na^+-dependent uptake of hexose and amino acids in LLC-PK_1 cells by amiloride and amiloride derivatives. *Fed. Proc.* **45,** 517. (Abstr.)
Cook, J. S., Shaffer, C., and Cragoe, E. J., Jr. (1987). Inhibition by amiloride analogs of Na^+-dependent hexose uptake in LLC-PK_1/C_{14} cells. *Am. J. Physiol.* **253,** C199–C204.
Crabbé, J. (1964). Stimulation of active sodium transport by the urinary bladder of amphibia under the influence of aldosterone. *In* "Water and Electrolyte Metabolism II" (J. de Graeff and B. Leijnse, eds.), pp. 59–68. Elsevier, Amsterdam.
Cribbs, D. H., Glenney, J. R., Kaulfus, P., Weber, K., and Lin, S. (1982). Interaction of cytochalasin B with actin filaments nucleated or fragmented by villin. *J. Biol. Chem.* **257,** 395–399.
Cuthbert, A., and Shum, W. (1974). Amiloride and the sodium channel. *Naunyn-Schmiedeberg's Arch. Pharmacol.* **281,** 261–269.
Cuthbert, A. W., Okpako, D., and Shum, W. K. (1974). Aldosterone, moulting and the number of sodium channels in frog skin. *Br. J. Pharmacol.* **138,** 129P. (Abstr.)
De Wolf, I., and Van Driessche, W. (1986). Voltage-dependent Ba^{2+} block of K^+ channels in apical membrane of frog skin. *Am. J. Physiol.* **251,** C696–C706.
Diamond, J. M. (1978). Channels in epithelial cell membranes and junctions. *Fed. Proc.* **37,** 2639–2644.
Dick, H. J., and Lindemann, B. (1975). Saturation of Na^+ current into frog skin epithelium abolished by PCMB. *Pfluegers Arch.* **355,** R72. (Abstr.)
Ding, G., Franki, N., Condeelis, J., and Hays, R. M. (1991). Vasopressin depolymerizes F-actin in toad bladder epithelial cells. *Am. J. Physiol.* **260,** C9–C16.
Dolphin, A. C. (1990). G protein modulation of calcium currents in neurons. *Annu. Rev. Physiol.* **52,** 243–255.
Drenckhahn, D., Schlüter, K., Allen, D. P., and Bennett, W. (1985). Colocalization of band 3 with ankyrin and spectrin at the basal membrane of intercalated cells in the rat kidney. *Science* **230,** 1287–1290.
Edelstein, N. G., Catteral, W. A., and Moon, R. T. (1988). Identification of a 33 kD cytoskeletal protein with high affinity for the sodium channel. *Biochemistry* **27,** 1818–1822.
Endoh, M., Manyama, M., and Tajima, T. (1985). Attenuation of muscarinic cholinergic inhibition by islet-activating protein in the heart. *Am. J. Physiol.* **249,** H309–H320.
Ercolani L., Stow, J. L., Boyle, J. F., Holtzman, E. J., Lin, H., Russell Grove, J., and Ausiello, D. A. (1990). Membrane localization of the pertussis toxin-sensitive G-protein subunits $\alpha_{i\text{-}2}$ and $\alpha_{i\text{-}3}$ and expression of a metallothionein-$\alpha_{i\text{-}2}$ fusion gene in LLC-PK_1 cells. *Proc. Natl. Acad. Sci. U.S.A.* **87,** 4635–4639.
Fain, J. N., Wallace, M. A., and Wojcikiewicz, R. J. H. (1988). Evidence for involvement of guanine nucleotide-binding proteins in the activation of phospholipases by hormones. *FASEB J.* **2,** 2569–2574.
Fimognari, G., Fanestil, D., and Edelman, I. (1967). Induction of RNA and protein synthesis in the action of aldosterone in the rat. *Am. J. Physiol.* **213,** 954–962.
Forscher, P., and Smith, S. J. (1988). Actions of cytochalasins on the organization of actin filaments and microtubules in a neuronal growth cone. *J. Cell Biol.* **107,** 1505–1516.
Frelin, C., Vigne, P., Barbry, P., and Lazdunski, M. (1987). Molecular properties of amiloride action and of its Na^+ transporting targets. *Kidney Int.* **32,** 785–793.

Frings, S., Purves, R. B., and MacKnight, A. D. C. (1988). Single-channel recordings from the apical membrane of the toad urinary bladder epithelial cell. *J. Membr. Biol.* **106,** 157–172.

Frizzell, R. A., and Schultz, S. G. (1978). Effect of aldosterone on ion transport by rabbit colon, *in vitro*. *J. Membr. Biol.* **39,** 1–26.

Frömter, E. (1972). The route of passive ion movement through the epithelium of *Necturus* gallbladder. *J. Membr. Biol.* **8,** 259–301.

Frömter, E., and Diamond, J. (1972). Route of passive ion permeation in epithelia. *Nature (London)* **235,** 9–13.

Garcia-Romeu, F. (1974). Stimulation of sodium transport in frog skin by 2-imidazolines (guanidinbenzimidazole and phentolamine). *Life Sci.* **15,** 539–542.

Garty, H., and Benos, D. J. (1988). Characteristics and regulatory mechanisms of the amiloride-blockable Na^+ channel. *Physiol. Rev.* **68,** 309–373.

Garty, H., and Lindemann, B. (1984). Feedback inhibition of sodium uptake in K-depolarized toad urinary bladders. *Biochim. Biophys. Acta* **771,** 89–98.

Garty, H., Yeger, O., Yanovsky, A., and Asher, C. (1989). Guanosine nucleotide-dependent activation of the amiloride-blockable Na^+ channel. *Am. J. Physiol.* **363,** F965–F969.

Geiger, B. (1985). Microfilament–membrane interactions. *Trends Biochem. Sci.* **10,** 456–461.

Geiger, B. (1989). Cytoskeletal-associated cell contacts. *Curr. Opinion Cell Biol.* **1,** 103–109.

Gilman, A. G. (1984). G proteins and dual control of adenylate cyclase. *Cell* **36,** 577–579.

Gilman, A. G. (1987). G proteins: Transducers of receptor-generated signals. *Annu. Rev. Biochem.* **56,** 615–649.

Goddette, D. W., and Frieden, C. (1986). The kinetics of cytochalasin D binding to monomeric actin. *J. Biol. Chem.* **261,** 15970–15973.

Gögelein, H., and Greger, R. (1986). Na^+ selective channels in the apical membrane of late proximal tubules (pars recta). *Pfluegers Arch.* **406,** 198–203.

Gögelein, H., and Van Driessche, W. (1981). The effect of electrical gradients on current fluctuations and impedance recorded from *Necturus* gallbladder. *J. Membr. Biol.* **60,** 199–209.

Gomperts, B. D., Barrowman, M. M., and Cockcroft, S. (1986). Dual role for guanine nucleotides in stimulus-secretion coupling. *Fed. Proc.* **45,** 2156–2161.

Goodman, D., Allen, J., and Rasmussen, H. (1971). Studies on the mechanism of action of aldosterone: Hormone-induced changes in lipid metabolism. *Biochemistry* **10,** 3825–3831.

Goodman, D., Wong, M., and Rasmussen, H. (1975). Aldosterone-induced membrane phospholipid fatty acid metabolism in the toad urinary bladder. *Biochemistry* **14,** 2803–2809.

Grinstein, S., and Furuya, W. (1984). Amiloride-sensitive Na^+/H^+ exchange in human neutrophils: Mechanism of activation by chemotactic factors. *Biochem. Biophys. Res. Commun.* **122,** 755–762.

Grinstein, S., Elder, B., and Furuya, W. (1985). Phorbol ester-induced changes of cytoplasmic pH in neutrophils: Role of exocytosis in Na^+/H^+ exchange. *Am. J. Physiol.* **248,** C379–C386.

Guharay, F., and Sachs, F. (1984). Stretch-activated single ion channel currents in tissue-cultured embryonic chick skeletal muscle. *J. Physiol. (London)* **352,** 685–701.

Hall, P. F. (1984). The role of the cytoskeleton in hormone action. *Can. J. Biochem. Cell Biol.* **62,** 653–665.

Hamill, O. P., Marty, A., Neher, E., Sakmann, B., and Sigworth, F. J. (1981). Improved patch-clamp techniques for high-resolution current recording from cells and cell-free membrane patches. *Pfluegers Arch.* **391,** 85–100.

Hamilton, K. L., and Eaton, D. C. (1985). Single-channel recordings from amiloride-sensitive epithelial sodium channel. *Am. J. Physiol.* **249,** C200–C207.

Hamilton, K. L., and Eaton, D. C. (1986a). Regulation of single sodium channels in renal tissue: A role in sodium homeostasis. *Fed. Proc.* **45,** 2713–2717.

Hamilton, K. L., and Eaton, D. C. (1986b). Single channel recordings from two types of amiloride-sensitive epithelial Na^+ channels. *Membr. Biochem.* **6,** 149–171.

Hartwig, J. H., Chambers, K. A., and Stossel, T. P. (1989). Association of gelsolin with actin filaments and cell membranes of macrophages and platelets. *J. Cell Biol.* **108,** 467–479.

Harwood, J. P., Low, H., & Rodbell, M. (1973). Stimulatory and inhibitory effects of guanyl nucleotides on fat cell adenylate cyclase. *J. Biol. Chem.* **248,** 6239–6245.

Helman, S. I., Cox, T. C., and Van Driessche, W. (1983). Hormonal control of apical membrane Na^+ transport in epithelia. Studies with fluctuation analysis. *J. Gen. Physiol.* **32,** 201–220.

Hescheler, J., Rosenthal, W., Trautwein, W., and Schultz, G. (1987). The GTP-binding protein, G_o regulates neuronal calcium channels. *Nature (London)* **325,** 445–447.

Hestrin-Lerner, S., and Hokin, L. E. (1964). Effects of hormones on Na^+ and H_2O transport and on phospholipid metabolism in toad bladder. *Am. J. Physiol.* **206,** 136–142.

Hillyard, S. D., Zeiske, W., and Van Driessche, W. (1982a). A fluctuation analysis study of the development of amiloride-sensitive Na^+ transport in the skin of larval bullfrogs (*Rana catesbeiana*). *Biochim. Biophys. Acta* **692,** 455–461.

Hillyard, S. D., Zeiske, W., and Van Driessche, W. (1982b). Poorly selective cation channels in the skin of the larval frog (stage ≤XIX). *Pfluegers Arch.* **394,** 287–293.

Holz, G. G., IV, Rane, S. G., and Dunlap, K. (1986). GTP-binding proteins mediate transmitter inhibition of voltage-dependent calcium channels. *Nature (London)* **319,** 670–672.

Ives, H. E., Yee, V. J., and Warnock, D. G. (1983). Mixed type inhibition of the Na^+/H^+ antiporter by Li^+ and amiloride. Evidence for a modifier site. *J. Biol. Chem.* **258,** 9710–9716.

Jelesma, C., and Axelrod, J. (1987). Stimulation of phospholipase A_2 activity by the $\beta\gamma$ subunits of transducin and its inhibition by the α subunit. *Proc. Natl. Acad. Sci. U.S.A.* **84,** 3623–3627.

Jorgensen, C. B., Levi, H., and Ussing, H. H. (1946). On the influence of the neurohypophyseal principles on the sodium metabolism in the axolotl (*Ambystoma mexicanum*). *Acta Physiol. Scand.* **12,** 350.

Kawamoto, S., and Hidaka, H. (1984). Ca^{2+}-activated, phospholipid-dependent protein kinase catalyzes the phosphorylation of actin binding proteins. *Biochem. Biophys. Res. Commun.* **118,** 736–742.

Kim, D., and Clapham, D. E. (1989). Potassium channels in cardiac cells activated by arachidonic acid and phospholipids. *Science* **244,** 1174–1176.

Kim, D., Lewis, D. L., Graziadei, L., Neer, E. J., Bar-Sagi, D., and Clapham, E. E. (1989). G-protein $\beta\gamma$-subunits activate the cardiac muscarinic K^+-channel via phospholipase A_2. *Nature (London)* **337,** 557–560.

Kleyman, T. R., Cragoe, E. J., Jr., and Kraehenburlhe, J. P. (1989). The cellular pool of Na^+ channels in the amphibian cell line A6 is not altered by mineralocorticoids. *J. Biol. Chem.* **264,** 11995–12000.

Koefoed-Johnsen, V., and Ussing, H. H. (1958). The nature of the frog skin potential. *Acta Physiol. Scand.* **42,** 298–308.

Kolb, H., Brown, C., and Murer, H. (1985). Identification of voltage-dependent anion channel in the apical membrane of a Cl^- secretory epithelium (MDCK). *Pfluegers Arch.* **403,** 262–265.

Kuhn, H. (1986). Proteins involved in the control of cyclic GMP phosphodiesterase in retinal rod cells. *In* "Membrane Control of Cellular Activity" (H. C. Luttgau, ed.), pp. 289–380. Springer-Verlag, Berlin.

Kurachi, Y., Hiroyuki, I., Sugimoto, T., Shimizu, T., Miki, I., and Ui, M. (1989). Arachidonic acid metabolites as intracellular modulators of the G protein-gated K^+ channel. *Nature (London)* **337,** 555–557.

Läuger, P. (1984). Conformational transitions of ionic channels. *In* "Single-Channel Recordings" (B. Sakmann and E. Neher, ed.), pp. 177–189. Plenum, New York.

Lassing, I., and Lindberg, U. (1985). Specific interaction between phosphatidylinositol 4,5-bisphosphate and profilactin. *Nature (London)* **314,** 472–474.

Lewis, S. A., Alles, W. P., and Clausen, C. (1986). Endogenous enzymes hydrolyze epithelial Na^+ channels. *Biophys. J.* **49,** 157. (Abstr.)

Li, J. H.-Y., and DeSousa, R. C. (1979). Inhibitory and stimulatory effects of amiloride analogues on sodium transport in frog skin. *J. Membr. Biol.* **46,** 155–169.

Li, J. H.-Y., and Lindemann, B. (1982). Chemical stimulation of Na^+ transport through amiloride blockable channels of frog skin epithelium. *J. Membr. Biol.* **75,** 179–192.

Li, J. H.-Y., Cragoe, E. J., Jr., and Lindemann, B. (1987). Structure–activity relationship of amiloride analogs as blockers of epithelial Na^+ channels: II. Side chain modifications. *J. Membr. Biol.* **95,** 171–185.

Lien, E. L., Goodman, D. B. P., and Rasmussen, H. (1975). Effects of an acetyl-coenzyme A carboxylase inhibitor and a sodium-sparing diuretic on aldosterone-stimulated sodium transport, lipid synthesis, and phospholipid fatty acid composition in the toad urinary bladder. *Biochemistry* **14,** 2749–2754.

Lifschitz, M. D., Schrier, R. W., and Edelman, I. S. (1973). Effect of actinomycin D on aldosterone-mediated changes in electrolyte excretion. *Am. J. Physiol.* **224,** F376–F380.

Light, D. B., McCann, F. V., Keller, T. M., and Stanton, B. A. (1988). Amiloride-sensitive cation channel in apical membrane of inner medullary collecting duct. *Am. J. Physiol.* **255,** F278–F286.

Light, D. B., Ausiello, D. A., and Stanton, B. A. (1989a). Guanine nucleotide-binding protein, $\alpha_{i\text{-}3}$, directly activates a cation channel in rat renal inner medullary collecting duct cells. *J. Clin. Invest.* **84,** 352–356.

Light, D. B., Schwiebert, E. M., Karlson, K. H., and Stanton, B. A. (1989b). Atrial natriuretic peptide inhibits a cation channel in renal inner medullary collecting duct cells. *Science* **243,** 383–385.

Light, D. B., Schwiebert, E., Fejes-Toth, G., Naray-Feges-Toth, A., Karlson, K., McCann, F., and Stanton, B. (1990a). Chloride channels in the apical membrane of cortical collecting duct cells. *Am. J. Physiol.* **258,** F273–F280.

Light, D. B., Corbin, J. D., and Stanton, B. A. (1990b). Dual ion channel regulation by cyclic GMP and cyclic GMP-dependent protein kinase. *Nature (London)* **344,** 336–339.

Lindemann, B. (1984). Fluctuation analysis of sodium channels in epithelia. *Annu. Rev. Physiol.* **46,** 497–515.

Lindemann, B., and Van Driessche, W. (1977). Sodium-specific membrane channels of frog skin are pores: Current fluctuations reveal high turnover. *Science* **195,** 292–294.

Lindemann, B., and Van Driessche, W. (1978). The mechanism of Na^+ uptake through Na-selective channels in the epithelium of frog skin. *In* "Membrane Transport Processes" (J. Hoffman, ed.), pp. 155–178. Raven, New York.

Loo, D. D., Lewis, S. A., Ifshin, M. S., and Diamond, J. M. (1983). Turnover, membrane insertion, and degradation of sodium channels in rabbit urinary bladder. *Science* **221,** 1288–1290.

Louvard, D. (1989). The function of the major cytoskeletal components of the brush border. *Curr. Opinion Cell Biol.* **1,** 51–57.

Lu, D., and Grinstein, S. (1990). ATP and guanine nucleotide dependence of neutrophil activation. Evidence for the involvement of two distinct GTP-binding proteins. *J. Biol.*

Chem. **265,** 13721–13729.

Manning, D. R., and Gilman, A. G. (1983). The regulatory component of adenylate cyclase and transducin. A family of structurally homologous guanine nucleotide-binding proteins. *J. Biol. Chem.* **258,** 7059–7063.

Mercier, R., Reggio, H., Devilliers, G., Bataille, D., and Mangeat, P. (1989). Membrane-cytoskeleton dynamics in rat parietal cells: Mobilization of actin and spectrin upon gastric acid secretion. *J. Cell Biol.* **108,** 441–453.

Mohrmann, I., Mohmann, M., Biber, J., and Murer, H. (1986). Sodium-dependent transport of P_i by an established intestinal epithelial cell line (CaCo-2). *Am. J. Physiol.* **250,** G323–G330.

Mohrmann, M., Cantiello, H. F., and Ausiello, D. A. (1987a). Inhibition of epithelial Na^+ transport by atriopeptin, protein kinase C and pertussis toxin. *Am. J. Physiol.* **253,** F372–F376.

Mohrmann, M., Cantiello, H. F., and Ausiello, D. A. (1987b). Renal epithelial cell growth can occur in the absence of Na^+/H^+ exchanger activity. *Am. J. Physiol.* **253,** C633–C638.

Molski, T. F. P., Naccache, P. H., Marsh, M. L., Kermode, J., Becker, E. L., and Sha'afi, R. I. (1984). Pertussis toxin inhibits the rise in the intracellular concentration of free calcium that is induced by chemotactic factors in rabbit neutrophils: Possible role of the "G proteins" in calcium mobilization. *Biochem. Biophys. Res. Commun.* **124,** 644–650.

Moran, A. (1987). Sodium-hydrogen exchange system in LLC-PK_1 epithelium. *Am. J. Physiol.* **252,** C63–C67.

Moran, A. and Moran, N. (1984). Amiloride-sensitive channels in LLC-PK_1 apical membranes. *Fed. Proc.* **43,** 447. (Abstr.)

Moscona-Amir, E., Henis, Y., and Sokolovsky, M. (1988). Guanosine 5′-triphosphate binding protein (G_i) and two additional pertussis toxin substrates associated with muscarinic receptors in rat heart myocytes: Characterization and age dependency. *Biochemistry* **27,** 4985–4991.

Myers, B. M., Cantiello, H. F., Schwartz, J. H., and Tauber, A. I. (1990). Phorbol ester-stimulated human neutrophil membrane depolarization is dependent on Ca^{2+}-regulated Cl^- efflux. *Am. J. Physiol.* **259,** C531–C540.

Nagle, W., and Crabbé, J. (1980). Mechanism of action of aldosterone on active sodium transport across toad skin. *Pfluegers Arch.* **385,** 181–187.

Nakamura, T., and Ui, M. (1985). Simultaneous inhibitions of inositol phospholipid breakdown, arachidonic acid release, and histamine secretion in mast cells by islet-activating protein, pertussis toxin. *J. Biol. Chem.* **260,** 3584–3593.

Neer, E. J., and Clapham, D. E. (1988). Roles of G protein subunits in transmembrane signalling. *Nature (London)* **333,** 129–134.

Nelson, D., Tang, J., and Palmer, L. (1984). Single-channel recordings of apical membrane chloride conductance in A6 epithelial cells. *J. Membr. Biol.* **80,** 81–89.

Nelson, W. J., and Veshnock, P. J. (1987). Ankyrin binding to (Na^+/K^+) ATPase and implications for the organization of membrane domains in polarized cells. *Nature (London)* **328,** 533–536.

Okabe, K., Yatani, A., Evans, T., Ho, Y.-K., Codina, J., Birnbaumer, L., and Brown, A. M. (1990). $\beta\gamma$ Dimers of G proteins inhibit atrial muscarinic K^+ channels. *J. Biol. Chem.* **265,** 12854–12858.

Omann, G., Allen, R. A., Bokoch, G. M., Painter, R. G., Traynor, A. E., and Sklar, L. A. (1987). Signal transduction and cytoskeletal activation in the neutrophil. *Physiol. Rev.* **67,** 285–322.

Ordway, R. W., Walsh, J. V., Jr., and Singer, J. J. (1989). Arachidonic acid and other fatty acids directly activate potassium channels in smooth muscle. *Science* **244,** 1176–1179.

Palmer, L. G. (1982). Ion selectivity of the apical membrane Na^+ channel in the toad urinary bladder. *J. Membr. Biol.* **67,** 91–98.
Palmer, L. G. and Frindt, G. (1986). Amiloride-sensitive Na channels from the apical membrane of rat cortical collecting tubule. *Proc. Natl. Acad. Sci. U.S.A.* **83,** 2767–2770.
Palmer, L. G., and Frindt, G. (1987). Effects of cell Ca^{2+} and pH on Na^+ channels from cortical collecting tubule. *Am. J. Physiol.* **253,** F333–F339.
Palmer, L. G., Edelman, I. S., and Lindemann, B. (1980). Current-voltage analysis of apical sodium transport in toad urinary bladder. Effects of inhibitors of transport and metabolism. *J. Membr. Biol.* **57,** 59–71.
Palmer, L. G., Li, J. H.-Y., Lindemann, B., and Edelman, I. (1982). Aldosterone control of the density of sodium channels in the toad urinary bladder. *J. Membr. Biol.* **64,** 91–102.
Paris, S., and Poyssegur, J. (1987). Further evidence for phospholipase C-coupled G protein in hamster fibroblasts. Induction of inositol phosphate formation by fluoroaluminate and vanadate and inhibition by pertussis toxin. *J. Biol. Chem.* **262,** 1970–1976.
Pfaffinger, P. J., Martin, J. M., Hunter, D. D., Nathanson, N. M., and Hille, B. (1985). GTP-binding proteins couple cardiac muscarinic receptors to a K^+ channel. *Nature (London)* **317,** 536–538.
Prat, A. G., Ausiello, D. A., and Cantiello, H. F. (1991). Actin filament organization controls Na^+ channel activity in A6 epithelial cells. *FASEB J.* (Abstr.) **5,** A690.
Rae, J. L., and Levis, R. A. (1984). Patch voltage clamp of lens epithelial cells: Theory and practice. *Mol. Physiol.* **6,** 115–162.
Rodbell, M., Birnbaumer, L., Pohl, S. L., and Krans, H. M. J. (1971a). The glucacon-sensitive adenyl cyclase system in plasma membranes of rat liver. V. An obligatory role of guanyl nucleotides in glucagon action. *J. Biol. Chem.* **246,** 1877–1882.
Rodbell, M., Krans, H. M. J., Pohl, S. L., and Birnbaumer, L. (1971b). The glucagon-sensitive adenyl cyclase system in plasma membranes of rat liver. IV. Binding of glucagon: Effect of guanyl nucleotides. *J. Biol. Chem.* **246,** 1872–1876.
Rosenthal, W., Hescheler, J., Trautwein, W., and Schultz, G. (1988). Control of voltage-dependent Ca^{2+} channels by G protein-coupled receptors. *FASEB J.* **2,** 2784–2790.
Sachs, F. (1987). Baroreceptor mechanisms at the cellular level. *Fed. Proc.* **46,** 12–16.
Saint Onge, D., and Gicquaud, C. (1990). Research on the mechanism of interaction between actin and membrane lipids. *Biochem. Biophys. Res. Commun.* **167,** 40–47.
Samuelsson, B. (1983). Leukotrienes: Mediators of immediate hypersensitivity reactions and inflammation. *Science* **220,** 568–575.
Samuelsson, B., and Funk, C. D. (1989). Enzymes involved in the biosynthesis of leukotriene B4. *J. Biol. Chem.* **264,** 19469–19472.
Sariban-Sohraby, S., and Fisher, R. S. (1990). Single channel activity by the amiloride binding subunit of the epithelial Na^+ channel. *Biophys. J.* **57,** 87a. (Abstr.)
Sariban-Sohraby, S., Latorre, R., Burg, M., and Benos, D. J. (1984). Amiloride-sensitive epithelial Na^+ channels reconstituted into planar lipid bilayer membranes. *Nature (London)* **308,** 80–82.
Sariban-Sohraby, S., Sorscher, E. J., Brenner, B. M., and Benos, D. J. (1988). Phosphorylation of a single subunit of the epithelial Na^+ channel protein following vasopressin treatment of A6 cells. *J. Biol. Chem.* **263,** 13875–13879.
Satoh, M., Nanri, H., Takeshige, K., and Minakami, S. (1985). Pertussis toxin inhibits intracellular pH changes in human neutrophils stimulated by *N*-formyl-methionyl-leucyl-phenylalanine. *Biochem. Biophys. Res. Commun.* **131,** 64–69.
Schellenberg, G. D., Anderson, L., and Swanson, P. D. (1983). Inhibition of Na^+-Ca^{2+} exchange in rat brain by amiloride. *Mol. Pharmacol.* **24,** 251–258.
Scherer, R. A., and Breitwieser, G. E. (1990). Arachidonic acid metabolites alter G protein-

mediated signal transduction in heart. Effects on muscarinic K^+ channels. *J. Gen. Physiol.* **96,** 735–755.

Schliwa, M. (1982). Action of cytochalasin D on cytoskeletal networks. *J. Cell Biol.* **92,** 79–91.

Schubert, B., VanDongen, A., Kirsch, G., and Brown, A. M. (1989). β-Adrenergic inhibition of cardiac sodium channels by dual G-protein pathways. *Science* **245,** 516–519.

Schwartz, M. A., and Luna, E. J. (1988). How actin binds and assembles onto plasma membranes from *Dictiostelium discoideum*. *J. Cell Biol.* **107,** 210–209.

Shimizu, Y. (1986). Chemistry and biochemistry of saxitoxin analogs and tetrodotoxin. *Ann. N.Y. Acad. Sci.* **479,** 24–31.

Siegl, P. K. S., Cragoe, E. J., Trumble, M. J., and Kaczorowdki, A. G. J. (1984). Inhibition of Na^+/Ca^{2+} exchange in membrane vesicle and papillary muscle preparations from guinea pig heart by analogs of amiloride. *Proc. Natl. Acad. Sci. U.S.A.* **81,** 3238–3242.

Sigworth, F. J. (1984). Electronic design of patch clamp. *In* "Single-Channel Recording" (B. Sackmann and E. Neher, eds.), pp. 3–35. Plenum, New York.

Small, J. V. (1989). Microfilament-based motility in non-muscle cells. *Curr. Opinion Cell Biol.* **1,** 75–79.

Smith, P. R., Saccomani, G., Angelides, K., Joe, E. N., and Benos, D. J. (1990). Amiloride-sensitive sodium channel is linked to the cytoskeleton in renal epithelia. *FASEB J.* **4,** A445.

Soejima, M., and Noma, A. (1984). Mode of regulation of the ACh-sensitive K^+-channel by muscarinic receptor in rabbit atrial cells. *Pfluegers Arch.* **400,** 424–431.

Soltoff, S. P., and Mandel, L. J. (1983). Amiloride directly inhibits the Na^+,K^+-ATPase activity of rabbit kidney proximal tubules. *Science* **220,** 957–959.

Sorota, S., Tsuji, T., Tajima, T., and Pappano, A. J. (1985). Pertussis toxin treatment blocks hyperpolarization by muscarinic agonists in chick atrium. *Circ. Res.* **57,** 748–758.

Stiernberg, J., LaBelle, E. F., and Carney, D. H. (1983). Demonstration of a late amiloride-sensitive event as a necessary step in initiation of DNA synthesis by thrombin. *J. Cell Physiol.* **117,** 272–281.

Stossel, T. (1982). The structure of cortical cytoplasm. *Philos. Trans. R. Soc. London, Ser. B* **299,** 275–289.

Tanuma, S., Kawashima, K., and Endo, H. (1988). Eukaryotic mono (ADP-ribosyl)transferase that ADP-ribosylates GTP-binding regulatory G_i protein. *J. Biol. Chem.* **263,** 5485–5489.

Taylor, A., Mamelak, M., Keaven, E., and Maffry, R. (1973). Vasopressin: Possible role of microtubules and microfilaments in its action. *Science* **187,** 347–350.

Trautwein, W., and Hescheler, J. (1990). Regulation of cardiac L-type calcium current by phosphorylation and G proteins. *Annu. Rev. Physiol.* **52,** 257–274.

Tsunoo, A., Yoshi, M., and Narahashi, T. (1986). Block of calcium channels by enkephalin and somatostatin in neuroblastoma × glioma hybrid. *Proc. Natl. Acad. Sci. U.S.A.* **83,** 9832–9836.

Ussing, H. H. (1949). The active ion transport through the isolated frog skin in the light of tracer studies. *Acta Physiol. Scand.* **17,** 1–37.

Ussing, H. H., and Zerahn, K. (1951). Active transport of sodium as the source of electric current in the short-circuited isolated frog skin. *Acta Physiol. Scand.* **23,** 110–127.

Van Driessche, W., and Borghgraef, R. (1975). Noise generated during ion transport across frog skin. *Arch. Int. Physiol. Biochim.* **83,** 140–142.

Van Driessche, W., and Hillyard, S. D. (1985). Quinidine blockage of K^+ channels in the basolateral membrane of larval bullfrog skin. *Pfluegers Arch.* **405,** s77–s82.

Van Driessche, W., and Zeiske, W. (1985). Ionic channels in epithelial cell membranes. *Physiol. Rev.* **65,** 833–903.
Vigne, P., Champigny, G., Marsault, R., Barbry, P., Frelin, C., and Lazdunski, M. (1989). A new type of amiloride-sensitive cationic channel in endothelial cells of brain microvessels. *J. Biol. Chem.* **264,** 7663–7668.
Volpi, M., Naccache, P. H., Molski, T. F. P., Shefcyk, J., Huang, C. K., Marsh, M. L., Munoz, J., Becker, E. L., and Sha'Afi, R. I. (1985). Pertussis toxin inhibits fMet-Leu-Phe but not phorbol ester-stimulated changes in rabbit neutrophils: Role of G proteins in excitation response coupling. *Proc. Natl. Acad. Sci. U.S.A.* **82,** 2708–2712.
Watkins, P., Kanaho, Y., and Moss, J. (1987). Inhibition of the GTPase activity of transducin by an NAD^+:arginine ADP-ribosyltransferase from turkey erythrocytes. *Biochem. J.* **248,** 749–754.
Welsh, M., Smith, P., Fromm, M. J., and Frizzel, R. A. (1982). Crypts are the site of intestinal fluid and electrolyte secretion. *Science* **218,** 1219–1221.
White, J. R., Naccache, P. H., and Sha'afi, R. I. (1982). Stimulation by chemotactic factor of actin association with the cytoskeleton in rabbit neutrophils. *J. Biol. Chem.* **258,** 14041–11047.
Will, P. C., Lebowitz, J. L., and Hopfer, U. (1980). Induction of amiloride-sensitive sodium transport in the rat colon by mineralocorticoids. *Am. J. Physiol.* **238,** F261–F268.
Yassin, R., Shefcyk, J., White, J. R., Tao, W., Volpi, M., Molski, T. F. P., Naccache, P. H., Feinstein, M. B., and Sha'afi, R. I. (1985). Effects of chemotactic factors and other agents on the amounts of actin and a 65,000-mol-wt protein associated with the cytoskeleton of rabbit and human neutrophils. *J. Cell Biol.* **101,** 182–188.
Yatani, A., and Brown, A. M. (1989). Rapid β-adrenergic modulation of cardiac calcium channel currents by a fast G protein pathway. *Science* **245,** 71–74.
Yatani, A., Codina, J., and Imoto, Y. (1987a). A G protein directly regulates mammalian cardiac calcium channels. *Science* **238,** 1288–1292.
Yatani, A., Codina, R. D., Sekura, L., Birnbaumer, L., and Brown, A. M. (1987b). Reconstitution of somatostatin and muscarinic receptor mediated stimulation of K^+ channels by isolated G_k protein in clonal rat anterior pituitary cell membranes. *Mol. Endocrinol.* **1,** 283–289.
Yatani, A., Mattera, R., Codina, J., Graf, R., Okabe, K., Padrell, E., Iyengar, R., Brown, A. M., and Birnbaumer, L. (1988). The G protein-gated atrial K^+ channel is stimulated by three distinct G_i alpha-subunits. *Nature (London)* **336,** 680–682.
Yechiel, E., and Barenholz, Y. (1985). Relationships between membrane lipid composition and biological properties of rat myocytes. Effects of aging and manipulation of lipid composition. *J. Biol. Chem.* **260,** 9123–9131.
Yechiel, E., Barenholz, Y., and Henis, Y. I. (1985). Lateral mobility and organization of phospholipids and proteins in rat myocyte membranes. Effects of aging and manipulation of lipid composition. *J. Biol. Chem.* **260,** 9132–9136.
Yin, H. L., Zaner, K. S., and Stossel, T. P. (1980). Ca^{2+} control of actin gelation. Interaction of gelsolin with actin filaments and regulation of actin gelation. *J. Biol. Chem.* **255,** 9494–9500.
Yorio, T., and Bentley, P. (1978). Phospholipase A and the mechanism of action of aldosterone. *Nature (London)* **271,** 79–81.
Yoshitomi, K., and Frömter, E. (1985). How big is the electrochemical potential difference of Na^+ across rat renal proximal tubular cell membranes *in vivo? Pfluegers Arch.* **405,** S121–S126.
Zeidel, M. L., Silva, P., Brenner, B. M., and Seifter, J. L. (1987). cGMP mediates effects of

atrial peptides on medullary collecting duct cells. *Am. J. Physiol.* **252,** F551–F559.

Zeiske, W., and Lindemann, B. (1974). Chemical stimulation of Na^+ current through the outer surface of frog skin epithelium. *Biochim. Biophys. Acta* **352,** 323–326.

Zeiske, W., and Lindemann, B. (1975). Blockage of Na^+ channels in frog skin by titration with protons and by chemical modification of COO^--groups. *Pfluegers Arch.* **355,** R71. Abstr.

Zeiske, W., Machen, T. E., and Van Driessche, W. (1983). Cl^- and K^+-related fluctuations of ionic current through oxyntic cells in frog gastric mucosa. *Am. J. Physiol.* **245,** G797–G807.

Zweifach, A., and Lewis, S. A. (1988). Characterization of a partially degraded Na^+ channel from urinary tract epithelium. *J. Membr. Biol.* **101,** 49–56.

Index

C

D

E

H

I

N

O

P

W

X

Y